D0611440

INTEGRATION OF ALTERNATIVE SOURCES OF ENERGY

INTEGRATION OF ALTERNATIVE SOURCES OF ENERGY

FELIX A. FARRET
M. GODOY SIMÕES

IEEE PRESS

A JOHN WILEY & SONS, INC., PUBLICATION

Published by John Wiley & Sons, Inc., Hoboken, New Jersey
Published simultaneously in Canada

Library of Congress Cataloging-in-Publication Data:

Farret, Felix A.
Integration of alternative sources of energy/by Felix A. Farret, M. Godoy Simões.
p. cm.
"A Wiley-Interscience publication."
Includes bibliographical references and index.
ISBN-13: 978-0-471-71232-9
ISBN-10: 0-471-71232-9
[1. Power resources. 2. Renewable energy sources.] I. Simões, Marcelo Godoy. II. Title.

TJ163.2.F37 2006
621.042–dc22 2005048975

Printed in the United States of America

10 9 8 7 6 5 4 3 2 1

To Alzira and Fares Farret, Mary and John Hilliard,
Gerry, Matheus, and Patrick, with whom I have
learned and to whom I owe so much. F.A.F.

For the next generation, to which Ahriel and Lira belong.
Hopefully, it will be powered by renewable energy. M.G.S.

CONTENTS

CONTRIBUTORS

Benjamin Kroposki is a senior engineer at NREL and leader of the Distributed Power Systems Integration Team. His expertise is in the design and testing of distributed power systems, and he has produced more than 30 publications in this area. Kroposki also participates in the development of distributed power standards and codes for IEEE, the International Electrotechnical Commission (IEC), and the National Electrical Code. He serves as secretary for IEEE P1547.1 and chairs IEEE P1547.4. Mr. Kroposki received his bachelor's and master's degrees in electrical engineering from Virginia Tech and is a registered professional engineer.

Thomas Basso is a senior scientist working at NREL under the NREL Distribution and Interconnection R&D area of the NREL Distributed Energy and Electricity Reliability Program. Prior at NREL, he conducted outdoor accelerated weathering of photovoltaic (PV) modules and was NREL project leader for PV management under the NREL/Department of Energy PV Advanced R&D Project. Before coming to NREL, he was a design engineer with a consulting engineering firm and worked for a manufacturer of air-cooled heat exchangers for the petrochecmical, refinery, and utility industries. He was also a standards engineer for the American Society of Mechanical Engineers and an instructor in the Mechanical Engineering Department of Northeastern University. He serves as secretary for IEEE SCC21, P1547, P1547.2, and P1547.3 and is a member of the IEC Joint Coordination Group for Decentralized Rural Electrification Systems, IEC TC8, the American Society of Mechanical Engineers, and the American Solar Energy Society.

Richard DeBlasio is technology manager of the NREL/Department of Energy Distributed Energy and Electricity Reliability Program, which includes Distribution and Interconnection R&D at NREL. Before joining NREL in 1978, he was with the U.S. Atomic Energy Commission in Washington, D.C. (1974–1978), and Underwriters' Laboratories (1972–1974). He was also a member of the technical staff at Stanford University (1965–1972). He is an electrical engineer; a senior member of IEEE; an IEEE SA Standards Board member; chair of IEEE SCC 21 on Fuel Cells, Photovoltaics, Distributed Power, and Energy Storage; and chair of the international standards committees IEC TC82 and Joint Committee on Decentralized Rural Electrification Systems.

N. Richard Friedman is chairman and CEO of the Resource Dynamics Corp., Vienna, Virginia, where he directs strategic business assessments, advises energy companies on customer marketing strategies, and identifies new business development ventures. He has a master's degree in engineering administration from George Washington University and a bachelor of science degree in electrical engineering from the Moore School at the University of Pennsylvania. He is a member of numerous technical and professional societies, including the IEEE Energy Policy Committee, IEEE Standards Coordinating Committee 21 (including acting as chairman of IEEE P1574.2), and the executive committee of the U.S. Combined Heat and Power Association.

Tom Lambert is the principal of Mistaya Engineering Inc. He received a B.Sc. in mechanical engineering from the University of Alberta and an M.Sc. in mechanical engineering from Colorado State University. He develops computer models and provides consulting services in the areas of micropower system design, wind resource assessment, and data visualization. Since 1997 he has developed, documented, and provided training and technical support for the U.S. National Renewable Energy Laboratory's HOMER and ViPOR models.

Paul Gilman is a project leader in the National Renewable Energy Laboratory's (NREL) International Program. He manages the program's training activities and has designed and facilitated training workshops for rural energy planners in Brazil, the Maldives, Mexico, and Sri Lanka, and for project developers in China. Paul is also on both the HOMER software development team and the Africa team of NREL's International Program. He is trained as an electrical engineer and has experience developing technical documentation for software and hardware products and has managed community development projects in West Africa.

Peter Lilienthal has been with National Renewable Energy Laboratory since February 1990. His position is senior economist with the International Programs Office. He has a Ph.D. in engineering-economic systems from Stanford University. He has been active in the field of renewable energy and conservation since 1978. This has included designing and teaching courses at the university level, project development of independent power projects, and consulting to industry and regulators. His technical expertise is in utility modeling and the economic and financial analysis of small power projects. Since 1993 he has been the lead analyst for NREL's International and Village Power Program and the developer of the NREL's Village Power Optimization software, HOMER, and ViPOR.

FOREWORD

Integration of Alternative Sources of Energy is an important work about a technology that has the potential to advance environmental goals and eventually support a sustainable future for society. Several countries—Denmark, Australia, Spain, Germany, the UK, and others—have begun the transition away from fossil fuels and nuclear energy as a response to increasing concerns about fuel supplies, global security, and climate change.

Alternative energy includes all sources and technologies that minimize environmental impacts relative to conventional hydrocarbon resources and economic issues related to fossil fuel resources. Therefore, fuel cells, natural gas, and diesel might be alternatives to coal or nuclear power. Renewable energy sources are those derived from the sun or other natural and replenishing processes, including solar (light and heat), wind, hydro (fall and flow), sustainable biomass, wave, tides, and geothermal energy. Throughout the book, the fundamentals of the technologies related to integration of such alternative and renewable energy sources are reviewed and described with authority, skill, and from the critical engineering point of view for the end user of energy.

In fifteen chapters the authors cover the principles of hydroelectric, wind, solar, thermal, and photovoltaic power plants. Induction generators—important electric machines for wind and hydropower generation—are described in detail. The chapters on fuel cells and biomass are of paramount importance at the current stage of the hydrogen economy. The literature on microturbines is scarce; therefore, the authors make important contributions to the technical description of these important devices, which have contributed, in the last few years, toward shifting traditional generation to distributed generation. A comprehensive evaluation of storage technology is complemented by the description of integrating control and association of sources into microgrids. There are two chapters authored by contributors that describe the standards and interconnection issues of alternative energy sources to the grid and principles of economic optimization. The book is complemented by three appendixes covering some important sources of power and heat related directly to the rational use of energy: diesel, geothermal, and Stirling engines.

I applaud the initiative taken by the authors in this timely book to cover more closely, the electrical rather than the mechanical aspects of energy sources. I am sure this work will contribute to our understanding of how to integrate alternative energy sources for home, commercial, rural, and industrial applications.

I strongly recommend this textbook to a wide audience, including engineering educators and students of electrical and mechanical engineering.

Marian P. Kazmierkowski

Warsaw University of Technology, Poland

PREFACE

Our goal in writing this book was to discuss the "electrical side" of alternative energy sources. From the beginning, we felt that this approach would be a challenge that would be very difficult to fulfill. Most of the current technical work explores just one or two types of alternative energy sources, but the integration of sources is our main objective. We also noticed that most of the works on this subject were concerned exclusively with those aspect of primary energy related directly to extraction and conversion of power rather than the processing of energy and delivery of a final product. This product—energy—should be ready to power a new reality and the dreams associated with renewable or alternative sources of energy. These sources of energy have a lot in common. However, when beginning to discuss hydro, wind, solar, and other sources of energy, their complexity is soon realized.

Not long ago, huge generating power plants dominated the entire field of energy production. It seemed that there was no way to develop and deploy small and dispersed alternatives. For several decades, small plants nearly vanished from the scene. This was a worldwide trend based on the argument that electrical efficiency and concentrated sites would be incontestably the best economic and rational choice for generating electricity. Throughout those decades, small rural communities and remote areas were simply outside the scope of the centralized model. Population growth and the development of nations soon made our society realize that more and more local and distributed energy production would be necessary for continuous industrial growth.

On the other hand, the availability of energy would not be enough, or would be so distant from consumption points that central power plants would have devastating effects on ecology, scenery, and quality of life. For a sustainable energy future, massive fossil fuel–powered plants are not economical; most of them waste more than 50% of the primary energy due to irrecoverable thermal losses. In addition, they demand the use of massive coolers and heat sinks to guarantee the operational conditions, quality, and stability of the final product. The energy must be transported throughout long and congested transmission lines, resulting in greater and greater waste, which is no longer reasonable. Small, dispersed generating units do have the possibility of adding representative amounts of energy to the network without noticeable long-term impact on the environment and economic investment.

Current computer and power electronics technology supports the integration and distributed generation of energy to sites that have so far been neglected. Our earth has sunlit deserts, windy remote locations, offshore sea sites, glaciers, and streams, with each environment contributing to the world energy frame work. They are not wastelands anymore.

Who can guarantee for how long we will have plentiful and available petroleum, in the right quantity for the demand? How can humankind believe in a sustainable future if the only admissible alternative is nuclear energy? How can we quantify only economic benefits to humankind from fossil fuel and nuclear power and neglect other important issues, such as health care costs avoided; air quality; space, sea, and river poisoning; sound and visual pollution; or an unpredictably safe future? Nobody wants to go to war to sustain our future.

In this book we review and organize all pertinent subjects in an orderly way. After a general introduction to the subject in Chapter 1, we review some general principles of thermodynamics in Chapter 2. Initially, we chose to explore the most common alternative energy sources, such as hydro, wind, thermosolar, photovoltaics, fuel cell power plants (fuel cell plants are expected to become widespread soon), and biomass. After we cover the basis of the primary sources, our approach is to show how to integrate them for electrical power production and injection into the main grid; these subjects are covered in Chapters 3, 4, 5, 6, 7, and 8. The means of interfacing the primary energy as microturbines, induction generators, and storage systems for electrical power are dealt with in Chapters 9 and 10. Chapter 11 is a special chapter on energy storage systems. Our main emphasis in Chapter 12 is on various means of integrating, transforming, and conditioning energy sources into more useful electrical applications rather than coping with the distribution of sources throughout the electrical network. Power transmission is left for the reader to study in related books. Problems related to system operation, maintenance, and management are tackled briefly in Chapter 13. This chapter also refers the reader to other, more specialized literature on the subject. In Chapter 14 we discuss the standards for interconnection. Finally, in Chapter 15 we give the reader the opportunity to learn more about the HOMER Micropower Optimization Model, which is a computer model developed by the U.S. National Renewable Energy Laboratory (NREL) to assist in the design of micropower systems and to facilitate comparison of power generation technologies across a wide range of applications. HOMER software can be downloaded from the NREL Web site.

This book is especially dedicated to those people who believe that there is a way to work for a clean, long-lasting, and beautiful world: to those students, engineers, and professionals who believe that engineering is decisive in its contribution to this journey and who dedicate their professional lives to this mission. Many of the users of this book will be in their senior years of undergraduate study or first graduate-level courses in energy; electrical, environmental, civil, chemical, or mechanical engineering; and agronomic sciences.

ACKNOWLEDGMENTS

The authors are very thankful to many people. This book would not have been possible without their contributions. We exchanged ideas with colleagues, professionals, and students, took their suggestions, and sometimes even gave them extra duties. The manuscript took almost two years to write; during this period, we interacted with people who understood the importance of this work and supported us firmly. We offer our utmost recognition and respect especially to Vladimir Popov, Burak Ozpineci, Luciane Canha, P. K. Sen, Edson H. Watanabe, José Antenor Pomilio, Fang Z. Peng, Ali Emadi, Ernesto Ruppert Filho, Bhaskara Palle, Sudipta Chakraborty, Peter Lilienthal, and Ben Kroposki. We thank the staff of John Wiley & Sons, especially Rachel Witmer and Angioline Loredo, for their full dedication to the project; they encouraged us and filled us with enthusiasm. Thanks to our families, who were deprived of our company and attention for many months; they never complained but rather, gave us their help and understanding. Thanks to Luzia Ornelas for the book cover design; the photograph of Earth is available thanks to NASA. Finally, we express our gratitude to our schools: the Federal University of Santa Maria and the Colorado School of Mines. Their institutional support encouraged cooperation in a very creative environment.

FELIX A. FARRET
M. GODOY SIMÕES

Santa Maria, Rio Grande do Sul, Brazil
Golden, Colorado, U.S.A.
Spring 2005

ABOUT THE AUTHORS

Felix A. Farret received his B.E. and M.Sc. in Electrical Engineering from the Federal University of Santa Maria in 1972 and 1986, respectively; specialist in instrumentation and automation by the Osaka Prefectural Industrial Research Institute–Japan; M.Sc. by the University of Manchester, UK, and Ph.D. from the University of London, UK, all in electrical engineering. He was operation and maintenance engineer for the Electrical State Company of Electrical Energy in the RGS state, Brazil, in 1973–1974. He has been working in an interdisciplinary educational background related to power systems, power electronics, nonlinear controls, and integration of renewable energies. He teaches at the Department of Electronics and Computation of the Federal University of Santa Maria, Brazil, committed to undergraduate and graduate activities and in research development, since 1974. He was a visiting professor at Colorado School of Mines, Engineering Division, U.S.A., in 2002–2003. He has published his first book in Portuguese titled *Use of Small Sources of Electrical Energy* (UFSM University Press, 1999), and co-authored the book *Renewable Energy Systems: Design and Analysis with Induction Generators* (CRC Press, 2004). Energy Engineering Systems is the focus of his present interests for distribution and industrial applications.

In more recent years, several technological processes in renewable sources of energy were coordinated by Dr. Farret and transferred to Brazilian enterprises such as AES-South Energy Distributor, Hydro Electrical Power Plant Generation of Nova Palma, RGE Energy Distributor, State Company of Electrical Energy and CCE Power Control Engineering Ltd. These processes were related to integration of micropower plants from distinct primary sources; voltage and speed control by load for induction generators; low power PEM fuel cell application and its model development. Injection of electrical power into the grid is currently his major interest. In Brazil, he has been developing several intelligent systems for industrial applications related to integration, location and sizing of renewable sources of energy for distribution and industrial systems, including fuel cells, hydropower, wind power, photovoltaics, battery storage applications and other ac–ac and ac–dc–ac links.

M. Godoy Simões received his B.Sc. degree from the University of São Paulo, Brazil, M.Sc. degree from the University of São Paulo, Brazil, and Ph.D. degree from The University of Tennessee, USA, in 1985, 1990, and 1995, respectively, and his D.Sc. degree (Livre-Docência) from the University of São Paulo in 1998. He is IEEE Senior Member and is with the Colorado School of Mines since 2000.

He published the first book in Portuguese about fuzzy modeling titled *Controle e Modelagem Fuzzy* (FAPESP and Edgard Blücher Publisher Company) and co-authored the book *Renewable Energy Systems: Design and Analysis with Induction Generators* (CRC Press, 2004).

Dr. Simões is a recipient of a National Science Foundation Faculty Early Career Development (CAREER) in 2002, the NSF's most prestigious award for new faculty members. Dr. Simões has been serving the IEEE – Institute of Electrical and Electronics Engineers in several capacities. He served as the Program Chair for the Power Electronics Specialists Conference 2005 and General Chair of the Power Electronics Education Workshop 2005. He is currently serving as IEEE Power Electronics Society Intersociety Chairman, Associate Editor for Energy Conversion, as well as Editor for Intelligent Systems of IEEE Transactions on Aerospace and Electronic Systems and as Associate Editor for Power Electronics in Drives of IEEE Transactions on Power Electronics. He was the founder and first chairman of the IEEE Denver Power Electronics Society chapter. He was actively involved in the Steering and Organization Committee of the IEEE International Future Energy Challenge for the competitions held in 2003 and 2005.

Dr. Simões has been working in research and educational activities for the development of intelligent control for high power electronics applications in renewable and distributed energy systems at Colorado School of Mines. He has been involved in educational projects at Colorado School of Mines in establishing a Minor in Energy and a Minor in Humanitarian Engineering, with a grant from Hewlett Foundation. He has also been involved in an interdisciplinary research, sponsored by National Science Foundation, on enhancing engineering responsibility with humanitarian ethics.

CHAPTER 1

ALTERNATIVE SOURCES OF ENERGY

1.1 INTRODUCTION

Since humankind's beginning, the ability to harvest and convert energy has been a means of survival. However, this has never been as evident as in recent years. When the industrial revolution in Europe caused an evolution of society and areas of larger population density, people realized that factors such as comfortable housing and energy could be important to the development of a country. Fossil fuels became essential products of a modern society, and new strategies were developed to guarantee their uninterrupted supply.

Since then, our population has grown—and so has its demands for industrial goods. On a planet of unaltered size and resources, this has had predictable consequences. Bloody wars, new frontiers, international agreements, and optimized use of resources are a few of the results. With the exhaustion of the planet's energy resources, the continuation of such effects can easily be foreseen.

Long ago, the need to generate large amounts of electrical energy and the realization that larger power plants were more efficient than smaller ones encouraged the construction of huge power plants. Examples include Itaipu Binational in Brazil, Guri in Venezuela, Sayano-Shushensk in Russia, and Churchill Falls in Canada. A more recent example is the plan for construction of the largest

Integration of Alternative Sources of Energy, by Felix A. Farret and M. Godoy Simões

(18 GW) hydropower plant in China, Three Gorges Dam. However, some of these areas are affected by immense floods, massive power transmission lines and towers, air pollution, modified waterways, devastated forests, large population densities, and wars. Because of this trend in development, distances to energy sources are increasing, material capacities are reaching their limits, fossil reserves are being exhausted, and pollution is becoming widespread. New alternatives must be devised if humanity is to survive today and for centuries to come.

1.2 RENEWABLE SOURCES OF ENERGY

Earth receives solar energy as radiation from the sun, and it receives it in a quantity that far exceeds humankind's use. By heating the planet, the sun generates wind, rain, rivers, and waves. Along with rain and snow, sunlight is necessary for plants to grow. The organic matter that makes up plants is known as *biomass*, and biomass can be used to produce electricity, transportation fuel, and chemicals. Plant photosynthesis (essentially, chemical storage of solar energy) creates a range of biomass products, from wood fuel to rapeseed, that can be used for heat, electricity, and liquid fuels.

Hydrogen can also be extracted from many organic compounds, as can water. Hydrogen is the most abundant element on Earth, but it does not occur naturally in a gas. It always combines with other elements, such as with oxygen to make water. Once separated from another element, hydrogen can be burned as a fuel or converted into electricity.

The sun also powers the evapotranspiration cycle, which allows water to generate power in hydro schemes—the largest source of renewable electricity today. Interactions with the moon produce tidal flows, which can produce electricity.

Although humans have been tapping into renewable energy sources (such as solar, wind, biomass, geothermal, and water) for thousands of years, only a fraction of their technical and economic potential has been captured and exploited. Yet renewable energy offers safe, reliable, clean, local, and increasingly cost-effective alternatives for all our energy needs.

Research has made renewable energy more affordable today than it was 25 years ago. Wind energy has declined from 40 cents per kilowatthour (kwh) to less than 5 cents. Electricity from the sun through photovoltaics (which literally means "light electricity") has dropped from more than $1 per kilowatthour in 1980 to nearly 20 cents/kwh today. Ethanol fuel costs have plummeted from $4/gallon in the early 1980s to $1.20 today.

Renewable energy resource development will result in new jobs and less dependence on oil from foreign countries. According to the federal government, the United States spent $109 billion to import oil in 2000. If we fully develop self-renewing resources, we will keep the money at home to help our economy.

There are some drawbacks to renewable energy development. An example is solar thermal energy, in which solar rays are captured through collectors (often, huge mirrors). Solar thermal generation requires large tracts of land, and this affects natural habitat. The environment is also affected when buildings, roads, transmission lines, and transformers are built. In addition, a fluid often used for solar thermal generation is toxic, and spills can occur. Solar or photovoltaic cells are produced using the same technologies as those used to create silicon chips for computers, and this manufacturing process also uses toxic chemicals. In addition, toxic chemicals are used in batteries that store solar electricity through nights and on cloudy days. Manufacturing this equipment also has environmental effects. So even though the renewable power plant does not release air pollution or use fossil fuels, it still has an effect on the environment.

In addition, there are production problems. All the solar equipment production facilities in the world make only enough cells to produce about 350 MW of electricity, just enough for a city of 300,000 people. California alone needs about 55,000 MW of electricity on a sunny, hot summer day. Producing that electricity by solar means would be about four times more expensive than if it were produced in a natural gas–fired power plant.

Wind power, too, has its downside, involving primarily land use. For example, the average wind farm requires 17 acres to produce 1 MW of electricity (about enough electricity for 750 to 1000 homes). However, farmers and ranchers can use the land beneath wind turbines. Wind farms can cause erosion in desert areas, and they affect natural views because they tend to be located on or just below ridgelines. Bird deaths also result from collisions with wind turbines and wires. This is the subject of ongoing research. But ultimately, combined with energy efficiency, renewable energy can provide everything fossil fuels offer in terms of energy services: from heating and cooling to electricity, transportation, chemicals, illumination, and food drying.

Energy has always existed in one form or another. It can be transformed into other forms of energy, but it cannot be created or destroyed. For example, the energy in a flashlight's battery becomes light energy when the flashlight is turned on. Food, the most natural stored chemical energy, resides in fat tissues and cells as potential energy. When the body uses that stored energy to do work, it becomes kinetic energy. Telephones transform a voice into electrical energy variations, which flow over wires or are transmitted through air. Other telephones change this electrical energy into sound energy through speakers. Cars use stored chemical energy in fuel to move, and they change chemical energy into heat and kinetic energy. Toasters change electrical energy into heat and light energy. Computers, television sets, and DVD players change electrical energy into coordinated types of mechanical movement and image and sound energy to reproduce the ambient of life.

In all such transformations of energy, intermediary transformations are involved. For example, consider the case for a home computer. Electricity allows self-organization of the main processor, according to a preestablished program, to convert ventilator movement to the cooling process for the main processing unit and the motherboard. The alternating current (ac) source power is converted into integrated direct current (dc) power to feed peripheral plates. After many electromagnetic processes, the monitor produces a luminous energy on-screen. Many processes and intermediary sources are integrated into a simple computer. They produce heat, light, movement, and circulation of electrical current to make it an impressively organized machine. This diversity of energy forms is an example of the changes happening in power systems.

1.3 RENEWABLE ENERGY VERSUS ALTERNATIVE ENERGY

A renewable energy source cannot run out and causes so little damage to the environment that its use does not need to be restricted. No energy system based on mineral resources is renewable because, one day, the mineral deposits will be used up. This is true for fossil fuels and uranium. The debate about when a particular mineral resource will run out is irrelevant in this context. A renewable energy source is replenished continuously.

Renewable energy sources—solar, wind, biomass (under specific conditions), and tides—are based directly or indirectly on solar energy. Hydroelectric power is not necessarily a renewable energy source because large-scale projects can cause ecological damage and irreversible consequences. Geothermal heat is renewable but must be used cautiously to guard against irreversible ecological effects.

There is no shortage of renewable energy because it can be taken from the sun, wind, water, plants, and garbage to produce electricity and fuels. For example, the sunlight that falls on the United States in one day contains more than twice the energy the country normally consumes in a year. California has enough wind gusts to produce 11% of the world's wind electricity.

Clean energy sources can be harnessed to produce electricity and process heat, fuel, and valuable chemicals with less effect on the environment than fossil fuel would cause. Emissions from gasoline-fueled cars and factories and other facilities that burn oil affect the atmosphere through the greenhouse effect. About 81% of all U.S. greenhouse gases are carbon dioxide emissions from energy-related sources.

At the International Climate Convention in Kyoto (1997), it was agreed that the developed nations of the world must reduce their greenhouse gas emissions. The European Union (EU) committed to reducing emissions of carbon dioxide (CO_2) by 8% from 1990 levels by the year 2010. The United States was to reduce emissions by 6% and Japan by 7% (see Table 1.1). These agreements are laid down in the Kyoto Protocol and aim for a society that uses renewable energies, not fossil fuels.

It is understandable that the world worries about emissions because our environment is unable to absorb them all. Table 1.2 lists some renewable sources of energy and their approximate production, or absorption, of CO_2 per kilowatthour.

Because every source is more or less intensive in what it produces, special measures have to be considered when considering global energy solutions. These include availability, capability, extraction costs, emissions, and durability. Table 1.3 shows indicators of renewable energy technologies, and Table 1.4 illustrates the intensity and frequency characteristics of some renewable sources.

TABLE 1.1 Kyoto's Pledged Emission Reductions in Selected Countries

Country	Emission Reduction (%)
Australia	−8
Canada	+6
Croatia	+5
EU	+8
Hungary	+6
Iceland	−10
Japan	+6
New Zealand	0
Norway	−1
Poland	+6
Russian Federation	0
Switzerland	+8
Ukraine	0
United States	+7

Source: Depledge, J., 2000. Tracing the origins of the Kyoto Protocol: an article by article textual history, UNFCCC/TP/2000/2, UNFCCC, Bonn.

Ironically, the atomic energy industry seems to be profiting from concerns about greenhouse gases and global climate change. Nuclear energy does not emit greenhouse gases, but its waste is stored in long-lasting containers and thrown into the sea in underground caves. Nevertheless, in the developed northern hemisphere, nuclear energy has little political or social support. The United States has not built a single reactor since the accident at Harrisburg, Pennsylvania, in 1979. In addition, there are no moves toward expanding nuclear power generation in any European Union member state that already has nuclear power stations. On the contrary, there is support for reduction in and closure of their atomic programs. Eight Western European countries (Denmark, Iceland, Norway, Luxembourg, Ireland, Austria, Portugal, and Greece) have never had a nuclear energy program and have instead

TABLE 1.2 Emission of CO_2 per Kilowatthour by Renewable Sources of Energy

Renewable Source	Emissions of CO_2/kWh (g)
Waste incineration	600
Biogasifier	−3800
Biomass	−4000
Photovoltaic cells	120
Wind turbine	10
Hydraulic power station	25
Nuclear power station	55
Gas-fired power station	400
Coal-fired power station	1160

Source: Ref. [1].

TABLE 1.3 Indicators of Renewable Energy Technologies

Renewable Energy Technology	Volatility (Approx. Time Variation)	Resource Availability	Range of Generation Cost (EU cents/kWh)	Preferred Voltage Level of Grid Connection (kV)
Biogas	Year	High	5.18–26.34	1.30
Biomass	Year	High	2.87–9.46	1.30, except cofiring
Geothermal electricity	Year	Low: country-specific	3.34–6.49	10.110
Large hydro power				
Run-of-river power plants	Months	Low	2.53–16.37	220.380
Storage power plants	Months	Low	Not considered	220.380
Small hydro power	Months	High	2.69–24.93	10.30
Landfill gas	Year	Low	2.50–3.91	1.30
Sewage gas	Year	Medium	2.85–6.24	1.30
Photovoltaics	Days, hours, seconds	High	47.56–165.32	<1
Solar thermal electricity	Days, hours, seconds	Low:country-specific	12.48–66.97	1.30
Tidal	12 hours	High	Not considered	10.380
Wave	Weeks	High	9.38–45.16	10.380
Wind				
Onshore	Hours, minutes	Low:country-specific	4.63–10.80	30.380
Offshore	Hours, minutes	Low:country-specific	6.09–13.39	110.380

Source: Ref. [1].

favored the alternative programs of renewable energy (see Figure 1.1). Outside Europe, only China, South Korea, Japan, Taiwan, and South Africa aspire to expand the share of nuclear power generated in their countries.

Today, the atomic energy industry is targeting developing countries, and the Kyoto Protocol is paving the way. The protocol provides for the use of "flexible instruments," which were introduced so that wealthy nations could achieve their emission reductions in other countries by paying royalties to compensate for pollution levels. One instrument is the Clean Development Mechanism (CDM). The CDM facilitates the financing of clean technologies (through investment in solar energy, wind turbines, hydroelectric power stations, and energy-saving

TABLE 1.4 Intensity and Frequency Characteristics of Renewable Sources

System	Major Periods	Major Variables	Power Relationship	Comment	Approximate Time Variation
Direct sunshine	24 h, 1 y	Solar beam radiance $G_b^*(W/m^2)$, beam angle from vertical q_z	$P \propto G_b \cos \theta_z$; $P_{max} \cong 1\,kW/m^2$	Daytime only, highly fluctuating	Hours to seconds
Diffuse sunshine	24 h, 1 y	Cloud cover, perhaps air pollution	$P < \sim 300\,W/m^2$	Significant energy, however	Day
Biofuels	1 y	Soil condition, solar radiation, water, plant species, wastes	Stored energy, $10\,MJ/kg$	Many variations, linked to forestry and agriculture	Year
Wind	1 y	Wind speed u_0 nacelle height above ground z, height anemometer mast h	$P \propto u_0^3 \dfrac{u_z}{u_h} = \left(\dfrac{z}{h}\right)^b$	Highly fluctuating, $b \approx 0.15$	Minutes to hours for wind farms
Wave	1 y	Reservoir height H_s, wave period T	$P \propto H_s^2 T$	High power density $\approx 50\,kW/m$ across wavefront	Week
Hydro	1 y	Reservoir height H, water volume flow rate	$P \propto HQ$	Established resource	Months
Tidal	12 h, 25 min	Tidal range R, contained area A, estuary length L, depth h	$P \propto R^2 A$	Enhanced tidal range if $L/\sqrt{h} \approx 36{,}000\,m^{1/2}$	12 h
Geothermal	None	Temperature of aquifer or rock formation, hence temperature difference from ambient	$P \propto (\Delta T)^2$	Very few suitable locations for electricity generation	None

Source: Ref. [2].

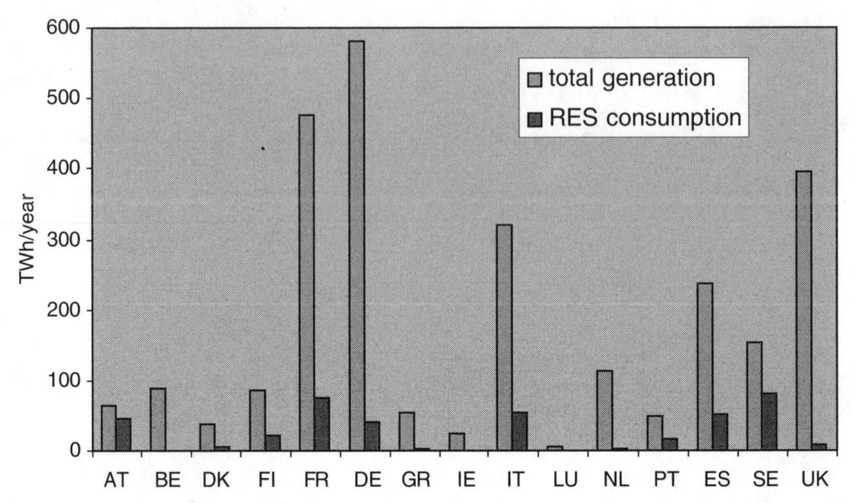

Figure 1.1 Electricity generation from renewable energy versus total electricity consumption in european countries, 2001. (From Ref. [2].)

technologies) and the transfer of these technologies from the northern to the southern hemisphere. Wealthy nations can use emission reductions achieved via the CDM to meet their Kyoto commitments, but the same cannot be said with respect to developing nations. Developing countries gain access to clean, endemic sources and compromise their future in much the same way as did the northern hemisphere. Is this the ideal win–win solution? The atomic energy industry claims that nuclear energy can be used as an effective solution in the struggle to prevent climate change.

From socioeconomic and environmental points of view, renewable energy increases supply security, has the lowest environmental effect of all energy sources, allows for local solutions, and offers sustainable energy development worldwide. Renewable energy also offers wider opportunities for investment, avoided fuel costs, CO_2 emissions savings, and new jobs. Generally speaking, renewable energy technologies are important because of the income that results from manufacturing, project development, servicing, and in the case of biomass, rural jobs and income diversification for farmers.

1.4 PLANNING AND DEVELOPMENT OF INTEGRATED ENERGY

Many studies show that the global wind resource technically recoverable is more than twice as much as the projection for the world's electricity demand in 2020. Similarly, theoretical solar energy potential (see Chapters 5 and 6 for solar thermal and solar photovoltaics, respectively) corresponds to almost 90,000,000 Mtoe (million tonne oil equivalent) per year, which is almost 10,000 times the world total primary energy supply [3]. The rapid deployment of renewable energy technologies, and their wider deployment in the near future, raise challenges and opportunities regarding their integration into energy supply systems.

The planning and development of integrated energy must consider the environment itself, the existence of energy sources, system needs, and local needs where it is desirable to install a renewable energy source. The capability of the grid supply, the electrical and mechanical behavior of the load, the distributed generation sources, and the effects on the regional economy define how successful the investment may be.

1.4.1 Grid-Supplied Electricity

The thickness, and hence cost, of conducting cables is inversely proportional to the voltage of power; therefore, high voltages are preferred for electricity supply along transmission and subtransmission lines. The practical limits relate to safety issues, especially sparking and insulation at high voltage. In practice, the voltage of long-distance transmission is 50 to 750 kV. Local area distribution is 6 to 50 kV, and supply to consumers is 100 to 500 V. Internally, in equipment, it is 3 to 48 V.

Grid electricity is converted from the primary source by:

- Moving wires in magnetic fields (Faraday effect)
- Photovoltaic generation with sunlight (photovoltaic effects)
- Chemical transformations as in fuel cells and batteries (electrochemical effects)

Transformation between transmission voltages is easiest and cheapest with alternating current (ac to ac). Transforming between direct currents (dc to dc) or between ac and dc is possible using electronic interfaces, which have become increasingly reliable and cheaper because of solid-state power electronics. Transmission with ac power has more loss per unit of distance than with dc power because of stray capacitances and inductances along lines, which increase current losses. Nevertheless, the ease of transformation means that the majority of power transmission is accomplished with high-voltage ac up to 350 km. The economic facilities of high-voltage dc transmission systems favor distances greater than 500 km.

To regulate power, voltage, speed, and frequency, each method depends on matching, instantaneously, load to generation. Generation is distinguished by its economic and physical ability to vary to match load. Examples are:

1. Base load generation (difficult or expensive to vary; e.g., nuclear power, large coal, and large biomass)
2. Peaking generation (easier to vary quickly but may be expensive; e.g., gas turbines and fuel cells)
3. Standby generation (easy to increase generation rapidly from off or idling modes; e.g., diesel, fuel cells, and gas turbines)

4. Intermittent generation (e.g., run-of-the-river hydro, wind, and most renewables, except biomass and geothermal)

Note that fuel cells can be used as peaking or standby generation; this depends on the availability of hydrogen or fuel gas and the maximum excess power that can be withdrawn from the cell without overly compromising its useful life. The same note is applied to reservoir hydropower, which may be either base (plenty of water) or peaking (limited water) load. Note also that intermittent does not mean unpredictable availability but guarantees that load always equals generation.

1.4.2 Load

It is important to realize that electricity users do not want electricity alone. They want a service, such as transportation (vertical or horizontal), lighting, water, welding, motor movement, communication, or warmth. The success of the electricity supply must be judged by the availability, quality, and cost of the service. The quality of a service (e.g., heating or water supply) tends to be measured by an intensive parameter (e.g., temperature or pressure) and the availability of that parameter. The cost of the service is measured by an extensive parameter (e.g., energy or kilowatthours) linked to its availability. The desire for the service presents a demand on the grid system, which power engineers see as load.

Ordinary consumers use the name of the service (e.g., television) for the function and never expect to use the word *load*. It is most efficient to use the word *demand* for the desire of consumers for service and the word *load* for the consequent electricity consumption. This subtle distinction is maintained in this book.

Satisfactory service can be maintained without the continuous consumption of electricity. For instance, water of a satisfactory temperature can be supplied from a previously heated tank. If the value of the intensive parameter is maintained (i.e., shower temperature), the consumer is satisfied even if electrical supply is interrupted. A demand that is satisfied by intermittent power is an *interruptible load*, also called a *switchable load*. If load and tariff management are used to optimize a power system (e.g., to increase the penetration of renewable energy), it is quite acceptable to use it to induce new trends in energy use.

1.4.3 Distributed Generation

Distributed generation (DG) is the application of small generators, typically 1 to 10 MW, scattered throughout a system to provide electrical energy closer to consumers. Current DG power sources include hydropower, wind, photovoltaics, diesel, fuel cells, and gas turbines. Renewable and other generators located downstream in a distribution network and involving small, modular electricity generation units close to the point of consumption are defined in this book as DG.

In this section we provide a qualitative analysis of the issues that drive the effects of DG on a transmission system. More technical details and deeper insights are dealt with in Chapter 13, where we also define the services that DG can provide to distribution systems. In this section, the transmission services that DG is technically capable of providing are identified, and guidelines are developed to enable DG to participate in markets for these services.

Studies, reports, and experts in the field of DG [3–6] refer broadly to the benefits that DG can provide to transmission and distribution systems. The amount of generation relative to system total load, or penetration, is the most important factor in determining the influence of DG on transmission operation. A single 2-MW generator may have a considerable effect on the operation of a distribution system but go entirely unnoticed on a transmission system. On the other end of the spectrum, if a fully mature DG market supplies 30% or more of total customer load, the effect and importance to transmission operation will be undeniable. A tougher question is: What are the effects at penetration levels between the two extremes, and how should they be treated in respect to system control and economic valuation? This is addressed by focusing on (1) localized transmission benefits that a relatively small penetration of well-sited DG can provide, and (2) the benefits to the larger transmission system that can feasibly be achieved by growing DG penetrations.

Among the services DG can provide to the distribution system are capacity support, contingency capacity support, loss reduction, voltage support, voltage regulation, power factor control, phase balancing, and equipment life extension. DG can be defined as generation located at or near a load. *Combined heat and power* (CHP) is associated with prime movers that provide shaft power to generators and encompasses two broad categories: reciprocating engines and turbines. (Fuel cells may soon make a significant entrance on the CHP stage as well, but they are not yet ready for prime time.) CHP systems (also known as *cogeneration*) are generally developed by a user to avoid the purchase of power from the grid or by the energy service provider that retails the power to the site.

CHP is considered a subset of DG and can be used when there is a potential for profitable use of thermal energy. CHP is an energy cascade that captures energy normally rejected as part of a process. In the traditional case, steam is raised with a boiler on-site, and power is purchased from the local utility. The thermal energy in the steam is then employed for another use.

1.5 RENEWABLE ENERGY ECONOMICS

To meet the demand for a broad range of services (e.g., household, commerce, industry, and transportation needs), energy systems are needed. An energy supply sector and the end-use technology to provide these energy services are also necessary. In the United States, European Union, Russia, and Japan, the electricity supply system is composed of large power units—mostly fossil-fueled and centrally controlled—with average capacities of hundreds of megawatts. Conversely, renewable energy sources are geographically distributed and if embedded in distribution networks, are often closer to customers and therefore subject to smaller losses.

In the power sector, most utilities have limited experience interconnecting numerous small-scale generation units with their distribution networks. Complicating matters, the possible level of renewables penetration depends on the existing electrical infrastructure. For example, transporting to land the power produced by a large offshore wind farm is (economically) possible only where sufficient electricity grid capacity is available. In some locations, new electricity infrastructures have been set up to provide high penetration levels of up to 100% electricity from renewables.

Distributed electricity generation, close to the end customer, differs fundamentally from the traditional model of a large power station that generates centrally controlled power. The DG approach is new and replaces the concept of economy of scale (using large units) with economy of numbers (using many small units), although it has yet to prove itself.

Far from being a threat, DG-based renewable energy can reduce transmission and distribution losses as well as transmission and distribution costs, provide consumers with continuity and reliability of supply, stimulate competition in supply, adjust prices via market forces, and be implemented in a short time and with scalonated resources because of its modular nature. The International Energy Agency alternative scenario (WEO, 2002; WEIO 2003) predicts savings of about 40% for the transmission grid and 36% for the distribution grid because of the increased use of DG. This is a significant and driving argument when recent blackouts in the United States, Brazil, and Italy are taken into account.

1.5.1 Calculation of Electricity Generation Costs

When calculating generation costs, a distinction must be made between existing and potential plants. For existing plants, the running costs (short-term marginal costs) are relevant only for the economic decision as to whether to use the plant for electricity generation. Conversely, for new capacities, long-term marginal costs are important [2].

Existing Plants Annual running costs are split into fuel costs and operation and maintenance (O&M) costs. *Fuel costs* are a function of the fuel price of the primary energy carrier and efficiency. *O&M costs* refer to electricity output, hence must be coupled with full-load hours. In general, one average operation time (full-load hour) is taken for each technology band. Analytically, generation costs for existing plants are given by

$$C = C_{\text{var}} = C_{\text{fuel}} + \tilde{C}_{\text{O\&M}} - R_{\text{heat}} = \frac{p_{\text{fuel}}}{\eta_{\text{el}}} + \frac{C_{\text{O\&M}}}{H_{\text{el}}} \times 1000 - p_{\text{heat}} \frac{\eta_{\text{heat}}}{\eta_{\text{el}}} \frac{H_{\text{heat}}}{H_{\text{el}}} \tag{1.1}$$

where C = generation costs per kilowatt [euros (€)/MWh]
C_{var} = running costs per energy unit (€/MWh)
C_{fuel} = fuel costs per energy unit (€/MWh)
$\tilde{C}_{\text{O\&M}}$ = operation and maintenance costs per energy unit (€/MWh)
R_{heat} = revenues gained from purchase of heat (€/MWh)

p_{fuel} = fuel price primary energy carrier (€/MWh$_{\text{primary}}$)
p_{heat} = heat price (€/MWh$_{\text{heat}}$)
η_{el} = efficiency—electricity generation
η_{heat} = efficiency—heat generation
H_{el} = full-load hours—electricity generation per annum (h/yr)
H_{heat} = full-load hours—heat generation per annum (h/yr)

The full-load hours represent the equivalent time of full operation for a year. This is calculated for a power plant by dividing the amount of electricity generated per year by the plant's nominal power capacity. For theoretical cost–resource curves, this reflects an important aspect: the suitability of sites. In the case of wind energy, the full-load hours are determined by the wind speed distribution and the rated wind speed of the machines. Knowing the expected full-load hours, the quantity of electricity to be generated can be calculated. Hence, costs per unit are determined. The number of full-load hours divided by the number of hours in a year (8765 hours, on average) equals the system capacity (dimensionless).

New Plants Electricity generation costs consist of variable costs and fixed costs. Generation costs are given by

$$C = C_{\text{var}} + \frac{C_{\text{fix}}}{q_{\text{el}}} = \left(C_{\text{fuel}} + \frac{C_{\text{O\&M}}}{H_{\text{el}}} \times 1000 - R_{\text{heat}} \right) + \frac{1000\, I(\text{CRF})}{H_{\text{el}}} \tag{1.2}$$

where C = generation costs per kilowatthour (€/MWh)
C_{var} = running costs per energy unit (€/MWh)
C_{fix} = fixed costs (€)
q_{el} = amount of electricity generation (MWh/yr)
$C_{\text{fix}}/q_{\text{el}}$ = fixed costs per energy unit (€/MWh)
C_{fuel} = fuel costs per energy unit (€/MWh)
$\tilde{C}_{\text{O\&M}}$ = operation and maintenance costs per energy unit (€/MWh)
R_{heat} = revenues gained from purchase of heat (€/MWh)
I = investment costs per kilowatt (€/kW)
CRF = capital recovery factor, $= \dfrac{z(1+z)^{\text{PT}}}{[(1+z)^{\text{PT}} - 1]}$
z = interest rate
PT = payback time of the plant (years)
H_{el} = full-load hours—electricity generation per annum (h/yr)

Fixed costs occur whether or not a plant generates electricity. These costs are determined by investment costs (I) and the capital recovery factor.

Investment Costs Investment costs differ according to technology and energy source. In general, investment costs per unit of capacity for renewable energy systems are higher than for conventional technologies based on fossil fuels. Also, differences exist among renewable energy technologies (e.g., investment costs per unit of capacity for small hydropower plants are generally at least twice those for wind turbines).

Investment costs decrease over time and are usually derived annually. It is usual to consider renewables as having zero fuel costs, apart from biomass (biogas, solid biomass, and sewage and landfill gas), so running costs are determined by operation and maintenance costs only. Therefore, the running costs for renewable energy systems are normally low compared with those of fossil-fuel systems.

Capital Recovery Factor The capital recovery factor allows investment costs incurred in the construction phase of a plant to be discounted. The amount depends on the interest rate and the payback time of the plant. For the standard calculation of generation costs, these factors may be set as follows for all technologies:

- Payback time (PT) of all plants: 15 years
- Interest rate (z): 6.5%

Different interest rates may be applied in any economical study. The interest rate depends on stakeholder behavior and is a function of a guaranteed political planning horizon of promotion scheme of technology of investor category.

Generation costs are calculated per unit of energy output, so fixed costs must be related to generation. Hence, fixed costs per unit of output are lower if the operation time of the plant—characterized by full-load hours—is high. Deriving generation costs for CHP plants is similar to calculating them for plants that produce electricity only. Both short-term marginal costs (i.e., variable costs) and fixed costs must be considered for new plants. Of course, variable costs differ between CHP and conventional electricity plants because the revenue from heat power must be considered in the former. In general, no taxes are included in the various cost components.

1.6 EUROPEAN TARGETS FOR RENEWABLES

Worldwide, several scenarios share the goal of sustainability in general or in the energy field. Thus, groundbreaking targets toward this goal are important for renewable energy and end-use energy efficiency. Such targets can guide policymakers during decision making and send important signals to investors, entrepreneurs, and the public. Case studies have demonstrated how concrete targets can lead to increased impact in various fields. In the case of renewable energies, policymakers formulate concrete policies and support measures to foster their development. Investors develop related strategies and renewable businesses as targets convince them that their investment will yield the returns projected.

Renewable power is available in many environmental energy flows, harnessed by a range of technologies. The parameters used to quantify and analyze these forms are listed in Table 1.4.

A study by C. Kjaer [7,8] in the EU emphasizes 10 requirements for any community-wide mechanism to create a sound investment climate for renewables:

1. Compatibility with the "polluter pays" principle
2. High investor confidence

3. Simplicity and transparency in design and implementation
4. High effectiveness in deployment of renewables
5. Encouragement of technological diversity
6. Encouragement of innovation, technological development, and lower costs
7. Compatibility with the power market and with other policy instruments
8. Facilitation of a smooth transition ("grandfathering")
9. Encouragement of local and regional benefits, public acceptance, and site dispersion
10. Transparency and integrity by protecting consumers and avoiding fraud and free riding

1.6.1 Demand-Side Management Options

In the transport sector, biofuels are just beginning to be developed in Europe. However, in some countries, such as Brazil, sugarcane and oily plants already play important roles in the energy matrix. In the transport sector, the integration of renewables requires the adaptation of an infrastructure that has grown over a century of development based exclusively on fossil fuels. Besides the gradual substitution of vehicles in circulation, it is necessary to develop a new supply chain for the production and distribution of biofuels. This will require substantial investment. However, development of the fossil fuel–based transport system also required investment, which was subsidized by the public sector in many countries.

In the heating sector, the full integration of renewable energy requires an adaptation of historical infrastructures. In many parts of Europe, it is already possible to construct buildings completely independent of fossil fuels or electricity for heating needs. This is achieved using state-of-the-art renewable heating and cooling applications linked with energy-efficiency measures and demand-side management (see Appendixes B and C).

A substantial economic restriction to the integration of renewable heating (i.e., solar thermal, biomass, and geothermal) is the long lifetime of buildings. The installation of renewable heating systems is more cost-effective during the construction of a building or when the overall heating system is being refurbished. This means that there is a small window of opportunity for cost-effective integration of renewable heating. If this opportunity is lost, a building will remain dependent on fossil fuels or electricity to cover its heating demand for decades. For this reason, it is essential that all possible measures be taken to ensure that renewable heating sources are installed in all new buildings. It is also necessary to promote the use of renewable heating whenever a conventional heating system is being modernized.

Renewable heating sources can also be used for cooling. An increasing number of successful systems are being installed, based mainly on solar thermal and geothermal energy. The growing demand for cooling is affecting electricity systems in Europe, and several countries are now reaching peak electricity demand in summer instead of winter. These problems can be mitigated by the development and commercialization of renewable cooling technologies.

The existing infrastructure and market dominance of conventional heating and cooling technologies create a substantial barrier to growth for renewable heating. Biomass heating and cooling can be competitive in areas where the fuel supply chain is well developed, but this is not yet the case in many parts of the world. Solar thermal systems can be good economic investments, but in many areas, users are not aware of this. In addition, most heating installers are trained only for conventional heating systems and therefore encourage customers to stick to conventional heating. The integration of distinct energy-intensity profiles can be considered in these cases.

The choices that millions of citizens make for their homes and offices are crucial to the future integration of renewable energies in the heating and cooling sectors. Raising awareness among the public and training the professionals involved (e.g., conditioning and climatization installers, building engineers, architects, and managers of heat-intensive buildings or devices) are therefore very important.

Increasing use of renewable energies must be accompanied by energy efficiency and demand-side management measures at the customer end. Renewable energy development and energy efficiency are strongly interdependent. The European Union has always stressed the need to renew commitment at the community and member-state levels to promote energy efficiency more actively. In light of the Kyoto agreement to reduce carbon dioxide emissions, improved energy efficiency, together with increased use of renewables, will play a key role in meeting the EU Kyoto target economically (see Table 1.1). In addition to a significant positive environmental effect, improved energy efficiency will lead to a more sustainable development and enhanced security of supply as well as many other benefits. An estimated economic potential for energy-efficiency improvement of more than 18% of present energy consumption still exists today in the EU as a result of market barriers, which prevent the satisfactory diffusion of energy-efficient technology and the efficient use of energy. This potential is equivalent to more than 1900 TWh, roughly the total final energy demand of Austria, Belgium, Denmark, Finland, Greece, and the Netherlands combined.

Special emphasis should be placed on urban areas, where a high proportion of energy is consumed. Urban areas are characterized by highly developed infrastructures, which do not always easily allow a rapid increase in renewable energy generation. The fact that electrical network infrastructures are generally overdimensioned in urban areas can, in some cases, allow a high penetration of photovoltaic generators and wind energy without changing the existing cabling, transformer stations, and so on. However, in general, the future energy infrastructure will need to be designed from the beginning to accommodate renewable energy effectively at a high level. The small contributions that every home makes by using energy derived directly from nature (as wind, heat, coolant, light, photovoltaic electricity, and clean air) will make the biggest difference in the end.

1.6.2 Supply-Side Management Options

The European renewable energy industry has already reached an annual turnover of €10 billion and employs 200,000 people. Europe is the global leader of renewable energy technologies, and the use of renewables has a considerable effect on the investments made in the energy sector. Renewable energy replaces imported fuels, with beneficial effects on the balance of payments. Although per unit of installed capacity, renewable energy technology is more capital intensive, when the external costs that have been avoided are taken into account, investing in renewables turns out to be cheaper for society than business-as-usual investments in conventional energy. Renewable energy technologies are often on a smaller scale than fossil fuel and nuclear projects, and they can be brought online more quickly and with lower risks. Finally, deployment of renewables creates more employment than do other energy technologies.

The development of smarter, more efficient energy technology over past decades has been spectacular. Technologies have improved, and costs have fallen dramatically (see Figure 1.2). The examples of wind and solar photovoltaics are striking. Investment costs for wind energy declined by around 3% per annum over the past 15 years. For solar photovoltaic cells, unit costs have fallen by a factor of 10 in the past 15 years (stimulated initially by the space program).

In the European Union, renewables already make up a significant share of total energy production. Germany, for example, has doubled its renewable output in the past five years to 8% of total electricity production. Denmark now gets 18% of its electricity from wind power alone and has created an industry with more jobs than

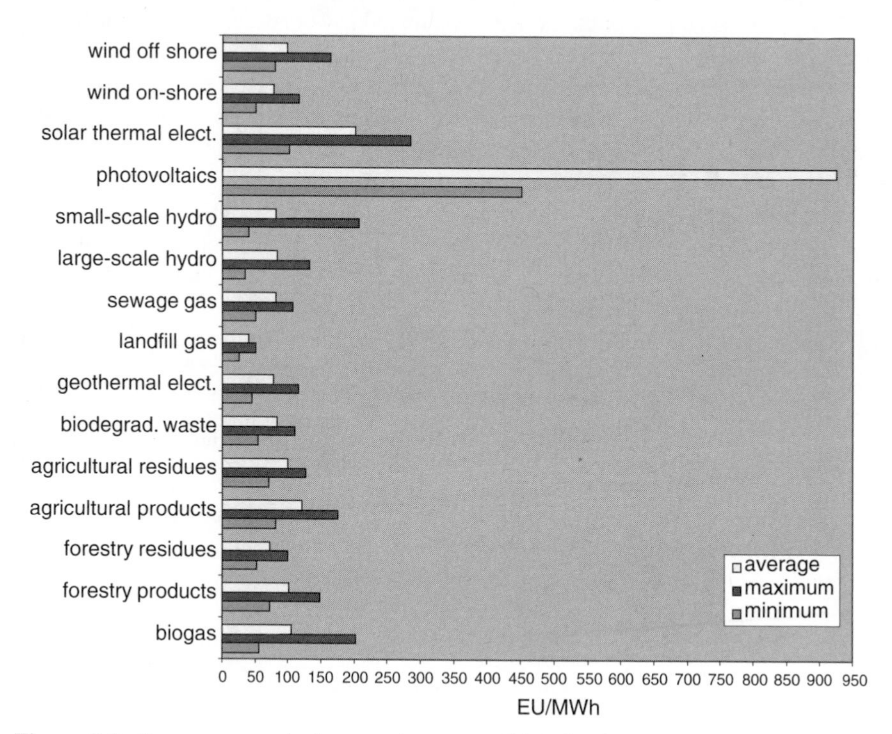

Figure 1.2 Long-run marginal generation costs (€/MWh) for renewable energy technologies in EU-15. (From Ref. [2].)

TABLE 1.5 Technical and Economical Characteristics of Selected Renewable Energy Technologies

	Unit Capacity (kW)	Electrical Efficiency (%)	Thermal Efficiency (%)	Lifetime (yr)	Full-Load Operating Hours	Investment Costs (€/kW)	Generation Costs (€ cents/ kWh)
Gas diesel engines	3–10,000	30–45	45–50	15	5000	450–1400	2.5–4.0
Microturbines	25–250	15–35	50–60	15	5000	1000–1700	3–4
Stirling engines	10–150	15–35	60–80	15	5000	1400–2200	4–5
Steam engines	0.5–10,000	15–35	40–70	15	5000	1500–3000	3–4
Wind power	0.1–5000	40–50	—	20	2500	900–1100	4–5
Fuel cells	0.5–2000	38–55	40–70	15	5000	2800–4400	5–8

Source: Vartiainen, E. et al., 2002; Gaya Group Oy, 2004; Obernberger, I., 2004.

in the electricity sector. Spain has leapt from using virtually no renewables a few years ago to become the second-biggest wind power country in Europe, with 6000 MW of capacity. Countries such as Finland, Sweden, and Austria have supported the development of very successful biomass power and heating industries through fiscal policies, sustained R&D support, and synergistic forestry and industrial policies. In addition to saving significant carbon dioxide emissions, equipment from all three countries is exported worldwide. Table 1.5 presents some technical and economical characteristics of selected renewable energy technologies, and Table 1.6 presents the EU-25 electricity production mix of 2002.

Going further, in 1991, the European Union and the United States launched ExternE, a joint project to assess the economic costs of externalities from the production and use of energy, and estimated that these costs amount to 1 to 2% of the EU's gross domestic product [8]. In July 2001, the European Commission issued a

TABLE 1.6 EU-25 Electricity Production Mix, 2002

Primary Source	Production (%)
Coal	30.4
Oil	6.1
Nuclear	31.5
Gas	17.0
Wind power	2.1
Large hydro	10.0
Other renewable energy	1.7
Other	1.2

Source: Ref. [7].

press release on the study's findings [9] which concluded that the "cost of producing electricity from coal or oil would double and the cost of electricity production from gas would increase by 30% if external costs such as damage to the environment and to health were taken into account."

1.7 INTEGRATION OF RENEWABLE ENERGY SOURCES

Integration of renewable energy sources involves integrating in a system any energy resource that naturally regenerates over a short period of time. This time scale is derived directly from the sun (such as for thermal, photochemical, and photoelectric energy), indirectly from the sun (such as for wind, hydropower, and photosynthetic energy stored in biomass), or from other natural movements and mechanisms of the environment (such as for geothermal and tidal energy). In the long term, renewable energies will necessarily dominate the world's energy supply system for the simple reason that there is no alternative. Humankind cannot survive indefinitely off the consumption of finite energy resources, concentrate supplies on some points on Earth, or carelessly spread its population over the world.

Today, the world's energy supply is based largely on fossil fuels and nuclear power. These sources of energy will not last forever and have proved to be a major cause of environmental problems. Environmental effects of energy use are not new, but it is increasingly well known that they range from deforestation to local and global pollution. In less than three centuries since the industrial revolution, humankind has burned away roughly half of the fossil fuels accumulated under Earth's surface during hundreds of millions of years. Nuclear power is also based on limited resources such as uranium, and the use of nuclear power creates such incalculable risks that nuclear power plants cannot be insured.

Renewable sources of energy are in line with an overall strategy of sustainable development. They help reduce dependence on energy imports and do not create a dependence on energy imports, thereby ensuring a sustainable security of supply. Furthermore, renewable energy sources can improve the competitiveness of industries, at least in the long run, and have a positive effect on regional development and employment. Renewable energy technologies are suitable for off-grid services; they can serve remote areas of the world without expensive and complicated grid infrastructure.

The ability to integrate electricity generated from renewables into grid supplies is governed by several factors, including:

- The variation with time of power generated
- The extent of the variation (availability)
- The predictability of the variation
- The capacity of each generator
- The dispersal of individual generators
- The reliability of plants

- The experience of operators
- The technology for integration
- The regulations and customs for embedded generation

Despite these difficulties, the experience of the past 25 years has shown that ever-increasing amounts of electricity from renewables can be integrated into grid supplies without significant financial penalty. The standard response of grid operators that are accustomed to large-scale centralized generation is that intermittent and dispersed renewable energy generation cannot be so integrated. However, given the requirement to accept specific renewable energy generation, the technology and methods have followed successfully. Examples include:

- Electrical safety equipment and grid-fault disconnectors
- Grid-linked inverters for photovoltaics, solar cells, and power from buildings
- Doubly-fed induction generators for variable-speed wind turbines
- Voltage reinforcement on rural power lines
- Cofiring of steam boilers with biomass
- Gas turbines for the output of gasifiers

The outstanding example of ever-increasing integration of renewable energy generation into the grid is Jutland, western Denmark. In the early 1980s, the limit for wind power exported to the grid was considered to be 20% of total supply. However by 2003, about 40% of annual electricity supply was from wind, and at times, significant areas were supplied totally by wind power. The reason for the change was the willing application of new technologies and practices.

Nevertheless, there are fundamental limitations for any renewable energy generation technology and plant; for instance, the sun never shines at night. Also, in the middle of large towns and cities, the surface roughness for wind move is not acceptable for small towers. So it is essential to integrate renewable energy generation options with control and storage such that they complement each other.

1.7.1 Integration of Renewable Energy in the United States

The United States currently relies heavily on coal, oil, and natural gas for its energy. Fossil fuels are nonrenewable: They draw on finite resources that will eventually dwindle and become too expensive or too environmentally damaging to retrieve. In contrast, renewable energy resources such as hydropower, wind energy, and solar energy are replenished constantly and will never run out.

As in any other place, most renewable energy in United States comes directly or indirectly from the sun. Sunlight, or solar energy, can be used directly for heating and lighting, to generate electricity, and for cooling as well as for a variety of commercial and industrial uses. The sun's heat also drives winds (whose energy is captured with wind turbines) and evaporates waters, turning them into rain or snow, which then flows into rivers or streams and whose energy may be captured in water dams.

Other renewable sources include geothermal energy, which is tapped from Earth's internal heat for electric power production and heating and cooling buildings, and the oceans' tides, which come from the gravitational pull of the moon and sun. In fact, ocean energy comes from a number of sources. In addition to tidal energy, there is the energy of the oceans' waves, which are driven by tides and winds. The sun also warms the surface of oceans more than it warms ocean depths, which creates a temperature difference that can be used as an energy source. All these forms of ocean energy can be used to produce electricity.

In contrast to fossil energy, renewable energy is an attractive source for several reasons: clean environment, long-lasting life, increased jobs, increased comfort, and industry and energy self-sufficiency through decreased dependence on other nations. An economy that uses less energy also produces less pollution, and an energy-efficient economy can grow without using more energy. Energy efficiency means using less energy to accomplish the same task. More efficient energy use results in less money spent on energy by homeowners, schools, government agencies, businesses, and industries. The money that would have been spent on energy can instead be spent on consumer goods, education, services, and products. From 1970 to 2000, U.S. energy consumption grew only 45% although the U.S. gross domestic product increased 160%. In other words, the energy used per dollar of gross domestic product decreased 44% from 1970 to 2000. By 1999, greenhouse gas emissions from energy use had risen 13% above 1990 levels. During that period, energy use increased 14.9%.

1.7.2 Energy Recovery Time

The cost of electricity depends entirely, or largely, on the size of power stations. Between 1960 and 1980, the ideal size of a station rose from 400 MW to 1000 MW. These days, 5 MW is regarded as ideal because small-scale power generation permits a flexible response to energy demand and return of capital. Small-scale units such as wind turbines, photovoltaic cells, fuel cells, and biogasification plants represent the future.

Regardless of the type of primary source, it takes energy to convert energy from one type into another. The lower the specific energy content, the more energy intensive is the conversion process. When the specific energy content is low, the energy process chain uses more energy than it generates in electricity. Most of the primary energy extracted today has a profitable content that makes conversion cost-effective.

However, if any energy were to gain momentum, a point would come when the specific energy conversion would no longer be cost-effective. The amount of time a power plant needs to operate before all the energy consumed in the chain has been earned back (and the power plant begins to produce net energy), the *energy recovery time*, is highly dependent on the specific energy content of the primary source. It is difficult to compare this figure with the energy recovery time for fossil fuel–powered power stations. A fossil fuel power station has to recover only the electricity used for construction and other constituent processes in the chain. In such a case, the recovery

TABLE 1.7 Recovery Time of Selected Sources of Energy

Alternative Source	Recovery Time (years)
Wind	0.62–0.90
Gas and oil	1
Photovoltaic system	1.5–3
Nuclear power station	10–18

Source: Ref. [3].

time for power stations fired by gas and oil is 0.09 of a full-load year (approximately 0.13 of a calendar year); for coal-fired power stations, it is 0.15 of a full-load year (approximately 0.21 of a calendar year) [8]. But unlike modern gas-fired power stations that generate and supply commercial heat, alternative sources of energy such as nuclear power plants, wind turbines, and photovoltaic systems can generate only electricity. All the energy used in the chain is recovered in the form of electricity, which increases recovery time considerably. As a frame of reference, assume that fossil fuel–fired power stations must recover the energy used in their construction only in the form of electricity. This results in a recovery time of 0.7 full-load year for gas- or oil-fired power stations, which is approximately one calendar year. Coal-fired power stations have a longer recovery time. Table 1.7 lists the recovery time of selected sources of energy.

Improvements in conversion yields and production methods will help reduce the recovery time for photovoltaic systems in the future. Photovoltaic technology is at a peak of development and at the moment is in the sharply rising section of the learning curve, which means that prices will fall significantly as more capacity is commercialized. It is conceivable that the recovery time for photovoltaics will drop to less than one year as technical progress continues. Nuclear energy, on the other hand, is a mature technology; the price of nuclear power will not decrease as more nuclear power stations are built. In the past, there were even cost hikes of approximately 14% a year until the mid-1980s. Since 1979, no new nuclear power stations were ordered in Organization of Economic Cooperation and Development countries, which ended the competitive time in which further price rises could occur. Clearly, the recovery time for nuclear power stations is much longer than that of other power stations and will never decrease. In contrast, the recovery time for photovoltaic systems, in particular, is certain to decrease if new technologies and materials are used.

Environmental issues such as the greenhouse effect have focused attention on fossil-fuel combustion and electricity generation around the world. In Australia, 47% of the annual emissions of greenhouse carbon dioxide come from fossil fuel–fired power plants [3–9]. As coal-based plants are retired, due to age and greenhouse concerns, there is an opportunity for renewable energy generation sources to grab a larger share of the global electric energy market.

Wind systems, solar systems, storage components, and complete energy systems are now commercially available from many suppliers [5] to fill niche markets.

Fundamental research (especially in the production of thin-film solar photovoltaic devices, hydrogen from sun-mirrored heat, and new forms of batteries) is occurring in many countries, and these activities are steadily reducing the cost of renewable energy systems. However, there are still issues to resolve before such systems gain a bigger portion of the electric energy marketplace. Such systems must lower the overall cost of delivered energy, gain acceptance by a conservative industry, and convince the industry's customers that renewable energy systems are safe, reliable alternatives to conventional grid-supplied power.

Another issue is that although the primary energy supply may be free, the cost of using wind and solar energy is not. This is because structures and energy collectors must be built and energy storage must be provided. Any initiative that increases the energy collected or stored will lead to a reduction in the price of energy delivered from a complete system [5,8]. *Balanced* and *optimized* are terms frequently used to indicate that a system is designed to size the renewable, storage, and fuel-based components to deliver minimal all-of-life costs in a specific site and for a specific customer loading pattern. Such a system operates to maximize renewable energy capture and to minimize all-of-life costs of components.

1.7.3 Sustainability

The Fifth Environmental Action Programme of 2000 established EU legislation and defined sustainable development as "that which meets the needs of the present without compromising the ability of future generations to meet their own needs" [10]. The policy objectives underlying this definition were to ensure compatibility between economic growth and efficient and secure energy supplies together with a clean environment.

Environmentally polluting by-products are produced by conventional energy generation, which also depends on finite energy sources that are gradually being depleted. However, energy is essential for socioeconomic progress in developing and industrialized countries, and the demand for energy will increase with global population.

Targets established in the EC white paper of 1997 foresee a 12% share of renewables in Europe's total energy consumption by 2010 (double the 1997 share). Individual targets for each renewable energy technology are also set. Annual growth rates between 1995 and 2001 show that one sector (wind) is far beyond the target and that others (i.e., hydro, geothermal, and photovoltaics) are in line with expectations. To reach the overall and sector targets (which is feasible), specific support actions have to be taken soon for technologies that lag behind, such as biomass and solar thermal. The deployment status of energy consumption by energy source in the United States is illustrated in Table 1.8.

Given the present state of market progress and political support, the expectation is that if strong additional support measures are adopted, the overall contribution of renewable energy to energy consumption in 2020 will be 20%. These estimates are based on a conservative annual growth scenario for the technologies. To reach the target, strong energy-efficiency measures have to be taken to stabilize

TABLE 1.8 U.S. Energy Consumption by Energy Source, 2003

Source	Production (%)
Coal	22.773
Natural gas	22.490
Petroleum	39.074
Nuclear	7.795
Hydroelectric	2.779
Biomass	2.865
Solar	0.063
Wind	0.108
Geothermal	0.314

Source: Renewable DOE-EIA Energy Annual, 2003.

energy consumption between 2010 and 2020. These novelties in the energy market have opened discussions about what would be necessary socially, politically, and economically for a country to adapt to new environmental surroundings. In particular, when the director of the National Aeronautics and Space Administration's Goddard Institute for Space Studies enlightened the U.S. Congress to the fact that human-induced global warming was detectable in the climate record, some skeptical members of Congress argued that the data were unclear and inconclusive. According to the Goddard Institute, the consequences predicted for global warming include worldwide floods, droughts, rising sea levels, category 5 hurricanes, and typhoons. These effects, though, were widely debated. Agreement was barely reached that deep reductions in carbon emissions cannot be made economically without the use of energy-efficiency and renewable energy technologies.

Since then, standards have become necessary to regulate new power interconnections with distribution systems. The Institute of Electrical and Electronics Engineers (IEEE) 1547 Standard for Interconnecting Distributed Resources with Electric Power Systems is the first in the 1547 series of planned interconnection standards [11]. There are major obstacles to an orderly transition to the use and integration of distributed power resources with electric power systems, as discussed in Chapter 15. Examples include the lack of uniform national interconnection standards and tests for interconnection operation and certification and the lack of uniform national building, electrical, and safety codes. Resolving this requires time to develop and promulgate consensus. The 1547 standard is a milestone for the IEEE standard-setting process and demonstrates a model for ongoing success for further national standards and for moving forward in modernizing the national electric power system.

Figure 1.3 depicts the 1999 U.S. energy flow in the net primary consumption given in quads and exajoules. A quad is 1 quadrillion (1015) Btu, and an exajoule is 1018 joules.

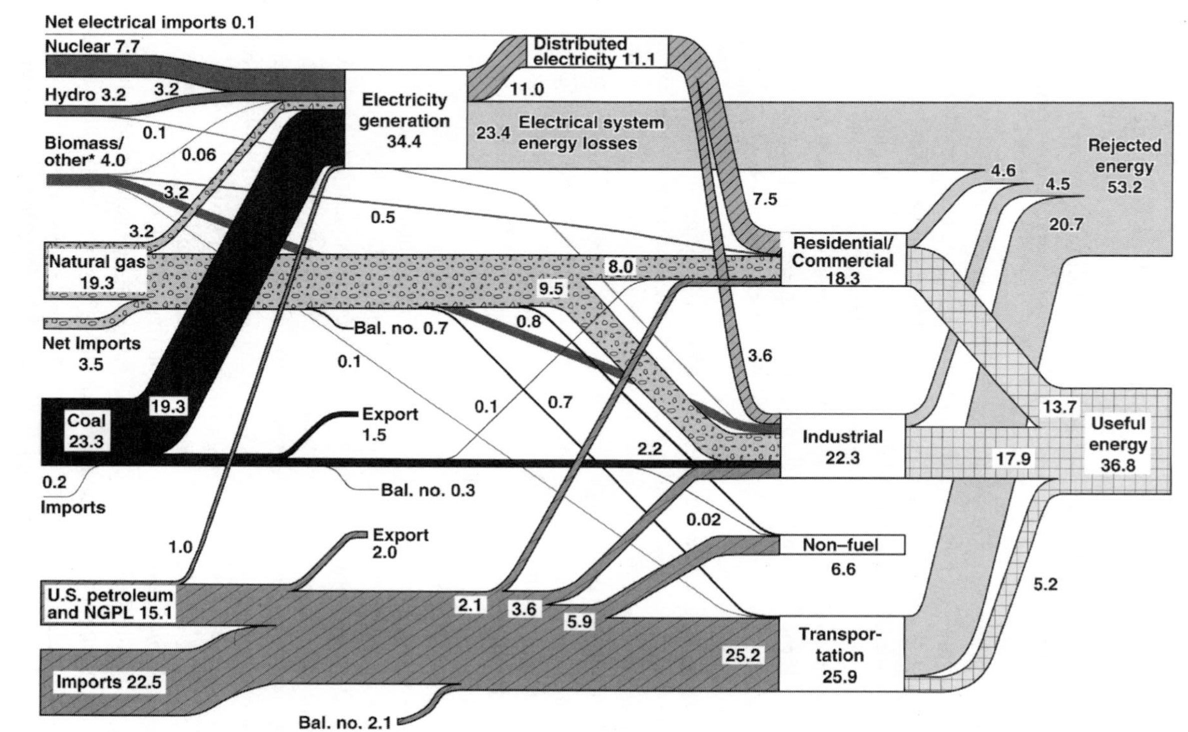

Figure 1.3 U.S. energy flow, 1999. Net primary resource consumption = 97 quads. (From Ref. [12].)

1.8 MODERN ELECTRONIC CONTROLS OF POWER SYSTEMS

Renewable and alternative energy sources must eventually be integrated with existing electric systems. Power electronics are a crucial enabling technology toward this end. Power electronics is a part of electronic application systems that encompasses the entire field of power engineering, from generation to transmission and distribution to transportation, storage systems, and domestic services. The progress of power electronics has generally followed microelectronic device evolution and influenced the current technological status of renewable energy conversion.

The power produced by renewable energy devices such as photovoltaic cells and wind turbines varies on hourly, daily, and seasonal bases because of variation in the availability of the sun, wind, and other renewable resources. This variation means that power is sometimes not available when it is required and that on other occasions there is excess power. The variable output from renewable energy devices also means that power conditioning and control equipment is required to transform this output into a form (i.e., voltage, current, and frequency) that can be used by electrical appliances. Therefore, energy must be stored and power electronics used to convert this energy.

Power-processing technology can be classified according to the energy, time, and transient response required for its operation. As the cost of power electronics falls, system performance improves. Applications are proliferating, and it is expected that this trend will continue with high momentum in this century. Modern industrial processes, transportation, and energy systems benefit tremendously in productivity and quality enhancement with the help of power electronics. The environmentally clean sources of power—such as wind, photovoltaics, and fuel cells—will be highly emphasized in the future because efficient-energy conversion from renewable sources depends heavily on power electronics.

In this book we are concerned with how alternative and renewable energy can be integrated electrically. Power electronics technology plays a major role in the injection of electrical power to the utility grid, as discussed in Chapter 12. If only photovoltaic and fuel cell systems are used, a dc-link bus could be used to aggregate them, and ac-power could be integrated through dc–ac conversion systems (inverters). If only hydro or wind power is used, variable-frequency ac voltage control can be aggregated into an ac link through ac–ac conversion systems. Ac–ac conversion systems can be created through several approaches discussed in this book.

Of course, alternative energy sources such as diesel and gas can also be integrated with renewables. They have a consistent and constant fuel supply, and the decision to operate them is based more on straightforward economics. Gas microturbines and diesel generators are commercially available with synchronous generators that supply 60 Hz, and a direct interconnection with the grid is typically easier to implement. When integrating and mixing these sources, a microgrid can be based on a dc- or ac-link structure. The design of such a microgrid must incorporate energy storage with seamless control integration of source, storage, and demand.

REFERENCES

[1] Wiertzstraat, Wise, *Coming Clean: How Clean Is Nuclear Energy*? Stichting GroenLinks in the European Union/The Group of the Greens/European Free Alliance, Brussels, Belgium, October 2000.

[2] G. Resch, H. Auer, M. Stadler, C. Huber, L. H. Nielsen, J. Twidell, and D. J. Swider, *Dynamics and Basic Interactions of RES with the Grid, Switchable Loads and Storage*, Work Package 1 within the 5th Framework Programme of the European Commission, supported by DG TREN, Contract NNE5-2001-660, October 2003.

[3] EREC (Integration of Renewable Energy Sources), Targets and benefits of large-scale deployment of renewable energy sources, presented at the Workshop on Renewable Energy Market Development Status and Prospects, Workshops in the New Member States, April–May 2004.

[4] http://www.energyquest.ca.gov/story/chapter19.html.

[5] http://www.nrel.gov/clean_energy/whatis_re.html.

[6] J. Diamond, California Energy Commission, consultant report prepared by Outside Energy Corporation, USA, January 2002.

[7] C. Kjaer, Position paper on the future of EU support systems for the promotion of electricity from renewable energy sources, news release, European Wind Energy Association, Brussels, Belgium, November 2004.

[8] Communication from the Commission to the Council and the European Parliament, *The Share of Renewable Energy in the EU*, report in accordance with Article 3 of Directive 2001/77/EC, *Evaluation of the Effect of Legislative Instruments and Other Community Policies on the Development of the Contribution of Renewable Energy Sources in the EU and Proposals for Concrete Actions*, SEC 366, May 26, 2004.

[9] Commission of Restructuring of British Energy, *A Quantitative Assessment of Direct Support Schemes for Renewables*, IP/04/1125, September 22, 2004; Working Group on Renewables and Distributed Generation, Ref. 2003-030-0741, Table 4, p. 22, January 2004.

[10] http://externe.jrc.es/overview.html.

[11] T. S. Basso and R. DeBlasio, *IEEE Standards for Interconnection*, P1547, IEEE Press, Piscataway, NJ, .

[12] G. V. Kaiper, *U.S. Energy Flow, 2000*, UCRL-ID-129990-00, Lawrence Livermore National Laboratory, Energy and Environment Directorate, Stanford, CA, February 2002.

CHAPTER 2

PRINCIPLES OF THERMODYNAMICS

2.1 INTRODUCTION

To analyze the conversion of energy from one form to another, the availability of energy to do work, and how heat will flow, it is important to understand the laws and applications of thermodynamics. In engineering thermodynamics, the emphasis is on processes involving transformations of work, heat, and internal energy by boilers, steam turbines, internal combustion engines, refrigerators, air conditioners, and other thermal and mechanical devices.

Fluids play an important role in modern storage and energy conversion systems. A system is *open* if it exchanges matter and energy with its surroundings; it is *isolated* if there is no interaction with its surroundings; and it is considered a *closed system* if it exchanges energy, but not matter, with its surroundings. Working fluid is the matter contained within the boundaries of such systems. Matter can be in solid, liquid, vapor (or gaseous), and plasmatic phases. The working fluid in applied thermodynamic problems is approximated by either a perfect gas or a substance that exists as liquid and vapor.

There is a need to supply energy to several sectors of human activity: (1) residential, for appliances and lighting, space heating, water heating, and air conditioning; (2) commercial, for lighting, space heating, office equipment, water heating, air

Integration of Alternative Sources of Energy, by Felix A. Farret and M. Godoy Simões

conditioning, ventilation, and refrigeration; (3) industrial, for water and steam boilers, direct process energy, and machine drives; and (4) transportation, for personal automobiles, light and heavy trucks, air transport, heat water recovery transport, pipe transport, rail transport, and so on. In all four sectors, in addition to primary energy sources, electrical energy, as produced from primary energy sources, is used extensively.

Thermodynamics can separate the possible from the impossible in thermal and mechanical processes. However, its primary use is in predicting the input energy required and output energy expected, fluid flows, temperature rise, and other property changes in a thermal system. Therefore, an understanding of thermodynamics principles is very important as they apply to the sustainability of modern society.

2.2 STATE OF A THERMODYNAMIC SYSTEM

The state of a system is the set of thermodynamic values known as *properties*; there are *extensive properties*, which depend on the amount of material (e.g., mass and volume), and *intensive properties*, which do not depend on the amount of material. Numerical computation involving physical quantities requires their units be given in a homogeneous system. The metric units adopted by the International Standards Organization (ISO) are called SI (Système International d'Unités). However, thermodynamic problems and industrial processes still adopt the U.S. Customary System (USCS) units. In U.S. Customary System, both force and mass are called pounds. Therefore, one must distinguish the pound-force (lbf) from the pound-mass (lbm). The pound-force is that force which accelerates 1 pound-mass at 32.174 ft/sec^2. Thus, 1 lbf = 32.174 lbm-ft/sec^2. The expression 32.174 lbm-ft/(lbf-sec^2), designated as g_c, is used to resolve expressions involving both mass and force expressed as pounds. For instance, the equation for Newton's second law would be written $F = ma/g_c$, where F is in lbf, m in lbm, and a in ft/sec^2. Similar expressions exist for other quantities, such as kinetic energy: $\text{KE} = mv^2/2g_c$ with KE in ft-lbf; potential energy: $\text{PE} = mgh/g_c$, with PE in ft-lbf; fluid pressure: $p = \rho gh/g_c$, with p in lbf/ft^2; and specific weight: $\text{SW} = \rho g/g_c$ in lbf/ft^3. In these examples, g_c should be regarded as a unit conversion factor. It is frequently not written explicitly in engineering equations, but it is required to produce a consistent set of units. Note that the conversion factor g_c [lbm-ft/(lbf-sec^2)] should not be confused with the local acceleration of gravity, g, which has different units (m/s^2) and may be either its standard value (9.807 m/s^2) or some other local value. If the problem is presented in USCS units, it may be necessary to use the constant g_c in the equation to have a consistent set of units.

Energy is an extensive property, and thermodynamic principles provide ways to calculate the change in energy for a closed system. The dimensions of energy are the same for work. Although the dimensions are the same, engineers may use the SI system where the unit is the joule (J), where $1\text{ J} = 1\text{ N}\cdot\text{m} = 1\text{ W/s}$. In the USCS system there are two main units, the British thermal unit (Btu), used in the study of heat, and the foot-pound-force (ft-lbf), used in the study of mechanical work. One

British thermal unit equals approximately the amount of energy required to raise the temperature of 1 pound-mass of liquid water 1 degree Fahrenheit at room temperature. Some properties important in thermodynamic problems are described next.

Pressure is a force acting on a unit area, with USCS units of pounds per square inch gauge (psig), but the "g" is generally assumed and thus omitted (i.e., psi). The SI unit is the pascal (Pa) or 1 N/m^2, and the conversion factor is 1 psi = 6894.8 Pa. Absolute pressure, the pressure above the lowest possible pressure, is the sum of the gauge pressure and the atmospheric pressure. Pressure-measuring devices generally measure gauge pressure rather than absolute pressure (i.e., gauge pressure is the difference between a referenced pressure and atmospheric pressure). By convention, measurements in psia always include the "a." A constant atmospheric pressure of 14.7 pounds per square inch absolute is usually assumed, corresponding to a gauge pressure of 0 psig. When pressures are very close to atmospheric pressure, as in much of the heating and air-conditioning field, one may follow the usual practice of expressing pressure in inches or meters of water. A pressure of 1 in. of water is the pressure at the bottom of a column of water 1 inch high. One psi is equivalent to 27.7 in. of water, and 1 atmosphere (atm) is equal to 1.01325×10^5 Pa.

Temperature is a measure of the average kinetic energy of a system, taken in practice as the relative hotness or coldness of a substance. Numerical values are given significance, however, through the use of well-defined temperature scales. The units of temperature can be defined in USCS units by degrees Fahrenheit (°F), with a conversion factor to Celsius as $T_C = \frac{5}{9}(T_F - 32)$. The unit of thermodynamic temperature or absolute temperature is the kelvin, the fraction 1/273.15 of the thermodynamic temperature of the triple point of water. The conversion of degrees Celsius to kelvin is $T_K = T_C + 273.15$. The thermodynamic temperature is used in the SI system because it is related to the energy possessed by matter. However, temperature itself is not a form of energy and should not be confused with the heat output of a device. A solar collector at near-zero flow conditions will generate very high temperatures. The energy collected for this condition is yet quite low compared to a high-flow, moderate-temperature condition.

Specific weight is defined as the weight per unit volume. The USCS units are pounds per cubic foot (lb/ft^3) and the SI unit of specific weight is kN/m^3. Because weight is a force, the specific weight is related to the density by Newton's second law.

Specific volume is defined as the reciprocal of the density. The specific volume v of a system is the volume occupied by a unit mass of the system, and $v = 1/\rho = V/m$ = volume / mass. The SI unit of specific volume is m^3/kg; other units are:

$$\begin{aligned}
1\ \mathrm{m^3/ton} &= 0.001\mathrm{m^3/kg} \\
1\mathrm{L/kg} &= 1\ \mathrm{dm^3/kg} = 0.001\ \mathrm{m^3/kg} \\
1\mathrm{cm^3/g} &= 0.001\ \mathrm{m^3/kg} \\
1\ \mathrm{in^3/lbm} &= 3.6175 \times 10^{-5}\ \mathrm{m^3/kg} \\
1\ \mathrm{ft^3/lbm} &= 0.0625\ \mathrm{m^3/kg}
\end{aligned}$$

The *weight rate of flow* can be defined as pounds per hour (lb/hr). Although volume flow rate is consistent with the SI system, several problems still use force per time; for example, the flow of water in an open channel can be expressed in units of volume per time, but industrial liquid flows are usually specified in gallons per minute (gal/min). Airflow is generally measured in cubic feet per minute (ft^3/min).

Velocity is the volume rate of flow F per unit area A, and the units vary with the application. In heat transfer the units may be feet per hour (ft/hr) for both liquids and gases, whereas in fluid mechanics the velocity of liquids is generally in feet per second (ft/sec), and the velocity of gases is in feet per minute (ft/min) or meters per second (m/s). The general relationship, with units of flow and area selected to give the desired units for the velocity, is

$$\text{velocity } V = \frac{\text{volume flow rate}}{\text{area, pipe or duct}} = \frac{F}{A}$$

Entropy is a quantity that measures the extent to which a system's energy is available for conversion to work. If a system undergoing an infinitesimal reversible change takes in a quantity of heat dQ at absolute temperature T, its entropy is increased by $dS = dQ/T$. The area under the absolute temperature–entropy graph for a reversible process represents the heat transferred in the process. For an adiabatic process, there is no heat transfer and the temperature–entropy graph is a straight line, the entropy remaining constant through the process. Any process in which there is no change in entropy occurs is said to be *isentropic*.

The *specific entropy* of a system is the entropy of the unit mass of the system and has the dimensions energy/mass/temperature. The SI unit of specific entropy is J/kg·K. Other units are

$$1\,\text{kJ/kg}\cdot\text{K} = 1000\,\text{J/kg}\cdot\text{K}$$
$$1\,\text{erg/g}\cdot\text{K} = 10^{-4}\,\text{J/kg}\cdot\text{K}$$
$$1\,\text{Btu/lb}\cdot{}^\circ\text{F} = 4186.8\,\text{J/kg}\cdot\text{K}$$
$$1\,\text{cal/g}\cdot{}^\circ\text{C} = 4186.8\,\text{J/kg}\cdot\text{K}$$

Enthalpy (from the Greek meaning "to heat") is not a specific form of energy but is useful in the analysis of open systems. It is defined as the mass of the system, m, multiplied by the specific enthalpy of the system, $H = mh$. It is measured in joules and is defined in terms of thermodynamic state functions (i.e., a state function, so enthalpy changes are path independent). The enthalpy of a body is the sum of its internal energy and the product of its volume and the pressure exerted upon it.

The *specific enthalpy* of a working fluid is defined in terms of its intensive or unit mass properties, given by $h = u + Pv$, where u is the specific internal energy (the amount of energy per unit of mass of a substance), P is the pressure, and v is the specific volume (m^3/kg). The SI unit of specific enthalpy is J/kg.

The *specific internal energy* is the internal energy of a system per unit mass of the system and thus has the same dimensions as energy/mass or enthalpy. To understand the concept of internal energy, we need to look at macroscopic and microscopic forms of energy. Potential energy and kinetic energy are *macroscopic forms* of energy. They can be visualized in terms of position and velocity of the objects. In addition to these macroscopic forms of energy, a substance possesses several microscopic forms of energy. *Microscopic forms* of energy include those due to the rotation, vibration, translation, and interactions among the molecules of a substance. None of these forms of energy can be measured or evaluated directly, but techniques have been developed to evaluate the change in the total sum of all these microscopic forms of energy. These microscopic forms of energy are collectively called *internal energy*, customarily represented by the symbol U. In engineering applications, the unit of internal energy is the British thermal unit (Btu), which is also the unit of heat. The specific internal energy measures the energy content of a system due to its thermodynamic properties, such as pressure and temperature. The change of internal energy of a system depends only on the initial and final states of the system and not in any way on the path or manner of the change. This concept is used to define the first law of thermodynamics.

Heat capacity is defined by the ratio of heat ΔQ required to raise the temperature of an object a small amount ΔT (i.e., $C = \Delta Q/\Delta T$). The specific heat c is the heat capacity divided by the mass m of the object; thus, $c = \Delta Q/m\,\Delta T$. Since an infinitesimal variation in entropy dS may be expressed as the ratio of an infinitesimal amount of the absorbed heat divided by the system temperature (i.e., $dS = dQ/T$), the following expression for entropy in terms of specific heat capacity holds:

$$dS = \frac{dQ}{T} = \frac{mc\,dT}{T} \tag{2.1}$$

When two phases (solid–liquid or liquid–gas) are present during heat transfer at constant pressure, a temperature change will not occur and the change in heat content of the substance must be obtained from tables of properties of the substance. The heat required to change the phase of an object with mass m is

$$Q = \pm\, mL \tag{2.2}$$

where L is the heat required for the phase change, the quantity L is the heat of fusion for a phase change from solid to liquid, or the heat of vaporization for a phase change from liquid to gas, or the heat of sublimation for a phase change from gas to solid.

The sign of heat depends on the direction of the phase change. When ice is heated from 0°F, to 32°F, the heat required is 0.49(32 − 0) = 15.7 Btu/lb, as shown in Figure 2.1. As more heat is added, the temperature remains at 32°F until all of the ice has been changed into water. The energy required to affect the change, termed the *latent heat of fusion*, is 144 Btu/lb. As more heat is added, at atmospheric pressure, the temperature rises to 212°F. The heat required to raise the water temperature is 1.0(212 − 32) = 180 Btu/lb. For further addition of heat, the temperature

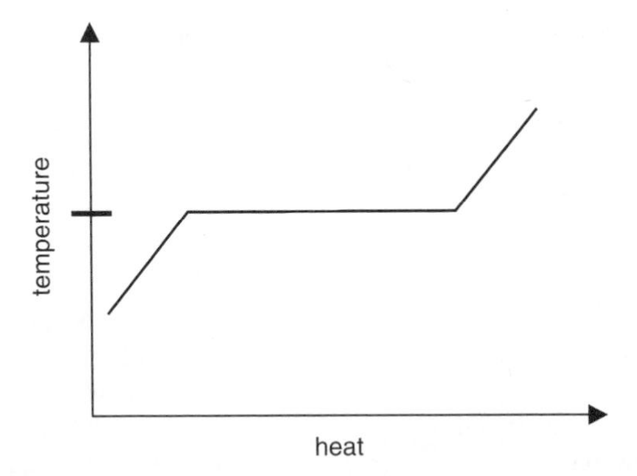

Figure 2.1 Temperature variation with phase change.

remains at 212°F until all the water has been converted into steam. The energy required for this change of phase, called the *latent heat of vaporization*, is 970 Btu/lb at an atmospheric pressure of 14.7 psi. Air contains various amounts of moisture as water vapor at very low pressures. The presence of water vapor has little effect on the specific heat of the air, but if water is added to the air (clothes drier) or condensed from the air (air conditioner, dehumidifier), an energy transfer of about 1060 Btu is required per pound of water. A brief list of values of specific heat is given in Table 2.1. The specific heat of gases (and to a much lesser extent, of liquids and solids) depends on the process by which heat is added or removed. The term *heat content* is no longer used in modern thermodynamics, although it is still common; the preferred term is *enthalpy.*

TABLE 2.1 Specific Weight and Specific Heat

	Thermodynamic Properties	
Substance	Specific Weight w (lb/ft^3)	Specific Heat C_p (Btu/lb-°F)
Air	0.075 (0 psig, 70°F)	0.24
Aluminum	165	0.22
Brick	125	0.20
Cast iron	450	0.12
Concrete	144	0.22
Flue gas	0.041 (0 psig, 500°F)	0.26
Glass	155	0.19
Ice	57.5	0.49
Steam	0.037 (0 psig, 212°F)	0.48
Steel	490	0.12
Stone	165	0.21
Water	62.5	1.00
White pine	27	0.67

TABLE 2.2 Higher Heating Value for Selected Fuels

Fuel	HHV
Anthracite coal	13,000 Btu/lb
Bituminous coal	13,500 Btu/lb
Lignite	6,500 Btu/lb
Peat	1,200 Btu/lb
Fuel oil	137,000 Btu/gal
Natural gas	1,050 Btu/ft^3
Propane	92,000 Btu/gal
Butane	102,050 Btu/gal
Wood, mixed varieties	7,200 Btu/lb
Hardwoods	25×10^{-6} Btu/cord
Softwoods	16×10^{-6} Btu/cord
Carbon (C)	14,600 Btu/lb
Hydrogen (H_2)	62,000 Btu/lb
Sulfur (S)	4,050 Btu/lb

The *heating value* represents the chemical energy stored in a fuel and is measured by burning a small sample of the fuel in an oxygen environment. The heat transferred to a water jacket surrounding the sample is termed the *higher heating value* (HHV) of the fuel. When a fuel is burned, the moisture present is converted into water vapor. The latent heat of vaporization must be subtracted from the HHV, and the result is reported as the *lower heating value* (LHV). For example, peat (decayed vegetable matter) is the first step in the transformation of plants and trees into coal. The HHV is very low, due to the high moisture content, so it must be dried before being used as a fuel. Lignite is a better fuel because geological conditions have reduced the moisture content, increased the carbon content, and raised the heating value to a usable level. Table 2.2 shows the chemical energy in some fuels, where the HHV is usually given in USCS units as Btu/lb for solids, Btu/gal for liquids, and Btu/ft^3 for gases. In the SI system it is termed *calorific value* and has units of J/kg or J/m^3. Bituminous and anthracite coals are products of the final stages of geologic formation and have heating values above 12,000 Btu/lb with moisture below 5%. The anthracites contain about 90% carbon and are found only in mountainous regions where thrust pressures were great. Petroleum contains about 85% carbon and 12% hydrogen. Natural gas can be found in conjunction with crude oil, but it can also be found independently. The primary component is methane, and the heating value varies from 1000 Btu/ft^3 (volume measured at 14.7 psia and 60°F) to 1100 Btu/ft^3. Because of this variation, it is often sold in the unit *therm* (1 therm = 100,000 Btu) rather than on a volume basis. Propane and butane are liquefied petroleum gases (LPGs). They are stored under pressure as liquids but are vaporized by heat transferred from their surroundings and burned as a gas. The pressure of propane storage is 189 psig at 100°F and 38 psig at 0°F. Butane is stored at lower pressures: −52 psig at 100°F and −7 psig at 0°F. The

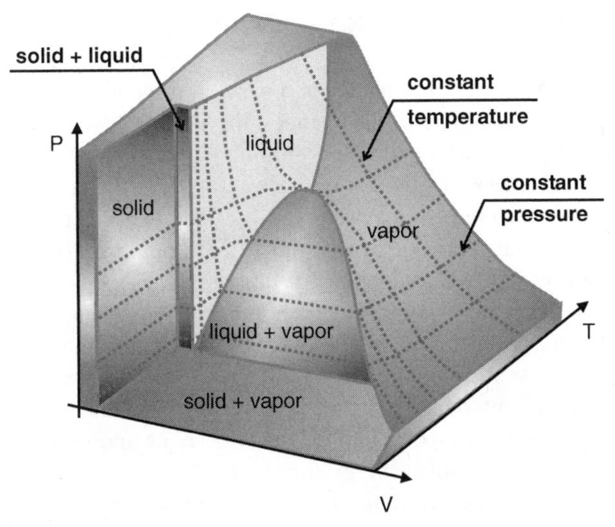

Figure 2.2 *P–V–T* surface for a substance.

heating value of wood averages about 7200 Btu/lb at 20% moisture. Wood is usually sold by the cord (i.e., a stack 8 ft × 4 ft × 4 ft) and has some variation in heating value, depending on the density of the wood.

An *intensive property* of simple and homogeneous thermodynamic systems is a function of two other intrinsic properties, so it can be represented in Cartesian three-dimensional space, plotted as a surface as in Figure 2.2. The thermodynamic properties of a pure substance can be related by the general relationship $f(P, V, T) = 0$, which represents a surface in (P, V, T) space. It is more a qualitative insight than a quantitative evaluation. The state of any pure working fluid can be defined completely simply by knowing two independent properties of the fluid. This makes it possible to plot the state changes on two-dimensional diagrams such as:

- A pressure–volume (*P–V*) diagram
- A temperature–entropy (*T–s*) diagram
- An enthalpy–entropy (*h–s*) diagram

When considering a transition between two points of a thermodynamic state, the following processes are often used:

Adiabatic	$dq = 0$	no heat transfer
Isentropic	$ds = 0$	constant entropy
Isothermal	$dT = 0$	constant temperature
Isochoric	$dV = 0$	constant volume
Isobaric	$dP = 0$	constant pressure

2.3 FUNDAMENTAL LAWS AND PRINCIPLES

When energy is transferred into or out of a system, principles of thermodynamics are used to describe the nature of the changes, the external effects in terms of work and heat, and the properties of the mass in the system, such as fluid pressure and temperature. The *law of conservation of energy* is used in the analysis of a system: Energy can be neither created nor destroyed (i.e., the total energy within the universe remains constant). Energy can be stored in a number of ways and can be transferred from one system to another by various processes.

The first law of thermodynamics is expressed as: The net heat added to a system minus the net work produced by the system must be equal to the increase in stored energy within the system. The first law can be used in a more general manner, considering in addition to work and heat, electrical energy and chemical energy as distinct energy forms that can move into or out of systems and must be balanced by corresponding changes in other energy forms.

The *second law of thermodynamics* is a straightforward application of a principle of physics which states that in a closed system, it is not possible to finish any real physical process with as much useful energy as it had originally (i.e., something is always wasted). This means that a perpetual-motion machine is impossible. The second law was formulated after nineteenth-century engineers noticed that heat could not pass by itself from a colder body to a warmer body. According to the philosopher of science Thomas Kuhn, the second law was first put into words by two scientists, Rudolph Clausius and William Thomson (Lord Kelvin). The *Clausius statement* is that "heat cannot itself pass from a cold to a hot body." The quantum physicist Richard P. Feynman mentioned in his lectures that French physicist Sadi Carnot discovered the second law 25 years earlier than Clausius and Thomson (that would have been before the first law of conservation of energy was discovered). The first law does not prohibit energy conversion at 100% efficiency, but such conversion is prohibited by the second law of thermodynamics.

The consequences of the Clausius statement impose strict limitations on a variety of thermodynamic processes. An air conditioner transfers energy from a cold body (the interior) to a hot body (the surroundings). The first law would permit the process to operate without input work as long as the heat gained by the interior was balanced by the heat rejected to the surroundings. The second law would prohibit such a device and imposes minimum limits on the input work required. The *Kelvin–Planck statement of the second law*, which is equivalent in consequences to the Clausius statement, is: It is impossible for any device to operate in a cycle and produce work while exchanging heat only with bodies at a single fixed temperature.

An important consequence of the second law is the *Carnot principle*, in which it is shown that with heat supplied at a temperature T_H and rejected at a temperature T_L, the maximum efficiency n at which the heat supplied can be converted into work is

$$\eta_{\max} = 1 - \frac{T_L}{T_H} < 1 \tag{2.3}$$

where T_L and T_H are both thermodynamic temperatures (in Kelvin). To calculate the efficiency in degrees Fahrenheit, the following expression can be used:

$$\eta_{max} = \frac{T_H - T_L}{T_H + 460} \tag{2.4}$$

In a steam boiler the efficiency of conversion of energy from the fuel to steam is limited by the first law to 100% (practical considerations set the limit at around 90%). If the steam is used for space or process heat, the limitation is again 100%, a value that can be approached as closely as desired. If the steam is used to produce work (and indirectly, electrical energy) with a heat engine, the maximum efficiency is limited by the second law.

2.3.1 Example in a Nutshell

In a steam electric power station, heat is supplied at 1000°F (steam temperature) and rejected to the condenser cooling water at a temperature of 60°F (Figure 2.3). What is the maximum possible efficiency of conversion?

$$\eta_{max} = \frac{1000 - 60}{1000 + 460} = 64\% \tag{2.5}$$

Thus, 36% of the energy supplied is lost to the cooling water. This loss, by the second law, cannot be reduced except by changes in operating temperatures. The actual efficiency is considerably less, and when combined with the boiler conversion

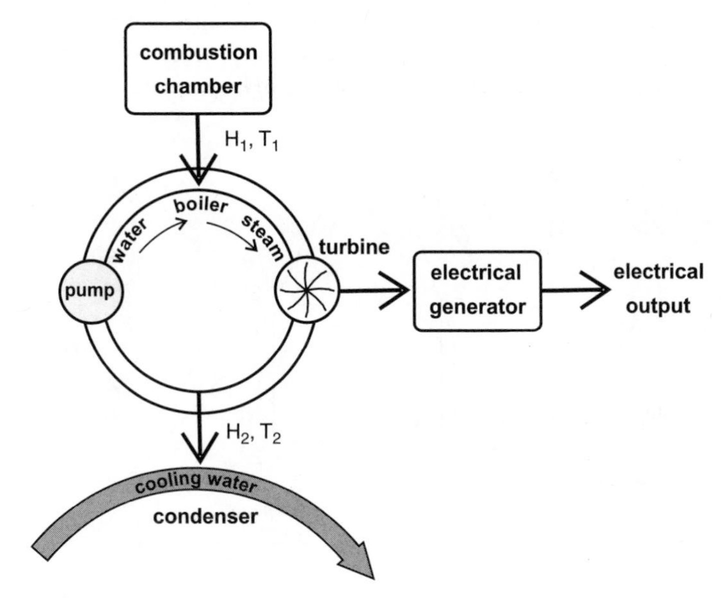

Figure 2.3 Steam electric power station maximum efficiency.

efficiency and losses in the conversion from shaft work to electrical energy, the overall efficiency of the modern steam electric station is around 35%. By the time the electrical energy is delivered to the point of use, less than one-third of the energy in the fuel remains.

A *heat engine* is defined as a device that converts heat energy into mechanical energy or, more exactly, as a system that operates continuously in which only heat and work may pass across its boundaries. The operation of a heat engine can best be represented by a thermodynamic cycle such as an Otto, Diesel, Brayton, Stirling, or Rankine cycle; for electrical power conversion the most relevant cycles are the Rankine (vapor–liquid system, typical of steam power plants) and the Brayton (gas turbine–based power plants).

The *Carnot cycle* is an idealized representation of the operation of a steam engine. To understand it, an ideal gas can be imagined to be confined in a cylinder by a piston, and allowed to absorb heat from a reservoir held at T_H or to reject heat to a reservoir held at T_L. The gas is led through a cyclical path (the Carnot cycle) depicted in Figure 2.4, which performs work on the surroundings. The overall effect is to convert heat into useful work. Assume absolute temperature or thermodynamic temperature (kelvin); the following transformations occur along the cycle:

- Reversible isothermal expansion at T_H, segment 2 to 3
- Reversible adiabatic expansion (from T_H to T_L), segment 3 to 4
- Reversible isothermal compression at T_L, segment 4 to 1
- Reversible adiabatic compression (from T_L to T_H), segment 1 to 2

Figure 2.5 shows the scheme of a simple power plant; it consists of a boiler, turbine, condenser, and pump. Fuel is burned in the boiler and heats the water to generate steam. The steam is used to rotate the turbine, which powers the generator. Electrical energy is generated when generator windings rotate within a magnetic

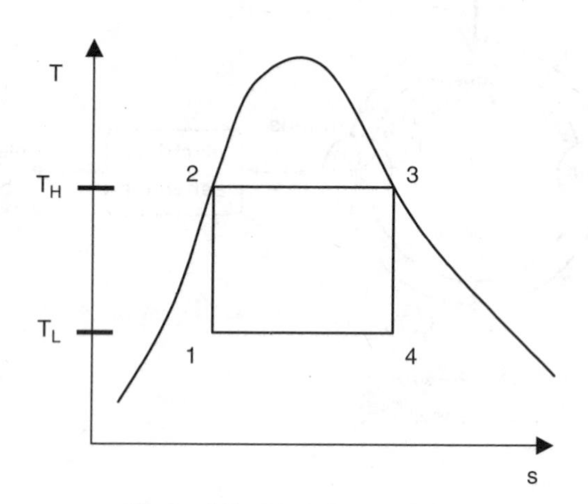

Figure 2.4 Ideal Carnot cycle.

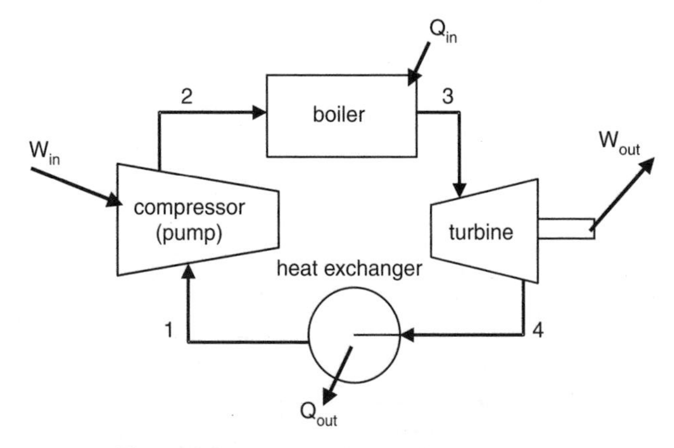

Figure 2.5 Scheme of a simple power plant.

field. After the steam leaves the turbine, it is cooled to its liquid state in the condenser. The liquid is pressurized by the pump before going back to the boiler. The power cycle balancing can be evaluated by the clockwise path on Figure 2.4 as follows:

- The heat absorbed by the working fluid during isothermal process 2–3 (high-temperature reservoir) is

$$q_{23} = T_H(s_3 - s_2) \Rightarrow Q_{\text{in}}$$

- During isothermal process 4–1 (low-temperature reservoir), the heat given up by the working fluid is

$$q_{41} = T_L(s_1 - s_4) = T_L(s_2 - s_3) = -T_L(s_3 - s_2)$$

- The work produced by the cycle is

$$W_{\text{cycle}} = q_{23} + q_{41} = T_H(s_3 - s_2) - T_L(s_3 - S_2)$$

- The thermal efficiency can be calculated by

$$\eta = \frac{\text{work produced by the cycle}}{\text{heat supplied to the cycle}} = \frac{q_{23} + q_{41}}{q_{23}} = \frac{T_H - T_L}{T_H} \tag{2.6}$$

The product of pressure and volume represents a quantity of work. This can be represented by the area below a *P–V* curve as plotted in Figure 2.6, where the area enclosed by the four curves represents the net work done by the engine during one cycle. By using the second law of thermodynamics it is possible to show that no heat engine can be more efficient than a reversible heat engine working between two fixed-temperature limits. Due to mechanical friction and other irreversibilities, no cycle can actually achieve this level of efficiency.

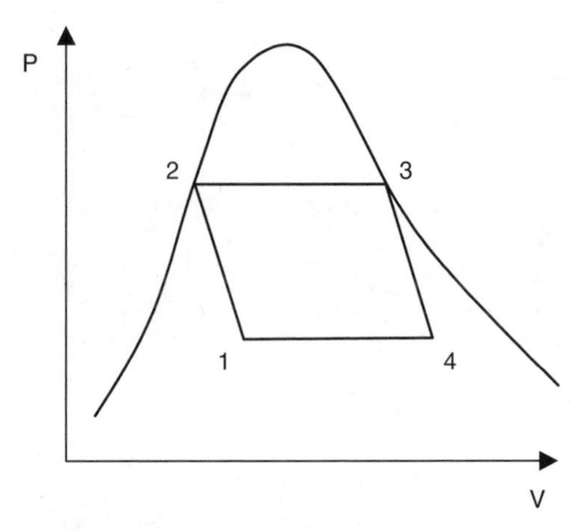

Figure 2.6 Pressure versus volume curve for an energy conversion cycle.

2.3.2 Practical Problems Associated with Carnot Cycle Plant

The Carnot cycle is not practical because of:

- The isentropic expansion in a turbine as in the line segment from 3 to 4. A major concern is related to quality of the steam inside the turbine because a high moisture content results in blade erosion.
- The isentropic compression process in a pump from 1 to 2. It is very difficult to design a condenser and transmission line system that precisely control the quality of the vapor to achieve an isentropic compression.
- The two-phase pump–compressor that is required.

A possible cycle to solve some of these problems could be proposed as depicted in Figure 2.7. However, this cycle requires compression (1–2) of the liquid at a very

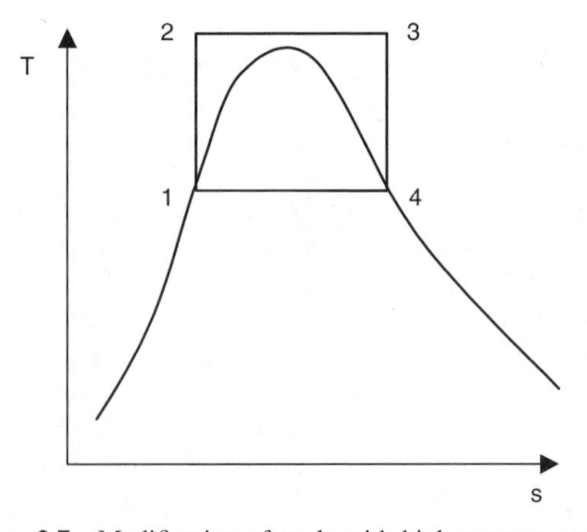

Figure 2.7 Modification of cycle with high-pressure steam.

high pressure (exceeding 22 MPa for steam), which is not practical. In addition, maintaining a constant temperature above the critical temperature is also difficult since the pressure will have to change continuously. A practical problem is a maximum temperature limitation for the cycle and constraints on metallurgical limitations imposed by materials used within the boiler and turbine.

2.3.3 Rankine Cycle for Power Plants

The cornerstone of modern steam power plants is the thermodynamic cycle proposed by W. J. M. Rankine, a Scottish engineer. The main components of a Rankine cycle steam plant are:

- A boiler that generates steam, usually at a high pressure and temperature
- A turbine that expands the steam to a low pressure and temperature, thereby producing work
- A generator driven by the steam turbine
- A condenser that cools the steam to a liquid so that it can be pumped back into the boiler
- A feed pump
- Feed heaters, which preheat the water before it enters the boiler
- A reheater, which is part of the boiler, reheats the steam after it has been partially expanded

Figure 2.8, a temperature–entropy diagram, illustrate the state changes for the Rankine cycle. Such a cycle avoids transporting and compressing two-phase fluid by trying to condense all fluid exiting from the turbine into saturated liquid before it is compressed by a pump. The actual power plant introduces regenerative and reheat modifications in the basic Rankine cycle.

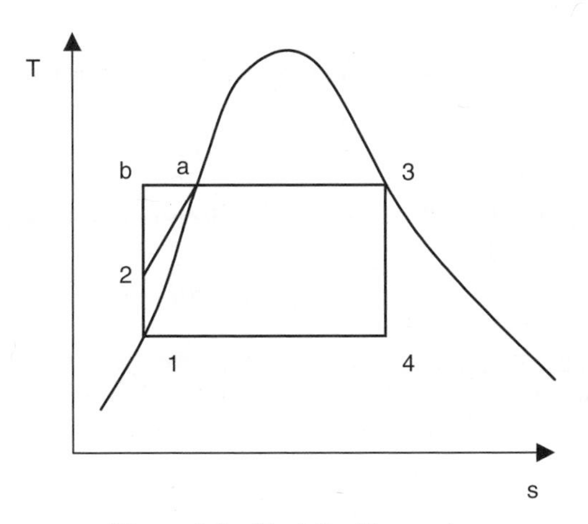

Figure 2.8 Ideal Rankine cycle.

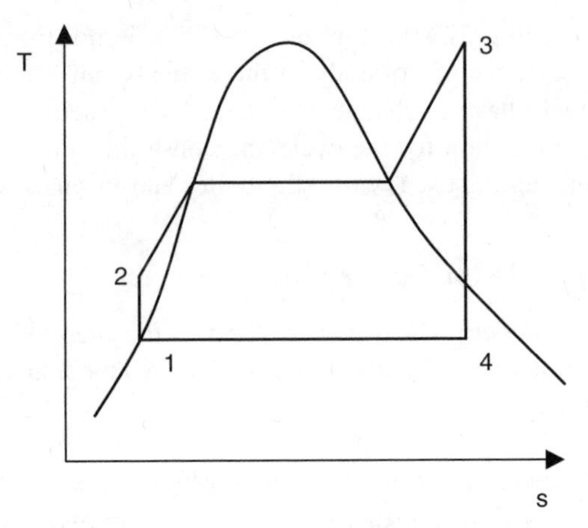

Figure 2.9 Rankine cycle with superheating.

It is evident in the *T-s* diagram of Figure 2.8 that the ideal Rankine cycle is less efficient than a Carnot cycle for the same maximum and minimum temperatures. The Rankine cycle work is represented by the area 2–*a*–3–4–1–2, which is less than the Carnot cycle work represented by the area 2–*b*–3–4–1–2. Figure 2.9 illustrates a typical superheated Rankine cycle, the following path describing the cycle:

- 1–2: Pump ($q = 0$); isentropic compression (pump)

$$W_{\text{pump}} = h_2 - h_1 = V(p_2 - p_1)$$

- 2–3: Boiler ($W = 0$); isobaric heat supply (boiler)

$$Q_{\text{in}} = h_3 - h_2$$

- 3–4: Turbine ($q = 0$); isentropic expansion (steam turbine)

$$W_{\text{out}} = h_3 - h_4$$

- 4–1: Condenser ($W = 0$); isobaric heat rejection (condenser)

$$Q_{\text{out}} = h_4 - h_1$$

By increasing the steam temperature, point 3 is shifted up and the dry saturated steam from the boiler is passed through a second bank of smaller-bore tubes within the boiler until the steam reaches the required temperature. Increasing the steam temperature increases the cycle efficiency and reduces the moisture content at the turbine exhaust end; sometimes two stages of reheating are connected in tandem. Three ways to improve the thermal efficiency are depicted in Figure 2.10.

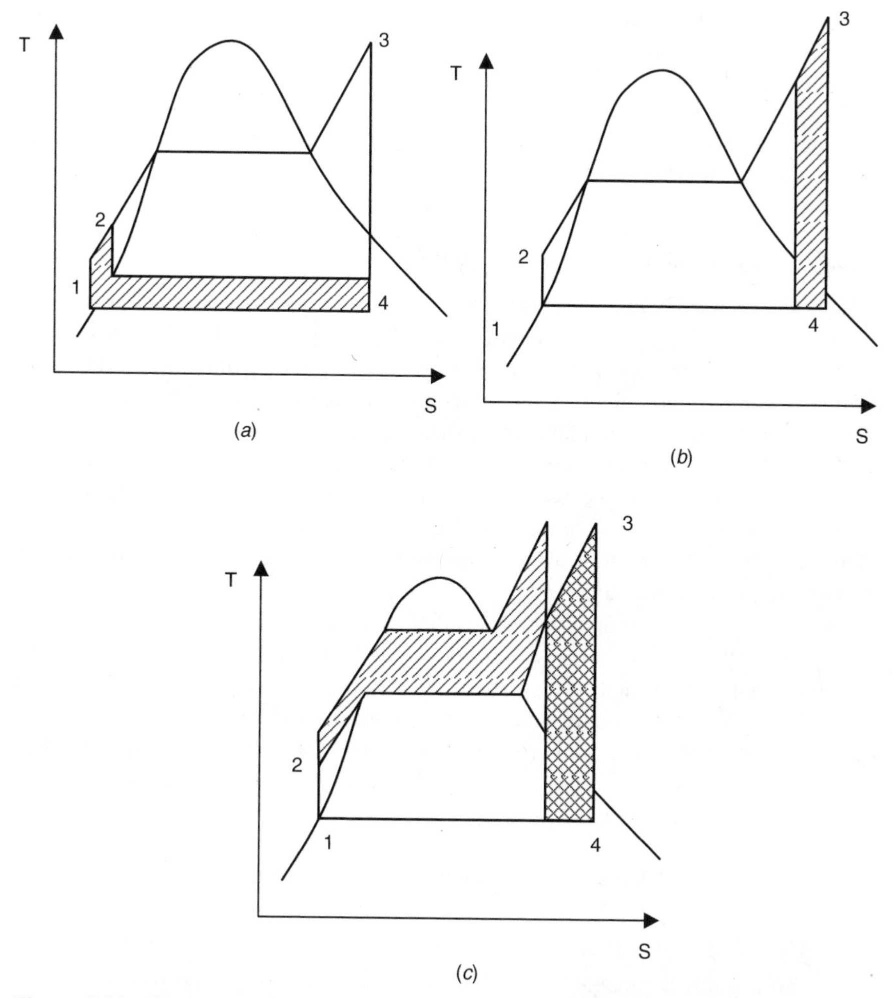

Figure 2.10 Thermal efficiency improvement with modified Rankine cycle: (*a*) lowering the condensing pressure (lower condensing temperature, lower T_L); (*b*) superheating the steam to higher temperature; (*c*) increasing the boiler pressure (increase boiler temperature, increase T_H).

The optimal way to increase the boiler pressure but not increase the moisture content in the exiting vapor is to reheat the vapor after it exits from a first-stage turbine and to redirect this reheated vapor into a second turbine. Regeneration helps to improve the Rankine cycle efficiency by preheating the feedwater into the boiler. Regeneration can be achieved by open or closed feedwater heaters. In open feedwater heaters, a fraction of the steam exiting a high-pressure turbine is mixed with the feedwater at the same pressure. In a closed system, the steam bled from the turbine is not mixed directly with the feedwater. Therefore, the two streams can be at different pressures, but such details are outside the scope of this book.

Fluids other than water and steam may be used in the Rankine cycle if they are considered more appropriate for a particular process. For example, ocean thermal conversion plants depend for their operation on the temperature difference between the hotter surface water (say up to 27°C) and the colder deep water (down to 4°C). The working fluid for these plants is ammonia because of its low boiling point, but the cycle used is the basic Rankine cycle.

2.3.4 Brayton Cycle for Power Plants

The basic gas turbine cycle was first proposed by George Brayton around 1870 and is currently used for gas turbine engine–based aircraft propulsion and electric power generation. Gas turbines may be used as stationary power plants to generate electricity as stand-alone units or in conjunction with steam power plants on the high-temperature side. In these plants the exhaust gases serve as a heat source for the steam. Steam power plants are considered external-combustion engines, in which the combustion takes place outside the engine. Figure 2.11 shows a gas turbine where air is drawn into a compressor, which raises the temperature and pressure. The high-pressure air proceeds into the combustion chamber, where the fuel is burned at constant pressure.

The resulting high-temperature gases then enter the turbine and expand to atmospheric pressure through a row of nozzle vanes. This expansion causes the turbine blade to spin, which then turns a shaft inside a magnetic coil. When the shaft is rotating inside the magnetic coil, electrical current is produced. The changes in properties of the working fluid brought about by the addition of fuel or the presence of combustion products are ignored. The following points are considered in an analysis of the Brayton cycle:

- The working substance is air which is treated as an ideal gas throughout the cycle.
- The combustion process is modeled as a constant-pressure heat addition.
- The exhaust is modeled as a constant-pressure heat rejection process.

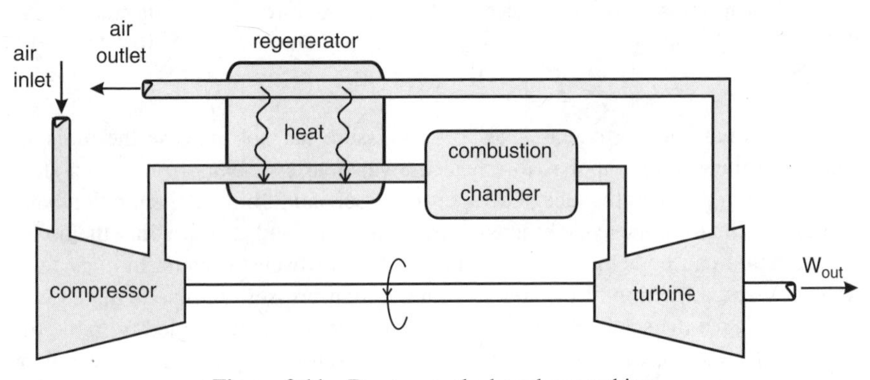

Figure 2.11 Brayton cycle–based gas turbine.

- The gas turbine is a rotatory device working at a nominal steady state.
- Spark ignition is used for startup, since air compressor output temperature is not high enough to ignite the fuel.

In the ideal air-standard Brayton cycle, air is assumed to follow the four processes depicted in Figure 2.12. The cycle path is composed of isentropic compression, constant-pressure heat input from the hot source, isentropic expansion, and constant-pressure heat rejection to the environment. In the combustion chamber heat is added to the gas at constant pressure, the density decreases, and the specific volume and temperature increase. Entropy is also increased, since combustion is not a reversible process. In the turbine the situation is the opposite of that in the compressor: The pressure decreases and specific volume increases. The temperature decreases and in an ideal expansion the entropy is constant. The encircled area in

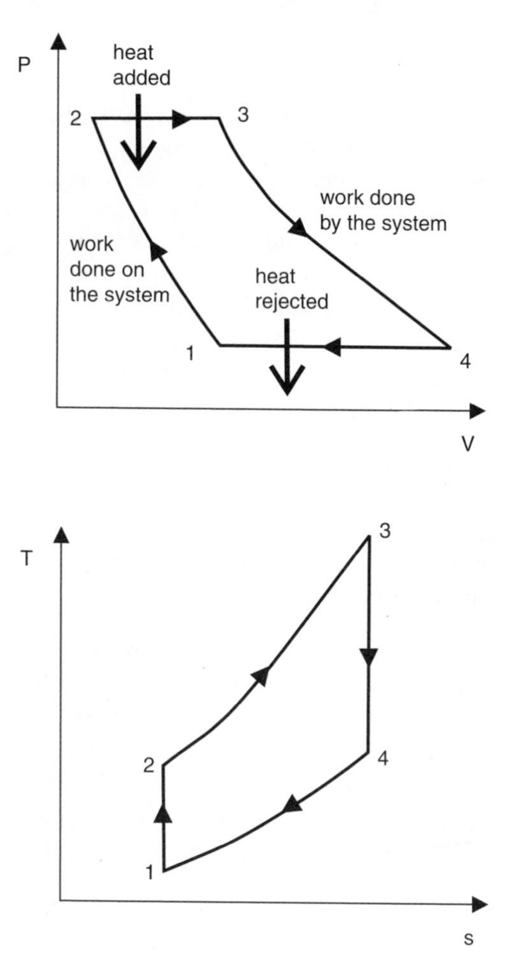

Figure 2.12 Ideal Brayton cycle.

the P–V diagram represents the net work produced by the gas turbine. The efficiency of the gas turbine is the ratio of net work produced and heat power added:

$$\eta = \frac{W_{\text{net}}}{Q_{\text{in}}} = 1 - \frac{1}{r_p^{(k-1)/k}} \tag{2.7}$$

where is r_p the pressure ratio in the turbine and k is the ratio of specific heats. The higher the gas turbine pressure ratio, the higher its efficiency. The equation is valid only for ideal gas turbines with no friction and reversible processes. A gas microturbine used for stationary power generation is a combination of a small gas turbine and a directly driven high-speed generator placed on the same shaft, without a gearbox and optimized mechanical design to minimize friction losses, usually applied in combined heat and power applications. Most designs have heat recuperation with excellent thermal efficiencies, good power output efficiencies, and high reliability. Chapter 8 covers microturbines in greater detail.

2.3.5 Energy and Power

Energy E may be defined as the "capacity to do work," such as the lifting of a weight. To lift a 10-lb weight a distance of 5 ft requires $5 \times 10 = 50$ ft-lb of energy. Energy may be converted from one form to another, stored, and transferred from one body to another. Energy takes a variety of forms: chemical, electrical, mechanical (work, kinetic, potential, flow), and thermal (heat, internal). Thermodynamics is a complex subject; most problems in energy conversion require detailed knowledge of the subject and careful attention to limitations of the fundamental relationships. In residential and commercial applications the range of working fluids is limited (air, water, flue gas). Flue gas is a mixture of air and gas products resulting from combustion flowing through a chimney. For most residential and commercial applications, the pressures are generally moderate and the processes involved are relatively simple. Consequently, it is possible to simplify the subject considerably while retaining the ability to solve most energy problems in this field. Residential and industrial applications are concerned with only three forms of energy: chemical, electrical, and heating. There are other forms of energy that may be of interest, but they are not usually involved in industrial problems.

- *Chemical energy* (CE) is energy such as that stored in fuels, usually identified by the term *higher heating value* and measured in British thermal units. (1 Btu is the energy required to increase the temperature of 1 lb of water 1°F).
- *Electrical energy* (EE) is a form of energy entering a system as electricity, measured in kilowatthours. (1 kWh is equal to 3413 Btu).
- *Heat* (Q) is a form of thermal energy that is transferred across system boundaries by a temperature difference, measured in Btu.

If a hot body is placed close to a cold body, heat flows from the higher- to the lower-temperature body, and the energy content of that body must increase. For most problems the energy change within the body can be treated as a change in

TABLE 2.3 Energy and Power Conversion Factors

Energy	Power
1 kWh = 3413 Btu	1 hp = 2545 Btu/hr
1 Btu = 778 ft-lb	1 kW = 3413 Btu/hr
1 hp-hr = 2545 Btu	1 hp = 0.746 kW
1 kWh = 1.34 hp-hr	1 hp = 33,000 ft-lb/min

the heat content of that body. When this energy is transferred to a solid, liquid, or gas, the temperature of the substance increases in proportion to the quantity of energy transferred. *Power* is the rate of energy transfer or conversion. Horsepower (hp), kilowatts (kW), and Btu/hr are units of power, not energy. Application of the first law involves energy balances. Usually, change of energy (energy rate) is used and is generally more convenient. Table 2.3 gives some useful energy and power conversion factors.

The mathematical statement of the first law is a powerful tool for solving energy problems and for a better understanding of energy transformations. The first law of thermodynamics as applied to residential and commercial energy problems is most conveniently stated thus: The energy added to a system equals the energy leaving the system plus the increase in stored energy within the system:

$$E_1 = E_2 + \Delta E_s \tag{2.8}$$

where the subscripts 1, 2, and s refer to energy entering, leaving, and stored, respectively. Expanding that equation for the forms of energy to be most concerned with and using energy rates gives

$$Q_1 + CE_1 + EE_1 + H_1 = Q_2 + \mathrm{CE}_2 + \mathrm{EE}_2 + H_2 + \Delta H_\mathrm{s} \tag{2.9}$$

where Q = heat transferred by temperature difference across system boundaries (Btu/hr)

CE = chemical energy (Btu/hr) = W (HHV of fuel in Btu/lb)

EE = electrical energy (Btu/hr) = 3413 kW

H = heat content of fluids crossing system boundaries (Btu/hr) = WC_pT

ΔH = change in stored heat content (Btu/hr) = $MC_\mathrm{p}(T'/\mathrm{t})$, where M is the weight of the storage material (lb) and T'/t is the change in temperature T (°F) in a time period t (hr)

2.4 EXAMPLES OF ENERGY BALANCE

2.4.1 Simple Residential Energy Balance

The following examples were considered in Ref. [1]. Suppose that the residence shown in Figure 2.13 is maintained at a constant temperature with no change in

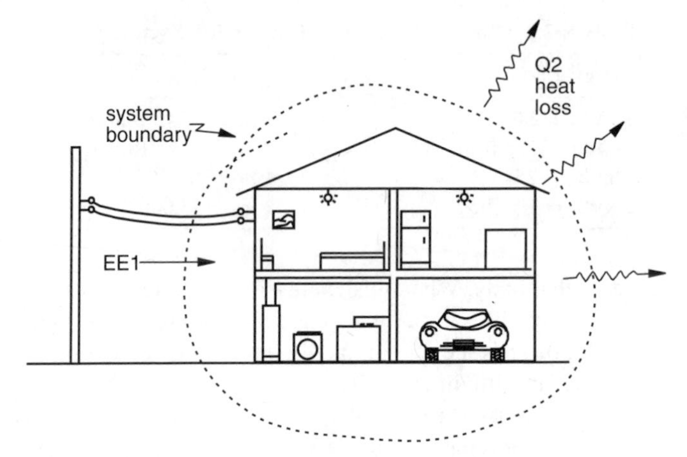

Figure 2.13 Residence at constant temperature.

energy stored and only two energy flows crossing the boundaries. Assume an average electrical power use of 1.5 kW. What is the heat loss from the house?

Use equation (2.9) with EE $= Q_2$ (all other terms are zero since there is no other energy flow outside the boundary):

$$1.5 \times 3413 = Q_2$$

so the heat loss is 5120 Btu/hr. Even though nothing was stated about how the energy was being used in the house, the energy balance tells us that 100% of all electrical energy entering the house is being converted into heat, no matter what the primary use. It may be for operating a refrigerator, stove, fan, radio, TV, lights, or a power tool. All of the electrical energy supplied to these devices must appear as heat after the overall energy conversion. The electrical demand of lights and appliances can easily exceed an average of 1 kW (3413 Btu/hr), perhaps one-third of the energy required to heat a small, well-insulated house at 30°F. This loss of indirect electrical heat is usually quite noticeable during a winter power outage. If an air conditioner or heat pump were being used, a third energy flow would be shown crossing the boundaries, and the results would be different.

2.4.2 Refrigerator Energy Balance

In using energy balances the boundaries can be drawn around any device, group of devices, or components of a device. The refrigerator shown in Figure 2.14 could be the system of interest. Electrical energy is flowing in, as is the heat gain Q_1 through the insulated walls. The heat flow Q_2 from the condenser is out of the system. Suppose that the electrical input averages 150 W (0.15 kW). Does the refrigerator add heat or subtract heat from the house? How many Btu/hr?

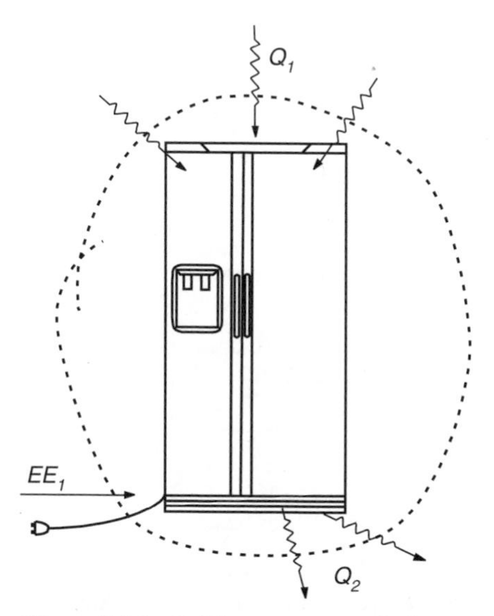

Figure 2.14 Refrigerator energy balance.

From equation (2.9) we have

$$EE_1 + Q_1 = Q_2$$
$$Q_2 - Q_1 = EE_1 = 3413(0.15) = 512\,\text{Btu/hr}$$

The net effect is to heat the house, since the heat given off by the condenser (Q_2) exceeds the cooling (Q_1) by 512 Btu/hr. The heat rejected by the condenser was taken as $Q_2 - Q_1$, which is the energy flow from the condenser to the surroundings due to a temperature difference. It is more likely that the condenser was cooled by a forced-air system (two airflows crossing the boundaries), so that the Q_2 term should have been omitted and the terms H_1 and H_2 added. The results would have been the same, since a heat balance of the air inside the condenser would show that $Q_2 = H_2 - H_1$. Where a fluid flows into and out of a system and information is not available or required on flow rates and temperatures, the simplifying substitution of Q for ΔH is desirable. The system boundaries could have been drawn to include only the compressor, condenser, and evaporator. The results would have been the same, since the heat gain by the evaporator equals the heat gain through the walls. Other boundaries could have been drawn but would not have yielded the desired information.

2.4.3 Energy Balance for a Water Heater

An electric water heater offers a good example of the use of stored energy and heat content in the energy balance (Figure 2.15). The 4.5-kW water heater shown is

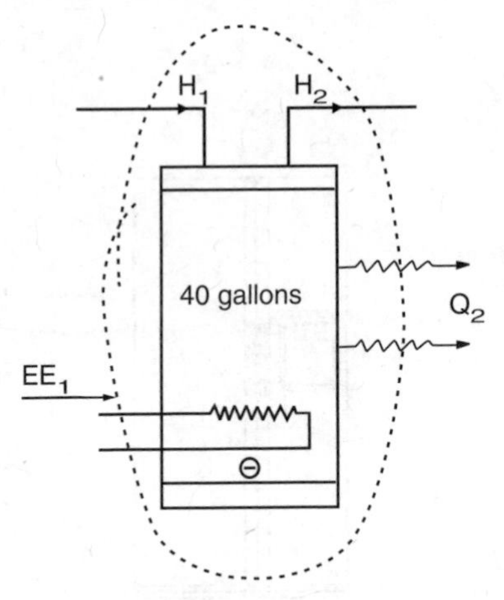

Figure 2.15 Water heater energy balance.

supplied with cold water at 60°F and delivers hot water at a thermostat setting of 130°F. Loss through the insulation (Q_2) is 300 Btu/hr. Assume 8.3 lb of water per gallon.

1. Can the unit supply the 130°F water continuously at a rate of 20 gal/hr (166 lb/hr)?
2. If the cost of electrical energy is 4 cents/kWh, what does it cost to heat 1 gallon of water at these conditions?
3. If the heater is initially at 70°F, how long will it take for the temperature to reach 130°F? (Neglect water use during the warm-up period.)

From equation (2.9):

$$H_1 + \mathrm{EE}_1 = H_2 + Q_2$$
$$H_1 = 166(1.0)(60) = 9960\,\mathrm{Btu/hr}$$
$$H_2 = 166 \times 1.0 \times 130 = 21,600\,\mathrm{Btu/hr}$$

so that $9960 + 3413(P_{\mathrm{kW}}) = 21,600 + 300$, and solving for the power required gives $P_{\mathrm{kW}} = 3.5\,\mathrm{kW}$. Since this is less than the 4.5 kW available, the 20-gal/hr rate is no problem. Since 20 gal/hr requires 3.5 kW, 1 gal requires 3.5/20 = 0.18 kWh and the cost is 0.18(0.04) = \$0.72. There is no flow under these conditions, but there is an increase in stored energy. From equation (2.9) once more:

$$\mathrm{EE}_1 = Q_2 + \Delta H_s$$
$$3413(4.5) = 300 + \frac{40(8.3)(1.0)(130 - 70)}{t}$$

Solving for t gives $t = 1.32\,\mathrm{hr}$.

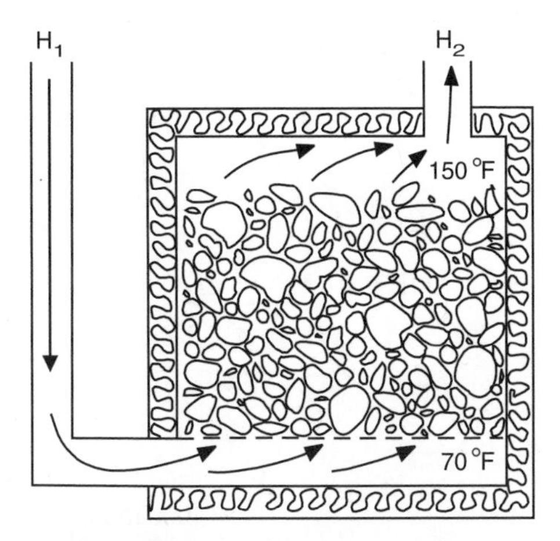

Figure 2.16 Rock bed energy balance.

2.4.4 Rock Bed Energy Balance

The 30,000-lb well-insulated rock bed shown in Figure 2.16 is used for energy storage in a solar air system. If the bed is initially at 150°F and the average airflow required to heat the building is 100 ft^3/min, calculate the average temperature of the rock bed after 24 hours of operation. Assume an ideal bed such that the outlet temperature remains at 150°F until the average temperature drops to the return air temperature of 70°F.

From equation (2.9),

$$H_1 = H_2 + \Delta H_s$$
$$H_1 = 100(4.5)(0.24)(70) = 7560\,\text{Btu/hr}$$
$$H_2 = 100(4.5)(0.24)(150) = 16,200\,\text{Btu/hr}$$
$$\Delta H_s = 30,000(0.21)\frac{T'}{24} = 262.5T'$$

Solving for the temperature change gives $T' = -33$°F. The negative sign indicates a temperature decrease. Since T' is the change in temperature, the final temperature of the bed is $150 - 33 = 117$°F. This is the average temperature of the bed—the top is still at 150°F and the bottom at 70°F. We could calculate the location of the line separating the two temperatures by equating the energy at the average temperature to the sum of the energy stored at the high and low temperatures. The results would indicate that the line was located 41% of the distance from the bottom to the top.

2.4.5 Array of Solar Collectors

In a 300-ft^2 array of solar collectors, the fluid absorbs energy at a maximum rate of 160 Btu/ft^2-hr. What airflow rate should be used to increase the air temperature by 50°F? For liquid-cooled collectors, what water flow would give a 10°F rise?

From the first law, the increase in heat content of the flowing fluid must equal the heat added:

$$H_2 - H_1 = Q_1$$

$$WC_p(T_2 - T_1) = Q_1$$

$$W = \frac{Q_1}{C_p \Delta T}$$

For air:

$$W = \frac{160(300)}{0.24(50)} = 4000 \text{ lb/hr}$$

$$\frac{4000}{4.5} = 889 \text{ ft}^3\text{/min}$$

For water:

$$W = \frac{160(300)}{1.0(10)} = 4800 \text{ lb/hr}$$

$$\frac{4800}{500} = 9.6 \text{ gal/min}$$

2.4.6 Heat Pump

The heat pump can be understood as an air conditioner. During the summer, the air on the room side is cooled while air outside is heated; during winter a cycle-reversing valve is used to switch the functions of the evaporator and condenser, and the refrigeration flow is reversed through the device. The capacity of air conditioners and heat pumps is measured in tons of refrigeration effect; 1 ton of capacity is equivalent to 12,000 Btu/hr. Figure 2.17 illustrates the heat pump concept. In the cooling mode the indoor heat exchanger functions as an evaporator, with heat flow from the room to the cold evaporating refrigerant. In the outdoor heat exchanger (condenser), heat is removed from the hot compressed vapor, and liquid refrigerant is returned through a restriction to the indoor unit. In the heating mode the roles of the heat exchangers are reversed, with the outdoor unit functioning as the evaporator and the indoor unit functioning as the condenser. In heating mode use,

$$Q_H = Q_C + \text{EE}_1$$

The useful heat delivered to the building is the sum of the electrical energy required for operation plus the heat transferred to the outdoor unit from the surroundings.

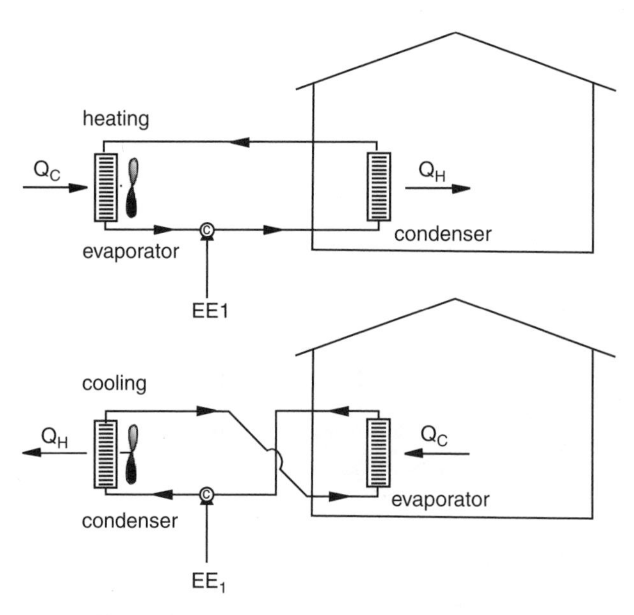

Figure 2.17 Heat pump efficiency evaluation.

The ratio of the useful energy output to the electrical energy input is determined by the coefficient of performance:

$$(\mathrm{COP})_{\mathrm{heat}} = \frac{Q_H}{\mathrm{EE}1}$$

In cooling mode use,

$$Q_C = Q_H - \mathrm{EE}_1 \quad \text{and} \quad (\mathrm{COP})_{\mathrm{cool}} = \frac{Q_C}{\mathrm{EE}_1}$$

2.4.7 Heat Transfer Analysis

Although thermodynamics is a powerful tool, there are practical problems that lack sufficient information to permit use of the first law. For example, industrial systems may recover waste heat in the flue gas from heaters and boilers as in Figure 2.18. For this analysis, heat transfer techniques are required to study conduction, radiation, and convection in order to estimate steady-state fluid temperatures. The energy engineer needs to understand a very interdisciplinary area for analysis, modeling, and design of heat and mass transfer processes, with application to common practical problems. A holistic perspective and integration of alternative energy sources with architectural projects considering climate aspects and natural solutions is also required for a successful sustainable energy system design.

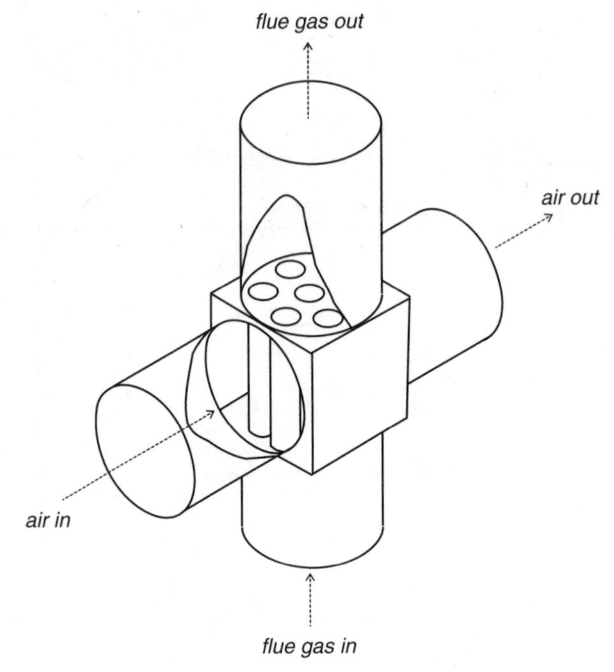

Figure 2.18 Flue gas heat recovery system.

2.5 PLANET EARTH: A CLOSED BUT NOT ISOLATED SYSTEM

Energy cannot be destroyed, so how has the public developed the perception that we have a shortage of energy? The scarcity is not in total energy but in energy forms or levels that are suitable for transfer to useful work or heat. When fuel is burned in a boiler, the products of the process (primarily steam and hot flue gases) have the same energy content as the fuel supplied. The flue gases mix with the atmosphere and are no longer available for useful work or heat. A portion of the energy in the steam can be utilized in the production of work or heat, but the end result of these processes will eventually appear as energy rejected to the environment, so that all of the original energy is, in effect, lost.

During the process of conversion of some form of energy into another intermediate form, which serves a useful purpose, a question arises as to the maximum efficiency of the conversion process. Is energy waste to the environment always a by-product of energy conversion? If so, is there a predictable or limiting portion of the energy converted that can never be recovered regardless of the skill or ingenuity of the designer?

The two laws of thermodynamics can be stated together: The total energy content of the universe is constant, and the total entropy is continually increasing. Entropy is thus a measure of the amount of energy that is no longer capable of performing work. An increase in entropy involves a reduction of the energy available to do work in the future. Increasing entropy can also be considered as a change from an ordered state to a disordered state. One example of increasing entropy is water falling over a dam. When the water is above the dam it has some potential energy due to gravity, which can be used to generate electricity or turn a wheel to perform some useful task. Once the water has fallen to the level below the dam, its total energy is the same (as the fall warms the water, increasing its thermal energy), but it no longer has the same capacity to do work. The water has moved from what is referred to as an *available* or *free energy state* (high-grade energy) to an *unavailable* or *bound energy state* (low-grade energy). This change in the energy state of the water as it falls over the dam is an increase in entropy. Unavailable energy is most often in the form of thermal energy.

The second law of thermodynamics tells us that the entropy of an isolated system tends to increase with time. An *isolated system* is one in which matter and energy cannot either enter or exit. The Earth is certainly not an isolated system because it receives a constant flow of energy from the sun and from outer space; however, with the exception of a few meteorites, artificial space, satellites, and some cosmic dust, Earth is a closed system when it comes to matter. The order (entropy) in a closed but not isolated system can be maintained over time, but only if there is an external source of energy. Thus, it is ultimately the energy of the sun that has allowed highly ordered and complex organisms and civilizations to develop on Earth.

From the beginning of life on Earth until the last several hundred years, all organisms, including human beings, utilized the available energy of the sun to provide the necessities of life. During this period of time, energy from the sun was used to maintain the entropy in our planetary system. Plants use sunlight directly, the process of photosynthesis converting the energy from the sun into plant matter, creating order. Animals and humans use sunlight indirectly by eating plants and other animals to survive. Humans in the past also used indirect solar energy in the form of wood for fuel and building materials, and flowing wind and water as well as animals for performing work. As long as life on Earth used sunlight in an evenly distributed manner and at a rate less than that of incoming solar radiation, order was maintained all over the planet.

As populations increased in certain parts of the world in the past several thousand years, humans started to cut down more trees, build up their houses and roads, and harvest more crops in a year than could be regenerated through the natural process of photosynthesis. Such practices led to deforestation and desertification of previously fertile land. Human population increases and the use of fossil fuels since the industrial revolution have greatly increased the rate at which humans are adding to the entropy of the planet.

Fossil fuels such as coal, oil, and hydrocarbons are complex and highly ordered chains of hydrocarbons, formed originally by photosynthesis from sunlight millions of years ago. When humans burn fossil fuels to provide energy to drive cars or generate electricity, the second law of thermodynamics states that a penalty must be extracted—and this penalty is an increase in entropy. In most populated parts of the world, energy from fossil fuels is consumed at an enormous rate. Fossil fuel consumption therefore results in a large increase in entropy, and this entropy shows up in the form of pollution.

The substances that human beings classify as pollutants have always been present on the planet, because Earth is a closed system. The reason that these materials cause a negative impact on the environment, therefore, is not that they exist, but that they have been dispersed or concentrated so much throughout the world's ecosystems in a very disordered fashion. To illustrate this, two pollutants of major concern will be considered, carbon dioxide (CO_2) and sulfur dioxide (SO_2). CO_2 and SO_2 are both given off when coal is burned to provide heat or to generate electricity. CO_2 is the most important greenhouse gas contributing to global warming, and SO_2 is the main cause of acid rain. Carbon and sulfur have no impact on the environment when they are locked up in the highly ordered form of coal. It is when fossil fuels are burned, and the combustion products are released in the atmosphere, that these compounds become pollutants.

Solar energy can be used to counterbalance increases in entropy caused by the use of fossil fuels. As long as the rate of entropy increase due to fossil fuels is less than the rate of entropy decrease from solar radiation, the net entropy on the planet will not increase. Since the amount of solar radiation reaching Earth is quite large, why cannot humans continue to use vast quantities of fossil fuels?

It is virtually impossible to channel solar radiation such that it can counteract the negative impact of pollution caused by fossil fuels. For example, it is difficult to imagine using the sun's energy to reduce the acidity of lakes and forest soils damaged by acid rain. A better solution would be to use solar radiation to power our societies directly. This would be more efficient than using fossil fuels to create our highly ordered civilization and then trying to eliminate the pollution and entropy by using sunlight. Renewable energy technologies such as solar energy (photovoltaic cells, active and passive solar heating, etc.), biomass (plant matter), wind power, and hydropower use sunlight either directly or indirectly, to produce the high-quality energy demanded by industrialized societies, as discussed in subsequent chapters.

REFERENCES

[1] C. B. Schuder, *Energy Engineering Fundamentals: With Residential and Commercial Applications*, Van Nostrand Reinhold, New York, 1983.

[2] A. W. Culp, Jr., *Principles of Energy Conversion*, McGraw-Hill, New York, 1979.

[3] S. W. Angrist, *Direct Energy Conversion*, Allyn & Bacon, Boston, 1982.

[4] F. P. Incropera and D. P. DeWitt, *Fundamentals of Heat and Mass Transfer*, 5th ed., Wiley, New York, 2001.

[5] K.-F. V. Wong, *Thermodynamics for Engineers*, CRC Press, Boca Raton, FL, 2000.

CHAPTER 3

HYDROELECTRIC POWER PLANTS

3.1 INTRODUCTION

With the present worldwide level of hydropower plant technology, self-production of electric power has gained greatly in popularity because it is associated with the fact that the majority of rural properties have rivers with water streams and small water heads that can be used as primary energy. That energy is commonly used for stationary machine drivers, generation of electricity, and water storage in elevated reservoirs. Small electric power plants can still be used with waterwheels or, depending on the flow or the available head, with turbines.

The first known small hydroelectric power plant was built in Northern Ireland in 1883 to supply energy for an electric train that began with two turbines of 52 hp (1 horsepower = 746 watts). It was preceded in 1882 by the first known turbine–generator set, with approximately 8 hp, driven from a gross water head of 10 m to supply some carbon-filament light bulbs.

In the same year, 1883, in Diamantina, Brazil, a hydroelectric energy scheme of 12 kW (two generators driven by waterwheels with an elevation of 5 m) was established. Ever since, the diversified evolution of the energy sector has been prodigious and has influenced all sectors of everyday life. With that diversity appeared several types of hydroelectric power plants.

Integration of Alternative Sources of Energy, by Felix A. Farret and M. Godoy Simões

Hydroelectric power plants can be driven from a water stream or accumulation reservoir. Run-of-river hydroelectric plants (those without accumulation reservoirs) are built along a river or a stream without lake formation for the intake of water. In this form the river course is not altered, and its minimum flow will be the same or higher than that of the turbine output power. The excess water should be diverted, and in this case, use of the water volume is not total. The costs are lower than those for reservoirs with less environmental impact, but as the plant is unable to store energy, there is no need for extensive hydrological studies.

The development of a hydroelectric power plant with an accumulation reservoir demands far more complex hydrological and topographical studies to determine the site elevation. Such data are of great importance, because they are used directly to establish the watercourse flow for calculation of the power to be produced by the power plant [1–4]. The water use is total, and the civil works have an effect on the environment proportional to the size of the plant. A reservoir system also has a higher generation potential than that of a water stream, compensating for the larger capital investment. The lake formed by water accumulation can be used for other purposes, such as recreation, creation of fisheries, irrigation, and urbanization.

Useful data for hydroelectric power plant projects might include:

- Local availability of materials necessary for the construction of a dam (when applicable)
- Labor availability for cost reduction
- Reasonable distance of the power plant from the public network, to allow future interconnection through a single-phase ground return or a single-phase, three-phase, or low-voltage dc link.

3.2 DETERMINATION OF THE USEFUL POWER

In general terms, ordinary turbines and waterwheels use the energy that can be evaluated as the sum of the three forms of energy given by *Bernoulli's theorem*. This expression remains constant for a given cross section and position in a channel:

$$\frac{v^2}{2g} + h + \frac{p}{\rho g} = \frac{P}{\rho g Q} \tag{3.1}$$

where v = water flow speed (m/s)
g = gravity constant (= 9.81)(m/s^2)
h = height of the water (m)
p = pressure of the water (N/m^2)
ρ = density of the water (kg/m^3) ($\cong$ 1000.00)
P = power (kg·m/s) (1 hp = 75 kg·m/s = 746 W)
Q = flow of the watercourse (m^3/s)

For ordinary modern turbines, the effective power at their input may be obtained from equation (3.1) (neglecting the terms v and p) for the potential energy in the watercourse as

$$P_t = \eta_t \rho g Q H_m \qquad (3.2)$$

where η_t is the turbine simplified efficiency (for standard turbines it is taken as 0.80) and H_m is the water head. The available flow of a watercourse (m^3/s) is expressed by

$$Q = Av \qquad (3.3)$$

Energy capture from the flow of large rivers is a very attractive idea. However, the initial enthusiasm may fade when the density of energy that can be obtained from the running water is taken into account. For example, the power density, of a 1.0-m/s stream, [by equation (3.1)], is approximately 500 W/m^2 for the river cross section. Complicating matters further, only 60% of that energy can theoretically be recovered for most small hydroelectric power plants, and almost always, machines using this type of energy are of rough manufacture whose efficiency rarely reaches 50%. That is, they use only 250 W/m^2 of the energy through the intercepted area across the water stream. As predicted by the combination of equations (3.1) and (3.3), the power available from a stream is proportional to the cube of the speed. If the stream speed is 2 m/s, the density would rise to 2000 W/m^2; and so on. However, such speeds are not usually very common in streams from which energy can be extracted.

In practice, useful power is the maximum power a site can offer through its topographical and hydrological characteristics of elevation and available flow in the watercourse. Notice that the useful power in the machine shaft should be the same as the maximum electric power that can be generated at that place. Depending on the geometric and material dimensions used in the machines and piping of the power plant, this electric power can be determined in kilowatts ($\eta = 0.6$, $\rho \cong 1000\,kg/m^3, g = 9.81\,m/s^2$), approximately, by

$$P = 6.0QH \qquad kW/m^2 \qquad (3.4)$$

where H is the gross head (m), that is, the difference between the levels of the crests of water in the reservoir (dam) and the river at the powerhouse site. The pump flow and pumping manometric height can be calculated, respectively, by

$$Q_b = 0.75Q \quad \text{and} \quad H_b = 0.55H \qquad (3.5)$$

Usually, the methods used for topography and hydrology works of small power plants are expedient as discussed below. There are references that provide more detailed routines for these calculations [3,4].

3.3 EXPEDIENT TOPOGRAPHICAL AND HYDROLOGICAL MEASUREMENTS

3.3.1 Simple Measurement of Elevation

Two very simple methods are used to measure the natural water head of a hydroelectric station. For the first, it is enough to use a carpenter level and two very flat inflexible metric rulers, one 3 to 4 m long and one 2 m long. The inferior tip of the smaller ruler is placed vertically at the level of the water (Figure 3.1). Then the larger ruler is placed on the ground and the carpenter level controls its horizontal position. It measures the height h_i and marks the level of a point i where the tip of the larger ruler rests on the ground. Restarting from that position, the operation is repeated between several pairs of elevation points, making sure that they were verified in all level differences, h_k. The height of the total head is the straight summation of the level differences, $h_k (k = 1, 2, \ldots, n)$:

$$h = h_1 + h_2 + h_3 + \cdots + h_n \tag{3.6}$$

For the second method, two metric rulers (about 2 m long) are used, together with a plastic tube (flexible, transparent, 1 cm in internal diameter, and at least 6 m long). The rulers are placed at two points vertically, between which the elevation is measured (Figure 3.2). With the help of the plastic tube (full of water), points of equal level are determined on each ruler, creating a horizontal plane of reference. Summation of all differences between the ground heights of every two points gives the total difference in level, h, as in equation (3.6).

3.3.2 Global Positioning Systems for Elevation Measurement

Global positioning systems (GPSs) provide specially coded satellite signals that can be processed in a GPS receiver, enabling the receiver to compute area, elevation, velocity, and time. Four GPS satellite signals are used to compute positions in three

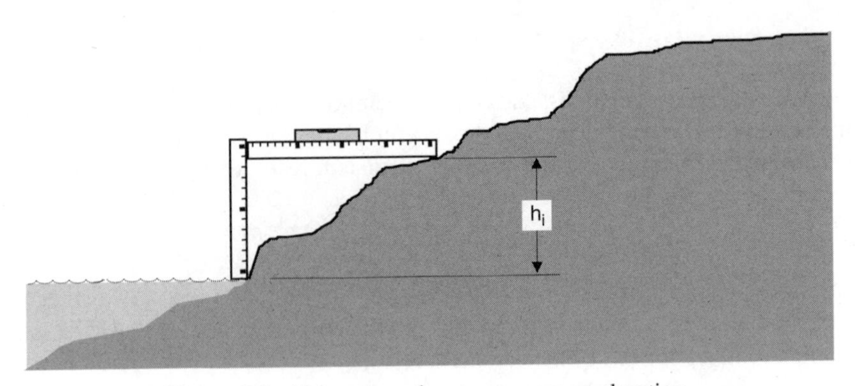

Figure 3.1 Using two rulers to measure an elevation.

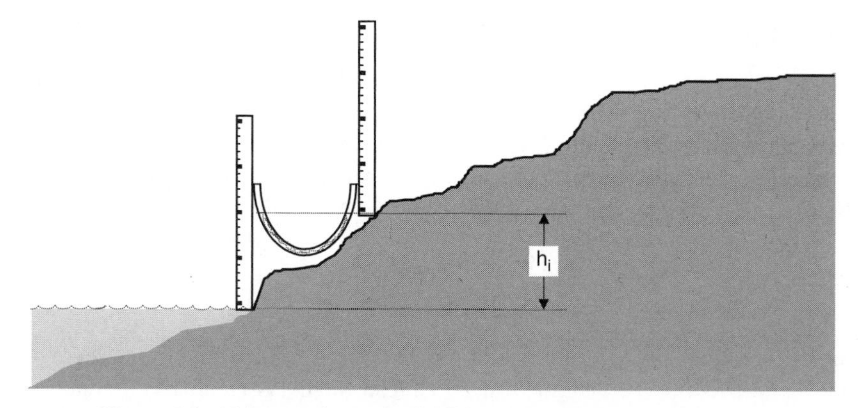

Figure 3.2 Using a ruler and a flexible tube to measure an elevation.

dimensions and time offset in the receiver clock. Although there are many thousands of civilian GPS users worldwide, the original system was designed for and operated by the U.S. military, funded and controlled by the U.S. Department of Defense.

The control segment consists of a system of tracking stations located around the world. The space segment of the system consists of GPS satellites, space vehicles (SVs), which send radio signals from space. The nominal GPS operational constellation consists of 24 satellites that orbit the Earth in 12 hours. There are often more than 24 operational satellites as new ones are launched to replace older satellites. The satellite orbits repeat almost the same ground track, as the Earth turns beneath them, once each day. The orbit altitude is such that the satellites repeat the same track and configuration over any point approximately every 24 hours (4 minutes earlier each day).

There are six orbital planes (with nominally four SVs in each), equally spaced (60° apart) and inclined at about 55° with respect to the equatorial plane. This constellation provides the user with between five and eight SVs visible from any point on Earth.

The master control facility is located at Schriever Air Force Base (formerly, Falcon AFB) in Colorado. These monitor stations measure signals from the SVs, which are incorporated into orbital models for each satellite. The models compute precise orbital data (ephemeris) and SV clock corrections for each satellite. The master control station uploads ephemeris and clock data to the SVs. The SVs then send subsets of the orbital ephemeris data to GPS receivers over radio signals.

The GPS user segment consists of GPS receivers and the user community. The receivers convert SV signals into position, velocity, and time estimates. Four satellites are required to compute the four dimensions x, y, z (position), and time. GPS receivers are used for navigation, positioning, time dissemination, and other research purposes.

3.3.3 Specification of Pipe Losses

There are four basic causes of losses in pipes conducting water or any viscous fluid from a higher potential point, say a water dam, to a lower potential point, such as a hydraulic turbine. The four causes are the specific mass ρ of the fluid (kg/m), the average speed v (m/s), the pipe diameter D (meters), and the viscosity coefficient η (kg/m·s), all related quantitatively by the *Reynolds number*:

$$N_R = \frac{\rho v D}{\eta} \tag{3.7}$$

For the specific case of water, which is the basis for all hydraulic turbines, $\rho = 1000\,\text{kg/m}$ and $\eta = 10^{-3}\,\text{kg/m}\cdot\text{s}$ at 20°C. The Reynolds number is a nondimensional number; it does not depend on a system of units. The number is used to establish the speed limit at which a fluid can run in a pipe without turbulent flow. It is common practice to establish N_R within the following limits: for $N_R = 2000$ and below, the flow is laminar; for N_R from 2000 to 3000, the flow is unstable; and for N_R above 3000, the flow is turbulent.

Part of the loss of energy in pipes is due to the roughness of their internal walls. However, it is important to note that turbulent flow is not caused by the roughness of the pipe but by the conditions at which the fluid runs in the pipe. That means that a smooth pipe can have turbulent flow and a rough pipe can have smooth flow, because when there is smooth pipe flow, the viscosity of the fluid forms a laminar sublayer covering the roughness of the pipe. As the speed of the fluid increases in transitional pipe flow, the sublayer is reduced in thickness, smoothing out some of the roughness. Finally, in rough pipe flow, the laminar sublayer is very thin and the roughness of the pipe is well exposed to the water flow, causing heavy pipe losses of otherwise useful kinetic energy. Pipe friction is proportionally inverse to the Reynolds number.

Pipe losses are extremely sensitive to the diameter of the pipe not only by affecting the Reynolds number inversely but by contributing to the increase in velocity of the fluid and the associated losses. It is a complex task to quantify these contributions mathematically, so it is usual in hydrology studies to unite all losses in a head equivalent h_i. The equivalent is defined as a term proportional to the pipe length and friction and to the kinetic energy given by Bernoulli's expression [equation (3.1)] and inversely proportional to the pipe diameter. This expression is known as the *Darcy–Weisbach formula*:

$$h_i = \lambda \frac{l}{D}\frac{v^2}{2g} \tag{3.8}$$

The friction factor λ is provided by the pipe manufacturer.

If equation (3.3) is used to obtain the square of the water velocity, it will be observed that the diameter affects equation (3.8) in its fifth power. The result from this equation has to be added linearly to the results from equation (3.6) to obtain the effective pipe head loss. The final value of practical interest is the effective hydraulic gradient, expressed as

$$\nabla h = \frac{h + h_i}{l} \tag{3.9}$$

3.3.4 Expedient Measurements of Stream Water Flow

Among the expedient methods used to measure the flow of a ditch, stream, or a river, the simplest is accomplished by floats, by rectangular or triangular spillways, or through measurement of the water resistivity difference between two points [3–5].

Measurement Using a Float In measurements with a float, a straight passage of watercourse with a uniform bed where the water flows calmly is chosen. If possible, it is measured a length L 10 m above the river course, marking the beginning and the end. Points of interest are then marked by two strings tied to stakes nailed on the margins perpendicular to the ditch or stream axis. Next, the float is placed some meters upstream from the beginning of the passage chosen and in the middle of the riverbed. The float can be a closed bottle ballasted with water at about one-third of its volume to keep it vertical. A chronometer measures the time, in seconds, that the float takes to travel the passage chosen.

The cross-sectional areas, limited by the levels of water crest and the ditch or stream bottom, should be determined for at least the initial and final points of the measurement. If the length of that passage is too long, it is advisable to determine the areas of one or more intermediate cross sections and use the average value. The flow Q is calculated by the formula

$$Q = \frac{0.8L\bar{A}}{t} \tag{3.10}$$

where 0.8 is a surface speed correction for the average speed across the section measured, L is the length of the passage used to measure the flow (m), $\bar{A}$ the average area of the cross sections m^2, and t the time of the float course (s). The speed correction coefficient for the superficial speed across the section is necessary because of to speed differences at various points of the river cross section, such as between waters closer to the margins and the riverbed as opposed to those in mid stream, due mostly to viscosity and friction.

Measurement Using a Rectangular Spillway Measurement using a rectangular spillway leads to more precise results than with a float, although it demands more work and is limited to cases where the morphological conditions of the watercourse allow its use. For this process, the watercourse is blocked by a board with a rectangular central aperture of known area, sufficient for the passage of all the water. The spillway width should be from one-half to two-thirds of the watercourse width. The aperture cuts should be chamfered in the direction of the water flow.

After all the fencing panel rifts and the spillway are secured, an upstream stake is nailed to the waterbed 1 or 2 m distant, with its top at the crest of the spillway (the lower side of the aperture). Once water is draining normally through the spillway, the height of the water level, H, is measured above the top of the stake. In this case, the water discharge can be calculated from

$$Q = 1.84(L - 0.2H)H^{3/2} \tag{3.11}$$

where L is the aperture width of the spillway (m; it should be larger than $3H$) and H is the head in meters of the upstream water level above the crest of the spillway at the location of the stake.

Measurement Using a Triangular Spillway A spillway with a triangular 'V-shaped' aperture is used when the discharges are very small (lower than 200 L/s) or when the stream has a smaller width than depth, which precludes the installation of a spillway of rectangular section. The installation measurement procedures for this type of spillway are the same as those noted for a rectangular spillway; however, the flow is calculated by

$$Q = 1.4H^{5/2} \tag{3.12}$$

where H is the head of the upstream water level (m) above the lower vertex of the spillway at the location of the stake.

Measurement Based on the Dilution of Salt in the Water Water flow can also be determined using the resistivity alteration caused by salt dilution along a river course [2]. In one method, salt is poured in the water all at once, and the variation in salt concentration in the course is measured downstream during a certain period of time. In a second method, a salt solution is poured in the water at a certain constant rate, and the concentration is measured downstream.

To implement the first method, approximately 3 kg of salt for every estimated m^3/s of water flow is needed, given by

$$Q = \frac{\bar{V}}{T} \quad m^3/s \tag{3.13}$$

where $\bar{V}$ is the volume of water at an average cross section of the river and the downstream length of the river used in the measurement, and T is the time period during which the readings are made. A mass M (kg) of salt is initially dissolved in a bucket of water and poured in the river all at once. About 50 m or more downstream, the water resistivity (and thus the conductivity) is measured periodically for 5 or 6 minutes such that it can be graphed (Figure 3.3). The mass of salt in a bucket, which has volume V_0, has an initial concentration given by

$$C_0 = \frac{M}{V_0} \quad kg/m^3 \tag{3.14}$$

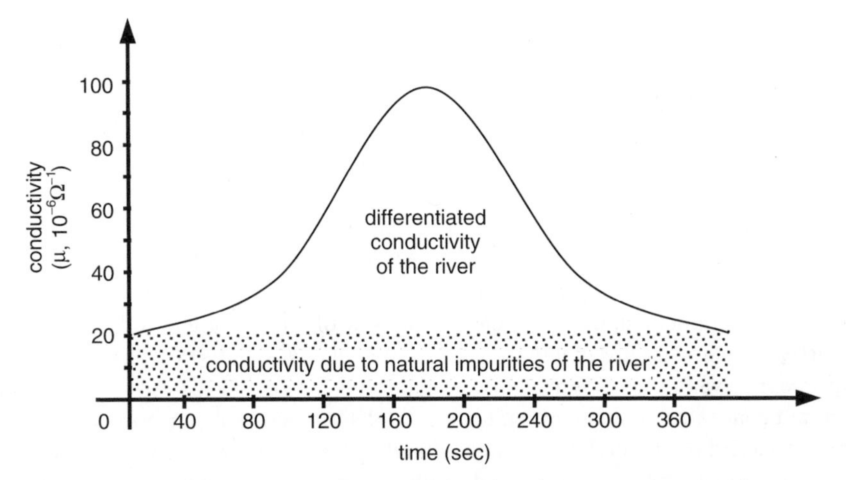

Figure 3.3 Typical curve of conductivity alterations at a point on a river.

where C_0 is the initial concentration of salt in the bucket, and V_0 is the bucket volume occupied by the salt solution.

All the salt (M) in the water stream would provide an average salt concentration of

$$\bar{C} = \frac{M}{\bar{V}} \qquad \text{kg/m}^3 \tag{3.15}$$

where $\bar{V}$ is the volume of water given by the average cross section of the river and the downstream length of the river used for the measurement. However, due to the proportionality k between the concentration of salt $\bar{C}$ and the water conductivity $\bar{\mu}(\Omega^{-1})$, it is possible to establish

$$k = \frac{\bar{C}}{\bar{\mu}} = \frac{C_0}{\mu_0} \tag{3.16}$$

Based on the classical equation for conductivity,

$$\mu_0 = \frac{1}{R_0}\frac{l}{S}$$

and on equation (3.16), it is easily shown that

$$\frac{R_0}{R} = \frac{\bar{\mu}}{\mu_0} = \frac{\bar{C}}{C_0} \tag{3.17}$$

The average conductivity of the water, $\bar{\mu}$, is proportional to the average salt concentration in the water, $\bar{C}$. This means that

$$\bar{\mu} = \frac{1}{T}\int_0^T \mu(t)\,dt = \frac{1}{T}A \tag{3.18}$$

where $A = \int_0^T \mu(t)\,dt$ is the area under the curve in Figure 3.3, discounting the natural impurities of the river. Combining equations (3.13), (3.15), and (3.18) yields

$$Q = \frac{\bar{V}}{T} = \frac{M/\bar{C}}{A/\bar{\mu}} = \frac{M}{kA} \tag{3.19}$$

Area A under the curve in Figure 3.3 is calculated in $(\Omega^{-1} \cdot s)$ and the scale factor k in kg/m$^2 \cdot \Omega^{-1}$ is obtained through measurement of C_0 the initial concentration of salt occupying volume V_0 in the bucket.

The second method, used to measure the flow of a more or less turbulent watercourse, is conducted by gradually pouring a bucket's initial salt concentration C_0 upstream [given by equation (3.14)] at a constant rate $q = V_0/T$ (m^3/s). The average salt concentration in the river, $\bar{C}$, may be determined using equation (3.15). For these measurements, a relatively long, straight passage of river (some 50 m long) is chosen, into which the salt is poured. With an ordinary multimeter or an ohmmeter (capable of measuring up to 0.01 Ω), the resistivity of the downstream water is measured at the end of the length chosen for testing. It is helpful to remember that the resistivity (or resistance) is the inverse of conductivity (or conductance) and is proportional to the concentration of salt. This method is based on the idea that the concentration of salt in the stream is reduced from C_0 to $\bar{C}$ and that Q is much larger than q. If those two assumptions are not met, a more precise form of the equation should be sought [2].

As the salt is being diluted in the water at a constant rate, its concentration in the water, and therefore the conductivity, will change as shown in Figure 3.4. This means that at a certain point downstream after some period of time (say, 200 s), the mixture will be complete, and the concentration of salt that changed from C_0 to $\bar{C}$ is measured indirectly through the conductivity of the water at that point [2]; that is,

$$\frac{\bar{V}}{V_0} = \frac{C_0}{\bar{C}}$$

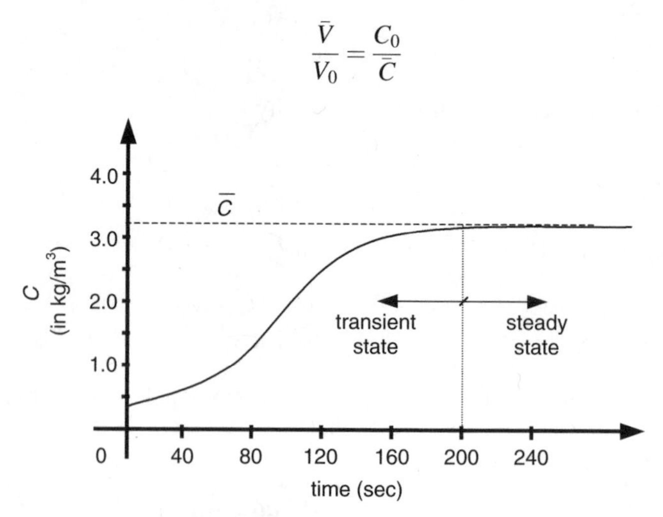

Figure 3.4 Variation of salt concentration in river flow.

As a result, during the steady-state concentration shown in Figure 3.4, the rate of high concentration of salt being poured into the water will be proportional to the volume of water $\bar{V}$ just passing downstream at a rate of $Q = \bar{V}/t$: that is,

$$Q = q\frac{C_0}{\bar{C}} \qquad \text{m}^3/\text{s} \tag{3.20}$$

This method is fast and simple and is precise to about 7%.

3.3.5 Civil Works

Civil works include construction of a dam (or water storage facility), a water intake (to divert a river from its normal course), power piping, a reservoir, forced piping, a balance chimney, a machine house, or a spillway (to return water to its normal course). Such conventional facilities cost a lot of money and require a civil engineer who is expert in the mechanics of soils, geology, and structures to set up the project and to follow it through the building stages [2–4]. To avoid unnecessary facilities when setting up a small power plant could necessitate employment of a specialist to minimize building expenses, and reduce time until the plant can be operated, and possibly, maintenance problems.

When sufficient height and water flow are available for a small power plant, a dam may not be needed. The water will be used in essentially the same proportion as it is available in the river. Such a system is known as a *run-of-river plant*. A portion of the stream flow is simply diverted to the powerhouse and, once used, is returned to the river. The large majority of small power plants are of the run-of-river type. This concept is sometimes misunderstood as being that of a turbine put directly into the river as a lower bucket type of waterwheel which would be the case for a flow turbine and not for a run-of-river turbine.

3.4 GENERATING UNIT

A generating unit is formed by a primary hydraulic machine and a generator together with such auxiliary equipment as a regulator (electromechanical, electro-electronic, or electronic), a water admission valve, an electrical command board, and an inertial flywheel. In Chapter 10 we discuss in general terms the equipment and electrical characteristics of small power plants. In this section we simply present some peculiarities of hydroelectric power plants.

3.4.1 Regulation Systems

Regulation systems maintain voltage and rotation at constant levels, thereby maintaining the frequency of the generating unit within the variation limits of the electric network demands. A centrifugal pendulum of the mechanical type (i.e., a servomechanism working under pressurized oil) commands conventional automatic speed regulators of the type used in small hydroelectric power plants, the simplest of which are the inertial flywheels (see Section 11.5).

When acquiring a regulator, its cost, the distance of the powerhouse to the load, and the load type should be taken into account. When considering cost, it is advisable to choose a regulator for turbine powers above 20 kW. In smaller turbines, regulation can be achieved by load adjustment, manual control of water flow, or through a simplified control for small integrated generating units [5–8]. Because voltage regulation is controlled by only one electric parameter, it is covered in more detail in Chapter 10.

3.4.2 Butterfly Valves

Butterfly valves are used in small hydroelectric power plants with metallic forced piping. They block water flow to the turbine during maintenance work and provide an additional resource for blocking the turbine in case of system failure. They are installed at the powerhouse by means of flanged connections, between the forced piping and the snail-shaped box housing the turbine.

3.5 WATERWHEELS

Waterwheels are quite primitive and simple machines, usually built of wood or steel, with shovels of steel blades fixed regularly around their circumference [6]. The water pushes the shovels tangentially around the wheel. The water does not exert thrust action or shock on the shovels as is the case with turbines. The water thrusting on the shovels develops torque on the shaft, and the wheel rotates. As one can infer, such machines are relatively massive, work with low angular speeds, and are of low efficiency, due to losses by friction, turbidity, incomplete filling of the buckets (or cubes), and other causes.

In most cases, waterwheels are handcrafted and a great variety of models exist [3,5,6]. The most common is the wheel with upper buckets, so called because it consists of fixed shovels around a wheel, filled with water coming in at the top of the machine (Figure 3.5). This type of wheel is also called hydraullic motor and it is driven by potential or gravity, because the water accumulates in the small compartments of the half-wheel out of the water flow. The weight of the water provokes a motor torque in the sense of its downstream drainage.

A second type of wheel is one with coulisse buckets, which uses potential and kinetic energy. Water drains through a mouthpiece tangent to the wheel, and the water thrust on the shovels develops a torque on the wheel's shaft, causing it to rotate.

In another type, known as a lower pushed wheel, water causes a thrust on the lower part of the wheel (see Figure 3.6). Such wheels can be mounted floating on a body of water or inserted into the water flow. This assembly is advantageous

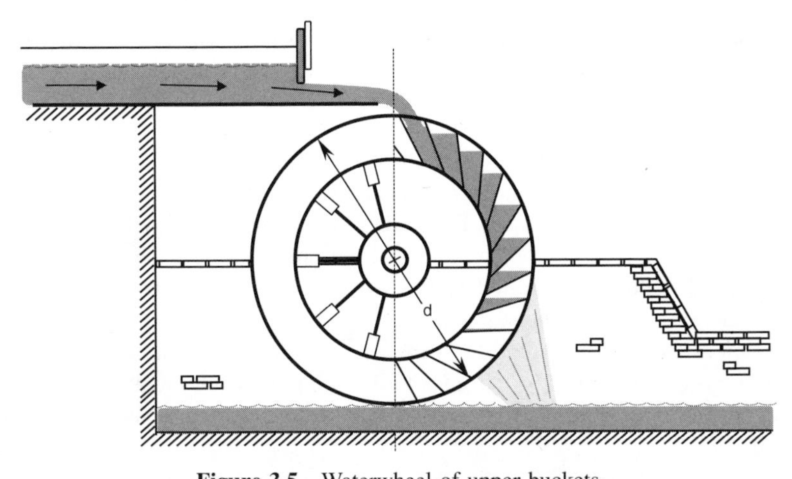

Figure 3.5 Waterwheel of upper buckets.

because it follows the potential variation of the stream and maintains almost constant torque.

In the case of bucket wheels, considering input and output at the same atmospheric pressure ($\Delta p = 0.0$), the effective shaft power [equation (3.1)] becomes

$$P_p = \eta \rho g Q H_m \tag{3.21}$$

where η is the efficiency (practical value $= 0.60$) and H_m is the head difference between crests of the water stream at the input and output of the channel.

Notice that the upper bucket wheel diameter is the minimum head of the water required to move it, whereas the lower bucket wheel is adapted for smaller heights,

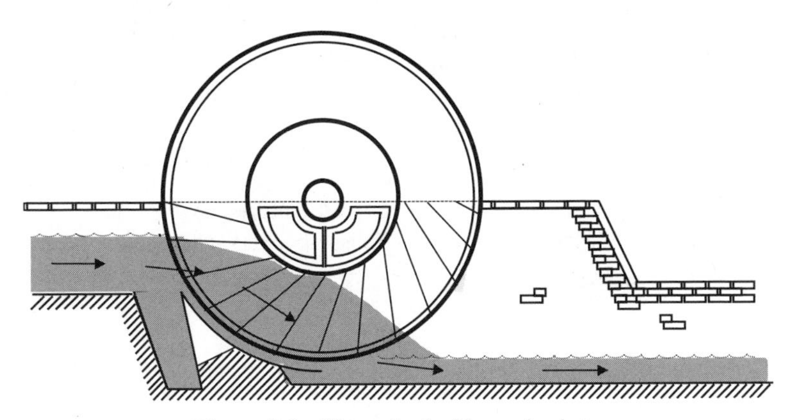

Figure 3.6 Waterwheel of lower buckets.

on the order of the wheel shovel length or less. In this case, the resulting kinetic energy power is given by

$$P_k = \frac{\eta \gamma A v^3}{2} \tag{3.22}$$

where η is the efficiency of the wheel and A is the normal cross-sectional area of the body of water intercepted by the waterwheel shovels. It should also be noted that waterwheels have a very low rate of rotation, so to produce electricity, they require speed multipliers (around 1:20), with accompanying appreciable losses.

The tangent speed of the water wheel is, approximately, that of the water in movement at the input mouthpiece of the wheel without load on the shaft. In other words, $v = k\sqrt{2gh_0} = 4.2\sqrt{h_0}$ (from 3 to 3.5 m/s), h_0 being the average height of the stream at the water admission channel (from 0.50 to 0.70 m). The kinetic energy is negligible.

The use of waterwheels for electric power generation presents some important advantages: (1) production and conservation goals are easily met; (2) operation is not affected by dirty water or by solids in suspension; and (3) their motor torque value is aided in a certain way with an increase in load because during the decreased rotation under load, the buckets have more time to fill with water and thus increase the moment applied on the machine shaft.

The largest disadvantage of waterwheels is their low angular speed of operation, which causes the generator to work with speed multipliers at high reduction rates, thus resulting in appreciable losses of energy. Another disadvantage is when the system needs any speed regulation, which could complicate such a simple energy-generating system.

3.6 TURBINES

Turbines are wheels driven by some fluid in movement that makes them rotate by action of the energy contained in them (in potential or kinetic form) on slats that can have several formats. The starting operation of the various types in use at present differs according to the form of energy used to drive them [4,9–11].

Several types of turbines are manufactured in a number of sites worldwide with projects and construction somehow perfected. They are more compact, made from melted metal, and usually operate at high rates of rotation and with high mechanical efficiency. Use of the following types of turbines is recommended:

- Pelton turbine
- Francis turbine
- Michel–Banki turbine
- Kaplan or propeller hydraulic turbine
- Deriaz turbine

The choice of turbine type depends on the application range, primarily on height and water flow and other criteria, such as maintenance, turbine cost and sensibility

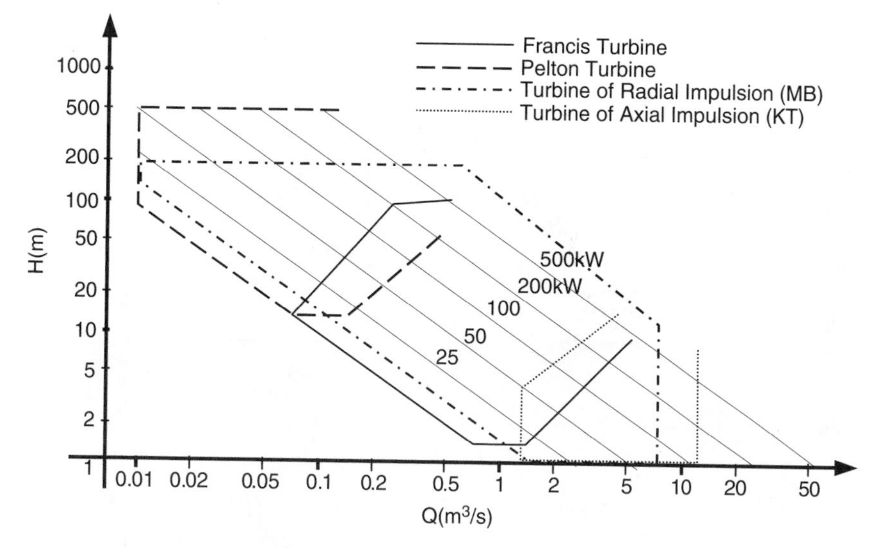

Figure 3.7 Application ranges of turbines.

TABLE 3.1 Head and Specific Speed Ranges for Various Types of Turbine Runners

Turbine Runner	Head Range (m)	n_s Range
Pelton	400–2000	0–30
Francis	50–500	20–120
Mixed-flow	20–80	120–180
Kaplan (vertical)	8–50	180–260
Bulb pit or tube (horizontal)	0–10	260–360
Michel–Banki	1.50–150	30–210

to materials in suspension. In Figure 3.7, the limits of operation recommended are represented for the aforementioned turbines, with their application fields a function of the head height, water flow, and power of each machine, as discussed in the following sections.

For general guidance, Table 3.1 lists various types of runners and the head associated with each type referred to by the specific speed, n_s, based on the optimum point, defined as

$$n_s = n\sqrt{\frac{Q}{\sqrt{H^3}}}$$

3.6.1 Pelton Turbine

A Pelton turbine (PT) is a turbine of free flow (action). The potential energy of the water becomes kinetic energy through injectors and control of the needles that

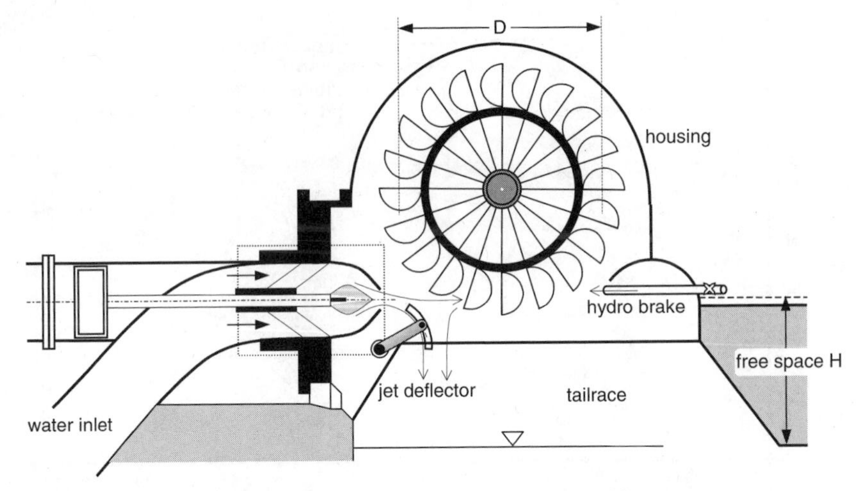

Figure 3.8 Cross section of a Pelton turbine.

direct and adjust the water jet on the shovels of the motive wheel. They work under approximate atmospheric pressure. Thus, the lower limit of the net height, H, is affected by the impact of the jet on the shovels of the motive wheel. In PTs equipped with multiple injectors, needles, and motive wheels of large diameter, the situation is more complex, necessitating better understanding of the manometric pressure.

In small hydroelectric power plants, operation of a PT can result in reasonable economy when it operates with flows above 30 L/s and heads from 20 m. The necessary combination of head with design flow to obtain a requisite PT power is shown in Figure 3.7. The wheel diameter of the Pelton crown of shovels, D (see Figure 3.8), as a function of the net head, H, the flow, Q, and the specific rotation, n_s, can be calculated by

$$D = \frac{97.5Q}{n_s^{0.9} H^{1/4}} \tag{3.23}$$

where $n_s = 240(a/D)z^{1/2}$ (rpm). Here a is the diameter of the water jet (meters) (Figure 3.9), D the diameter of the rotor circle at the point of jet incidence (i.e., the diameter of the crown of shovels at the impact points) (meters), and z the number of jets. The diameter of the jet at the injector output can be calculated by

$$a = 0.55Q^{1/2}H^{1/4} \tag{3.24}$$

where H is the head.

Most modern Pelton designs include a flow splitter (Figure 3.9). This is a way of reducing the axial thrust that would be caused by water escaping a shovel. By splitting the shovel, the water jet exerts equal force on both sides of the shovel, thus reversing the water at almost 160°, to avoid striking back on the incoming water.

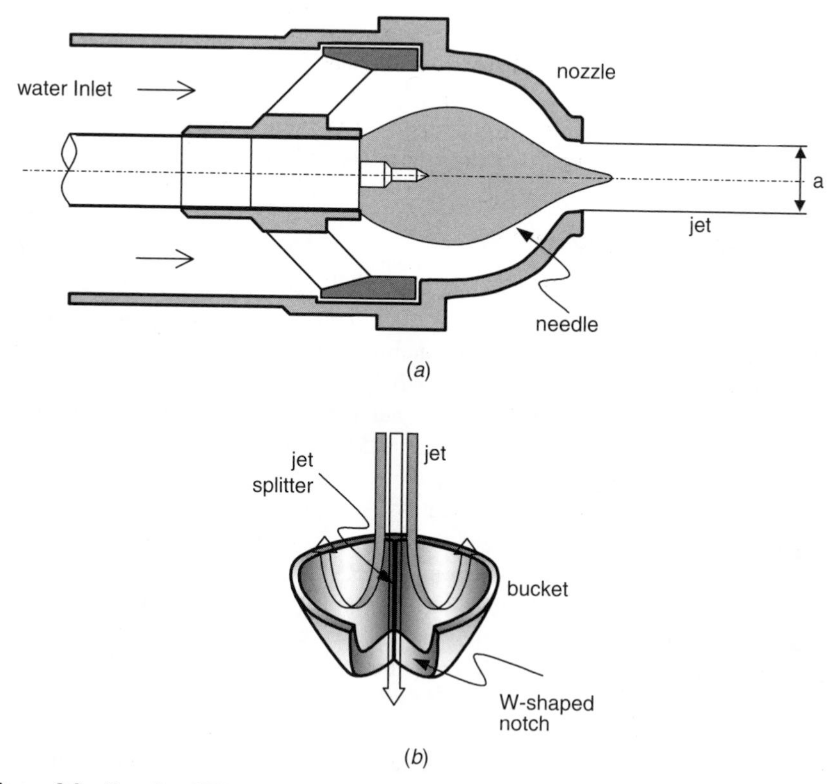

Figure 3.9 Details of PT jet control action: (*a*) injector nozzle, needle, and jet diameter; (*b*) bucket and jet splitter.

The injector's aperture has a smaller diameter according to manufacturer specifications. To optimize turbine efficiency and ensure that the water exits freely after the incidence of the jet on the shovels, the Mosonyi condition should be observed:

$$\frac{D}{a} > 10 \tag{3.25}$$

The optimal diameter of the rotor circle is obtained as

$$D_{\mathrm{opt}} = 5.88Q^{1/2}z^{-9/2}H^{-1/4}$$

As indicated in Figure 3.8, in many situations there is a normal need to deflect some amount of incoming water to resolve transient load situations. Such situations are related to the control of water flow during sudden appreciable load changes; to ease synchronization of the generator with other machines of the network; or in extreme cases, during heavy load rejection. The action of the deflector is to minimize the shock intensity of a sudden reduction in the speed of the water in the pipelines, called *water hammer*. These phenomena are caused by abnormal generator load rejection. With water hammer, a very high increase in upstream pressure may damage the pipelines and the turbine itself. A counter jet of water, known as a *hydro brake*, may also be used to help reduce the rotor speed in such operations, as illustrated in Figure 3.8. Maintenance and repair of Pelton turbines are expeditious, due to the relatively simple construction of the rotor and its easy access.

3.6.2 Francis Turbine

The field of application of Francis turbines (FTs) in small hydroelectric power plants is from 3 to 150 m of height and from 100 L/s of design flow of the motive water (Figure 3.7). According to local conditions, the turbine shaft can be horizontal or vertical. The horizontal position is the most suitable, because it eases direct connection of generators in conventional manufacturing (synchronous or asynchronous). Vertical shafts are difficult to repair and maintain, and more space is needed above the machine to connect the generators, resulting in a greater weight, which increases costs, and in addition, necessitates some means of erection.

Francis turbines are probably used most extensively because of their wide range of suitable heads, characteristically from 3 to 600 m in extreme cases. At the high-head range, the flow rate and output must be large; otherwise, the runner becomes too small for reasonable fabrication. At the low-head end, propeller turbines are usually more efficient unless the power output is kept within certain limits.

PTs have the advantage of a better efficiency curve than that of FTs (Figure 3.10). That curve is more horizontal for $Q_{\text{effective}}/Q_{\text{design}} = 25\%$, which is explained by the small variation in flow speed as a function of the flows themselves, and wear caused by sensitivity to materials in suspension in the moving water. Wearing caused by fine grains in suspension is less for Francis and Michel–Banki turbines than for PTs, as shown in the following sections. The following disadvantages of FTs are observed with respect to PTs;

1. Disassembly and assembly of an FT for maintenance and repair are very difficult.
2. The turbine is very sensitive to cavitation.
3. The efficiency curve of the turbine is not optimum, particularly at flows much smaller than the design flows (Figure 3.10).
4. The turbine is more sensitive to materials in suspension dragged by the water, which causes possible wear and consequent efficiency reduction.
5. The fast closure times of this turbine cause more intense water hammer, to cope with which the tube specifications must be overdimensioned.
6. The FT is not stable for operation at power 40% lower than its maximum power.
7. Complex mechanisms of control demand expert work during maintenance and recovery.

The advantage of an FT over a PT is due to the suction tube, which makes usable the totality of the elevation between upstream and downstream used for generation

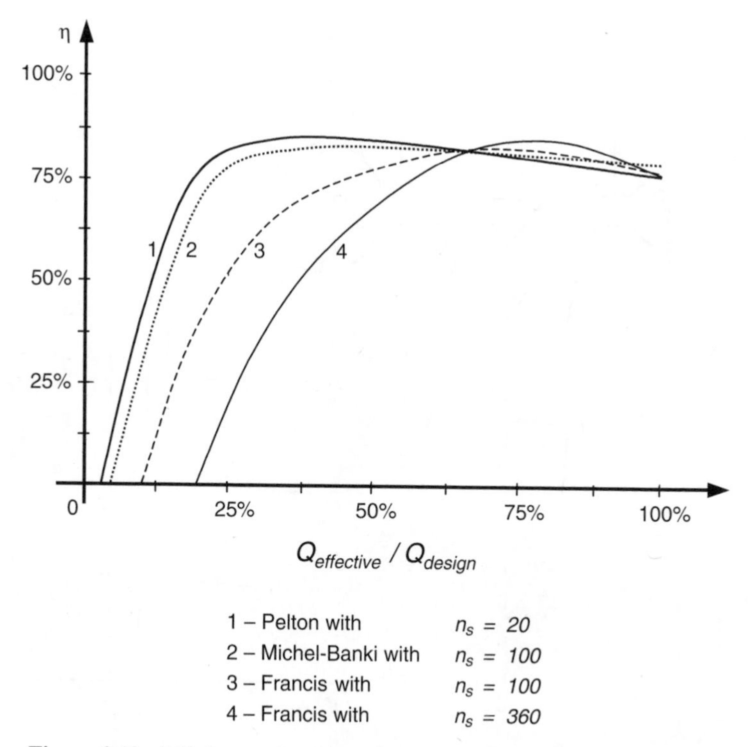

Figure 3.10 Efficiency of turbines for small hydroelectric power plants.

of hydroelectric energy. According to Bereshnoi, the diameter of the rotor can be determined by the relationship

$$D = \frac{(0.16n_s + 35.1)H^{1/2}}{n} \tag{3.26}$$

where n_s is the rated specific rotation (rpm) and n is the number of revolutions (effective speed of rotation) (rpm). The speed of specific rotation is given by

$$n_s = \frac{n(1.36P_t)^{1/2}}{H^{5/4}} \tag{3.27}$$

where P_t is the turbine power (kW) and H is the height of the water head (meters). Empirically, it can established that for heads between 10 and 50 m, the rotation of a Francis turbine is between 900 and 1200 rpm, and for heads higher than 50 m, the range of values is 1200 to 1500 rpm.

Having thus verified the estimated diameter of the wheel, one can determine the approximate dimensions of the turbine set (the snail-shaped box), with details shown in Figures 3.11 and 3.12. Table 3.2 presents the main proportions for

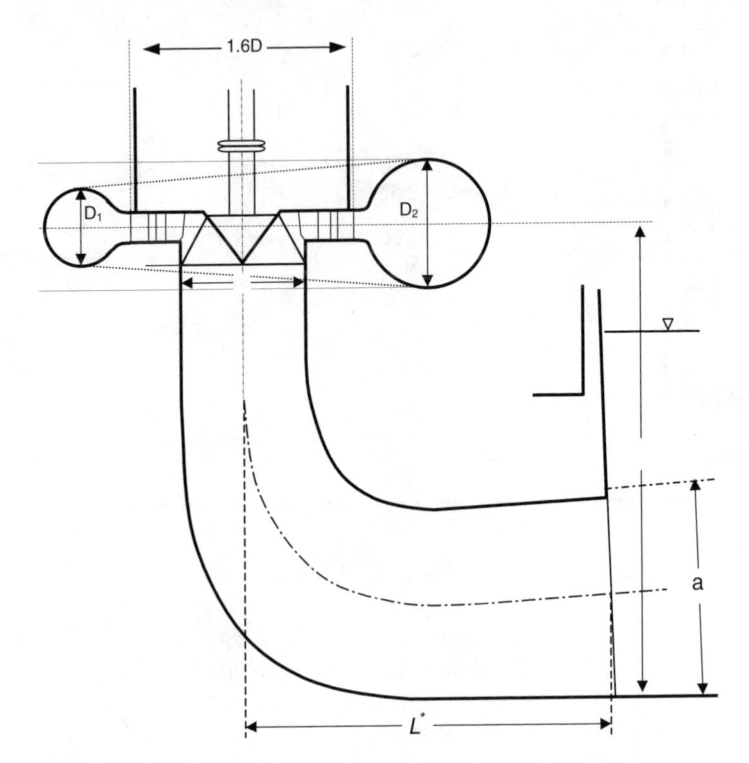

Figure 3.11 Dimensions of a spiral Francis turbine.

construction of a Francis turbine. In the table notice that a distinction is made between the length of vertical and horizontal aspiration tubes, L_v and L_h, respectively.

In small hydroelectric power plants, the turbines are generally already equipped with a tube of vertical aspiration; in such cases there is no need for a concrete aspirator tube.

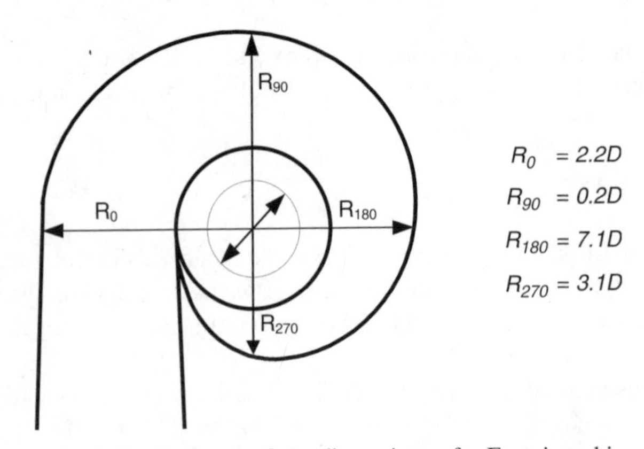

Figure 3.12 Estimate of the dimensions of a Francis turbine.

TABLE 3.2 Approximate Dimensions for a Francis Turbine

Description	Variable	Dimension (m)
Smaller diameter	D_1	$0.6D$
Larger diameter	D_2	$1.0D$
Height of the aspirator tube	a	$1.2D$
Width of the aspirator tube	b	$3.0D$
Area of the aspirator tube	ab	$3.6D^2$
Length of the vertical aspiration	L_v	$(5 - n_s/200)D$
Length of the horizontal aspiration	L_h	$2.2D$
Maximum height	A	$(3.4 - n_s/400)D$

3.6.3 Michel–Banki Turbine

Michel–Banki (also called *radial thrust*) turbines currently manufactured have a capacity of up to 800 kW . Their flows vary from 25 to 700 L/s (according to the machine dimensions), with head heights in the range 1 to 200 m. The number of slats installed around the rotor varies from 26 to 30, according to the wheel circumference, whose diameter is from 200 to 600 mm.

A Michel–Banki turbine (MBT) can be installed with an output free from the water or with a suction tube, in which the entire water elevation is used. The MBT set is shown in Figure 3.13, where the path followed by the water flow around the turbine rotor can also be observed.

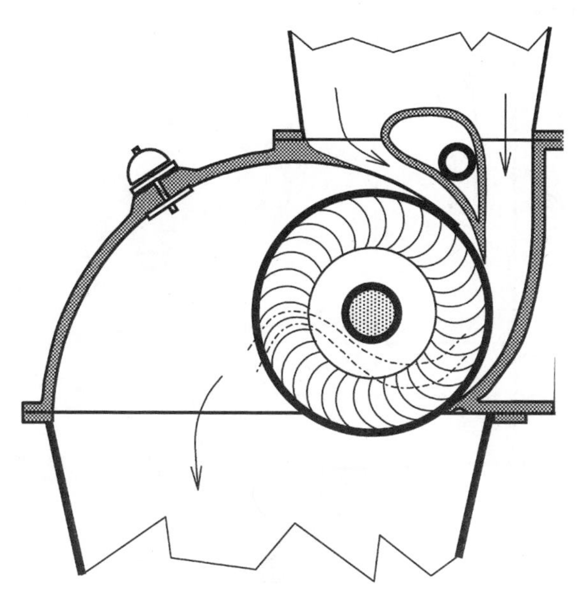

Figure 3.13 Michel–Banki turbine with vertical input.

Compared with other types of turbines, the MBT, with its rotor division in cells (longitudinal segments of the rotor) in the proportion 1:2, presents a big advantage. In other words, this multicell turbine can be operated at one- or two-thirds of its capacity (in the presence of low or average flows) or at full capacity (in the presence of design flows: that is, three-thirds).

By that disposition of variable flow capacities, any water flow is useful with optimum efficiency. The efficiency curves of the MBT are almost horizontal (Figure 3.14). Due to this tuning possibility, the turbine can be operated even at 20% of its full power. The tuning device for variable flows works with rotatory slats, which can also serve as elements of turbine closure when the head does not exceed 30 m, as illustrated in Figure 3.14 for the Michel–Banki–Ossberger type. This is an interesting feature in many areas of recent electrification, where there are only conditions of low perceptual use of the turbine and where the demand for energy increases gradually. In these cases, a small hydroelectric power plant will reach its full potential of energy production only after some time. The major advantages of the MBT are fast assembly of the machine set, less demand for civil structures, and easy access to all elements of the equipment during maintenance.

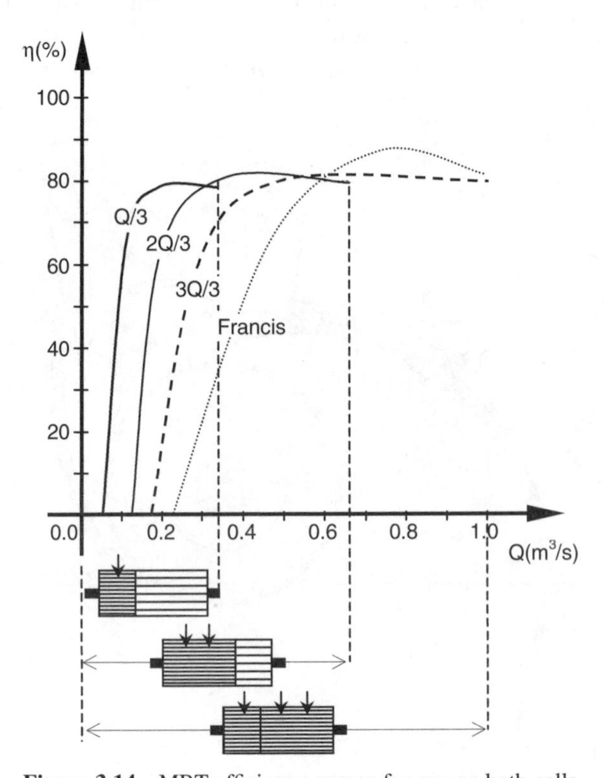

Figure 3.14 MBT efficiency curves for one or both cells.

3.6.4 Kaplan or Hydraulic Propeller Turbine

A Kaplan turbine (KT) is related to low heads or water stream, limited power, and optimum flow variations during the year and is recommended for heads from 0.8 to 5 m, approximately. To start a Kaplan turbine, a vacuum pump fills a siphon with water and forms an elevation between upstream and downstream. To stop the turbine, it is enough to stop the water flow through the relief valve on the distributor's upper part.

In addition to being lower in cost than conventional types of low-head turbines (such as bulb and bulb-well types), the Kaplan turbine has the advantage of maintaining its electromechanical parts out of the water. This feature eases routine inspection and maintenance and adds safety in case of floods. As its installation does not demand water reservoirs or prominent civil structures, the impact on the environment is negligible. According to the flow type (i.e., regulated or variable), the wheel shovels are fixed or adjustable, respectively (Figure 3.15). The efficiency of the turbine becomes higher if adjustable shovels are used because they adapt better to changes in the watercourse.

Generators with output powers below 100 kW can be driven by either a pulley or a belt. The turbine wheel should be installed so that the lowest point of the wheel is above the upstream maximum level. That height is calculated for each project, as a function of cavitation, related directly to the downstream level of operation. The selection of a Kaplan turbine that adapts best to the flow and head of a watercourse can be based on a graph similar to the one in Figure 3.16.

A small power plant using a Kaplan turbine does not dispense some auxiliary services: the electric motor of the vacuum pump and the electric motor drivers of the wheel shovels, when applicable (approximately, from 0.5 to 3.0 kW). A rotation

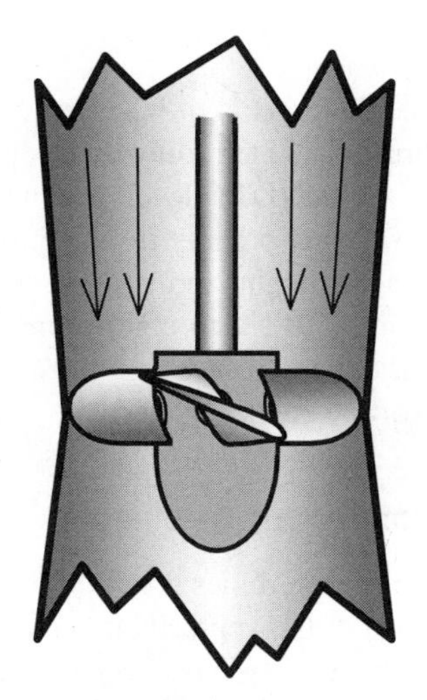

Figure 3.15 Kaplan turbine.

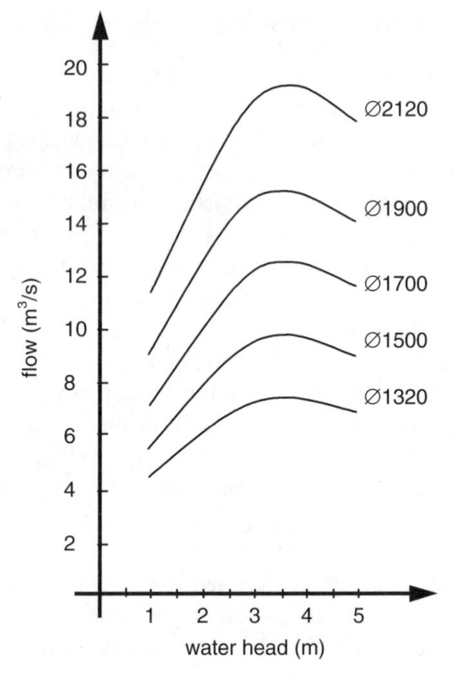

Figure 3.16 Flow–head curves for the Kaplan turbine.

multiplier should also be installed for high-speed generators. All of this contributes to additional losses.

3.6.5 Deriaz Turbines

When a turbine is neither radial nor axial, it is known as being of mixed flow, diagonal or semiaxial. The thrusting water follows an approximately conic surface around the runner. A Deriaz turbine (DT), developed in the 1960s, can reach a capacity of up to 200 MW. Its flows vary broadly from 1.5 to 250 m^3/s (according to the machine dimensions), with head heights in the range 5 to 1000 m. The runner diameter may be up to 7000 mm with six to eight runner blades. Diagonal turbines operate very economically as either turbines or pumps. They resemble a fast Francis turbine except for the size of the runner blades.

3.6.6 Water Pumps Working as Turbines

The inverse use of water pumps as turbines for small hydroelectric power plants has become quite popular because of the appreciable reduction in facility costs [3,12]. These pumps, usually of small capacity, have been used for many years in industrial applications to recover energy that would otherwise be lost. Throughout the world, pump–turbines have been used in small power plants with energy storage. They present the following advantages:

1. They cost less because they are mass produced for other purposes (i.e, as water pumps for buildings and residences).
2. Their acquisition time is minimal because they have a wide variety of commercial standards and are available in hardware stores and related shops.

However, they have a few disadvantages: They have slightly reduced efficiency compared to the same head height used for water pumping, and they are sensitive to the cavitation characteristics and operating range.

The main difference between the operations of those machines as a pump or as a turbine is that the fluid (water) flowing to the pumps is determined by the head height, of which there is not efficient control. On the other hand, a hydraulic turbine has efficient flow control, which is one of the reasons for its higher cost.

The use of pump–turbines in industries is justified by the need to dissipate some form of rotating energy produced in excess [12]. If the water flow already had to be controlled for some other reason, it can also be used to produce another form of energy when the constant flow is diverted for the turbine. Constant power is generated, which can be injected into the network or used to produce energy separately whenever a load controller is available. When there is a lot of variation in the water flow, the issue gets complicated. The problem can be solved by the use of another pump–turbine with a differentiated capacity (for instance, 1:2) to restrict the range of variation of the water flow in each, or by electronic control by the load (see Chapter 10) [13,14].

3.6.7 Specification of Hydro Turbines

To get a good purchase price for a turbine, several manufacturers should be consulted. Comparing values demands good specifications (as close as possible to that desired), so that small differences in the product characteristics do not unduly influence the final price. Superdimensioning should be avoided, as it dissuades manufacturers from offering equipment better adapted to each case. The following items should be available at the time of purchase:

1. Buyer's identification (name, profession, address, telephone, etc.)
2. Cost estimation or detailed specification
3. Sketch of the installation site for the power plant (if possible, with pictures), additional elements available in the nature, and usable technical characteristics

For a hydro turbine it is necessary to specify:

1. Available average, minimum, and maximum heights

2. Available flow for generation with its duration curve, site water storage capacity, information about historical variations, and so on
3. Water quality (materials in suspension, abrasiveness, and other useful characteristics)
4. Present and future energy needs of the generator or turbine, according to the effective capacity of the site (otherwise, any specification becomes useless)
5. Possible need for an isolation valve for the turbine
6. Specifications for the type of regulator required
7. Specifications for the generator (see Chapter 10)
   a. Type (induction, synchronous, or direct current)
   b. Electric output (alternating or direct current, maximum power, voltage, number of phases, frequency)
   c. Climate (winds, temperature, and humidity)
   d. Operating mode (manual, automatic, semiautomatic, or remote control, stand-alone or for injection in the network; in this last case, is there a sale, purchase, or exchange agreement with the electric power company?)
   e. Other indispensable equipment (control panels, protection and drivers, single- or three-phase transmission and distribution lines)
8. Location of the powerhouse, if there is one, with upstream head, distance from the water intake, dimensions, drainage height (minimum, maximum and average)
9. Specification of the diameter, length, configuration, and manufacturing material if a feeder of water is used under pressure for the turbine
10. Access to the site (boat, bike, highway, rails, walking path, or inhospitable access by land) for transport of equipment and maintenance planning
11. Existence of animals, fish, or any other life that should be preserved

REFERENCES

[1] Eletrobrás/DNAEE, *Small Hydroelectric Power Plants Handbook*, 1985.

[2] A. Brown, Stream flow measurement by salt dilution gauging, *ITIS*, November 1983.

[3] A. R. Inversin, *Micro-hydropower Sourcebook*, National Rural Electric Cooperative Association, International Foundation, Washington, DC, 1986.

[4] H. Lauterjung, Inventory of offer of hydropower plants, in *Taller de Microcentrales Hidroelectricas, 1990: Proceedings of Unesco-GTZ-PLL*, November 1990, pp. 89–188.

[5] W. G. Ovens, *A Design Manual for Water Wheels*, Vita, Zingem, Belgium, 1975.

[6] Z. Souza, R. D. Fuchs, and A. H. M. Santos, *Hydro and Thermo Electrical Power Plants*, Electric Brazilian Power Plants, Federal School of Engineering of Itajubá, Minas Gerais, Brazil, and Edgard Blücher, São Paulo, Brazil, 1983.

[7] G. P. Schreiber, *Hydroelectric Power Stations*, Vol. 3, Engevix Studies and Projects of Engineering, Edgard Blücher, São Paulo, Brazil, 1987.

[8] Z. Souza, *Hydroelectric Power Plants: Sizing of Components*, Edgard Blücher, São Paulo, Brazil, 1992.

[9] A. J. Macintyre, *Hydraulic Motive Machines*, Guanabara, Rio de Ganiero, Brazil, 1983.

[10] The small notable, catalog, MTV-1527, Betta Hydraulic Turbines, Franca, Brazil, 1991.

[11] *Força Energética*, Vol. 1, No. 2, July–August 1992.

[12] W. Bolliger, J. A. Menin, and S. Pumps, Éxito of las bombas como turbinas, *Sulzer Technical Review* (Spanish edition), pp. 27–30, February 1997.

[13] C. D. Mello, Jr. and F. A. Farret, Improvements in the structure and operation of an experimental micro power plant with electronic control by the load, in *Proceedings of the 12th CIGRÉ National Seminar on Production and Electric Power Transmission*, October 1993, Vol. 2, pp. 46–51.

[14] E. L. Henn, *Máquinas de fluído* (Machines of Fluid), Universidade Federal de Santa Maria, Santa Maria, Brazil, 2001.

CHAPTER 4

WIND POWER PLANTS

4.1 INTRODUCTION

Continuing development of modern society is contingent on the sustainability of energy. The vulnerability of the current energy chain, reliant on nonrenewable fossil fuel resources, will provoke a collapse in our society with the exhaustion of its natural reserves. That is why wind power energy is strongly advocated in projects and studies where the following factors are considered:

- High cost of hydro- and thermoelectrical generation
- Areas with fairly high average wind speeds (>3 m/s)
- Need to feed remote loads, where a transmission network is uneconomical
- Nonexistence of rivers or other energetic hydroresources in close proximity
- Need for renewable, nonpolluting energy

Wind power energy is derived from solar energy, due to uneven distribution of temperatures in different areas of the Earth. The resulting movement of air mass is the source of mechanical energy that drives wind turbines and the respective generators. Across Europe there are several leading wind energy markets: Germany, Spain, Denmark, Great Britain, Sweden, the Netherlands, Greece, Italy, and France.

Integration of Alternative Sources of Energy, by Felix A. Farret and M. Godoy Simões

There are several European companies exporting products and technology. U.S. wind resources are also appreciable. U.S. wind resources are large enough to generate more than 4.4 trillion kWh of electricity annually. There are sufficient wind intensities for power generation on mountainous areas and deserts, as well as in the midlands, spanning the wind belt in the Great Plains states. North Dakota alone is theoretically capable (if there were enough transmission capacity) of producing enough wind-generated power to meet more than one-third of U.S. demand. According to the Battelle Pacific Northwest Laboratory, wind energy can supply about 20 % of US. electricity, with California having the largest installed capacity. In South America, Brazil has vast wind potential because of its extended Atlantic shore. Particularly in the Brazilian northeast states, there are scarce hydroelectric resources. However, the integrated generation of wind and solar power is a powerful alternative [1–3]. The Brazilian wind power potential is evaluated at 63 trillion kWh/yr.

4.2 APPROPRIATE LOCATION

To select the ideal site for placement of wind power turbines, it is necessary to study and observe the existence of enough wind to make extraction of energy possible at a desired rate. Although flat plains may have steady strong winds, for small-scale wind power the best choice is usually along dividing lines of waters (i.e., the crests of mountains and hills). In those geographical locations, there is good wind flow perpendicular to the crest direction. Some basic characteristics to be observed for defining a site are:

- Wind intensities in the area
- Distance of transmission and distribution networks
- Topography
- Purpose of the energy generated
- Means of access

4.2.1 Evaluation of Wind Intensity

The energy captured by the rotor of a wind turbine is proportional to the cubic power of the wind speed. Therefore, it is very important to evaluate the historical wind power intensity (W/m^2) to access the economical feasibility of a site, taking in account seasonal as well as year-to-year variations in the local climate [3–5]. To estimate the wind power mechanical capacity, P, Bernoulli's equation (defined in Chapter 3) is used with respect to the mass flow derivative of its kinetic energy, K_e:

$$P = \frac{dK_e}{dt} = \frac{1}{2} v^2 \frac{dm}{dt} \tag{4.1}$$

Figure 4.1 Typical three-blade wind rotor.

The mass flow rate per second is given by the derivative of the quantity of mass, *dm/dt*, of the moving air that is passing with velocity v through the circular area A swept by the rotor blades, as suggested in Figure 4.1. According to equation (3.2), for any average flow ($\bar{Q} = A\bar{v}$) of a fluid, the upstream flow of mass can be given in terms of the volume of air V as

$$\frac{dm}{dt} = \rho\frac{dV}{dt} = \rho A\bar{v} \tag{4.2}$$

where $\rho = m/V$ is the air density in kg/m^3 (= 1.2929 kg/m^3 at 0°C and at sea level) and A is the surface swept by the rotor or blades (m^2).

The effective power extracted from wind is derived from the airflow speed just reaching the turbine, v_1, and the velocity just leaving it, v_2. The number of blades must also be considered. Equation (4.2) considers the average speed $(v_1 + v_2)/2$ passing through area A of the rotor blades (i.e., effectively acting on the rotor blades). Therefore, equation (4.2) becomes

$$\frac{dm}{dt} = \frac{\rho A(v_1 + v_2)}{2} \tag{4.3}$$

Also, there is a difference in the kinetic energy (usually expressed in kg $\cdot$ m/s = 9.81 W) in the wind speed just reaching and just leaving the turbine. A net wind mechanical power of the turbine is imposed by this kinetic energy difference, which may estimated by equation (4.1) as

$$P_m = \frac{dK_e}{dt} = \frac{1}{2}(v_1^2 - v_2^2)\frac{dm}{dt} \qquad \text{W/m}^2 \tag{4.4}$$

Combining equations (4.3) and (4.4), the following power is calculated:

$$P_m = \frac{dK_e}{dt} = \frac{1}{4}\rho A(v_1^2 - v_2^2)(v_1 + v_2)$$

whose upstream wind speed, v_1, is used to give

$$P_m = \frac{1}{4}\rho A v_1^3 \left(1 - \frac{v_2^2}{v_1^2}\right)\left(1 + \frac{v_2}{v_1}\right)$$

or

$$P_m = \frac{1}{2}\rho C_p A v_1^3 \tag{4.5}$$

where $C_p = (1 - v_2^2/v_1^2)(1 + v_2/v_1)/2$ is the power coefficient or rotor efficiency as expressed by Betz.

If C_p is considered a function of v_2/v_1, the maximum of such a function can be obtained for $v_2/v_1 = \frac{1}{3}$ as $C_p = 16/27 = 0.5926$. This value is known as the *Betz limit*, the point at which if the blades were 100% efficient, the wind turbine would no longer work because the air (having given up all its energy) would stop entirely after passing its blades. In practice, the collection efficiency of a rotor is not as high as 59%; a more typical efficiency is between 35 and 45%. To achieve the best designs, several sources of data, such as meteorological maps, statistical functions, and visualization aids, should be used to support the analysis of the wind potential.

Meteorological Mapping There are meteorological maps showing curves connecting the same intensity of average wind speed. Although such maps may yield valuable information, they are not sufficient for a complete analysis because the meteorological data acquisition stations do not look only for a determination of the wind potential for power energy use. Therefore, a final decision on the site will be made only after a convenient local selection procedure and thorough data acquisition [2,3].

Local wind power is directly proportional to the distribution of speed, so that different places that have the same annual average speed can present very distinct values of wind power. Figure 4.2 displays a typical wind speed distribution curve for a given site. That distribution can be made monthly or annually. It is determined through bars of occurrence numbers, or percentile of occurrence, for each range of wind speed over a long period of time. It is usually observed as a variation of wind

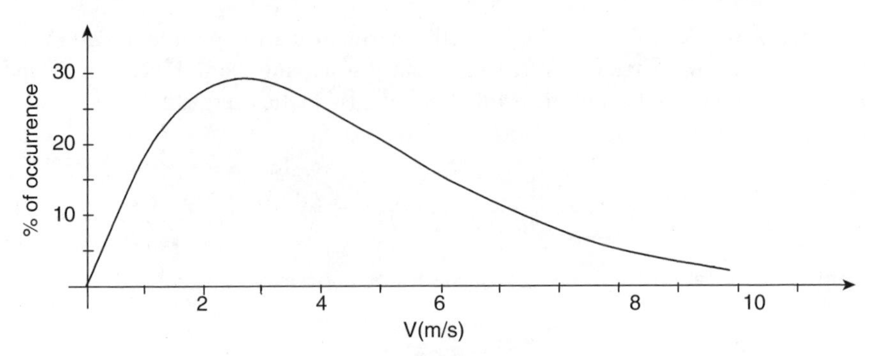

Figure 4.2 Wind speed distribution.

linked with climate changes in the area. It is typical in temperate climates to have summer characterized as a season of little wind and winter as a season of stronger winds.

When the wind speed is lower than 3 m/s (referred to as a *calm period*), the power becomes very limited for the extraction of energy and the system should be stopped. Therefore, for power plants, calm periods will determine the time required for energy storage. As discussed next, power distribution varies according to the intensity of the wind and with the power coefficient of the turbine. A typical distribution curve of power assumes the form shown in Figure 4.3. Sites with high average wind speeds do not have calm periods, and there is not much need of storage. However, high wind speed may cause structural problems in a system or in a turbine.

The worldwide wind pattern is caused by differences in temperature around the globe as well as by Earth's rotation. The temperature difference causes a huge convection effect above the ground surface, which makes the airflow move upward on the equator and go downward at about 30° north and south latitudes. This air

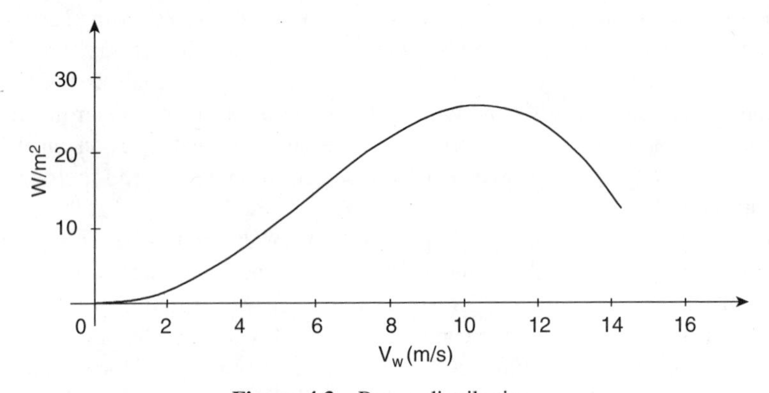

Figure 4.3 Power distribution.

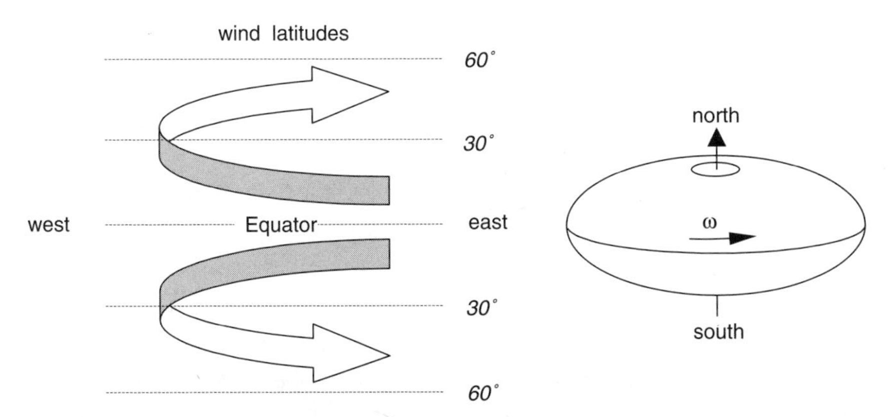

Figure 4.4 Global wind patterns.

convection is distorted by the tangent speed differences of Earth's spin such that the wind tends to move east to west. Horizontal wind speed near the equator is very low because winds move more upward. The most pronounced intensities of wind go from west to east between latitudes 30° and 60° north and south of the equator, forming two natural wind tunnels by a motion caused by the circulation of east to west winds below the 30° latitudes (see Figure 4.4). Usually, west wind speeds are higher than east wind speeds. This general tendency is affected strongly by the ground formation (mostly by valleys and mountains). As a result, it is very useful to represent wind variation by a probabilistic vector of determined intensity and direction.

Weibull Probability Distribution To establish a probability distribution, it is important to establish the duration curve of the wind speed for every hour of the day, every day of the year, a total of 8760 pieces of data. Figure 4.5 shows a curve

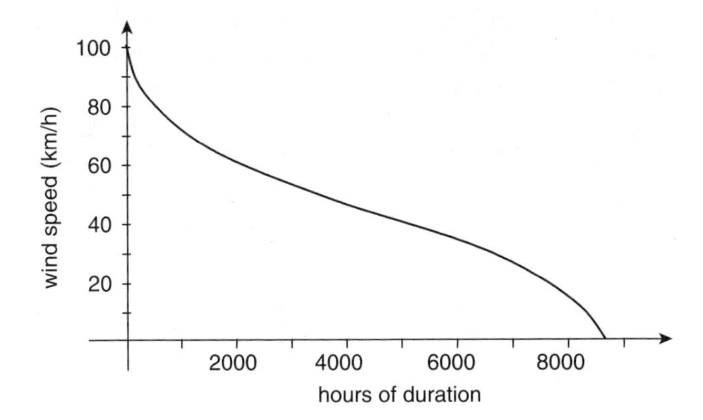

Figure 4.5 Curves of accumulated duration of wind speed.

formed by points marked in several speed ranges, in accordance with the number of hours accumulated for a particular wind intensity and above that range [4–7].

Wind movement around the Earth is a random phenomenon. Therefore, a large sample of wind data taken over many years should be gathered, to increase confidence in the data available. This is not always possible, so shorter periods are often used. The data are often averaged over the calendar months and can be described by the Weibull probability function, given as [4]

$$h(v) = \frac{k}{c}\left(\frac{v}{c}\right)^{k-1} e^{-(v/c)^k} \quad \text{for} \quad 0 < v < \infty \tag{4.6}$$

The Weibull function expresses the fraction of time the wind speed is between v and $v + \Delta v$ for a given Δv. In practice, most sites around the world present a wind distribution for k (shape factor) within the range 1.5 to 2.5. For most of them, $k = 2$, a typical wind distribution found in most sites, is known as the *Rayleigh distribution*, given by

$$h(v) = \frac{2}{c}\left(\frac{v}{c}\right) e^{-(v/c)^2} \tag{4.7}$$

Factor c, known as the *scale factor*, is related to the number of days with high wind speeds. The higher c is, the higher the number of windy days. This parameter is enough to represent the wind speed for most practical cases.

Weather repeats in seasons from one year to the next. For the purpose of wind variation studies, this period is usually taken as divided into the total number of hours (8760 hours/year). The unit of h in equation (4.6) may be stated as a percentage of hours per year per meter per second.

Equation (4.7) can be plotted for various parameters c as shown in Figure 4.6. The values of h are the number of hours in a year that the wind speed is within the interval from v to $v + \Delta v$ divided by the speed interval Δv. This plot provides a useful and realistic view of the average speed.

There is a predominant wind speed throughout a year. The predominant speed can be considered in at least three different ways, with very distinct implications. One could be defined as a single speed at which the wind blows most of the time. It is a simplistic method that does not take into account secondary predominant speeds, which may contain a lot of energy. A second way of defining wind speed is by the average speed experienced by h throughout the year, given by

$$V_{\text{av}} = \frac{1}{8760}\int_0^{\infty} hv\, dv \tag{4.8}$$

In this case, if $c = V_{\text{av}}$, the Rayleigh distribution given by equation (4.7) becomes

$$h(v) = \frac{2v}{V_{\text{av}}^2} e^{-(v/V_{\text{av}})^2} \tag{4.9}$$

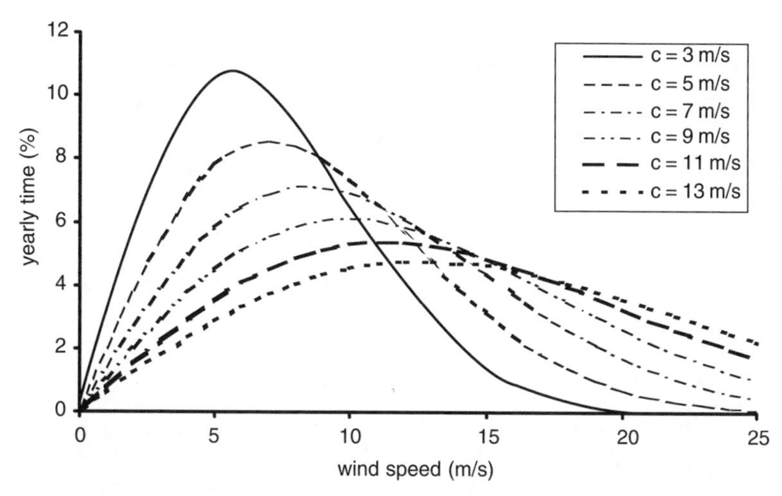

Figure 4.6 Rayleigh distribution of wind speed (parameter c).

A third and more appropriate definition is to use the concept of *root mean cube speed*, V_{rmc}, analogous to the root mean square (effective value). This definition is based on the idea that yearly average power varies with the cube of the wind speed experienced by h; that is,

$$V_{\mathrm{rmc}} = \sqrt[3]{\frac{1}{8760}\int_0^{\infty} hv^3\,dv} \tag{4.10}$$

Table 4.1 lists some popular formulas for the calculation of wind speed, root mean cubic power, and energy density over the number n of speed samples during the year, usually with $\rho_i = 1.225\,\mathrm{kg/m^3}$ at $t = 15°\mathrm{C}$.

If we assume that the power coefficient C_p in equation (4.5), is no higher than 0.5 for high wind speeds and two-blade turbines, we can conveniently take this value as the maximum practical rotor efficiency. Therefore, the maximum output power for the root mean cubic of the wind speed is

$$P_{\mathrm{rmc}} = \frac{\rho}{4} A V_{\mathrm{rmc}}^3 \qquad \text{watts} \tag{4.11}$$

TABLE 4.1 Common Wind Speed, Root Mean Cubic Power, and Energy Density Formulas

Average Speed (m/s)	Root Mean Cubic Speed (m/s)	Cubic Power Density (W/m^2)	Cubic Energy Density (Wh/m^2/yr)
$V_{\mathrm{av}} = \frac{1}{n}\sum_{i=1}^{n} v_i$	$V_{\mathrm{rmc}} = \sqrt[3]{\frac{1}{n}\sum_{i=1}^{n} v_i^3}$	$P_{\mathrm{rmc}} = \frac{1}{2n}\sum_{i=1}^{n} \rho_i v_i^3$	$E_{yw} = \frac{8760}{2n}\sum_{i=1}^{n} \rho_i v_i^3$

The total annual energy at the site can be obtained from equation (4.11) by multiplying it by 8760 hours. So the energy density per year and per swept area of the blades is given by

$$E_{yw} = 8760 \frac{P_{\mathrm{rmc}}}{A} = 8760 \frac{\rho}{4} V_{\mathrm{rmc}}^3 \qquad \mathrm{Wh/m^2} \tag{4.12}$$

The Rayleigh wind and energy density distribution is depicted in Figures 4.7*a* and *b* for $c = 7\,\mathrm{m/s}$ and $c = 10\,\mathrm{m/s}$, respectively. Note that for a minor change in

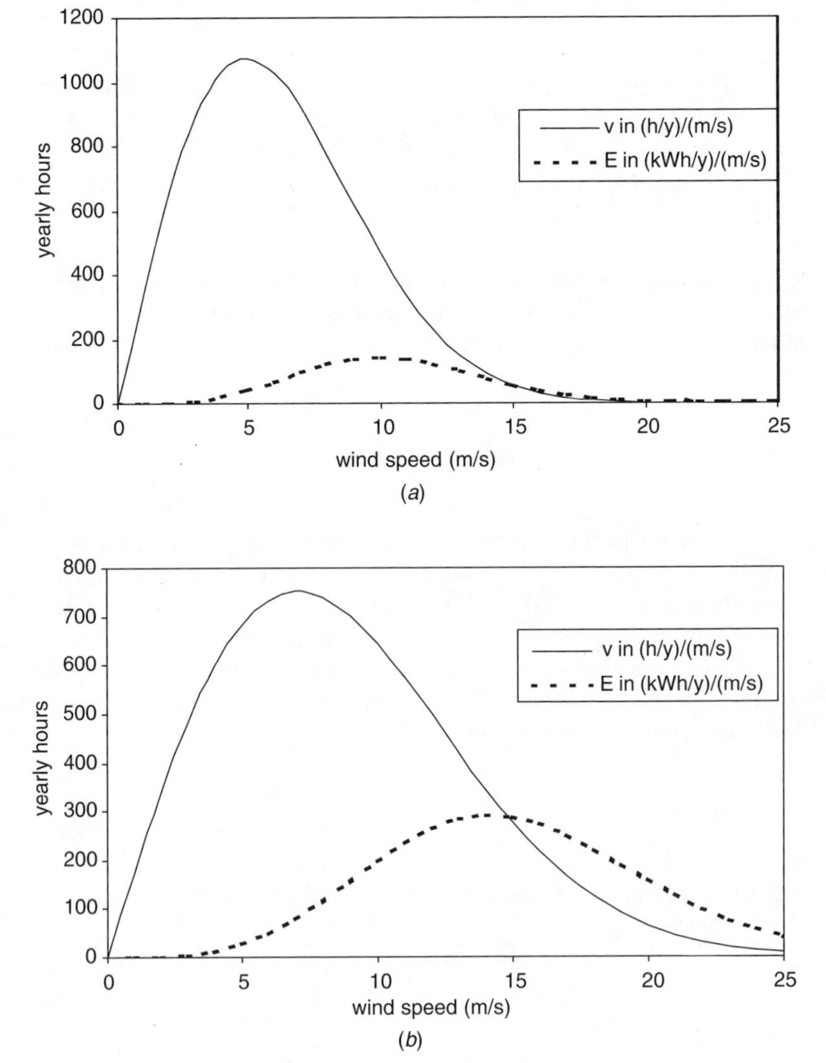

Figure 4.7 Rayleigh wind and energy density distribution ($k = 2$): (*a*) $c = 7\,\mathrm{m/s}$; (*b*) $c = 10\,\mathrm{m/s}$.

the wind speed, there is a significant increase and shift in the power density along the speed axis. Such changes would not be noticed much if only the yearly maximum or average speeds were used for those estimations, because they retain almost the same value in the two distributions represented.

Based on these curves and on the load to be fed, it is possible to establish the turbine–generator power that would have the best utilization factor for a given wind condition. It is evident that the longer the wind behavior in an area is observed, the more appropriate will be the design of a small power plant. Variable-speed control is recommended for a good wind energy system, due to the cubic relation of the yearly energy to the wind speed.

Analysis of Wind Speed by Visualization If a site being evaluated is not close to meteorological stations, a good practical suggestion is to observe the existing trees. Their deformation level will serve as a good indicator of the wind speeds in the area. The intensity of the wind increases with height; therefore, trees of larger span are reached by more intense winds, which can harm their growth. The following levels of tree deformation are recognized:

- *Brush*. When branches are on the lee side (the side exposed to wind), especially in the absence of leaves the winds are weak.
- *Flag*. Branches are the lee barriers (the static trunks in their original position, free to windward), in other words, the side from which the wind blows.
- *Lie down*. The wind is strong enough to produce permanent deformations in trunks and branches.
- *Shear*. The wind is always strong, to the point of breaking branches, giving the impression that they were cut uniformly.
- *Rug of trees*. The wind energy is so strong that it limits the growth of trees to some inches of the soil, giving the impression that the trees form a rug.

For small wind power installations, a visual assessment of the wind intensity is described in Table 4.2, which is an adaptation of the Beaufort table [3]. The International Committee of Meteorology adopted such a table in 1874; wind measurement using anemometers did not begin until 1939.

Technique of the Balloon Another very simple method to measure wind speed is by measurement of the speed of a balloon allowed to move freely in the air [3,4]. Two points of a known reasonable distance on the ground are selected. The time required for the balloon to pass between the two points in a horizontal line is recorded, and the wind speed is inferred from that.

4.2.2 Topography

The analysis of topography is very important when choosing a site. This is an important point when mountainous locations are under consideration. For those

TABLE 4.2 Estimation of Wind Speed (Beaufort Table)

Degree	Classification	Effects of the Wind on Nature	Speed (m/s)
0	Calm	Everything is still. Smoke goes up vertically.	0.00–0.30
1	Almost calm	Smoke is dispersed. Weather vanes are still. Wind is felt on the face.	0.30–1.40
2	Breeze	Wind is felt on the face. The noise of leaves agitated by the wind is heard. Weather vanes move.	1.40–3.00
3	Fresh wind	Leaves and small branches of trees are agitated constantly. Flags are stretched out.	3.00–5.50
4	Moderate wind	The wind lifts dust and paper from the ground. Small tree branches are agitated.	5.50–8.00
5	Regular wind	Small trees with leaves begin to balance.	8.00–11.00
6	Wind mildly strong	Large branches move, electrical lines whistle. It begins to be difficult to walk against the wind.	11.00–14.00
7	Strong wind	Entire trees are agitated. It is definitely difficult to walk against the wind.	14.00–17.00
8	Very strong wind	Branches of trees break. It requires a great effort to walk.	17.00–21.00
9	Windstorm	Tiles are lifted.	21.00–25.00
10	Gale	Trees are torn down. There is construction damage.	25.00–28.00
11	Storm	The wind assumes characteristics of a hurricane (rarely happens far away from coasts).	28.00–33.00
12	Hurricane	The air is full of solid particles and drops of water. The sea is entirely whitish.	33.00–36.00

areas, the wind speed increases on the front side of the mountain and decreases on the opposite side. In plane areas, the position of tree barriers should be observed with respect to the wind, as their proximity to the place chosen for turbine installation is critical. There should be a maximum of free space between tree barriers and the power plant, to avoid any slowing down of the wind (see Section 4.3).

With increased altitude, the movement of a wind draft takes a more complex form, due to different land shapes. Thus, it is necessary to have a reasonable knowledge of the characteristics of the soil to be occupied. The roughness of the ground surface causes wind shear, which changes significantly with altitude. Some data collected at Merida Airport in Mexico show that the wind speed can be four to five times higher at an altitude of about 450 m with respect to the ground surface, and then it starts to decrease again [4]. At about 100 m from the surface, many places on Earth have a wind speed sufficient for wind energy, but the costs of such high installation severely limit a decision in favor of using such heights.

4.2.3 Purpose of the Energy Generated

Wind power turbines can be sited for electric power generation, pumping of water, grinding of grains, and other uses. For electricity generation, a turbine should be installed at the highest places or places of maximum wind. For pumping water, the installation is obviously constrained by the availability of water. Valleys can behave like wind tunnels, thus becoming good places for the installation of pumping systems [4].

High-scale electric power generation demands installation of several turbines. The number of units will depend on the total capacity of each unit, economical sizing of the turbine [cost per capacity (kW)], and the effects of interference among the turbines.

4.2.4 Means of Access

A turbine site should have good access conditions, especially in cases of energy generation on larger scales, to ease installation, operation, and maintenance services. The site should be close to roads, highways, or even railways, since there is heavy machinery and equipment to be transported. There should be suitable safety conditions against vandalism and theft. There is a great impact on a project's total cost if highways, security systems, and other facilities are needed.

An economically favorable factor is that the high-speed characteristics of the wind necessary for power plant installations lower the costs for the right-of-way because generally, these lands are impracticle for other uses, such as agricultural.

4.3 WIND POWER

As discussed in Section 4.2, only part of the full energy available from the wind can be extracted for energy generation, quantified by the power coefficient, C_p. The power coefficient is the relationship of the power extraction possible to the total amount of power contained in the wind [3,8–12]. Figure 4.8 relates the power coefficient to the tip speed ratio, λ, defined as the relationship between the rotor blade tip speed and the free speed of the wind for several wind power turbines. As stated earlier, in quantitative terms, the tip speed ratio is defined as $\lambda = v/V = \omega R/V$, where ω is the angular speed of the turbine shaft, v the blade tangent tip speed, and R the length of each blade. It emphasizes the importance of knowing the purpose for which the energy will be used, to allow determination of the best selection for wind power extraction.

From equation (4.5), the turbine mechanical power can be given by

$$P_t = \frac{(C_p \rho A V^3)}{2} \qquad \mathrm{kg \cdot m/s} \tag{4.13}$$

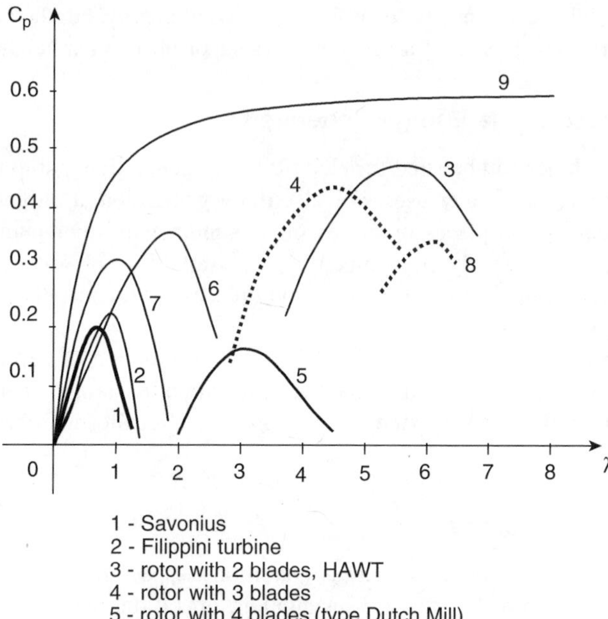

1 - Savonius
2 - Filippini turbine
3 - rotor with 2 blades, HAWT
4 - rotor with 3 blades
5 - rotor with 4 blades (type Dutch Mill)
6 - rotor with 6 blades
7 - rotor with multiple blades, of farm or spiked
8 - Darrieus turbine
9 - Theoretical turbine (*cp* = *16/27*)

Figure 4.8 Output characteristics of various turbines.

The air density ρ in Table 4.1 can be corrected by the gas law ($\rho = P/RT$) for every pressure and temperature with the following expression:

$$\rho = 1.2929 \frac{273}{T} \frac{P}{760} \tag{4.14}$$

where P is the atmospheric pressure (mm Hg), R is Reynold's constant, and T is the Kelvin absolute temperature. Under normal conditions ($T = 296\,\text{K}$ (25°C) and $P = 760\,\text{mmHg}$), the value of ρ is 1.192 kg/m^3, and at $T = 288\,\text{K}$ (15°C) and $P = 760\,\text{mmHg}$, it is 1.225 kg/m^3.

For equation (4.14) we can assume a temperature decrease of approximately 1°C for every 150 m. The influence of humidity can be neglected. If just the altitude h (in meters) is known (10,000 ft = 3048 m), the air density can be estimated by the two first terms of a series expansion like

$$\rho = \rho_0 e^{-[(0.297/3048h)]} \approx 1.225 - 1.194 \times 10^{-4}\,\mathrm{h} \tag{4.15}$$

TABLE 4.3 Speed and Power Loss Measured at the Margin of Trees

Porosity (%)	Loss (%)	Distances in the Direction Against the Wind in Tree Width				
		5%	10%	15%	20%	30%
20	Speed	16	7	4	3	2
	Power	41	18	12	8	6
40	Speed	20	9	6	4	3
	Power	49	25	17	13	9
Height of the area of turbulent flow (in tree heights)		1.5	2.0	2.5	3.0	3.5
Width of the area of turbulent flow (in tree widths)		1.5	2.0	2.5	3.0	3.5

Source: Refs. [3] and [12].

So, from equation (4.13), the turbine torque is given by

$$T_t = \frac{P_t}{\omega} = \frac{\rho ARV^2 C_T}{2} \tag{4.16}$$

where the torque coefficient is defined as $C_T = C_p/\lambda$.

If $S = 1\ \text{m}^2$ and $\rho = 1.2929\ \text{kg/m}^3$, the maximum wind potential can be obtained from equation (4.13) (without taking into account the aerodynamic losses in the rotor, the wind speed variations in several points in the blade sweeping area, the rotor type, etc.) as being

$$\frac{P}{A} = 0.5926(0.6464)V^3 = 0.3831\text{V}^3 \tag{4.17}$$

where P/A is the wind power per swept area (W/m^2) and V is the wind speed (m/s). The coefficient in equation (4.17) is usually quite small because of losses and uneven distribution of the wind on the blades, and it may be approximated by

$$\frac{P}{A} = 0.25V^3 \tag{4.18}$$

In general, obstacles such as trees, wind barriers, human construction, and forests decrease wind speed (i.e., they present some *porosity* with respect to the wind). *Porosity* is defined as the relationship between the open area and the total perpendicular area to the direction of wind flow. Porosity may be expressed as noted in Table 4.3.

4.4 GENERAL CLASSIFICATION OF WIND TURBINES

For reasons of stability and high torque, today's wind turbine engineers avoid building large machines with an even number of rotor blades. A rotor with any number of blades (and with at least three blades) can be considered approximately like a circular plate when calculating the dynamic properties of the machine [10–12]. A rotor with an even number of blades will cause stability problems in a machine with a stiff structure; at the very moment when the uppermost blade bends backward (because it gets maximum power from the wind), the lowermost blade passes into the wind shade in front of the tower. With an odd number of blades, this phenomenon is minimized.

Although one-blade wind turbines exist, their commercial use is not widespread, because the problems noted for the two-blade design apply even more strongly to one-blade machines. In addition to higher rotational speed, noise, and visual intrusion problems, they require a counterweight to be placed on the other side of the hub from the rotor blade to balance the rotor. Compared to a two-blade design this feature adds weight to the generating system without generating additional power.

Two-blade wind turbine designs save the cost of one rotor blade and its weight, but they increase the blade base traction by $\frac{3}{2}$. They require higher rotational speed (with respect to the larger number of blades) to yield the same energy output. This makes its market acceptance difficult and causes more noise and visual intrusion. In recent years, several traditional manufacturers of two-blade machines have switched to three-blade designs.

The rotor of two- and one-blade machines must be sufficiently flexible to tilt, to avoid too-heavy shocks in the no-wind position of the turbine when the blades pass the tower. The rotor is therefore fitted on an axis perpendicular to the main shaft that rotates along with it. This arrangement may require additional shock absorbers to prevent the rotor blade from hitting the tower.

Most modern wind turbines are three-blade designs with the rotor position maintained on the wind side of the tower using electrical motors in their yaw mechanism. This design, usually called the *classical Danish concept*, tends to be a standard when other concepts are judged. The vast majority of turbines sold in world markets are of this design. It was introduced with the renowned Gedser wind turbine. Another characteristic is its use of an induction generator.

For very simple types of wind power battery charger [5], the wind vanes are just a set of blades coupled directly to a fixed-inertia flywheel and to a dynamo (or generator) shaft with permanent ceramic magnets on the rotor. On the other hand, high-speed wind power turbines are better for bulky generation of electricity and are relatively less costly. However, most turbines are quite a bit more complex.

In general, wind turbines can be divided into two groups: horizontal shaft and vertical shaft (with or without accessories). Horizontal-shaft turbines include:

- Blade type with one, two, or three blades
- Multiple-blade, farm, or spiked type, which can work with wind coming from the front or back (many variations exist, much as the multirotor type)
- Double opposite blade type, which can use sails in place of blades

Vertical-shaft turbines are subdivided based on the following working principle: some use drags or friction, and some use lifting (as in an airplane wing). Some turbines utilize both principles.

Drag turbines are machines whose surface executes movements in the wind direction; in other words, they work with the force of the wind drag acting on them. The following types stand out:

- Savonius: single or multiple blades, with or without eccentricity (when the rotation shaft is or is not shifted with respect to the shaft that contains its gravity center)
- Blade, paddles, or oars
- Cup

Lifting turbines are machines whose rotor movement is perpendicular to the wind direction, and they are moved by the lifting action of the wind. Among the lifting turbines are the triangular Darrieus, the Darrieus Giromill, and the Darrieus–Troposkien.

4.4.1 Rotor Turbines

Blade turbines have high rotation with high efficiency. They have automatic regulation of turn speed through the attack angle as a function of the wind speed. The blades present variable sections and great strength against mechanical stresses. This type of turbine needs some orientation mechanism with direct action on turbines at low loads and with indirect action on turbines at higher loads [2,3].

Blade turbines are those most used for electricity generation, feeding batteries or injecting energy directly into the grid. They can be used to pump water as they allow high-rotation driving. They possess the disadvantage of needing towers of great height for installation, on which will be generators, control equipment, and the transmission system.

Blade turbines can vary with respect to the conventional model, but such variations do not indicate notable differences in performance. One is the multirotor, with several turbines mounted on the same tower and interlinked by shafts. Another is the sail turbine, which follows the conventional model with sails rather than blades. These are made of special fabric (sails of embarkation), usually in four or six pieces. They are low in cost, but their efficiency is also lower. They operate at low rotation and high torque.

4.4.2 Multiple-Blade Turbines

Turbines with multiple blades compensate for their low operational rotation. They are simple to manufacture and have high torque. Such a turbine is usually mounted on a horizontal shaft for low power uses. They have lower efficiency than that of blade turbines and are generally used to pump water. They demand very high towers. Multiple-blade turbines of the farm type are manufactured with metallic foils of uniform curved profile. In addition to the usual multiple-blade models, there is a cup type used in the construction of anemometers and toys.

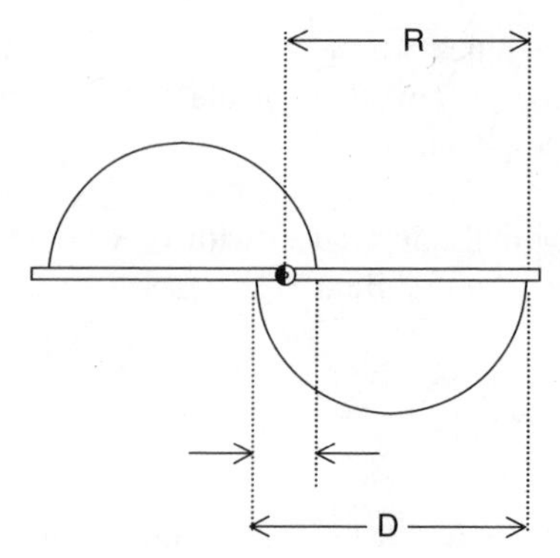

Figure 4.9 Cross section of a Savonius turbine.

4.4.3 Drag Turbines (Savonius)

Drag turbines operate on the principle of the friction caused by wind on the turbine blades. A Savonius turbine represents this design. This turbine model is of simple construction and consists of two parts of a barrel cut in the middle and fastened opposite each other by one of their opposed longitudinal edges (Figure 4.9). Savonius turbines are in wide use in rural areas for water pumping, attic ventilation, and agitation of water to prevent freeze-up during the winter. They are used little in electricity generation. The best configuration for the half-barrels is given by the relationships

$$R = D - 0.5S \tag{4.19}$$

$$S = 0.1D \tag{4.20}$$

The torque of a Savonius turbine is due to the difference in pressure between the concave and convex surfaces of the blades and by the recirculation of wind coming from behind the convex surface. Its efficiency reaches 31%, but it presents disadvantages with respect to the weight per unit of power, because its constructional area is totally occupied by material. A Savonius rotor needs 30 times more material than is needed by a rotor of the conventional type.

For the rudimentary installation of a Savonius turbine, metallic barrels of 200 L capacity with an H-shaped wood structure are used. The useful power can be determined for several wind speeds through the dimensions of the barrel:

$D = 0.60\,\text{m}$ (diameter of each half-barrel)
$h = 0.85\,\text{m}$ (height of the barrel)
$R = 0.57\,\text{m}$ (radius projection exposed to the wind)

To calculate the area exposed to the wind:

$$A = 2Rh = 2(0.57)(0.85) = 0.96\,\text{m}^2$$

TABLE 4.4 Output Power (Watts) of a Savonius Turbine According to the Wind Speed

Number of Barrels, n	1	2	3	4
Wind Speed, V (m/s)	Total Height of Barrels, h (m)			
	0.85	1.70	2.55	3.40
2	0.70	1.40	2.10	2.80
4	5.52	11.04	16.56	22.08
6	18.66	37.32	55.98	74.64
8	44.24	88.48	132.72	176.96

With this result and assuming a power coefficient C_p of 0.15 (as defined in Section 4.3), from equation (4.13) the power for n barrels is (see Table 4.4)

$$P = 0.6AC_pV^3n = 0.6(0.96)(0.15)V^3n = 0.0864V^3n \qquad \text{watts} \tag{4.21}$$

4.4.4 Lifting Turbines

Lifting turbines operate through the lifting effect produced by wind. The most common models are the triangular (delta) Darrieus turbine, the Giromill turbine, and the Darrieus–Troposkien turbine. The triangular Darrieus turbine has straight blades and variable geometry. The straight blades present disadvantages due to their natural bending; however, the variable geometry alters the power coefficient. Therefore, the rotation tends to be constant.

The Giromill turbine has straight, vertical blades and a rotating movement around its shaft. It is used for large-scale systems because it is considered a low-load system—those operating with only a few hundreds of watts. It has an automatic mechanism that maintains the attack angle position that supplies the best working conditions.

Among vertical shaft turbines, the Darrieus–Troposkien turbine is best suited to wind power plants. Studies of this turbine have proved its economic benefits and constructive simplicity. It is the most used turbine for electric power generation and it is discussed below in more details.

Darrieus–Troposkien Turbine The Darrieus–Troposkien turbine consists of blades, a shaft, a tower, and guy wires. The blades are curved with glide sections. The rotor is vertical and connects the top to the bottom of the blade. The tower is fixed on the ground by solid foundations that sustain the shaft. The lower part of the shaft is attached to the tower by rollers. In the upper part, the rollers are fixed to guy wires to keep the turbine in its upstraight position. The other extremities of the guy wires are fixed on the ground.

Rotor The rotor has curved blades with section glides of aerodynamic profile fixed on the shaft extremity of the rotor. The blades can be made of aluminum, fiberglass, steel, or wood. The rotor shaft can be tubular or latticed with an external cover to improve the aerodynamics.

Lifting The rotor support is made of one tower and guy wires. The tower sustains the rotor through bearings and rollers. It can still shelter parts of the system, such as a gearbox, generators, or pumps. The guy wires are fixed through bearings with rollers at the upper part of the rotor shaft in one of the extremities; in the other, they are fastened to the ground. Guy wires are essential for turbines driving small loads.

Speed Multipliers The system of speed multiplication used in turbines to reach generating speed is carried out through gearboxes of parallel or perpendicular (smaller losses) shafts. A system of belts can also be a good solution. The cost of the multiplier depends on the multiplication rate. The cost of the generator increases with reduced rotation, as the number of poles or turns per coil must then be increased. Thus, the optimum value of the multiplication rate is a trade-off between the speed multiplication rate and the number of poles. Speed multiplication also causes a representative percentage of the total losses of a wind energy system. In extremely small systems, this loss may represent about 20% of the total loss.

Braking System Braking systems have both safety and maintenance purposes since a turbine must have some mechanical speed limitation. Sizing will determine the best system to be adopted (hydraulic, electromagnetic, or mechanic).

Starting System The Darrieus–Troposkien turbine type needs a special starting system, for which it may adapt an electric motor connected to the network, a dc motor fed by batteries recharged by a generator connected to the turbine itself or to an auxiliary turbine. For instance, a Savonius turbine can be coupled to the shaft of the Darrieus itself, primarily for small-load turbines.

Generation System The electricity generated can feed the grid directly or be stored in batteries. Due to the large variations in rotation, the use of induction generators is recommended for stand-alone mode up to 10 kW (e.g., for places difficult to access). In these cases, self-excitation capacitors will be needed (see Chapter 10). Larger generators can be used to feed the grid directly.

4.4.5 System TARP–WARP [7]

The trend in present wind power technology is to increase the diameter of the rotor shaft as much as possible to have a larger sweeping area for higher generated output power. Contradicting this tendency, new technologies of energy generation from the wind are being proposed, such as the Toroidal Accelerator Rotor Platform (TARP), which combines generation and transmission. This concept of wind capture is based on the Wind Amplified Rotor Platform (WARP). Small wind turbines are grouped and installed in modules, to conform to the needs of distributed generation. This entire platform consists of a determined number of TARPs piled up one on top of another.

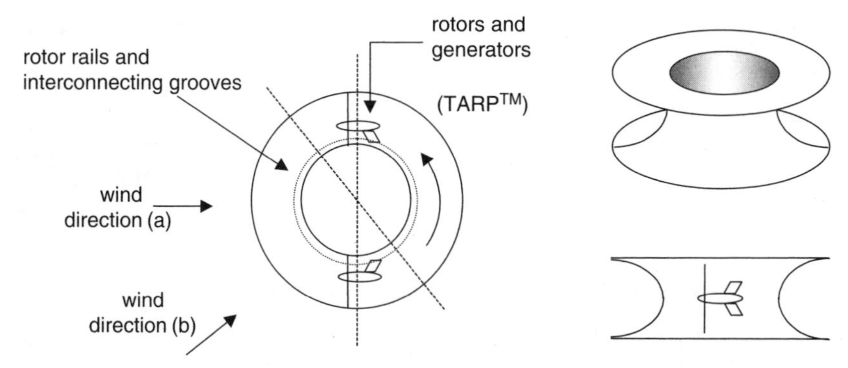

Figure 4.10 Effect of wind direction on the direction of wind-powered rotors.

Some toroidal form of aerodynamic turbine shelter characterizes the working principle of the TARP. Wind accelerates around the shelter, amplifying the density of wind power energy available. Each TARP structure supplies a field of increased outlying flow in all directions, impelling two wind power rotors of small diameter disposed about 180° from one another around the channel of toroidal flow so formed on the shelter.

TARP rotors have a typical diameter of 3 m or less and are coupled directly to the generator through a system of brakes but without a speed multiplication gearbox. As shown in Figure 4.10, if the wind changes from direction (a) to direction (b), a torque will form on the rotating wheel shelter that moves until it is balanced in the new wind direction. The wind power structure also serves as a support, as protection, and as housing for the turbine controls and other internal subsystems. This configuration differs dramatically from the traditional single rotor with a mounted horizontal shaft in a tower, and it is claimed to be of high efficiency [6].

The design just described overcomes the traditional configuration of wind turbines through an odd combination of distribution/transmission, superior performance, easy operation, easy maintenance, high readiness, and reliability. It needs little land area, has a better appearance, has less interference and less electromagnetic noise in TV transmission, and reduces the mortality rate of birds. The estimated cost of a kilowatthour is from 2 to 5 cents, depending on the wind power resources. Such systems still have high installation costs (due to the need for higher tower heights) and are not economically feasible for small power applications. Figure 4.11 shows the technological evolution of wind power turbines.

4.4.6 Auxiliary Equipments

Auxiliary equipments are devices used to improve the efficiency of wind power turbines and are most common for blade turbines. They include solar wind generators, confined vortexes, diffusers, wind concentrators, plane guides, deflectors, Venturi tubes, heating towers, and rotor accelerating shelters [11].

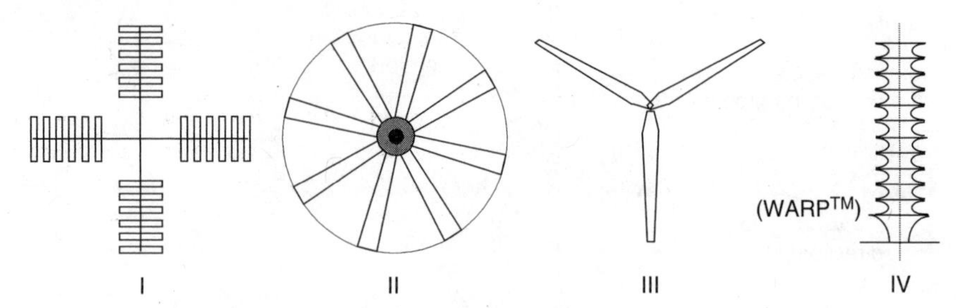

Figure 4.11 Technological generations of wind turbines.

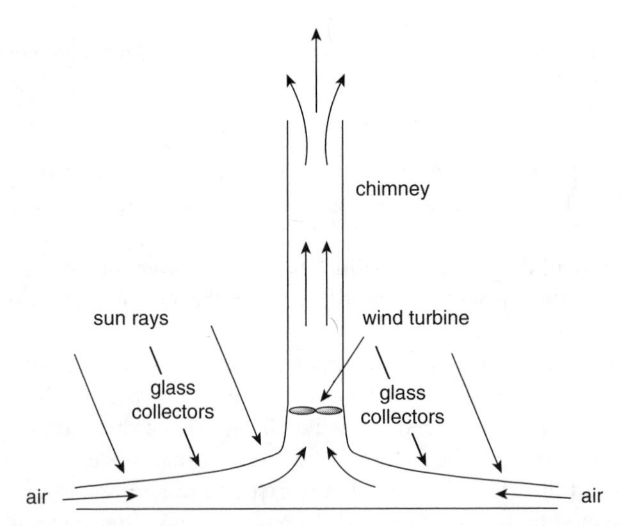

Figure 4.12 Principle of solar power towers. (From Refs. [13–15].)

Solar wind generators are special auxiliary equipment which are able to store the heat irradiated by the sun on black surfaces protected against convection effects. Warm-air circulation is guided by the chimney effect and ends up crossing a turbine when in its ascending movement. The confined vortex consists of a tower where, in its interior, the effects of a tornado are reproduced through the orientation of free wind heating. In a Spanish prototype, the tower sits in the center of a 7-km (4-mile)-radius circular glass building, as shown in Figure 4.12. Under the glass, the sun warms the air. As the warm air rises, it is drawn through turbines at the base of the tower, thus generating renewable electricity.

4.5 GENERATORS AND SPEED CONTROL USED IN WIND POWER ENERGY

As discussed in Section 4.3, the power of a wind generator varies with the cube of the wind speed; that is, if the wind speed doubles, the power will increase eight

TABLE 4.5 Production of High-Power Turbines in Europe

Manufacturer	Capacity (kW)	Blade Diameter (m)	Tower Head Mass (tons)	Swept Area 1000 m^2	kg/m^2	Prototype	Series Production
Enercon	4.5	112	500	9.8	51	2002	2004
GEWind Energy	3.6	104	280	8.5	33	2002	2003
Nordex	5.0	115		10.4		2004	
Neg Micon	4.2	110	214	9.3	23	2003	2005
Prokon	5.0	116	290	10.6	27	2004	2006
REPower	5.0	125	350	12.1	29	2004	2005
Vestas	3.0	90	102	6.4	16	2002	2004
WinWind	3.0	91	150	6.5	23	2004	2005

Source: Adapted from Energimagasinet, May 2003.

times. The wind generator efficiency is affected directly by the wind speed, whose rated value in turn depends on rotor design and rotation. Again, the rotation is expressed as a function of the blade tip speed for a given wind speed. Therefore, to maintain a fixed generator speed on a high efficiency level is virtually impossible. This is the main reason why two-blade rotors (see Figure 4.8) can be used only for higher-power turbines. Table 4.5 lists major manufacturers of large turbines in Europe through May 2003.

With respect to speed control, there are three different methods of safe power control in wind turbines: stall regulation, pitch control, and active stall control. Around two-thirds of wind turbines are passive stall regulated, with the machine rotor blades bolted onto the hubs at a fixed attack angle. In these cases the rotor blade is twisted when moving along its longitudinal axis to ensure that the blade stalls. By *stall* we mean that turbulence on the side of the rotor blade not facing the wind is created gradually rather than abruptly when the wind speed reaches its critical value. If stall control is applied, there are no moving parts in the rotor and a complex control system is needed, but stall control represents a very complex aerodynamic design problem and related design challenges in the structural dynamics of the entire wind turbine to avoid stall-induced vibrations.

On pitch control, an electronic controller checks the output power several times per second. When the output power becomes too high, it sends an order to the blade pitch mechanism, which immediately turns the hydraulic drive of the rotor blades slightly out of the wind direction, and vice versa when the wind speed drops again. The rotor blades are kept at an optimum angle by the pitch controller. In active pitch control, the blade pitch angle is adjusted continuously based on the measured parameters to generate the desired output power up to the maximum limit, as in conventional regulated pitch control. Electric stepper motors are also used for this purpose.

There are four speed bands to be considered in the operation of a wind turbine (see Figure 4.13). The first band goes from zero to the minimum speed of generation (cut-in). Below this speed the power generated just supplies the friction losses.

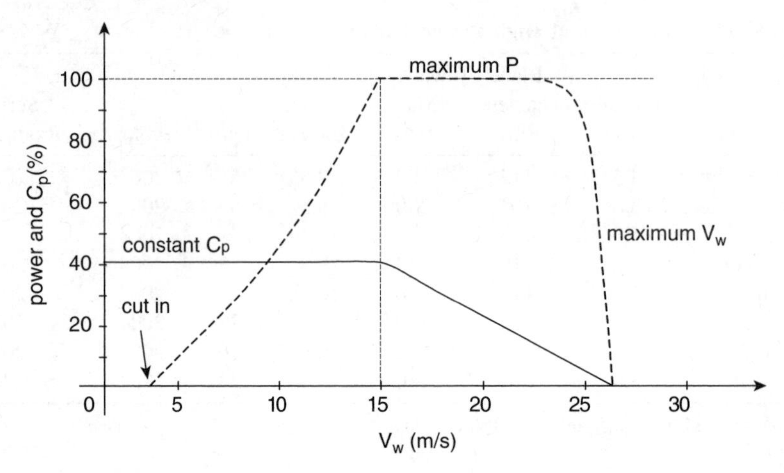

Figure 4.13 Speed control range for wind turbines.

The second band (optimized constant C_p) is the normal operation maintained by a system of blade position control with respect to the direction of wind attack (pitch control). In the third band, for high-speed winds, the speed is controlled such as to maintain a maximum constant output power (constant power) limited only by the generator capacity. Above this band (at wind speeds around 25 m/s), the rotor blades are aligned in the direction of the wind, to avoid mechanical damage to the electrical generator (speed limit).

The simplest way of driving a wind power turbine is, of course, by not using any special control form. In this case, the speed of the turbine is constant (as it is connected directly to the public grid), which operates at a fixed frequency or under a constant load. Power generated this way cannot be controlled since it is altered by the speed of the winds. So just a fraction of that total energy could be captured with extremely strong winds, and it would be uneconomical to size the power plant to the maximum power of the winds. There are benefits to incorporating some control in a wind power turbine, such as an increase in the energy capture from nature and a reduction in dynamic loads [8,9]. Furthermore, the speed actuators allow flexible adjustment of the operating point, and the project's safety margins can be decreased.

The more commonly used form of speed and power control for wind generators is control of the attack angle of the turbine blades, where the generator is connected directly to the public grid. Another control form which has received a lot of attention in the last few years is the use of generators driven by turbines with fixed attack angles of the blades, connected to the public network through a power electronic interconnection. Control occurs on the load flow, which in turn, acts on the turbine rotation [2,5,10]. As the rotor speed must change according to the wind intensity, the speed control of the turbine has to command low speed at low winds and high speed at high winds, so as to follow the maximum power operating point as indicated in Figure 4.14.

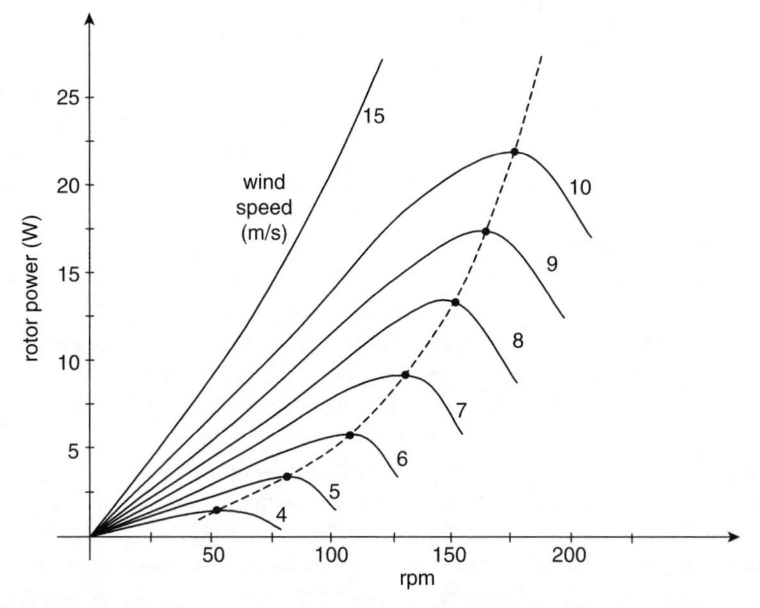

Figure 4.14 Turbine rotation versus power characteristic related to wind speed.

In addition, to invest in wind turbines, the following ancillary equipment (discussed in Chapter 10) are necessary:

1. Yaw control, for maximum wind intensity
2. A support tower, to hold the wind capture system and the generator
3. A speed multiplier, necessary for the low operating rotations usually found in wind power turbines (up to approximately 500 rpm) when it is known that 60-Hz commercial generators (the pattern throughout the United States) with two poles need 3600 rpm and those with four poles need 1800 rpm
4. Voltage and frequency or rotation control
5. Voltage-raising and voltage-lowering transformers to match the generation, transmission, and distribution voltages at normal consumption levels (usually, 480V, 127/220 V or 220/380 V)
6. An electric distribution network for consumers
7. Protection systems for overcurrent, overspeed, overvoltage, atmospheric outbreaks, and other anomalous forms of operation

4.6 ANALYSIS OF SMALL GENERATING SYSTEMS

Several types of generators can be coupled to wind power turbines: parallel and compound generators, dc and ac types, and especially, induction generators (see Chapter 10).

Once the installation site of a wind power plant has been selected, the next steps are to select the turbine rating, the generator, and the distribution system. In general, the distribution transformer is sized to the peak capacity of the generator according to the available distribution network capacity. As a practical rule, the output characteristics of a wind turbine power do not exactly follow those of the generator power, and they must be matched in the most reasonable way possible. Based on the maximum speed expected for a wind turbine, and taking into account the cubic relationship between wind speed and power, the designer must select the generator and gearbox to match these limits. The most sensitive point is the correct selection of the rated turbine speed for the power plant. If it is too low, generation for high-speed winds will not be possible. If it is too high, the power factor will be too low. There is an iterative design process on the match of the characteristic of commercially available wind turbines and generators with regard to cost, efficiency, and the maximum power generated. The maximum value of C_p should occur at approximately the same speed as that of the maximum power in the power distribution curve. So the tip speed ratio must be kept optimally constant at the maximum speed possible to capture the maximum wind power. This feature suggests that to optimize the annual energy capture at a given site, it is necessary that the turbine speed (the tip speed ratio) should vary to keep C_p maximized, as illustrated in Figure 4.15 for a hypothetical turbine. Obviously, the design stress must be kept within the limits of the turbine manufacturer's data, since torque relates to the instantaneous power as $P = T\omega$.

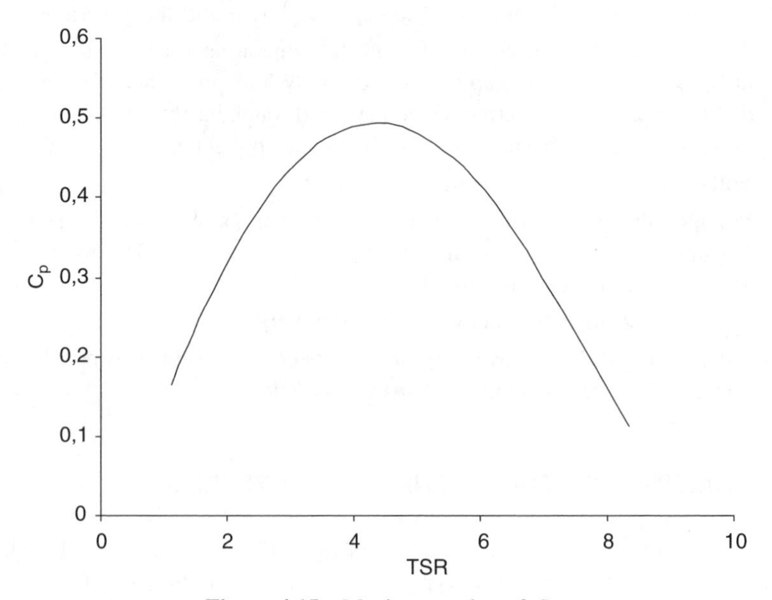

Figure 4.15 Maximum value of C_p.

It should be pointed out that in the case of parallel generators connected to batteries, small rotation increments are linked to large increments in the output current. When the terminal voltage of the battery stays constant (increasing the nominal rotation by a determined percentage), the current suffers an increase of k times its initial value. As a result, the power also suffers an increase of k times its initial value.

If the voltage drop of the generator occurs at higher rates than the current increment, a dc compound generator of three brushes should be used rather than a parallel generator. For a compound generator, the design is executed so that the decrease in the characteristic is at the same rate as the number of demagnetization coil turns. For three-brush generators (due to the increase in coil current), both field voltage and magnetic field are reduced.

Due to the advantages of working as a motor or generator (and its low cost with respect to other generators), the induction machine offers enhanced conditions for wind power plants. Only an induction generator connected to an infinite bar is considered in this chapter; other types are discussed more fully in Chapter 10.

The aerodynamic efficiency of a wind power machine of very small size varies, at most, between 40 and 45%, but in practice it is, on average, 35%. The efficiency of a dc generator is on the order of 55 to 60%. Therefore, the total efficiency of a small system (a few hundred watts) is on the order of 20%. One of the main causes of loss is the excitation demanded by the generator. In that case it is advisable to choose permanent magnet-based dc generators. It is possible to eliminate the switches in alternators of six or more poles by the use of permanent magnets. It is also advisable to use rectifiers to charge batteries. The generator should present voltage and frequency proportional to the rotation so as to provide current regulation by the inductive reactance of the circuit.

In conclusion, the designer needs the following specifications before purchasing a wind power turbine:

1. Wind intensities in the area and the duration curve
2. Topography
3. Purposes of the energy generated
4. Present and future energy needs for the generator and turbine according to the area's wind capacity (otherwise, any specification becomes useless)
5. Determined need for a turbine isolation valve
6. Specifications for the type of regulator
7. Specifications for the generator (see Chapter 10) according to:
   a. Type: induction, synchronous, or direct current
   b. Electric output: alternating or direct current, maximum power, voltage level, number of phases, frequency
   c. Climate: temperature and humidity
   d. Automation level: manual, automatic, semiautomatic, or remote control

   e. Operation: stand-alone or for injection in the network (in which case a sale, purchase, or exchange commercial agreement should exist with the electric power company)
   f. Number of phases: single- or three phase transmission and distribution
8. Other equipment: control panels, protection, and drives
9. Location of the machine house (if there is one, distance and dimensions)
10. Access possibilities to the site: boat, bike, highway, rails, walking distance, suitability for land transportation of equipment and maintenance planning

The cost of wind power plants has been changing in the last few years. By the year 2004, they cost around $0.50 per watt and $0.06 per kilowatthour.

REFERENCES

[1] S. Adeodato and W. Oliveira, A riqueza dos melhores ventos (The richness of the best winds), *Globo Ciência*, Vol. 63, pp. 20–25, 1996.

[2] D. P. Sadhu, *Estudos sobre energia eólica* (Studies About Wind Energy), Department of Mechanical Engineering, UFRGS, Pôrto Alegre, Brazil, 1983.

[3] J. S. Rohatgi and V. Nelson, *Wind Characteristics: An Analysis for Generation of Wind Power*, Alternative Energy Institute, West Texas A&M University, Canyon, TX, 1994.

[4] R. R. Barcellos, Use of wind energy in RS using a turbine of vertical shaft, M.S. thesis, PPGEM-UFRGS, Pôrto Alegre, Brazil, 1981.

[5] M. R. Patel, *Wind and Solar Power Systems*, CRC Press, Boca Raton, FL, 1999.

[6] G. C. Chang and F. Hischfeld, *Mechanical Engineering*, Vol. 100, No. 2, pp. 38–45, 1982.

[7] A. L. Weisbrich, S. L. Ostrow, Padalino, and J. P. Warp, WARP: a modular wind power system for a distributed electric utility application, *IEEE Transactions on Industry Applications*, Vol. 32, No. 4, 1996.

[8] *The Ruthland Windcharger: User's Handbook for the Series 910*, Marlec Engineering Company, Northants, England, 1995.

[9] C. A. Portolann, F. A. Farret, and R. Q. Machado, Load effects on dc–dc converters for simultaneous speed and voltage control by the load in asynchronous generation, *Proceedings of the IEEE International Conference on Devices, Circuits and Systems*, Caracas, Venezuela, 1995, pp. 122–127.

[10] P. Novak, T. Ekelund, I. Jovik, and B. Schmidtbauer, Modeling and control of variable-speed wind-turbine drive-system dynamics, *Proceedings of the IEEE Control Systems Society Conference*, 1995, pp. 28–38.

[11] R. M. Hilloowala and A. M. Sharaf, A ruled-based fuzzy logic controller for a PWM inverter in a stand alone wind energy conversion scheme, *IEEE Transactions on Industry Applications*, Vol. 32, No. 1, 1996.

[12] W. Hughes, D. K. McLaughlin, and R. Ramakumar, *Applications of Wind Energy Systems*, 1976.

[13] R. N. Meroney, Wind in the perturbed environment: its influence on WECS, presented at the American Wind Energy Association Conference, Boulder, CO, 1977.

[14] J. Ross, The EnviroMission, Armadale, Victoria, Australia, http://www.aie.org.au/pubs/enviromission.htm, accessed February 22, 2005.

[15] EnviroMission, http://www.wentworth.nsw.gov.au/solartower/, accessed March 17, 2005.

CHAPTER 5

THERMOSOLAR POWER PLANTS

5.1 INTRODUCTION

The sun is a perennial, silent, free, and nonpolluting source of energy and is responsible for all lifeforms on the planet. Its use for energy generation can be direct or indirect. Indirect solar energy is related primarily to wind power, hydropower, photosynthesis, sea tidal energy, and to the microbiological conversion of organic matter into liquid fuels (the subjects of other chapters in this book). The sun does not reach directly only one point on Earth during each day. Also, its intensity does not stay unchanged when available, and a utilization factor is fundamental in the definition of how economically feasible it may be to harness the sun's energy.

Direct solar energy is used to heat water (domestic, industrial, or commercial uses), to cool and air condition, to dry agricultural products, for distillation (mainly for the production of salt or brine by evaporation for seawater), and for electric power generation [1–5]. Thermal solar energy is most appropriate for areas of the planet that form the *solar belt*, those areas about 30° to the north or south of the equator, where direct solar radiation is very high throughout the year.

Two main types of solar technology exist for the conversion of solar energy into electricity. The first is related to the transformation of solar light directly into electricity that is done through modules consisting of photovoltaic cells, the subject of

Integration of Alternative Sources of Energy, by Felix A. Farret and M. Godoy Simões

Chapter 6. In this chapter we deal with solar heat only and discuss basic elements of the use of solar energy in water heating and in indirect production of electric power in power plants. The second technology type is adapted more for large-scale applications using solar radiation directly. This is the largest source of energy, usable primarily in areas along the solar belt.

5.2 WATER HEATING BY SOLAR ENERGY

A solar energy conversion process may be divided into three phases: reception, transfer, and accumulation [2]. Heat capture is accomplished directly or through collecting plates. A typical example of a collecting plate consists of a blackbody with a large radiation index of absorption. A very conventional example of the use of solar energy is a flat collector for water heating (see Figure 5.1). These collectors are basically formed by a box of insulating material, usually of fiberglass and resins of polyester, isolated internally with phenolic glass wool and a blackbody that covers a largely copper exchanger [3,4]. To increase thermal resistance and minimize losses, a crystal glass about 4 mm thick, perfectly isolated with glass wool or silicon, covers the device.

A practical system of water heating with solar collectors is illustrated in Figure 5.2. Cold water from the cold reservoir reaches the base of the solar collector pipes, which absorb heat. Through thermal expansion and natural or forced convection, it returns to the reservoir. The water flow continues in this cycle, and the temperature increases gradually after each passage through the collector pipes [1–5].

The equation used for practical implementation of a flat solar plate is [1]

$$Q = FA[I(ab) - U(T_i - T_a)] \tag{5.1}$$

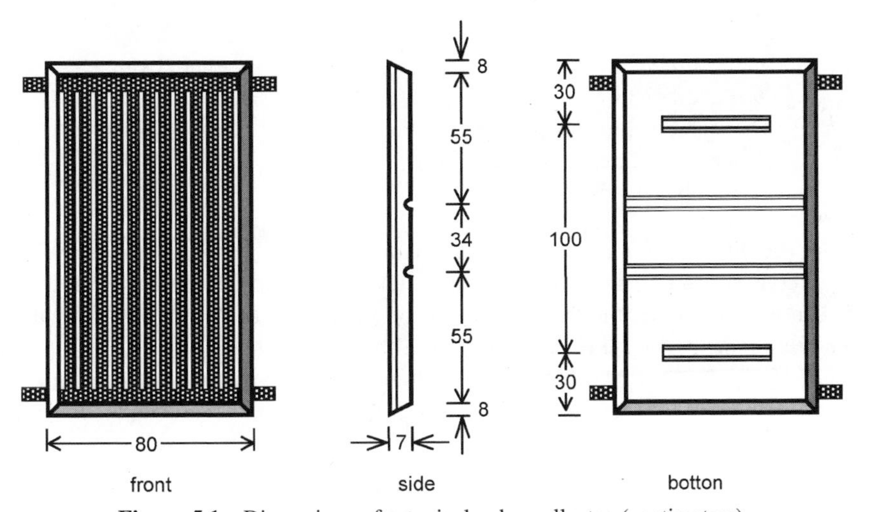

Figure 5.1 Dimensions of a typical solar collector (centimeters).

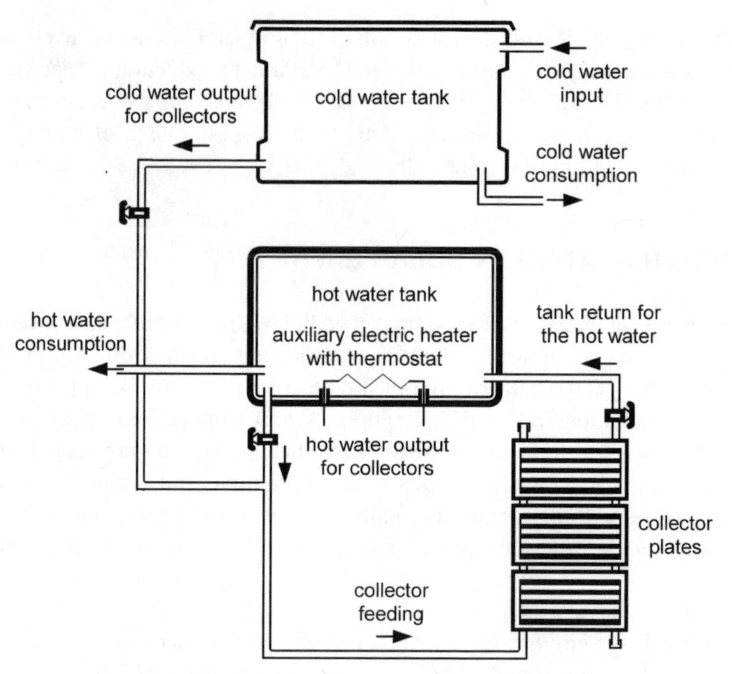

Figure 5.2 Heating water with solar collectors.

where Q = energy extracted by the plate (W)
F = efficiency factor of heat removal from the plate
A = plate area (m^2)
I = rate of the incident to absorbed solar radiation per unit of area of plate surface (W/m^2)
a = coefficient of solar transmittance of the transparent covers
b = coefficient of solar absorption of the plate sheet
U = coefficient of energy loss of the plate ($W/°C \cdot m^2$)
T_i = temperature of the fluid (°C)
T_a = ambient temperature (°C)

Typical dimensions (centimeters) of a solar collector are also shown in Figure 5.1. These collectors are manufactured carefully, taking into account the variations in the incidence of radiation, the temperature of the fluids, and the ambient temperature. The useful energy for the collector is defined by

$$Q_u = AGc_p(T_i - T_o) \tag{5.2}$$

where G is the fluid volume per unit of collector area, c_p the specific heat of the collector fluid (in the case of water, 4190 J/kg · °C), and T_o the output temperature of the fluid. The instantaneous efficiency of the collector, η, is defined as

$$\eta = \frac{Q_u}{AI} \tag{5.3}$$

For water heating by solar energy, the use of flat collectors can convert the sun's radiation into heat. Generation of water steam through solar energy, at average temperatures between 150 and 200°C, has countless applications. The direct heating of air in driers of grains and seeds are also very usual nowadays, with the advantage that these less sophisticated systems can compete with those using conventional fuels. The adoption of any solar process helps to avoid losses in the volume of cropped grains (of up to 50%), due to deficient storage, exposing them to humidity. In any case, solar collectors can avoid burning fossil fuels in some industries, such as those for food transformation, and can contribute to a reduction of electric and gas consumption in residences [6–10].

To avoid damage to the ambient, the use of conductive solar plates implies the need for an evaluation of the environmental conditions of the site at which the plates will be used. For instance, sites with high-speed winds can make the use of solar plates unfeasible.

5.3 HEAT TRANSFER CALCULATION OF THERMALLY ISOLATED RESERVOIRS

Units for accumulation of hot water have double walls, isolating them completely against temperature losses. To increase their durability, accumulators should be immune to corrosion. Some commercial modules are built of stainless steel AISI 304 and special resins of polyester for hot water and can resist temperatures up to 280°C and pressures of 7 atm. A simplified method of heat transfer calculation in thermally isolated reservoirs in steady and transient states is presented next.

Steady-State Thermal Calculations Heat goes from the high-temperature side, T_{high}, to the low-temperature side, T_{low}, according to

$$P = \frac{T_{\text{high}} - T_{\text{low}}}{R_{\text{th}}} \tag{5.4}$$

where P is the steady-state power dissipated in the reservoir [watts (or joules/second)], T the temperature (°C), and $R_{\text{th}} (= R)$ the thermal resistance (°C/W), whose components depend on the manufacturing material (pipelines and reservoir) (R_{ps}) and the reservoir ambient (R_{ra}). Equation (5.4) for the thermal calculation may be generalized as

$$T_{\text{high}} = T_{\text{low}} + RP \tag{5.5}$$

The thermal resistance R can be used to avoid the definition of what the heat distribution is exactly in a multilayer reservoir. The calculation above obtains average values for steady-state temperatures when the loss is constant and the temperature is already stabilized at a certain value. The thermal storage capacity should be taken into account to establish the conditions for high time constants during transient and/or heat charge or discharge conditions.

Transient-State Thermal Calculations At an instant $t < t_0$, the external temperature of the reservoir is the lowest temperature, T_{low}. At instant t_0, power transference begins, limited by the thermal capacity of the reservoir, which prevents the temperature from rising abruptly (exponentially). To express the instantaneous difference in temperature, it is defined as $\Delta T = Z_{th}P$, where Z_{th} is the reservoir thermal impedance, which varies with time. The equivalent thermal circuit, including the thermal capacity C, is illustrated in Figure 5.3, with its transient response to a sudden power step illustrated in Figure 5.4. The amount of heat Q stored in the

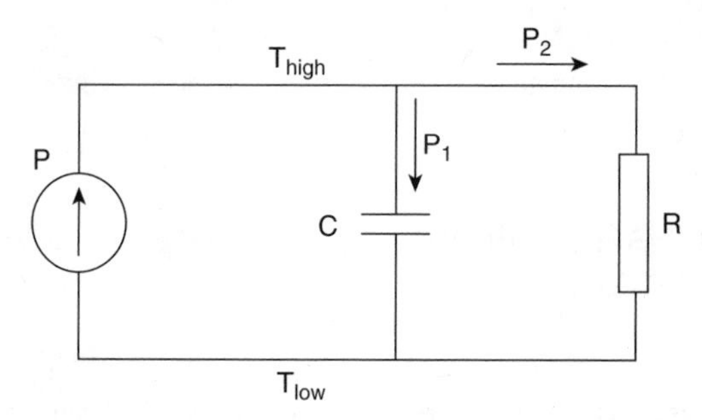

Figure 5.3 Reservoir equivalent thermal circuit.

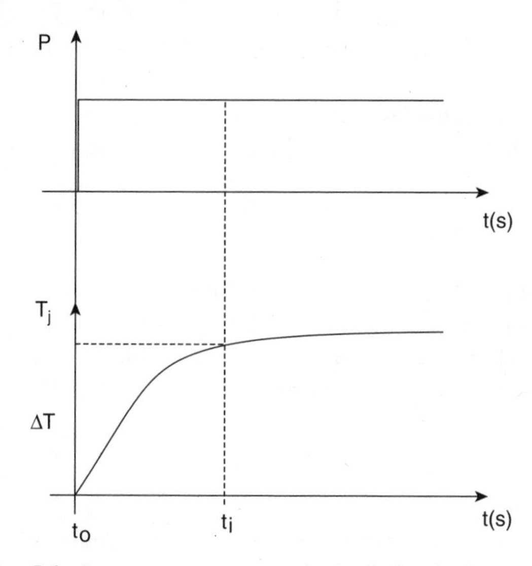

Figure 5.4 Instantaneous temperature variation in the reservoir.

reservoir mass m as a whole is

$$\Delta Q = c_p m \, \Delta T = C \Delta T = \int P_1 dt \tag{5.6}$$

where c_p is the specific heat of the reservoir mass m, $C = c_p m$ is the heat capacity of the reservoir as a whole, and P_1 is the power necessary to maintain the temperature difference ΔT across the walls. From the heat balance given in equations (5.4) and (5.6) and Figure 5.3 comes

$$\Delta T = T_{\text{high}} - T_{\text{low}} = RP_2 = \frac{1}{C} \int P_1 dt \tag{5.7}$$

where P_2 is the power diverted for heat storage in the reservoir walls. Differentiating equation (5.7) with respect to time yields

$$R \frac{dP_2}{dt} = \frac{P_1}{C}$$

As $P = P_1 + P_2$, if we set $P/C = P_1/C + P_2/C$ and isolate P_1/C in equation, (5.7), we obtain

$$\frac{P}{C} = R \frac{dP_2}{dt} + \frac{P_2}{C} \quad \text{or} \quad \frac{dP_2}{dt} + \frac{P_2}{RC} = \frac{P}{RC}$$

whose solution is $P_2 = P(1 - e^{-t/RC})$. However, as $\Delta T = T_{\text{high}} - T_{\text{low}} = RP_2 = RP(1 - e^{-t/RC})$, we may use

$$\frac{\Delta T}{P} = R(1 - e^{-t/RC}) = Z_{\text{th}} \tag{5.8}$$

This is a very simplified solution where the reservoir manufacturer supplies the value of Z_{th}. Otherwise, a field evaluation has to be made.

The concept of transient thermal impedance, Z_{th}, is used for applications with repetitive thermal loading and unloading conditions at high power, that is, when the reservoir works under different states of heat variation. The heat path conditions to take the heat out of the interior side of the reservoir to ambient are too complex to be expressed precisely by a simple exponential form like that in Figure 5.4. This figure for the variation of Z_{th} is useful only for homogeneous materials such as copper pipes, tank walls, or the like, where the heat spreads quickly. The same does not occur in the path of liquid in a tank where the heat goes to the tank walls, pipes, isolation, drafts, and internal layers of air, since the losses are not uniform. The larger the transient load, the smaller the mass that experiences a temperature increase; one cannot give a single value to its thermal storage capacity. In summary, the method presented above is very simplified; even so, it is useful in daily practice.

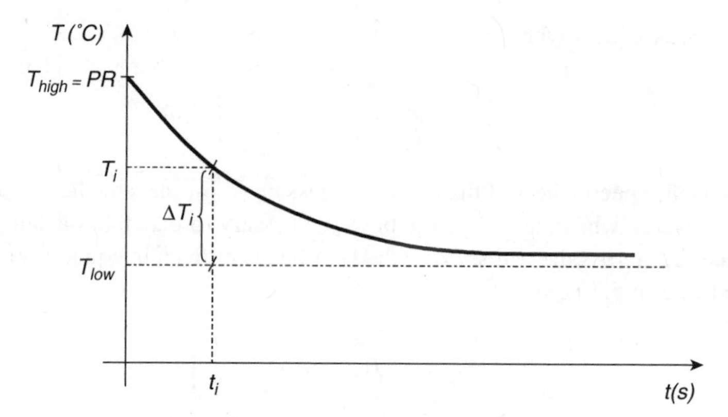

Figure 5.5 Temperature decaying for measurement of C_i.

Practical Approximate Measurements of the Thermal Constants, R and C in Water Reservoirs One practical way of measuring the thermal constants R and C could be as follows. An electrothermal element is immersed in the middle of the thermal fluid in a reservoir such that an external steady-state power V^2/R being supplied to it can be measured, as depicted in Figure 5.4. After some period of constant power to the thermal element, the internal temperature will reach a stable value to be measured simultaneously with the ambient temperature close to the tank. The value of R can then be calculated by equation (5.4).

To calculate C, we may continue the experiment from the previous test to evaluate R and its last measured temperature, except that now the heating active element is switched off. The temperature will begin to decrease exponentially, as in Figure 5.5, and its values at periodic intervals can be logged in a table or graph.

The thermal capacity may be evaluated from equation (5.8) for $T_{i=0} = T_{\text{high}} = PR$ and put into a more convenient form:

$$C_i = -\frac{t_i}{R \ln(\Delta T_i / PR)} \tag{5.9}$$

There will be as many values of C_i as the instants of measurement of T_i, and a good average value could be obtained from there. A high standard thermal isolation should guarantee warm water conservation for very long periods. Large differences in temperature between the inner and outer surfaces of the reservoir will increase its losses.

5.4 HEATING DOMESTIC WATER

Heating a volume of water demands temperature elevation (i.e., an increase in energy). For domestic purposes, that amount of energy is linked to the individual habits of each occupant of a residence. Recorded data show that, on average, each member of a family uses approximately 100 L of water a day. The amount of energy used to heat water each month is then estimated as

$$L = NP(100)(T_w - T_m)\rho_w c_p \qquad (5.10)$$

where L = energy used for water heating per month
N = number of days per month
P = number of people
T_w = acceptable minimum temperature for the hot water (60°C)
T_m = ambient temperature of the water
ρ_w = water density (1.0 kg/L)

Equation (5.10) can be used to predict the amount of energy utilized to heat water in any given period.

Example To estimate the average monthly energy use to heat water [joules (W/s)] for a four-person family, for water at an ambient temperature of 11.0°C, equation (5.10) yields

$$\begin{aligned} L &= (30\,\text{days})(4\,\text{people})(100\,\text{L/day})(60^\circ - 11^\circ)(1.0\,\text{kg/L})(4190\,\text{J/kg}\cdot{}^\circ\text{C}) \\ &= 2.464 \times 10^9\,\text{J} \end{aligned}$$

5.5 THERMOSOLAR ENERGY [8–10]

For applications of electric power generation using solar thermal energy on a large scale, the best known technologies are the parabolic trough, parabolic dish, solar power tower, and hydrogen production. For smaller uses and remote locations, parabolic dishes are better known, due to their great development potential.

Solar thermal power uses various means to generate heat, with the water being converted into steam that will be used to drive a conventional steam turbine to produce electricity (Figure 5.6). Sometimes a fossil fuel is used as a backup, so that the power plant can continue to produce energy when solar energy is not available [11,12].

Solar energy is sometimes considered a land-intensive technology, due to the amount of land used in such facilities. However, the amount of energy produced by a solar thermal power plant for a given land area is larger than that produced by a large electrical thermopower plant of the same proportions. Also, solar thermal power plants use desert lands that are arid or semiarid, as in the Brazilian northeast, the northern Sahara, and Central America. Another positive aspect of solar thermal power is that diversifying energy reserves results in less dependence on fossil fuels, whose prices float widely on the international market and are subject to increase as their supply sources become exhausted.

The environment seems to be the major beneficiary of such clean technologies. Emissions of carbon dioxide in the production of energy create about 50% of the

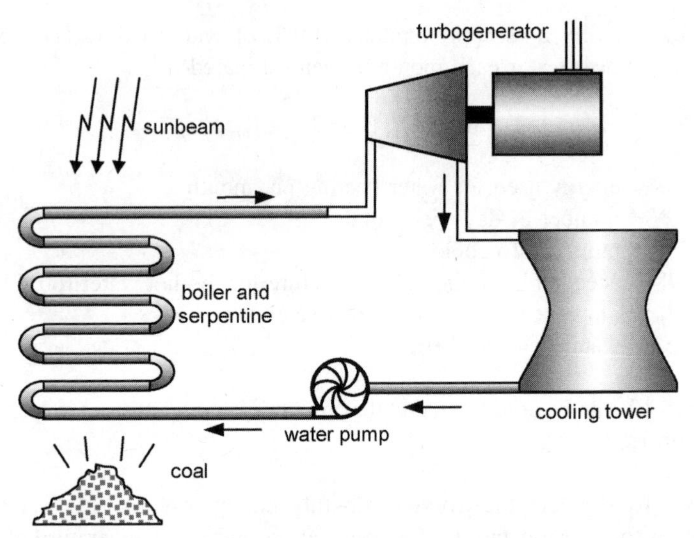

Figure 5.6 Conversion of solar heat and fossil fuel into electricity.

gases responsible for the greenhouse effect. The hybrid thermal solar power plants in operation in California, where fossil fuel is used only as reserve power, help in the reduction of carbon dioxide, nitrogen oxide, and sulfuric dioxide. A typical installation of 80 MW using solar troughs reduces carbon dioxide emissions into the atmosphere by 4.7 million tons and avoids using 2 million tons of coal during 25 useful years of trough life [13–16].

5.5.1 Parabolic Trough

A parabolic trough uses a type of clear oil in pipes that absorbs heat reflected off the trough. The parabolic troughs are long, trough-shaped reflectors that focus the sun's energy on a pipe running along the mirror's curve, as shown in Figure 5.7. Heat-absorbent oil is used inside the conducting pipeline to carry the thermal energy to the water in the boiler heat exchanger, whose temperature can reach about 400°C. Heat energy from the oil is transferred through a heat exchanger to boil water to dry steam of high pressure that drives the turbine of an electric generator. The cross section of the troughs is parabolic, to ensure conversion efficiency. In larger systems, the design of the trough includes a rotating shaft that allows each group to follow the sun from east to west. Solar farms reunite several long and parallel lines of groups of parabolic troughs to concentrate and maximize the sunbeams on heat-absorbing pipelines.

The sunlight incident on the absorber is maximized by the high reflectance coefficient of the parabolic reflector, which is positioned as high as possible. The peak optical efficiency of a parabolic trough is in the range 70 to 80%, but only something like 60% is practically useful since there are other losses, such as those due to heat losses in the solar field piping. Typical solar thermal power plants can

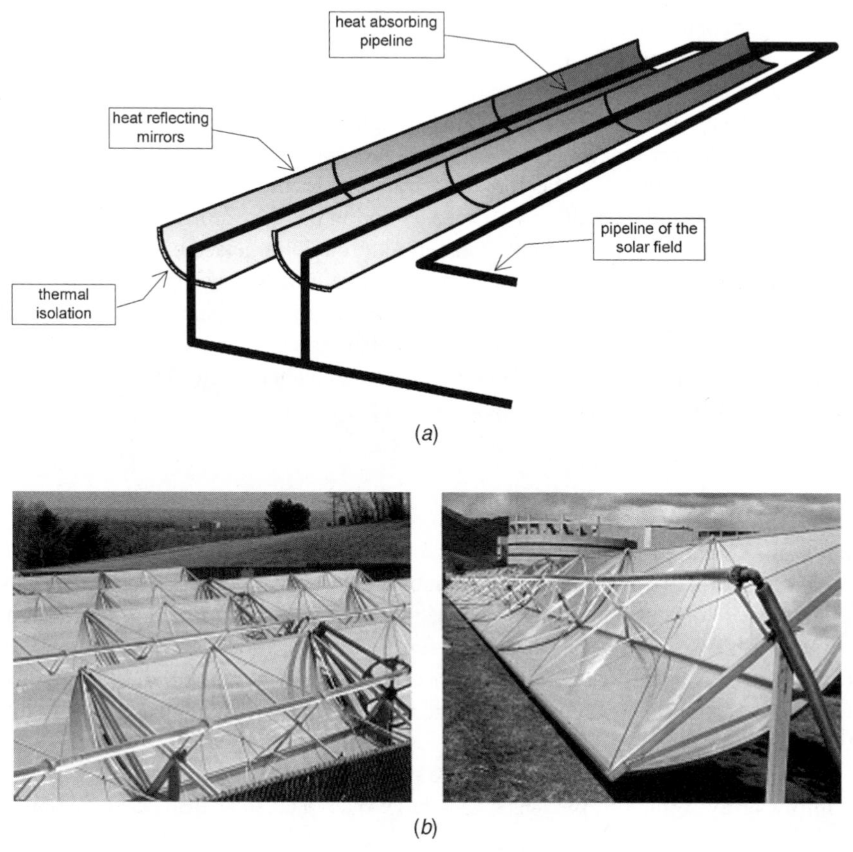

Figure 5.7 Parabolic troughs: (*a*) disposition of more than one solar array; (*b*) operational solar array. (Photo courtesy of Jefferson County Detention Facility in Golden, Colorado.)

supply energy at full load in 2000 to 3000 of the 8760 annual hours. Some losses are always present in the receiver of a trough concentrator, which is usually made of a metal absorber sometimes surrounded by glass tubes with antireflexive properties. The absorber is coated with a selective surface to filter out infrared radiance from the incoming light, to make way for light in the visible range. The intensity of the sun may be multiplied by a concentration ratio in the range 20 to 100.

Although silver reflectors have the higher reflectance, aluminum is preferred for the structures of the trough system because silver is more expensive and more difficult to protect against the corrosive effects of the outdoor environment. It is also important to keep the reflectors clean since dirt will degrade the reflectance of light from the parabola. Larger systems may have to use yaw controllers (precise to a fraction of a degree), whose values are almost negligible in these cases, but as their use depends on the size of the trough collector, small systems should not use them. Typical uses of parabolic solar troughs are to generate power for hot water, space heating, air conditioning, steam generation, industrial process heating, desalination, and electrical power generation.

Solar heating design makes possible annual savings of up to 70% of the overall hot water heating bill. Also, commercial trough systems can compete with natural gas and can deliver energy at a steady cost for more than 20 years. Little maintenance is needed since the collectors track the sun continually during the day to heat a closed-loop circulation of an antifreeze solution such as propylene glycol. Heat from the solar collectors is usually transferred through immersed copper coils to hot water storage tanks. The still expensive manufacturing technology of trough systems is being enhanced through research on parabolic trough technology that uses new methods and designs for power plant integration of solar technology with fuel gas that result in lower investment costs, cleaner energy, higher efficiency, and lower costs for production of electricity.

5.5.2 Parabolic Dish

For remote stand-alone applications, parabolic dishes with yawing engines seem to be a very attractive solution. It is the most modular of all the options presently available for concentrating sun rays in a single point, and can produce higher temperatures. It goes commercially from a few watts to several kilowatts at temperatures as high as 800°C. For higher power capacities, several units can be combined to produce the desired output power. Figure 5.8 is an example of such a configuration for a domestic application.

Solar dish–engine systems convert the energy from the sun into electricity using a mirror array formed in the shape of a large dish. The solar dish has a parabolic

Figure 5.8 Parabolic dish.

shape, which focuses the sun's rays onto a receiver. The receiver transmits the energy to an engine, typically a Stirling- or Brayton-cycle engine, to generate electric power at very high efficiency (see Appendix C). This high efficiency is due to the high concentration ratios achievable with parabolic dishes and the small size of the receiver, whose highest values are achievable at higher temperatures. Tests of prototype systems and components at locations throughout the United States have demonstrated net solar-to-electric conversion efficiencies as high as 30%, significantly higher than for any other solar technology.

Some interesting experiments with larger solar dishes have been reported in several countries, such as Australia with CSIRO in collaboration with industry partner Solar Systems Pty Ltd., to demonstrate the integration of solar thermal energy and methane gas (see Figure 5.9). It has been claimed that the system can produce a range of solar-enriched fuels and synthesis gas (CO and H_2). These gases can be used as power generation fuel gas, as metallurgical reducing gas, or as chemical feedstock (e.g., in methanol production). The dish built and operated by CSIRO as a demonstration facility designed to process 44 kW thermal of natural gas, including the process chemistry, reactor design, and power generation potential to prove the value of solar-thermal technology as an economical option for large-scale energy delivery [7,11–13].

Figure 5.9 CISRO solar dish system in Australia.

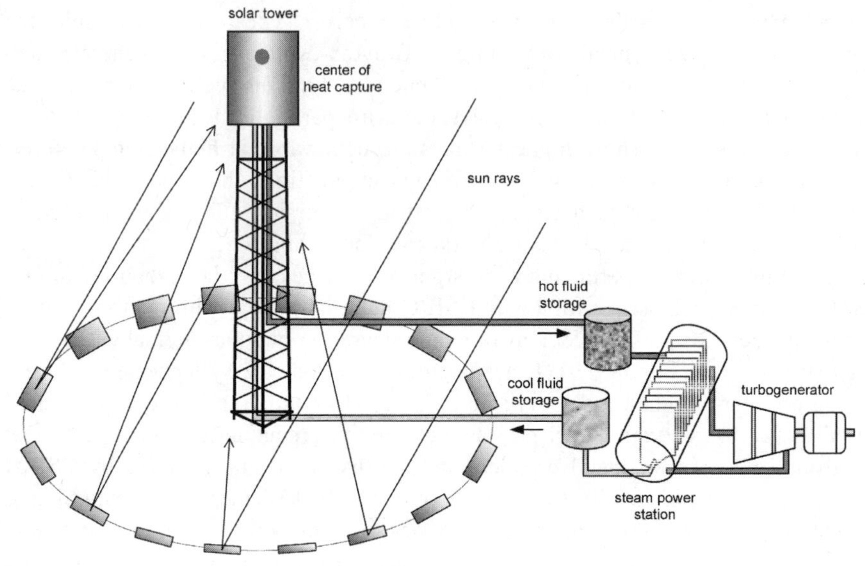

Figure 5.10 Solar power tower.

5.5.3 Solar Power Tower

The principle of capturing and concentrating the solar light on reception mirrors is used in an energy tower. In this case, the mirrors are distributed in a convenient form and not in arrays, as is the case with parabolic troughs (Figure 5.10). In the center of the tower circle is placed a receiver containing a fluid that can be water, air, oil, liquid metal, molted salt, or diluted salt. The position of the mirrors is heliostatic. The warm fluid goes from the receiver to the tower block and then to a steam turbine. As the sun heats salt in the receiver, it goes to the hot storage tank. The hot salt is then pumped through a steam generator and the steam drives a turbine–generator to produce electricity. The same salt is finally returned to the cold storage tank to be used again in the concentrator and closes the cycle. This technology is not yet well established, but the tower of energy can supply temperatures higher than those supplied by any other heat concentrators used to date.

Test designs in the United States using diluted salt as a transfer means to substitute for fossil fuel for thermal storage were built to work under complementary demand. These energy towers are considered good prospects for the long term, due to their high efficiency and low cost in the production of electricity, especially for larger units (100 to 200 MW). Power plants as big as 200 MW are being considered in a formal joint venture and profit-sharing agreement between SolarMission Technologies, Inc. and Sunshine Energy (Aust) Pty Ltd. In the case of small facilities, the intermediary heated fluid using these technologies (e.g., oil) can be used directly for heating poultry houses, greenhouses, and grain driers.

5.5.4 Production of Hydrogen

In the production of hydrogen, steam reforming for natural gases and synthesized hydrogen from either gasoline or methanol are being considered for fuel distribution supply in fuel cell–powered vehicles. However, in solar belt regions, the use of an environmentally friendly option for a solar-thermal distributed hydrogen process for these vehicles will probably need to be based on the already existing infrastructure to generate the hydrogen required.

For this purpose, sunlight can be concentrated on small solar reactors to achieve ultrahigh temperatures: 1700°C or higher. The chemical reaction rates at these temperatures are enormous. These reactors are made of one outer quartz tube and two inner concentric graphite tubes. The graphite used in the concentric tubes shown in Figure 5.11 are of different consistencies. The external graphite tube is solid graphite; the internal graphite tube is porous graphite. Sunlight has to pass through the quartz tube to heat the solid external graphite tube, which, in turn, radiates and heats the internal porous graphite tube to the necessary reaction temperature [17–19]. The natural gas to be used in these reactors must pass through a hydrogenerator and onto a zinc oxide bed to remove mercaptans and hydrogen sulfide from the stream. The chemical reaction is given by

$$C_xH_y + \text{concentrated sunlight} \Rightarrow C + 2H_2 + \text{unreacted } C_xH_y$$

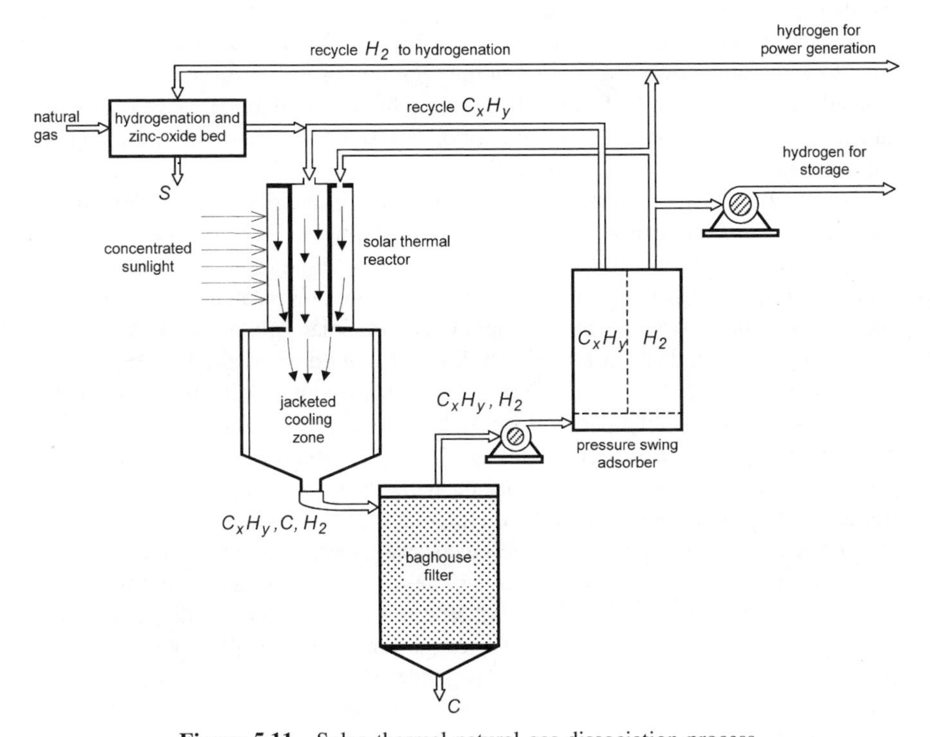

Figure 5.11 Solar–thermal natural gas dissociation process.

Single-pass conversions reported by researchers from the University of Colorado say that 70% of natural gas can be purified by this method. The hydrogen can be compressed and stored, and the carbon black can be sold into the carbon black market, kept in storage, or used to feed a highly efficient carbon-conversion fuel cell. The hydrogen that results from this process can be used in stationary fuel cells or for storage. Certainly, the cost of compression and storage of hydrogen will exceed 50% of the capital of a solar–thermal plant in a distributed service station (DSS).

5.6 ECONOMICAL ANALYSIS OF THERMOSOLAR ENERGY

There are structural problems related to financing solar energy, primarily because of the high installation costs, lack of government financial assistance, amount of time needed to gain the return of initial installation cost, and the belief that only large power plant installations can manage the world's energy problems. There are, however, real opportunities to be sought by people with long-term vision.

Three general sources of capital are available for a renewable power project: equity, debt, and grant financing. An equity investment is the purchase of ownership in a project. A debt investment is a loan to the project. For example, in the purchase of a house, the person buying the house is the equity investor, and the mortgage is the debt [19–23]. Since most solar power projects cannot compete with conventional fossil power technologies today, a number of organizations have grant financing to help buy down the noneconomic portion of the project. These grants typically are made available to help account for environmental, developmental, and economic externalities that would not otherwise be accounted for during a normal competitive project selection [20].

One example is the contribution of solar–thermal electricity generation to climate change, which can be assessed by calculating the greenhouse gas costs (GGCs) incurred during the provision of services and production of materials needed for the construction and operation of solar power plants. Both process and input/output analyses are able to yield GGCs for solar-only plants with a typical uncertainty of about 25%, at a moderate level of input data detail. The uncertainty of the GGCs for hybrid systems decreases with increasing fossil share. The GGCs of utility-sized solar-only parabolic trough, central receiver, and parabolic dish plants range from 30 g to 120 g $CO_2 - e$/kWhel. Furthermore, GGCs vary with plant size and, most important, depend on whether a fossil-fueled backup or heat storage system is chosen to increase the plant's capacity factor.

Comparisons with other renewable electricity generation technologies must be judged with care because of differences in the assessment boundary and methodology, and in the power plant's location, lifetime, capacity factor, dispatchability, and other characteristics. The parabolic trough option is typically applied to grid-connected plants and is commercially available with over 9 billion kilowatthours operational to date; the solar collection efficiency is up to 60%, with peak solar-to-electrical conversion efficiency up to 21%. The main disadvantage is the low temperature with moderate steam quality. The parabolic dish is typically applied to stand-alone applications or small off-grid installations. It is modular with very high conversion efficiency with peak solar-to-electrical conversion of about 30%. The drawback is the low-efficiency combustion in hybrid systems. The central receiver is for grid-connected plants with high solar collection efficiency: up to 46% at operating temperatures close to 550°C. However, the technology is not yet commercially proven to allow precise capital cost projections.

REFERENCES

[1] Hot water for solar energy, *Força Energética*, Vol. 1, pp. 14–16, May–June 1992.

[2] Solar energy: electricity or hot water, *Força Energética*, Vol. 2, No. 3, pp. 25–26, March 1993.

[3] W. L. Hughes, *Energy in Rural Development: Renewable Resources and Alternative Technologies for Developing Countries*, Advisory Committee on Technology Innovation, Board on Science and Technology for International Development, Commission on International Relations, National Academy of Sciences, Washington, DC, 1976.

[4] D. S. Ward, S. Karaki, G. O. G. Löf, C. C. Smith, M. Z. Lowenstein, C. B. Winn, M. E. Larson, and I. E. Valentine, *Solar Heating and Cooling of Residential Buildings: Sizing, Installation and Operation of Systems*, U.S. Department of Commerce (EOA), Solar Heating Application Laboratory, Colorado State University, Fort Collins, CO, 1977.

[5] *A energia do sol* (Solar Energy), catalog, Heliodinâmica, São Paulo, Brazil, 1990.

[6] N. Hadjsaid. J. F. Canard, and F. Some, Computer applications in power systems, *IEEE Power Engineering Society Magazine*, Vol. 12, No. 2, April 1999.

[7] L. M. Auerbach, *A Home Site Power Unit: Methane Generator*, Report on Alternative Energy Systems, Madison, WI, 1974.

[8] M. R. Patel, *Wind and Solar Power Systems*, CRC Press, Boca Raton, FL, 1999.

[9] A. W. Culp, *Principles of Energy Conversion*, McGraw-Hill, Toronto, Ontario, Canada, 1979.

[10] KanEnergi, *New Renewable Energy: Norwegian Developments*, Research Council of Norway in cooperation with Norwegian Water Resources and Energy Directorate, Hanshaugen, Oslo, Norway, December 1998.

[11] http://www.astm.org/.

[12] http://nssdc.gsfc.nasa.gov/space/model/atmos/us_standard.html.

[13] M. Romero, M. J. Marcos, R. Osuna, and V. Fernández, Design and implementation plan of a 10-mW solar tower power plant based on volumetric-air technology in Seville (Spain), in *Proceedings of the Solar 2000 Solar Powers Life—Share the Energy*, Madison, WI, June 17–22, 2000.

[14] J. Ross, On a mission of towering heights, http://www.aie.org.au/pubs/enviromission.htm, November 2004.

[15] M. Werder and A. Steinfeld, Life cycle assessment of the conventional and solar thermal production of zinc and synthesis gas, a solar process technology, *Energy*, Vol. 25, pp. 395–409, 2000. Paul Scherrer Institute, Villigen PSI, Switzerland; Swiss Federal Institute of Technology, Department of Mechanical and Process Engineering, Institute of Energy Technology, Zürich, Switzerland.

[16] R. Kistner and H. W. Price, Financing solar thermal power plants Paper RAES99-7727, *Proceedings of the Conference on Renewable and Advanced Energy Systems for the 21st Century*, Lahaina, Maui, HI, April 11–15, 1999.

[17] A. Weimer, J. K. Dahl, K. Buechler, A. Lewandowski, and C. Bingham, Solar-thermal processing of methane to produce hydrogen and syngas, *American Chemical Society Journal on Energy and Fuels*, Vol. 15, No. 5, pp. 1227–1232,

[18] V. Quaschning and F. Trieb, Solar thermal power plants for hydrogen production, in *Proceedings of the Hypothesis IV Symposium*, Stralsund, Germany, September 9–14, 2001, pp. 198–202.

[19] A. Chambers, C. Park, R. T. Baker, and N. M. Rodriguez, Hydrogen storage in graphite nanofibers, *Physical Chemistry*, Vol. 102, pp. 4253–4256, 1998.

[20] C. Dey and M. Lenzen, Greenhouse gas analysis of electricity generation systems, in *Conference Proceedings of ANZSES Solar 2000*, Grifith University, Queensland, Australia, November 29–December 1, 2000, pp. 658–668.

[21] *Força Energética*, Vol. 1, No. 2, July–August 1992.

[22] *The Leading Position in a Growing Market*, Advanced Power Systems catalog, Neste, 1996.

[23] Intermediate Technology Development Group, *Handson: It works*, TVE-Series, Schumacher Centre for Technology and Development, Rugby, Warwickshire, England, 1999.

CHAPTER 6

PHOTOVOLTAIC POWER PLANTS

6.1 INTRODUCTION

Solar light is converted directly into electricity with modules consisting of many photovoltaic solar cells. Such solar cells are usually manufactured from fine films or wafers. They are semiconductor devices capable of converting incident solar energy into dc current, with efficiencies varying from 3 to 31%, depending on the technology, the light spectrum, temperature, design, and the material of the solar cell (see Table 6.1). A solar cell could be understood simply as a battery of very low voltage (around 0.6 V) continually recharged at a rate proportional to the incident illumination. The series–parallel connection of cells allows the design of solar panels with high currents and voltages (reaching up to kilovolts). In order to implement a full electric power system, it is necessary to include power electronic conditioning equipment, energy storage and monitoring plus protection devices.

The most attractive features of solar panels are the nonexistence of movable parts, very slow degradation of the sealed solar cells, flexibility in the association of modules (from a few watts to megawatts), and the extreme simplicity of its use and maintenance. In addition, solar energy is a very relevant source, with characteristics such as: It is autonomous, its operation does not pollute the atmosphere (i.e., it does not harm any ecosystem), and it is an inexhaustible and renewable

Integration of Alternative Sources of Energy, by Felix A. Farret and M. Godoy Simões

TABLE 6.1 Commonest Materials Used in PV Modules

Type	Theoretical Efficiency		Practical Tests,	Modules	
	cm^2	η (%)	η (%)	cm^2	η (%)
Monocrystalline silicon (Si)	4	29	23	100	15–18
Polycrystalline silicon (Si)	4		18	100	12–18
Amorphous silicon (a-Si)	1	27	12	1000	5–8
Gallium arsenide (GaAs)	0.25	31	26		
Copper indium–selenide (CIS)	3.5	27	17		
Cadmium telluride (CdTe)	1	31	16		

Source: Ref. [1].

source with great reliability. However, up to now, manufacturing costs have been a major impediment to its widespread use. Power electronics has been the great enabling technology in the diversification of solar energy applications. This chapter covers basic notions of the use of solar electrical energy for direct and indirect energy production in electrical power plants.

6.2 SOLAR ENERGY

The sun is of great importance for the planet Earth and the ecosystem of our society. Rays emitted by the sun, gamma rays, reach the terrestrial orbit a few minutes after they leave the sun surface, crossing approximately 150 million kilometers. Clouds reflect about 17% of sunlight back into space, 9% is scattered backward by air molecules, and 7% is actually reflected directly off the surface back into space. Therefore, the travel through the atmosphere decreases the radiation at Earth's surface to about 35% less than the level in the stratosphere. At noon on a clear day, the luminous power at the ground level is approximately 1000 watts per square meter (or 1 sun = $1000\ W/m^2$).

The photovoltaic (PV) industry, in conjunction with the American Society for Testing and Materials (ASTM) (http://www.astm.org) and U.S. government research and development laboratories, developed and defined two standard terrestrial solar spectral irradiance distributions. These two spectra define (1) a standard direct normal spectral irradiance (ASTM E-891) and (2) a standard total (global, hemispherical, within a 2π steradian field of view of the tilted plane) spectral irradiance (ASTM E-892). The direct normal spectrum is the direct component contributing to the total global (hemispherical) spectrum [2,3].

The spectra represent terrestrial solar spectral irradiance on a surface of specified orientation under a set of specified atmospheric conditions (Figure 6.1). These distributions of power (watts per square meter per nanometer of bandwidth) as a function of wavelength provide a single common reference for evaluating spectrally selective PV materials with respect to performance measured under varying natural and artificial sources of light with various spectral distributions. The conditions

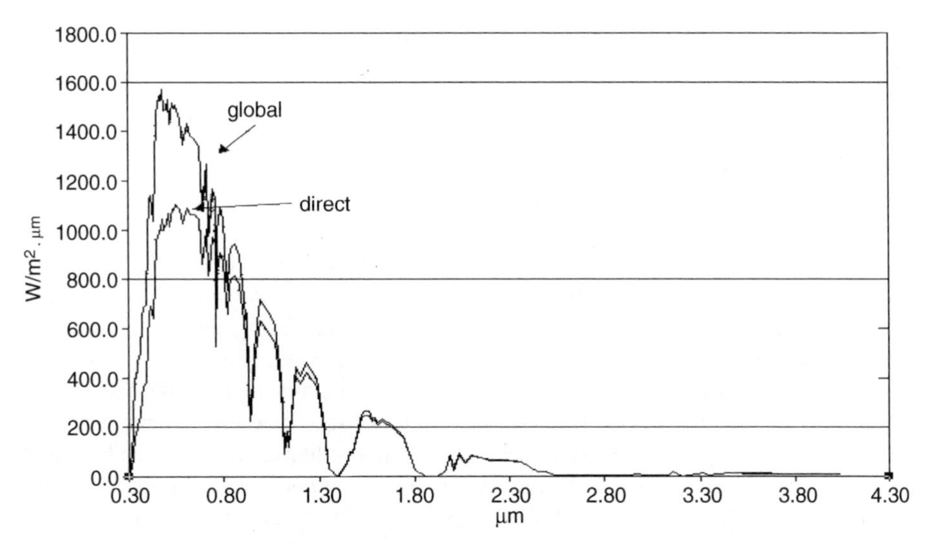

Figure 6.1 Direct and global radiation, 37° tilt (ASTM E-892 and E-891).

selected were considered to be a reasonable average for the 48 contiguous states of the United States over a period of one year. The tilt angle selected is approximately the average latitude for the continental United States.

Global solar radiation also includes thermal energy, whose use consists of domestic and industrial use of heat, for either heating fluid or for transformation to another form of energy, as discussed in Chapter 5. The *photovoltaic effect* consists of the direct transformation of the radiant energy into electricity without the intermediate production of heat.

Several countries in the tropical regions are privileged by their solar energy potential (e.g., Brazil has about 2500 MW, five times that of the United States). So there are enormous possibilities for the use of solar energy as a thermal and photovoltaic source.

Semiconductor characteristics define the electrical energy conversion efficiency of solar cells. The material energy gap, manufacturing quality of the cell material, and technology employed in the manufacturing process sum up to a final efficiency. Usually, the conversion efficiency of a commercial solar cell (in other words, the relationship between the electric power generated and the power of the incident radiation on the semiconductor) is of 10 to 15%. For instance, at noon on a clear day, a photovoltaic module can generate the average of 100 W/m^2.

Crystalline silicon and hydrogenated amorphous silicon are currently the most appropriate semiconductor materials for photovoltaic conversion. However, it is already possible in laboratories to produce solar cells of crystalline silicon with up to 25% conversion efficiency. Hydrogenated amorphous silicon is obtained from 10 to 12% of efficiency in the laboratory and 7 to 8% in modules produced industrially. Manufacturing costs are less than for the crystalline silicon cell. Monocrystalline silicon cells are efficient, reliable, durable, and expensive. Less expensive technologies are polycrystalline silicon of a low degree of purity or hydrogenated amorphous silicon [1,4,5].

6.3 GENERATION OF ELECTRICITY BY PHOTOVOLTAIC EFFECT

Semiconductor materials have bands of allowed and forbidden energy in their spectrum of electronic energy (the energy gap). Inside the allowed band, there are valence and conduction bands, separated by such an energy gap. The electrons occupy the valence band and can be excited in the conduction band by thermal energy or by absorption of photons with energy quantum higher than the energy gap. The bandwidth of the energy gap is characteristic for each semiconductor.

When an electron passes from one band to other, it leaves in its place a hole that can be considered a positive charge. When voltage is applied across the semiconductor, the electrons and their holes contribute to the electrical current, since the presence of that electric field makes those particles move in opposite directions with respect to each other. Therefore, an electrostatic potential inside the material is created to separate positive from negative charges. Now, when the semiconductor is illuminated, it behaves like a battery; in other words, the charges accumulate in opposite areas of the chip. If an external wire connects the two areas, there will be a circulation of electric current. In solar cells made of crystalline silicon, the processes of controlled and selective contamination of the semiconductor material originate the electric field [5–8].

The equivalent electrical circuit of a single cell is represented in Figure 6.2, from where the following equation can be derived:

$$I_o = I_\lambda - I_d - I_p \tag{6.1}$$

where I_λ is the photon current, which depends on the light intensity and its wavelength; I_d the Shockley temperature-dependent diode current; and I_p the PV cell leakage current. The photon current is proportional to the illumination intensity

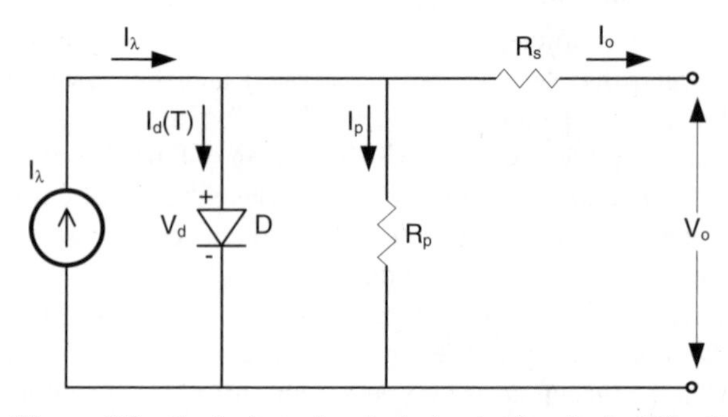

Figure 6.2 Equivalent electrical circuit of a single PV cell.

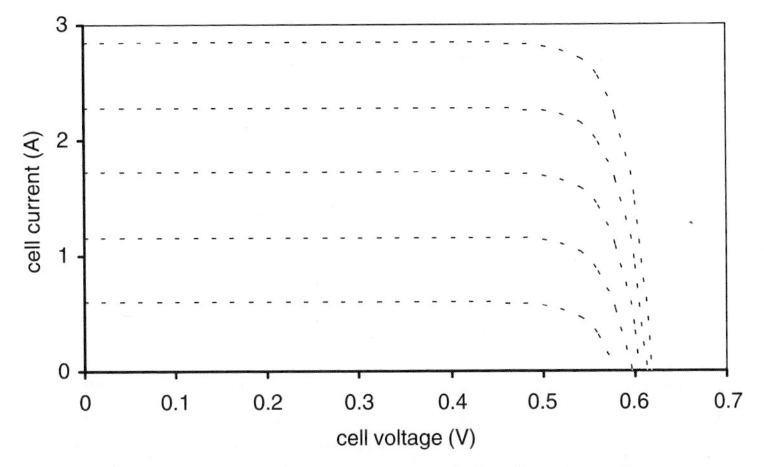

Figure 6.3 Typical *I–V* characteristic of a solar cell.

and depends on the light wavelength, λ. The parameters of this current are related to the cell short-circuit current, I_{sc}, and to the cell open-circuit voltage, V_{oc}. The short-circuit current may be obtained from the I–V characteristic for a given solar cell (Figure 6.3) when the cell output voltage is $V_o = 0$. In its turn, the open-circuit voltage is obtained for zero output current, $I_o = 0$ (no load).

In addition, if the photon current is known for certain standard illumination intensity, say $L_s = 1.0$ sun and a prescribed value of $I_{\lambda o}$, it is possible to obtain the photon current approximately for every other level L through the expression

$$I_\lambda = \frac{L}{L_s} I_{\lambda o} \tag{6.2}$$

The Shockley diode current is given by the classical expression

$$I_d = I_s(e^{qV_d/\eta kT} - 1) \tag{6.3}$$

where I_s = reverse saturated current of the diode, typically 100 pA for the silicon cell

k = 1.38047×10^{-23} J/K is the Boltzmann constant

q = 1.60210×10^{-19} C is the electron charge

V_d = diode voltage (volts)

η = empirical constant

T = $273.2 + t_C$ is the absolute temperature given as a function of the temperature (°C), t_C, generally taken as $T = 298$ K (i.e., 25°C)

q/kT = 38.94452 C/J for $t_C = 25$°C; or in a more general way, for any temperature, $q/k = 11605.4677$ C.

A parallel resistance R_p can represent the internal losses, or leakage current, across the Shockley diode. These values usually range between 200 and 300 Ω. Also, there is a series resistance between the photon current source and the load across the photovoltaic cell terminals, R_s. The usual value of this resistance is very small (0.05 to 0.10 Ω), reflecting directly on the manufacturing quality of the PV cells.

Under these circumstances, equation (6.1) becomes

$$I_o = I_\lambda - I_s(e^{qV_d/kT} - 1) - \frac{V_d}{R_p} \tag{6.4}$$

The diode voltage is $V_d = I_o(R_s + R_L) = V_o(1 + R_s/R_L)$, being a function of the load resistance, R_L, and for this reason, of the cell output power. Therefore, substituting V_d from Figure 6.3 and using equation (6.4), more precise results for the output current can be obtained from

$$I_o = \frac{R_p}{R_p + R_s + R_L}[I_\lambda - I_s(e^{qV_d/kT} - 1)] \tag{6.5}$$

For large values of the load resistance, R_L, with respect to the parallel resistance, R_p, of the cell, differences in the output current values may become remarkable.

It is important to use an empirical factor, η, in the exponential term of either equation (6.4) or (6.5), so that they can be adjusted to the practical data from the manufacturers. The adjusted forms of these equations are, respectively,

$$I_o = I_\lambda - I_s(e^{qV_d/\eta kT} - 1) - \frac{V_d}{R_p} \tag{6.6}$$

or

$$I_o = \frac{R_p}{R_p + R_s + R_L}[I_\lambda - I_s(e^{qV_d/\eta kT} - 1)] \tag{6.7}$$

As shown in Figure 6.3, the diode voltage and the open-circuit voltage, V_{oc}, change with the load current. For the very particular case when $I_o = 0$ in an illuminated panel, the single-cell open-circuit voltage may be obtained from equation (6.6) as

$$V_{oc} = V_d|_{I_o=0} = (I_\lambda - I_d)R_p = [I_\lambda - I_s(e^{\eta qV_{oc}/kT} - 1)]R_p$$

where the logarithmic simplified form is

$$V_{oc} = \frac{kT}{\eta q}\ln\left(1 + \frac{I_\lambda}{I_s} - \frac{V_{oc}}{I_s R_p}\right)$$

This is a transcendental equation that can be solved numerically. However, as V_{oc} is on the order of 0.6 V, R_p is on the order of 300 Ω and has opposite sign. Then

$$V_{oc} \cong \frac{kT}{\eta q} \ln \frac{I_\lambda}{I_s} \tag{6.8}$$

As the series resistance of a PV cell is very small (on the order of 0.1 Ω), when the cell is short-circuited, practically the only opposition to the current through is R_s. So the short-circuit current can be given by

$$I_{sc} \cong I_\lambda \tag{6.9}$$

The output power of the cell is the product of V_o by the output current given by equation (6.7), which gives

$$P_o = \frac{V_o R_p}{R_p + R_s + R_L} [I_\lambda - I_s(e^{qV_o(1+R_s/R_L)/\eta kT} - 1)] \tag{6.10}$$

The maximum power may be obtained by differentiating equation (6.10) with respect to V_o and setting the derivative equal to zero to find the external load voltage V_{om} for the maximum output power of the cell, which must satisfy

$$I_s e^{qV_{om}(1+R_s/R_L)/\eta kT} = \frac{I_\lambda + I_s}{1 + V_{om}^{q(1+R_s/R_L)/\eta kT}}$$

Another way of doing that is by plotting equation (6.10) and obtain its maximum graphically. Nevertheless, the maximum power may be given by [16].

$$P_m = \frac{V_{om}(I_s + I_\lambda)}{1 + \eta kT/qV_{om}(1 + R_s/R_L)} \tag{6.11}$$

If the incident flux power, P_i, on the cell is known, the conversion efficiency for the maximum power becomes

$$\eta_m = \frac{P_m}{P_i} = \frac{V_{om}(I_s + I_\lambda)}{P_i(1 + kT/\eta q V_{om})(1 + R_s/R_L)} \tag{6.12}$$

6.4 DEPENDENCE OF A PV CELL CHARACTERISTIC ON TEMPERATURE

Solar cells are influenced by temperature in two ways: current gain varies with illumination; in addition reverse-biased diode current and voltage are also affected. These effects are taken into account in the terms $V_T = kT/q$, I_λ, and I_s of Shockley's equation. For the saturation current, a somehow precise expression could be given approximately by [8–11]

$$I_s(T) = KT^m e^{-V_{GO}/\eta V_T}$$

where K = a constant independent on temperature
m = 2 for germanium and $m = 1$ for silicon
η = 1 for germanium and $\eta = 2$ for silicon
V_{GO} = 0.785 V for germanium and $V_{GO} = 1.21\ V$ for silicon and it is known as the equivalent voltage of the prohibited band (eV)
V_T = $kT/q = T/11,605.4677$ is an equivalent voltage to the junction temperature
T = 298°C.

Taking the derivative of the expression above in its logarithm form yields [8–10]

$$\frac{d(\ln I_s)}{dT} = \frac{dI_s/I_s}{dT} = \frac{m}{T} + \frac{V_{GO}}{\eta T V_T}$$

It is a common practice to establish the PV panel parameters with respect to standard (STC) and normal (NOC) operating conditions (Table 6.2). Under standard conditions with the parameters given, the theoretical logarithm per unit variation dI_s/I_s with temperature is about 8%/°C for silicon and 11%/°C for germanium [1]. In commercial diodes these variations are smaller because surface leakage is not as important effect for these diodes as for photovoltaic cells. Therefore, the theoretical values will be smaller, typically 7%/°C for both silicon and germanium diodes. These 7%/°C values may be interpreted as $(1.07)^{10} \cong 2.0$ (i.e., the reverse saturation current doubles for every 10°C). With respect to I_s, some references give this value as if the diode reverse saturation current I_{s0} for temperature T_0 changes at another temperature T [3]:

$$I_s(T) \cong I_{s0} \times 2^{(T-T_0)/10}$$

A more practical approach to this issue is the use of temperature coefficients such as a positive temperature coefficient (PTC) for the current and a negative temperature coefficient (NTC) for the voltage. For the short-circuit current,

$$I_{sc} = I_{sc0}(1 + \alpha \Delta T)$$

TABLE 6.2 PV Panel Parameters

Parameter	Standard Conditions	Normal Conditions
Solar illumination (W/m^2)	1000	800
Cell temperature (°C)	25	42
Solar spectral irradiance	ASTM E-892	ASTM E-892

where α is the positive temperature coefficient of the current. For a typical single-crystal silicon PV cell, α is about 500 µunits/K.

On the other hand, if temperature is increased and the voltage is fixed, the diode current will increase. By taking the derivative from Shockley's equation, it is possible to find an expression for the variation of the diode voltage with its saturation current and, consequently, its variation with the temperature as a difference of two terms. By balancing these two terms through reduction of voltage and increase of current, it may be kept a constant current for both silicon and germanium diodes. Under this condition, it is usually necessary that

$$\frac{dV_d}{dT} \cong -2.5\,\text{mV}/^\circ\text{C}$$

However, this relationship also decreases with an increase in temperature. So it becomes necessary to use a more precise expression for this variation, such as the one given by [3]

$$\frac{dV_d}{dT} = \frac{V_d - (V_{GO} + m\eta V_T)}{T}$$

With the usual values, it may be stated that

$$\frac{dV_d}{dT} \cong -2.1\,\text{mV}/^\circ\text{C for germanium} \quad \text{and} \quad \frac{dV_d}{dT} \cong -2.3\,\text{mV}/^\circ\text{C for silicon}$$

The practical approach would suggest something like $\beta = 5$ µunits/K as the negative temperature coefficient for the cell open-circuit voltage. So the temperature corrected for the open-circuit voltage is

$$V_{\text{oc}} = V_{\text{oc0}}(1 - \beta\,\Delta T)$$

6.5 SOLAR CELL OUTPUT CHARACTERISTICS

Figure 6.4 shows very useful characteristics typically used for control design of power electronic systems connected to photovoltaic arrays: the output power versus the load voltage and the output power versus the load current at different illumination and temperature levels. Voltage-based power control is clearly better, as suggested by Figure 6.4*a*, where the family of curves are smoother and evenly spaced [5,7,9,11].

Cell temperature also plays an important role in the efficiency of solar panels. Figure 6.4*c* and *d* show that for a cell illuminated by 0.6 sun under two different temperatures, $t_1 = 50^\circ$C and $t_2 = -25^\circ$C, the cell short-circuit current decreases whereas the open-circuit voltage increases.

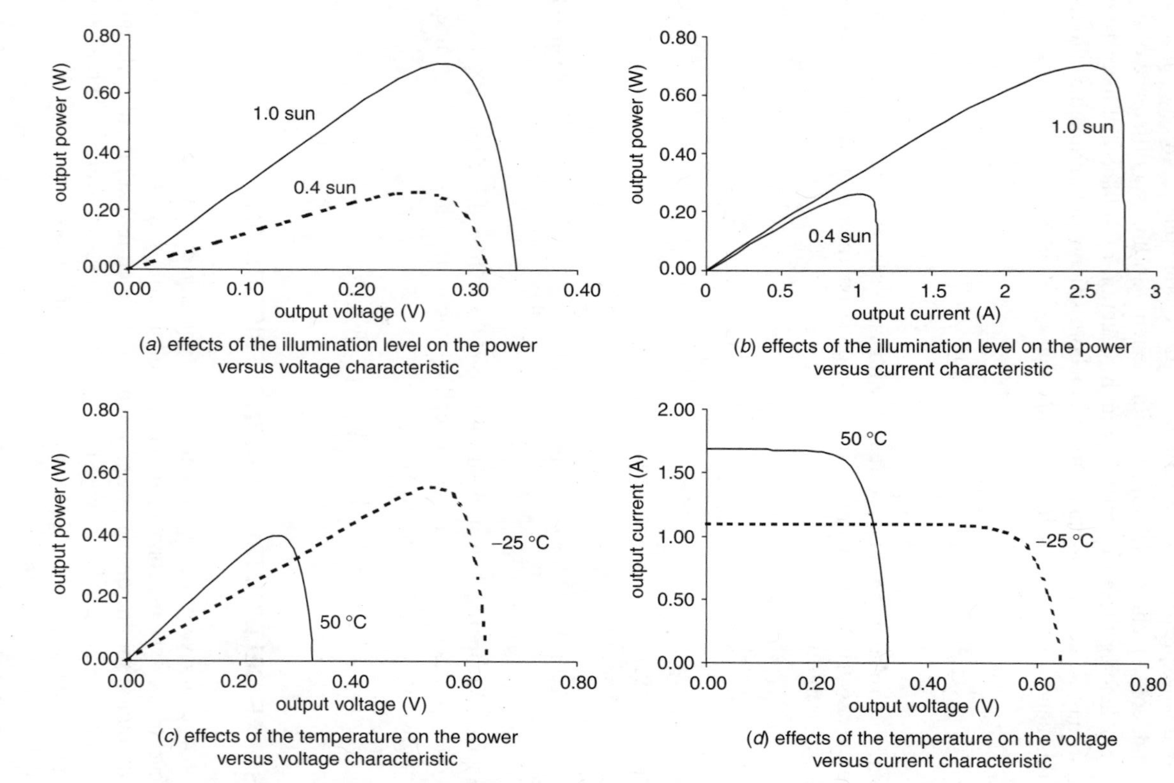

(*a*) effects of the illumination level on the power versus voltage characteristic

(*b*) effects of the illumination level on the power versus current characteristic

(*c*) effects of the temperature on the power versus voltage characteristic

(*d*) effects of the temperature on the voltage versus current characteristic

Figure 6.4 Effects at 0.6 sun of illumination and distinct temperature on output power.

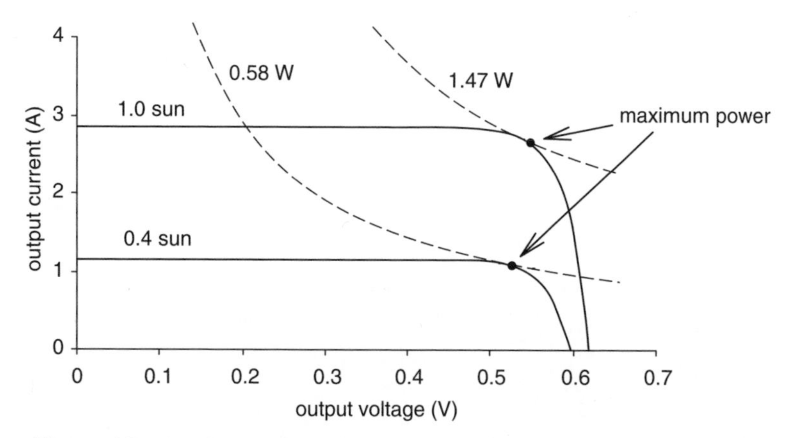

Figure 6.5 Conditions of maximum power for various illumination levels.

Equation (6.2) supports the fact that the output power of a PV cell increases with the illumination level, generating a locus of maximum power. Figure 6.5 represents the maximum output power locus for an individual PV cell. Comparing this figure with Figure 6.4, it is interesting to observe that neither the maximum voltage nor the maximum current are the same for different levels of illumination at the maximum power point. As the output power is given as

$$P_o = V_o I_o$$

then the peak power may be given from the following condition:

$$\frac{dP_o}{dV_o} = V_o + I_o \frac{dV_o}{dI_o} = 0$$

that is,

$$\frac{dV_o}{dI_o} = -\frac{V_o}{I_o} \tag{6.13}$$

The meaning of this expression is that the dynamical internal resistance of the source should match the external load resistance, leading to special power peak tracking control approaches such as those presented in detail in Chapter 11.

6.6 EQUIVALENT MODELS AND PARAMETERS FOR PHOTOVOLTAIC PANELS

Figure 6.6 represents a more detailed model of a photovoltaic cell. Typical values of capacitor C are very low ($\approx$10 pF) and are not usually taken into account in dc

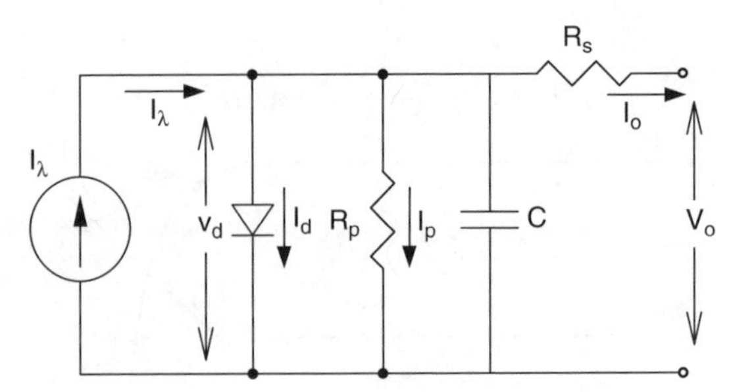

Figure 6.6 Equivalent detailed circuit of a single photovoltaic cell.

current analysis. The other parameters for a silicon photovoltaic cell are typically [5,6,12,13] R_s in the range 0.01 to 1.0 Ω and R_p in the range 200 to 800 Ω. The illumination current is proportional to the radiant luminous intensity of the sun given in lumens (footcandles) (luminous flow) or photocandles (luminous intensity), where 1 lumen = 1.496 exp (−10) W; 1 lumen/ft^2 = 1 footcandle (fc) = 1.609 exp (−12) W/m^2; and 1 sun = 1000 W/m^2.

Dark-current electric parameters of a photovoltaic panel may be taken from laboratory tests.

6.6.1 Dark-Current Electric Parameters of a Photovoltaic Panel

Three parameters need to be selected from the PV characteristic representing three measurement points on the diode curve, as suggested in Figure 6.7. An easy way to obtain the parameters of the circuit in Figure 6.6, also known as dark-current electric parameters, may be established from the panel in a moderately dark condition, that is, deactivating the internal source current of the circuit. With an external voltage

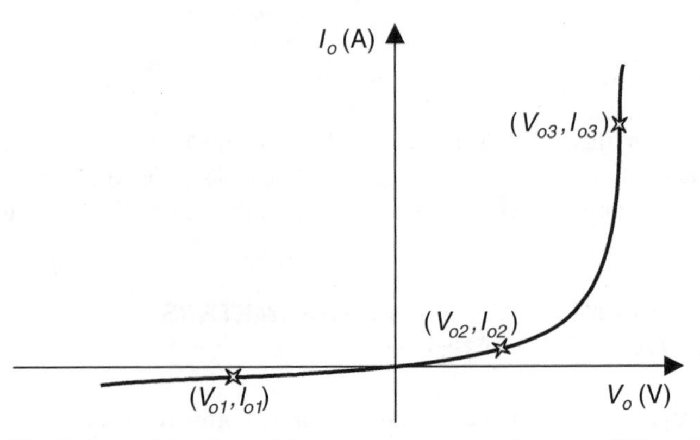

Figure 6.7 Suggested location of three measurement points on a PV panel curve.

source, one has to measure the points (V_{o1}, I_{o1}), (V_{o2}, I_{o2}), and (V_{o3}, I_{o3}) as shown in Figure 6.7 and define

$$\begin{aligned} V_{d1} &= V_{o1} - I_{o1}R_s \\ V_{d2} &= V_{o2} - I_{o2}R_s \\ V_{d3} &= V_{o3} - I_{o3}R_s \end{aligned} \tag{6.14}$$

The selection of these points should cover the most significant portions of Shockley's equation. For better results during the tests, the following locations of points on the curve are suggested: (1) close to the maximum reverse voltage of the cell, (2) close to the rated current of the cell, and (3) close to a tenth of the rated current of the cell. From the set of equations (6.14),

$$R_s = \frac{V_{d1}}{I_{o1} - I_{d1}} = \frac{V_{d2}}{I_{o2} - I_{d2}} = \frac{V_{d3}}{I_{o3} - I_{d3}} \tag{6.15}$$

can be established, from where, together with equation (6.3), we obtain

$$\begin{aligned} I_s &= \frac{(V_{d2}/V_{d1})(I_{o1} - I_{o2})}{(V_{d2}/V_{d1})[\exp(qV_{d1}/\eta kT) - 1] - [\exp(qV_{d2}/\eta kT) - 1]} \\ &= \frac{(V_{d3}/V_{d1})(I_{o1} - I_{o3})}{(V_{d3}/V_{d1})[\exp(qV_{d1}/\eta kT) - 1] - [\exp(qV_{d3}/\eta kT) - 1]} \end{aligned} \tag{6.16}$$

All terms to the right of equations (6.15) and (6.16) are functions of R_s of the measured pairs of values and constants. So from equation (6.16) we define a function of R_s as

$$\begin{aligned} f(R_s) = &\frac{(V_{d2}/V_{d1})(I_{o1} - I_{o2})}{(V_{d2}/V_{d1})[\exp(qV_{d1}/\eta kT) - 1] - [\exp(qV_{d2}/\eta kT) - 1]} \\ &- \frac{(V_{d3}/V_{d1})(I_{o1} - I_{o3})}{(V_{d3}/V_{d1})[\exp(qV_{d1}/\eta kT) - 1] - [\exp(qV_{d3}/\eta kT) - 1]} \end{aligned} \tag{6.17}$$

whose numerical solution for $f(R_s) = 0$ gives two distinct values for R_s. One is discarded for physical reasons, such as when R_s is back-substituted into equation (6.15), it may give a negative solution and distinct values for R_p with extremely large and distinct values of I_s. Once the right solution for R_s and R_p is selected, these values are replaced in equation (6.16) to give two identical solutions for I_s.

The limitations of this method are the correct knowledge of the empirical constant η, which is not usually supplied by the manufacturer, and the extended precision for the calculation because of the exponential functions involved. When a number of individual cells are in a series or parallel arrangement in a PV panel, this number has to be included in the equations of the foregoing method, as discussed below.

6.6.2 Model of a PV Panel Consisting of *n* Cells in Series

When integrating a group of PV cells in series, it is assumed that the cells are identical, resulting in the equivalence evolution of Figure 6.8. This assumption is usually well accepted if it is taken into account that individual cells inside the panel are manufactured with very similar characteristics, to avoid circulation of internal currents among the cells. With this in mind, it is acceptable to make the assumption

$$I_{\lambda 1} = I_{\lambda 2} = \cdots = I_{\lambda n} \qquad R_{s1} = R_{s2} = \cdots = R_{sn}$$
$$I_{d1} = I_{d2} = \cdots = I_{dn} \qquad R_{p1} = R_{p2} = \cdots = R_{pn}$$
$$V_{d1} = V_{d2} = \cdots = V_{dn}$$

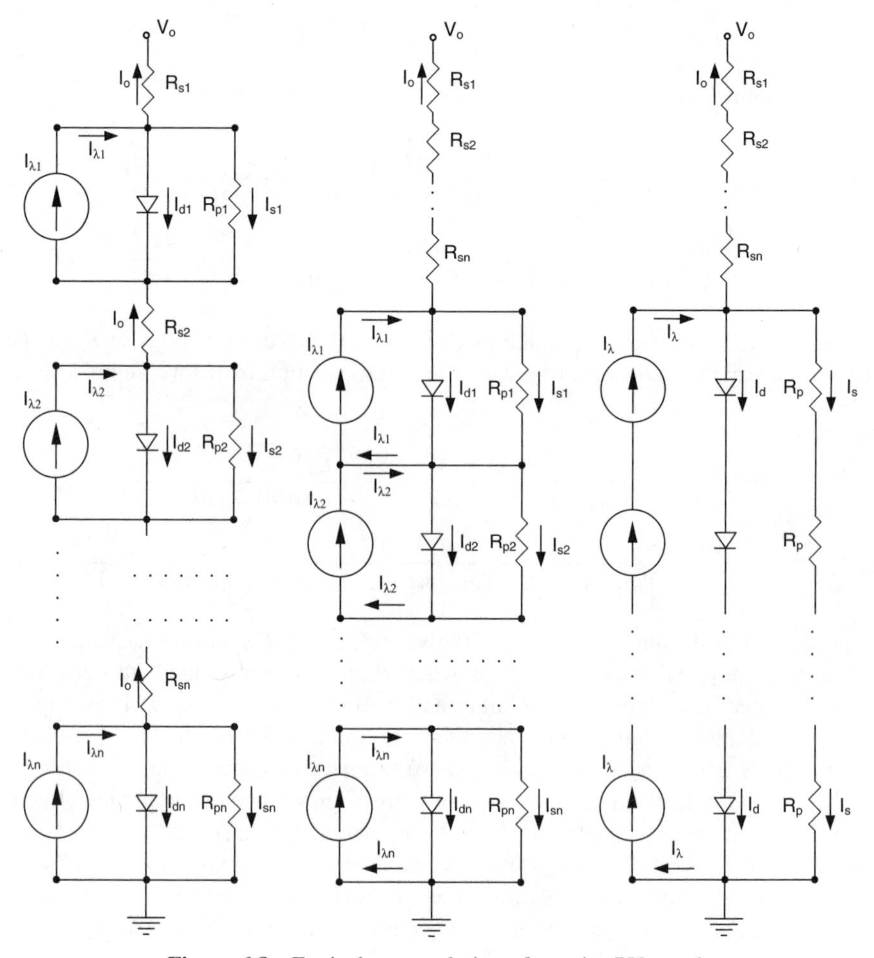

Figure 6.8 Equivalence evolution of a series PV panel.

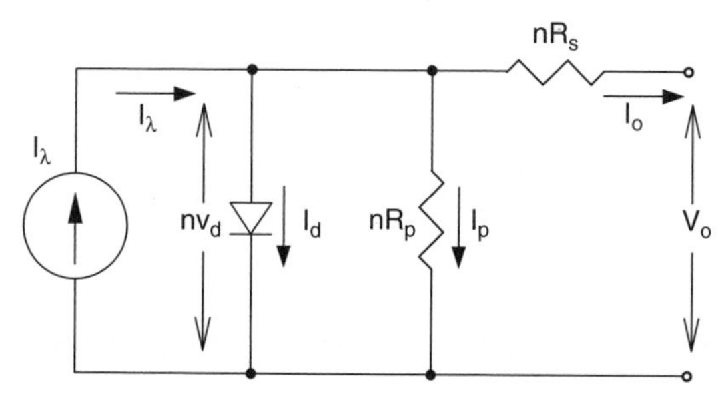

Figure 6.9 Equivalent model of a PV panel with identical cells in series.

The respective output voltage and current of the group are

$$V_o = nV_{oi} \qquad I_o = I_{oi}$$

where V_{oi} and I_{oi} are, respectively, the average voltage and current in the individual cell i.

A more compact form of the equivalent circuit of Figure 6.8 is shown in Figure 6.9. As shown in Figure 6.8, there are four relevant parameters: the efficiency, η, the series resistance, R_s, the parallel resistance, R_p, and the current saturation of the internal diode, I_s. The efficiency of a solar panel may be given for the maximum electrical power by

$$\eta = \frac{P_{\text{electrical}}}{P_{\text{illumination}}} \times 100\%$$

As an example, suppose that the manufacturers data sheet indicates the following: effective illuminating area $= 0.136\,\text{m}^2$, $V_{mp} = 15.6\,\text{V}$, and $I_{mp} = 1.4\,\text{A}$ for the standard test conditions (1000 W/m^2, 25°C, airmass 1.5); then

$$\eta = \frac{15.6(1.4)}{1000(0.136)}(100\%) = 15.74\%$$

In Figure 6.9 and from Shockley's equation it is possible to observe that the voltage across the terminals of an individual cell can be given by

$$V_{oi} = \frac{kT}{q}\ln\left(\frac{I_d}{I_s} + 1\right) - I_o R_s$$

So the output voltage of a panel with identical cells in series will be given by

$$V_o = nV_{oi} = \frac{nkT}{q}\ln\left(\frac{I_d}{I_s} + 1\right) - nI_o R_s$$

Shockley's equation for the current through the diodes then assumes the following form:

$$I_d = I_s\left\{\exp\left[\frac{q(V_o + nI_oR_s)}{nkT}\right] - 1\right\}$$

6.6.3 Model of a PV Panel Consisting of *n* Cells in Parallel

Assuming that the PV cells of a panel are connected in parallel across common terminals *a* and *b* as in Figure 6.10. In these conditions, there is no direct interaction among current sources or individual diode voltages because they are considered exactly the same. Therefore, this equivalent model may assume a more compact form, as shown in Figure 6.11. Tests of this equivalent model can be seen as a test carried out on a single panel with individual output voltage and current for the same cell:

$$V_o = V_{oi} \qquad I_o = nI_{oi}$$

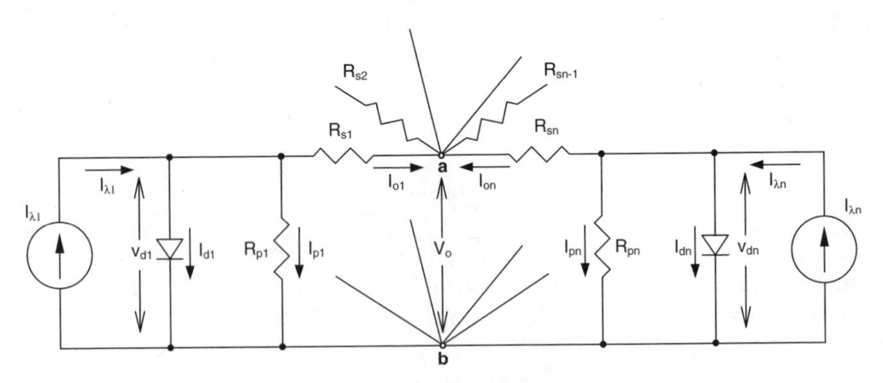

Figure 6.10 Group of identical PV cells connected in parallel.

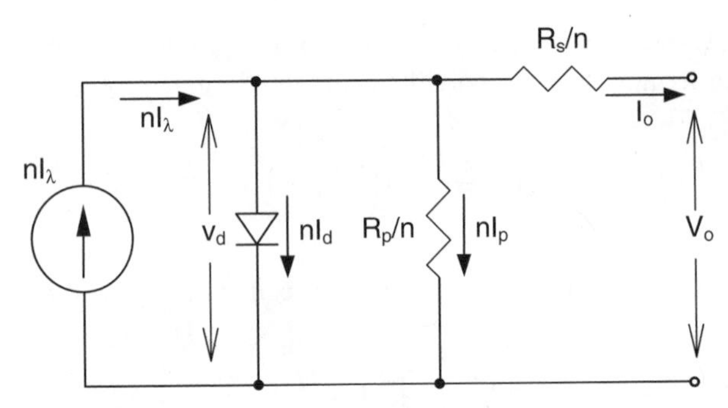

Figure 6.11 Equivalent model of a photovoltaic panel with cells in parallel.

The method just described can be extended to equivalent models of series–parallel combinations of PV panels or other arrangements of interest.

6.7 PHOTOVOLTAIC SYSTEMS

A photovoltaic electric system consists of photovoltaic modules, an electronic converter, a controller, and a bank of batteries. Each module is made of photovoltaic cells connected in series, parallel, or series/parallel, as part of a panel supported by a mechanical structure. A power converter may be used to regulate the flux of energy from the panel to the load so as to assure energy quality within certain limits. The electricity converted from solar modules should be stored in banks of batteries of high capacity to make the overall system feasible economically. The bank of batteries should store enough energy to minimize the night and/or low-light-intensity periods.

Maintenance of the modules should be limited to a wipe with a damp cloth, always using maximum care. Getting leaves or deposits on the panels should be avoided because these are very sensitive structures. The electric system should be installed in the following way:

1. Provide equipment grounding with a copper pole ground busbar of at least 2 m in length.
2. Connect the negative conductor, usually black in color with an alligator claw, to the ground bar.
3. Connect the other end of the negative cable, black in color, to the battery's negative terminal.
4. Connect the negative conductor of the solar module to the negative pole of the electric fence equipment.
5. Connect the positive conductor of the solar module to the equipment's positive pole.
6. Connect the positive cable of the electric fence to the positive terminal of the battery.
7. Connect the equipment's electric discharge conductor to the galvanized wire of the fence.

Several other aspects should also be considered at the design stage of a photovoltaic-based power supply system [14–18]:

1. The site should be open, without constant shadows for the solar light.
2. Use one or more solar modules.
3. Use structures or a support of aluminum.
4. Use a load controller.
5. Use a bank or set of batteries.
6. Consider using an indicator of the level of battery charge.
7. Use points of load consumption.

6.7.1 Illumination Area

Solar electric power generation is based on incident sunlight rather than heat. Therefore, for satisfactory efficiency, it is necessary that the solar panels stay exposed directly to sunshine most of the day, not exposed to tree shadows or buildings or other sorts of shielding. On cloudy days there may be some electric power generation at lower levels.

As discussed in Section 6.6, a solar panel consists of several parallel strings of many cells connected in series. These parallel arrays should match each other as closely as possible so as to avoid internal current circulation through the cell series. Any unbalance in these internal voltages should be avoided. Although all precautions are taken during manufacture of the units, external causes may be a source of imbalance over the string voltages. A very common problem with photovoltaic panels is to have an unexpected shadow over its surface. Common causes of shadow over solar panels are birds flying around, planes, its own structure, passersby, stray leaves, and wind-carried materials. Such transient effects can be analyzed using the circuit of Figure 6.12. Under no light, the photocurrent is zero ($I_\lambda \cong 0$), and only the internal diode, R_s and R_p, remains as a path for the current impinged from other active cells. The voltage of the healthy cells will supply current to the shadowy string cells, dissipating internally valuable energy. In a series group of these cells under shadow, if more cells are shadowed, they will go beyond a certain critical voltage value ($I_\lambda \rightarrow 0$), and the current through the association of all these internal resistances will act as a heat source, decreasing the overall efficiency of the solar panel.

To avoid the effect of having partially shadowed cell arrays, most modern solar panels are manufactured so as to divide the strings into several substrings with bypass diodes under the total string current, as illustrated in Figure 6.12 for a shadowed PV cell group. This improves performance because the string power still flows through the bypass diodes.

6.7.2 Solar Modules and Panels

Solar modules are a group of cells that transform the radiant light in dc electric power, usually in standard connections of 12, 24, or 48 V. They are usually composed of mono crystalline silicon cells, protected by antireflexive glass and by a special synthetic material. The number of modules to be used in a system is determined by the electric power consumption needs. There are several commercial arrays for many purposes, as can be seen in Figure 6.13.

6.7.3 Aluminum Structures

Light aluminum structures act as supports for the mechanical assembly of solar modules. Their purpose is to position the solar cells, preventing movement of the modules and maximizing the sun's radiation.

The normal line to the panel face must be directed to the geographical south on northern hemisphere, or the north in the southern hemisphere. The latitude of the

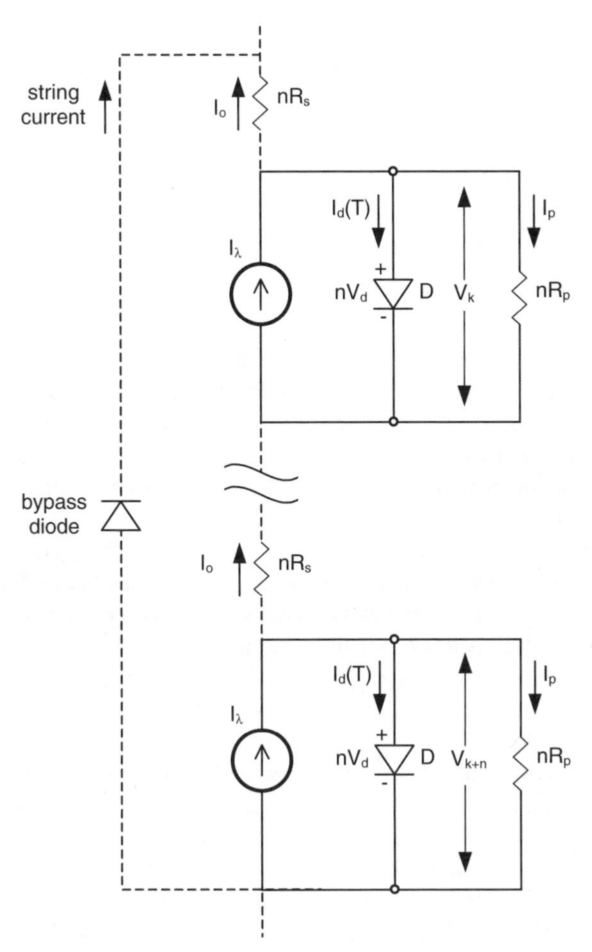

Figure 6.12 Bypass diode in a series PV cell connection.

Source: Siemens Solar Industries [17]

Figure 6.13 Several arrays of commercially available CSi solar panels.

TABLE 6.3 Panel Angle Based on Terrestrial Latitude

Latitude (deg)	Panel Angle (deg)
0–15	15
20	20
25	25
30	35
35	40

site where the solar power plant is to be installed determines the inclination of the panel. Besides, it is more advantageous to tilt the panel perpendicularly to choose the maximum radiation. The maximum energy would be obtained if the perpendicular position of the panel follows the sun in its daily and seasonal variations of inclination, although that may not be very practical because of the high cost of mechanical trackers.

It should also be noted that the sun's inclination varies from April to September from 5° to 23° in latitude above and below the same perpendicular line of the site's latitude. In industrial and other cases where as much power is needed in the winter as in the summer, it is better to set the inclination of the panel to the angle of the latitude of the site. Otherwise, the position selected for the panel should follow the seasonal power needs of the facilities. Table 6.3 indicates the suggested inclinations for solar panels based on terrestrial latitude.

A simple method of determining the best inclination for a given location is the use of a shadow diagram of a perfectly vertical stem tip (for best and easy results, use a bricklayer's plumb line) on the soil. Stem tip shadow positions are marked on the ground, let's say, from 10:00 to 15:00 hours. Then a line is drawn connecting these points. The perpendicular to this line will show the north–south direction being the rod shadow positioned to the north sides if the place is in the northern hemisphere. Note that there will be a small direction deviation in relation to magnetic north.

6.7.4 Load Controller

The load controller is discussed in more detail in Chapter 12. It optimizes and conforms to the transfer of photovoltaic energy generated by the modules in order to match consumption needs. It also protects the battery against excessive overcharging and overdischarging.

6.7.5 Battery Bank

Banks of batteries accumulate the electric power generated during the day, making it available whenever necessary. They usually consist of lead–acid deep-cycle batteries, preferably sealed and possibly with no need of water replacement (although those batteries are more expensive), with a nominal voltage rated at the dc-link voltage. Further details of battery banks, maintenance, and sizing are provided in Chapter 11. A battery bank also requires a healthy monitoring system to indicate the state of charge and to support the maintenance.

6.8 APPLICATIONS OF PHOTOVOLTAIC SOLAR ENERGY

There are tremendous opportunities for photovoltaic applications, such as for residential power and illumination, public illumination, camping, stroboscopic signaling, fluvial and marine embarkations, electric fences, telecommunications, telephone booths in remote areas, water supply, conservation of food and medicines, control of plagues, and general electric power supply [2,3,19,20].

6.8.1 Residential and Public Illumination

Several options are manufactured for the residential and public illumination market. They are available in a varied number of solar modules that provide illumination in a level and amount adapted for each need. There are commercial fluorescent and dulux lighting systems ranging from 9 to 22 W and low-power incandescent illumination lamps from 45 to 110 W. Traditional manufacturers are Philips, Osram, and General Electric.

A typical home lighting commercial kit consists of a 9 W/11 W/15 W energy-efficient compact fluorescent light luminary, solar photovoltaic module, battery, and a charge controller unit. The PV panel charges the battery during the daytime, and the energy thus stored in the battery is used to power the lights whenever required. The bulb controller very often has a built-in high-efficiency electronic inverter to reduce energy losses and noise and provide more lighting time. A charge controller unit is used to control the charging of the battery. It comes with electronics, which controls overcharging and overdischarging of the battery, which in turn increases battery life.

Solar energy can provide a good illumination solution for streets, blocks and parks. Electric power is obtained from solar panels and stored in a system of charged batteries through a load controller. Such systems are usually installed individually on top of each illumination pole with high-efficiency compact bulbs. Solar photovoltaic street light systems use the efficient 11 W/18 W compact fluorescent light providing illumination equivalent to an 60 W/75 W/100 W/150 W incandescent light bulb. These systems are equipped with high-efficiency, high-frequency electronic inverters to reduce energy losses and noise and provide lighting for a longer period. The system does not require any external power supply since the solar panel is inherently safe for public use and requires less frequent maintenance.

A truly portable solar lantern is also possible with an energy-efficient compact fluorescent bulb, battery, and electronics all placed in a compact housing and packaged along with a PV module. This solar lantern is equipped with a high-efficiency electronic inverter and a charge controller with deep discharge and overcharge protection to enhance battery life without need of maintenance.

Also, solar energy is used as a practical and economical source of illumination in camping areas, maintaining the charge of the battery system or being transformed to alternative energy at compatible voltage patterns for the most varied domestic means available in the market [19–21]. For fluvial and marine embarkations and similar types of movable loads, the generated energy is indicated for those cases where there is need of a simple, compact, light, and efficient energy system without need of fuels.

6.8.2 Stroboscopic Signaling

Photovoltaic systems allow electric power provisioning and signaling in remote places. They are used for warning signals in places of difficult access or where no electric grid connection is available. They are suitable for signaling air and marine traffic, highway traffic, and in the signaling of towers, antennas, buildings, and so on. They have been widely employed in developed countries such as the United States and in the load supply in battery banks and in signaling buoys of ports, lights, and oceanographic rafts. Their principal advantages are their low maintenance needs, good reliability, and moderate operational cost, alleviating the need for periodic technical inspections.

The modern high towers built to integrate mobile telephone networks all around the world benefit from the use of power modules composed of solar panels, a controller, and batteries. These modules are lightweight, long-lasting, autonomous, efficient, and suitable for many other applications that use high tower supports, such as for TV, radio, and satellite installations.

Photovoltaic systems are also used for signaling in road infrastructure. These systems, involving a solar panel, electronic controllers, and a backup battery, are often installed in remote areas where only limited skilled labor is available for maintenance. The battery must be able to withstand daily shallow cycles and seasonal deep cycles of operating temperatures from −30 to +50°C. There are batteries designed especially for these applications.

6.8.3 Electric Fence

Solar energy can be useful for cattle confinement, eliminating the need to change batteries or to install an electric network close to a pasture. Even in the absence of solar light for long periods, the simplest systems can maintain autonomy for up to a week. Twisted flexible copper wires are used in electric fences. The use of very thin wires and fence wires themselves should be avoided, because they cause unnecessary losses and can easily suffer rust or oxidation (isolation). Electric supports for a fence can be fixed directly on poles with the aid of handles. It is necessary to adjust the system for the best possible inclination angle of the panel with respect to the latitude. Batteries should be installed inside plastic, wood, or concrete cabinets to protect them from weather, humidity, and sludge. After connecting the battery terminals, they should be covered with grease or vaseline to avoid sulfating.

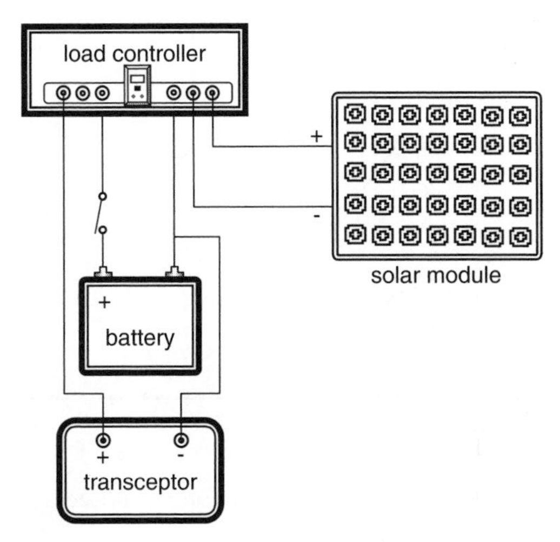

Figure 6.14 Telecommunication systems fed by solar modules.

6.8.4 Telecommunications

Photovoltaic systems for telecommunication applications must supply energy with long autonomy, to overcome long absences of sunlight. In many countries, telephone booths along highways use solar panels to avoid long passages of the feeding electric lines. The telephone signals are transmitted via radio transceptors, whose modules are fed through load controllers (see Figure 6.14). There are many installations like this in the world: for example, the electrification in Morocco of 22 control stations of the Mahgreb–Europe gas pipeline from the Algeria–Morocco border to the Strait of Gibraltar. The following equipment was installed at each station: a 8.5-kWp solar generator, 4000-Ah stationary batteries, a regulation and control system, and a galvanized steel mounting structure. Another example is the Radiolink network of the Dirección General de Telecomunicaciones (Spanish Telecommunications Authority), which covers all of Spain. Installed at each station were a b-kWp solar generator, 4000-Ah stationary batteries, a regulation and control system, a galvanized steel mounting structure integrated in the cover, and other pieces of equipment.

6.8.5 Water Supply and Micro-Irrigation Systems

Solar pumping systems are designed especially for water supply and irrigation in remote areas where no reliable electricity supply is available. Rural applications of solar energy are very valuable, despite the high cost [5,6]. Small villages and towns far from the electric grid may invest in irrigation systems or potable water supply powered by water pump and solar modules such as the ones recommended by the California Energy Commission (http://www.consumer-energycenter.com) in the Californian Emerging Renewables Program. The flow

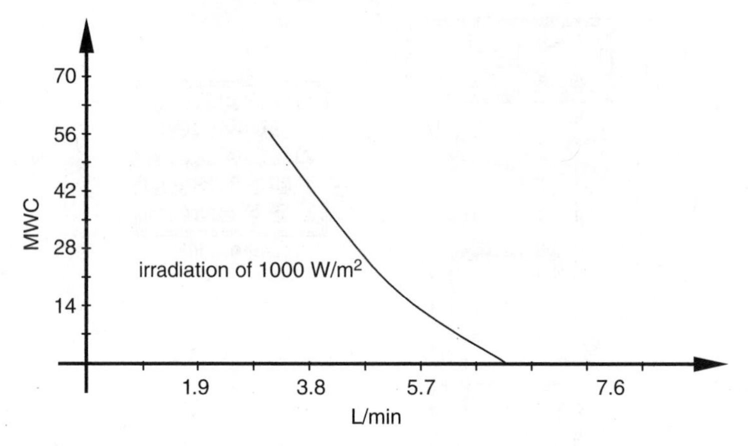

Figure 6.15 Pressure–flow characteristic for a surface water pump.

characteristic is inversely proportional to the pumping head, as in any conventional system of water pumping. For a surface pump, a performance curve such as that shown in Figure 6.15 may be used. An example is illustrated of a hydropump reaching a flow level of 5.7 L/min for 14 m of water column (MWC) for typical days of $1000\,W/m^2$ radiation (i.e., noonday sun). Therefore, the maximum suction head should not exceed 1.5 m. The solar surface water pumping system is usually a solar photovoltaic system, with an energy-efficient dc surface centrifugal pump coupled with about a 1-kW PV panel array. This system is typically used to provide potable water, water for crop irrigation, and water for livestock. For the example shown in Figure 6.16, the connection of a surface pump to solar modules is very simple.

Submersible pumps manufactured from stainless steel and bronze with external diameters of up to 100 mm (see Figure 6.17) can be used. The flow is also inversely proportional to the pumping heads. For instance, in Figure 6.18, for a flow of 125 L/h (2.08 L/min), a head of 32 m of water column can be reached. For higher efficiency with solar panels, a current amplifier should be used.

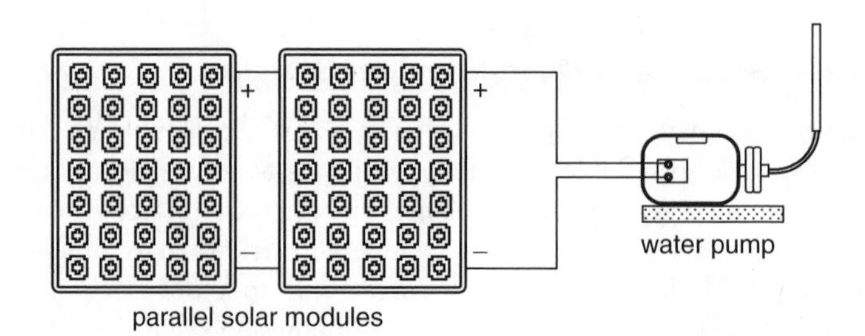

Figure 6.16 Surface water pump using solar modules.

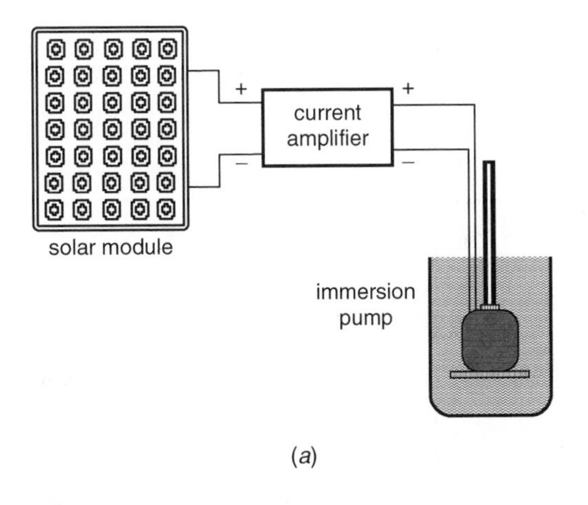

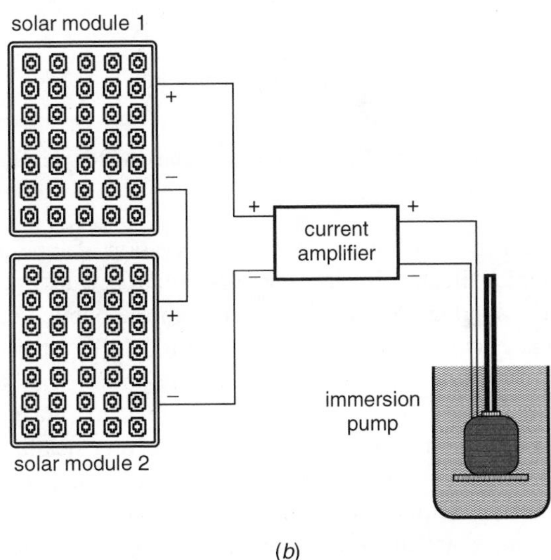

Figure 6.17 Powering a submersible pump with solar modules: (*a*) operation at 12 V; (*b*) operation at 24 V.

6.8.6 Control of Plagues and Conservation of Food and Medicine

Electric, audible, or burning high-voltage traps can easily be installed at the focus points of plagues. Photovoltaic energy provides a very cost-effective, clean, and silent autonomy for this kind of application.

Photovoltaic refrigerators are designed especially for low consumption, fed by solar energy. The starting operation is similar to that of an ordinary refrigerator except for the power source, which is fed from 12-V batteries of 300 or 600 Ah for 75 L (from 4 to 7°C) or 150 L (from 0 to −4°C) for four or six solar modules, respectively, working for 24 hours of the day.

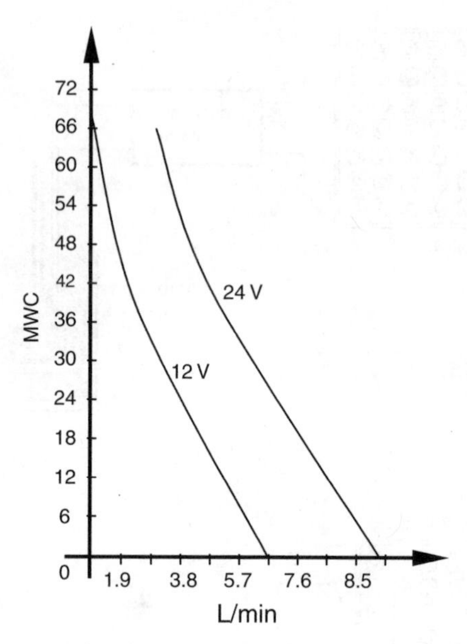

Figure 6.18 Characteristic of water pumps of the submersible type fed by solar panels.

The great interest in solar panels for the conservation of food and medicines is due to their hygienic, silent, reliable, and self-operation, guaranteeing their value independent of any shortage of electric power from the public network [7–10]. In remote areas, solar energy is associated with the reliability of batteries for emergency and security systems and for the data and information systems required in medical centers, intensive care units, operating theaters, and in similar applications where lighting interruption can be a matter of life and death. In addition, the building's regular UPS system and auxiliary generator set, with its associated engine-starting battery, requires the same level of reliability. Various types of metering systems are installed to measure all types of fluids, electricity, gas, and water necessary in the conservation of food and medicines.

6.8.7 Hydrogen and Oxygen Generation by Electrolysis

Reliance on imported oil from unstable regions of the world has forced the pursuit of hydrogen as a near-future nonfossil energy alternative. Electrolysis by solar or wind energy the most practical technical means to eliminate fossil fuels from the energy production cycle, especially when these energies come from the sun. High temperatures and/or electricity can also help in this process, enabling countries to use their own renewable electricity and water to produce emission-free hydrogen.

There are several advantages to producing hydrogen by electrolysis from solar panels: no moving parts; plug and play; installation at the point of use; virtually maintenance-free; no transportation; and direct production of high-purity hydrogen and oxygen, among others. When employing renewable energy sources for water electrolysis, it is vital to prevent electrode degradation induced by fluctuating energy input. Mechanical stability and process efficiency suffer when input power varies. The lifetime of standard electrodes in this operational mode lasts only a few hours. For this reason, electrolyzers are conventionally operated at constant rated power. During shutdown, standard electrodes will corrode without a protective voltage. This requires an electricity backup and causes energy losses. Several other losses are present in the process efficiency of water electrolysis: the low ripple-controlled rectifier, electrolyzer, distribution, compression, and others. There are several companies worldwide that supply equipment for this purpose, some of them claiming reduced production costs without transportation, storage, or compressor requirements (see, e.g., http://www.electrolyser.com). In Section 7.8 we describe a detailed study regarding when electrolysis may become a convenient alternative process for electrical energy storage.

6.8.8 Electric Power Supply

Solar energy can supply ac power of 110, 127, or 220 V at 60 or 50 Hz (depending on the inverter) or dc power at 12, 24, or 48 V. That energy can be used for small domestic electric equipment [7,9,10]. The available energy is obtained from a battery that is constantly recharged by solar modules. The load is fed directly in dc current or it is transformed, through switched power inverters (forced commutation), to a sinusoidal alternating current, rectified, and stabilized in voltage and frequency (see Figure 6.19).

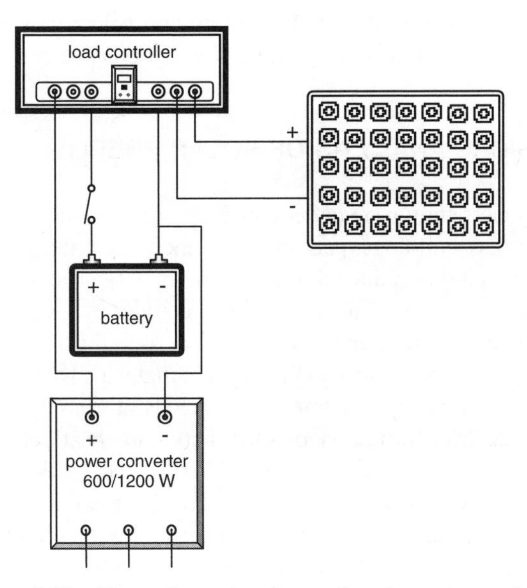

Figure 6.19 Three-phase electric supply using solar modules.

Typically, the solar power available at 12:00 a.m. on a clear and sunny day is 1 sun = 1000 W/m^2, with an average for 365 days of 174 W/m^2. That means that for an electric power supply one could expect

$$P(W) = 174\eta A \tag{6.18}$$

where η is the efficiency of the photovoltaic cells and A is the effective panel area (m^2).

6.8.9 Security and Alarm Systems

Terrorist attacks have emphasized the importance of protecting water power supplies against sabotage and similar attacks. The German municipal authority introduced appropriate legislation with its technical notification W1050 in March 2003. Solar power remote monitoring is used to protect property, to detect leakage, and for automatic remote reading of meters, all powered by a single solar panel.

Electronic surveillance of a plant and premises cannot replace mechanical protection and regular control. As part of an operational security concept, possible impairment of a water or gas supply can be averted by early detection through remote supervision of signaling contacts and key cylinders and the use of motion or sabotage detectors at the site.

Automatic remote reading of water meters is another application of solar panels. With manual readings it is common for safety reasons to use two-person teams, but this method introduces reading errors, is high in labor cost, and makes less data available. Remote reading allows new data to enter the system every 15 or 20 minutes, giving an exact and instantaneous profile of consumption. An alarm signal can warn of a minimum level limit, a high consumption velocity, or faulty pipes. This system requires the attention of only one or two workers once a year.

6.9 ECONOMICAL ANALYSIS OF SOLAR ENERGY

Normalization, under study in many countries, would limit the use of electric heaters and ovens in new buildings, perhaps recommending a tariff for overuse during peak periods, and trying to induce the replacement of electric heaters and ovens by other modes of generation. The restrictions could be softened for residences endowed with demand controllers or thermal reservoirs for the storage of hot water during hours of low demand (early in the day). In that way, heating needs would not concentrate power during only certain short periods of the day.

It is known that transformation of electricity into heat represents an overall reduced efficiency (on the order of 5%). That is due to the various processes through which electricity passes until it is available in an appropriate way to the user [5,6]. Electricity has more worthy uses: for illumination, to drive motors, and in other productive processes. Without reducing the comfort level of the population, the consumption of electricity for heaters and ovens could be better used in those more productive processes. However, it is difficult to compare the equipment cost that provides economic advantages of absolute reliability for the society as a whole when the user just analyzes the result in kilowatthours of consumption paid in the effective tariffs. Even so, growing energy prices demonstrate significant advantages for the use of direct solar energy for heating of water and home appliances.

The cost of the initial investment in thermal solar energy is high, and the solar field absorbs around 50% of the total costs of facilities. Thermal solar systems now in commercial operation are designed to integrate the solar energy in power plants fed by fossil fuel, aiming at a reduction in the cost of fuel. The factors that work against increased investment in this area are still the low cost and smaller scale of fossil fuel power plants (coal or oil), despite all the clear disadvantages (see Chapter 13).

The cost of electricity obviously differs from country to country, but has risen from an average of $49 per megawatt in 1995 to about $100 per megawatt in 2005. Such data demonstrate the great value of solar energy use in the future. At the present time, however, the use of solar energy demands equipment and advanced technologies, which results in a high-cost enterprise. At the time of publication of this book, the installation cost for solar power plants was about $3.50 per watt and the operating cost was $0.20 per kilowatt.

REFERENCES

[1] KanEnergi, *New Renewable Energy: Norwegian Developments*, Research Council of Norway in cooperation with Norwegian Water Resources and Energy Directorate, Hanshaugen, Oslo, Norway, December 1998.

[2] http://nssdc.gsfc.nasa.gov/space/model/atmos/us_standard.html.

[3] http://www.astm.org/.

[4] I. Chambouleyron, Solar electricity, *Science Today*, Vol. 9, No. 54, pp. 32–39, June 1989.

[5] R. Messenger and J. Ventre, *Photovoltaic Systems Engineering*, CRC Press, Boca Raton, FL, 2000.

[6] *Eletricidade diretamente do sol* (Direct electricity from the sun), catalog, Soleco Do Brasil, Santa Maria, Brazil, 1997.

[7] M. R. Patel, *Wind and Solar Power Systems*, CRC Press, Boca Raton, FL, 1999.

[8] R. Boylestad and L. Nashelsky, *Electronic Devices and Circuit Theory*, 6th ed., Prentice Hall, Upper Saddle River, NJ, 1996.

[9] J. Millman and C. C. Halkias, *Electronic Devices and Circuits*, McGraw-Hill, New York.

[10] J. Millman and C. C. Halkias, *Integrated Electronics: Analog and Digital Circuits and Systems*, McGraw-Hill, New York, 1972, pp. 58, 797.

[11] R. R. Spencer and M. S. Ghausi, *Introduction to Electronic Circuit Design*, Prentice Hall, Upper Saddle River, NJ, 2003.

[12] *A energia do sol* (Sun energy), catalog, Heliodinâmica, São Paulo, Brazil, 1990.

[13] *Força Energética*, Vol. 1, No. 2, July–August 1992.

[14] W. H. Lee and W. G. Scott, *Distributed Power Generation: Planning and Evaluation*, Marcel Dekker, New York, 2000.

[15] M. H. J. Bollen, *Understanding Power Quality Problems: Voltage Sags and Interruptions*, Series on Power Engineering, IEEE Press, Piscataway, NJ, 2000.

[16] A. W. Culp, *Principles of Energy Conversion*, McGraw-Hill, Toronto, Ontario, Canada, 1979.

[17] *Solar Energy: The Alternative Source That Spares Energy of the Net*, Catalog, Siemens, 2002.

[18] W. L. Hughes, *Energy in Rural Development: Renewable Resources and Alternative Technologies for Developing Countries*, Advisory Committee on Technology Innovation, Board on Science and Technology for International Development, Commission on International Relations, National Academy of Sciences, Washington, DC, 1976.

[19] Solar energy: electricity or hot water, *Força Energética*, Vol. 2, No. 3, pp. 25–26, March 1993.

[20] *The Leading Position in a Growing Market*, Advanced Power Systems, catalog, Neste, 1996.

[21] L. B. Dos Reis, *Geração de energia elétrica* (Generation of Electrical Energy), Manole, São Paulo, Brazil, 2003.

CHAPTER 7

POWER PLANTS WITH FUEL CELLS

7.1 INTRODUCTION

The operating principles of fuel cells were demonstrated initially in 1839 at the Royal Institution of London by an English barrister and physicist, Sir William Grove, who showed the reversibility of water electrolysis. The first practical application of fuel cells is credited to Francis T. Bacon of Cambridge University. In 1950 Bacon published groundbreaking results of an alkaline cell prototype. Fuel cells then became known worldwide when the National Aeronautics and Space Administration (NASA) used them in the Apollo program during the 1950s and later in the Gemini program. Obviously, fuel cells were a very convenient technology for the space program in not being polluting, producing electricity and heat, and having as by-product potable water from hydrogen, exactly what scientists wanted for a spaceship. In the past few years fuel cells have appeared as the most promising innovation in the market of alternative energies for stationary, portable, and automotive applications, as a natural energy conversion system from hydrogen stored from electrolysis. What appeals most about fuel cells is their construction, which can be clean and compact, their need for only a few movable parts, their modular technology, and the fact that they do not inflict on the environment emissions of sulfur and nitrogen oxides (SO_x and NO_x) [1–3].

Integration of Alternative Sources of Energy, by Felix A. Farret and M. Godoy Simões

Present interest in fuel cells is enormous. Numerous companies and research centers throughout the world are working on many developments related to fuel cell energy systems: Ballard Generation Systems, Global Thermoelectric, Fuel Cell Technologies (Canada), Sulzer Hexis (Switzerland), UTC Fuel Cells, Schatz Energy Research Center and Energy Partners, M-C Power, General Motors, Siemens–Westinghouse Corporation, GE Power Systems, Teledyne Energy Systems, H-Power, Avista, Ida Tech/North West Power Systems, and Plug Power in the United States; Toshiba, Mitsubishi Electric Corporation, and Ebara Corporation in Japan; ECN in Holland; Nuvera Fuel Cells in Italy; Rolls-Royce in England; and MTU, DaimlerBenz, Dornier, and Buderus Heiztechnik in Germany, among others. All this interest in fuel cells supports the idea that direct combustion of fossil fuel is declining in importance. Current efforts toward commercial and regular production of fuel cells are intensive, primarily to improve its performance as related to the space between electrodes. This space is critical for the compactness of this energy source, as well as for the removal of sulfur and carbon monoxide, particularly in the PEM (proton exchange membrane) and SOFC (solid oxide fuel cell) types. These compounds contaminate the platinum catalysts, thus degrading fuel cell performance with time.

Fuel cell characteristics differ from those of current dominant technologies for distributed generation in electric power systems, which are based on internal combustion engines using reciprocal primary movers or steam turbines. Such technologies are widely used at present and offer cheap, reliable energy with satisfactory heat use and efficiency. However, these types of machines cannot definitively find place in a planet concerned with its own survival. Characteristics such as noise, vibration, and emission of pollutant gases (e.g., NO_x, CO_x) have not yet found the most appropriate optimization formula of acceptance. Because of that, frequent maintenance should be planned, thus increasing the cost of small units, which already have low overall efficiency, on the order of 30% [4–6]. Fuel cells seem to be a good option despite not yet having a mature enough technology to be a feasible solution for the world market.

7.2 THE FUEL CELL

Fuel cells are electrochemical devices sometimes compared to conventional automobile batteries. However, there is a fundamental difference in that fuel and oxidizer are supplied to fuel cells continually so that they can generate continuous electric power. Therefore, a fuel cell is important as an energy vector, combining with the stored fuel as a potential energy source. Whereas fuel cells are based purely on the electrochemical reaction of hydrogen or other fuel gases, in ordinary batteries the reactants are either consumed or must be regenerated through electric recharge, as in automotive lead–acid batteries [1,2]. The hydrogen combines with the oxygen inside a combustion-free process, liberating electric power in a chemical reaction that is very sensitive to the operating temperature. Despite initial use in aerospace applications, new experimental materials have been contributing significantly to improved use, efficiency, and cost of fuels for terrestrial applications. A very significant figure that shows a lot of improvement is the amount of platinum catalyst needed in the electrodes. It has decreased in the last few years from 28 mg/cm^2 of electrode area to about 0.1 mg/cm^2 [3].

There are different types of fuel cells classified according to their operation at low or high temperatures. Low-temperature PEM fuel cells (up to about 100°C) are available and ready for mass production. They are used in applications where high-temperature cells would not be suitable, such as in commercial and residential power sources as well as in electric vehicles. The most compact systems are appropriate for powered electric vehicles and are available from 5 to 100 kW. High-temperature (about 1000°C) SOFCs are typically used for industrial and large commercial applications and operate as decentralized stationary units of electric power generation. Because of such high operating temperatures, great heat dissipation is expected to be recovered, and integration with gas-powered microturbines is very successful. A 70% overall efficiency is reported in such applications, representing a great contrast with respect to the energy generation in coal power plants, whose efficiency is between 35 and 40%.

Figure 7.1 shows the principles of an operating fuel cell. They are powered by the flow of a fuel gas such as hydrogen, which could be pure or derived form (e.g., methanol, coal gas, natural gas, gasoline, naphtha) and that of an oxidizer (e.g., oxygen, air). After the hydrogen passed through a porous anode separated by an electrolyte of a porous cathode, the resulting ion meets the oxygen to form water. As a result of this reaction, the accumulated electrons in the anode form an electrical field and find their way through an external conductive connection

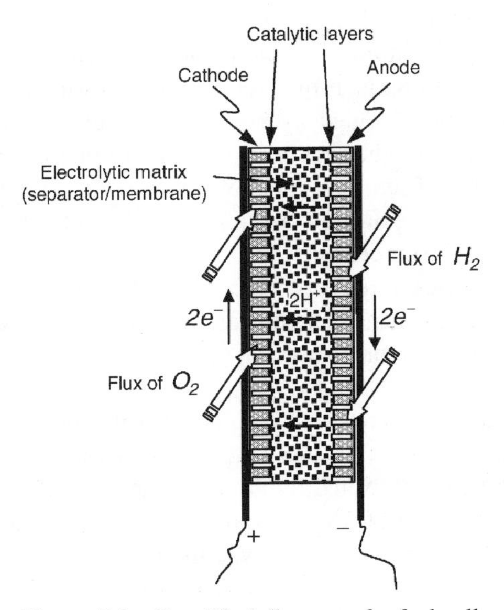

Figure 7.1 Simplified diagram of a fuel cell.

between the anode and the cathode. An electrical current is established through an external load. As in any other chemical reaction, heat is produced which has to be dissipated in some way. As a result of the gas reaction, there is by-product generation of heat and water, in contrast with the conventional mixture and burning of fossil fuel.

From a physiochemical point of view, fuel cells are the electrolysis antipode process in the sense that the first consumes H_2 and O_2 and produces electricity, heat, and water, whereas the second consumes electricity and water and produces heat, H_2, and O_2. This duality widely motivates practical possibilities for fully clean and sustainable systems, as discussed in Chapter 2.

7.3 COMMERCIAL TECHNOLOGIES FOR GENERATION OF ELECTRICITY

Hydrogen and pure oxygen are not yet commercially available in the quantities required for cost-effective use in fuel cells. There is abundant oxygen in the air and abundant hydrogen in several organic materials, particularly natural gas. It is available at low temperature but demands a fuel processor, or reformer, before reaching the cell. The reformer converts, methane, for example (added in a steam) to a reformed mixture of hydrogen, sulfur, ammonia, carbon dioxide, and carbon monoxide. The fuel processor still has to accomplish a catalytic endothermic steam reform: that is, the reaction of water steam with fuel hydrocarbon to produce hydrogen, thereby increasing complexity and losses, leading to an efficiency of around 35 to 48%. Purification of hydrogen is still required because particulates may degrade the fuel cell membrane life. Although natural gas is a fossil fuel, it can bridge the current economic status toward sustainable production of hydrogen within a couple of decades. One issue that needs to be evaluated is the cost comparison of microturbines powered by natural gas with fuel cells running from reformed hydrogen. The fuel cell choice makes sense only when combined with other applications that are interconnected to hydrogen storage and production. Therefore, fuel cells will be feasible economically when combined with hydrogen-powered electric vehicles, electrolyzers, and small-scale renewable energy sources.

One of the peculiarities of fuel cells to be used as commercial sources of energy is the need to consider the benefits and restrictions of a very complex market concerned with the following four points:

Cost: Projected at $2 to $3 per watt, the cost of fuel cells is much higher than the cost of generators, batteries, and other forms of alternative energy. A recent very positive trend motivated by the manufacture of fuel cell–powered cars allows a projection of $0.79 per watt in full-scale production. Competitiveness is a little more difficult with river and maritime applications, equipment for telecommunications, and recreation vehicles, which already have a reasonable market share [4–7].

Product availability: Customers want to have products to try out and compare with other options; otherwise, it becomes very difficult to convince people to use them.

Fuel: With the exception of large boats that use diesel or gasoline, several other industry segments use propane. However, if hydrogen is available for automobiles, which is a much larger market, this will easily be extended to boats.

Safety: A certain apprehension exists on behalf of potential buyers of recreation vehicles and small boats about replacing propane with hydrogen.

Power generation faces a highly competitive market under deregulation policies. Therefore, economic factors must convey a proper decision as to the type of generator to be used for each application. The main economical factor is maximization of the commercial power made available to consumers with minimum maintenance at the highest levels of efficiency and minimum capital investment [4,6,7]. The only item difficult to analyze in fuel cell applications is still a possibly lower capital investment. Many countries, including Germany and England, have been investing effectively in clean generation policies by subsidizing initial production and use of fuel cells. For a market with such diversified demands, several fuel cell technologies are in development [8–11]. The main research lines being considered at the time of publication of this book are:

- Proton exchange membrane or solid polymer fuel cells (PEMFCs or SPFCs)
- Phosphoric acid fuel cells (PAFCs)
- Alkaline fuel cells (AFCs)
- Molten carbonate fuel cells (MCFCs)
- Solid oxide fuel cells (SOFCs)
- Direct methanol fuel cells (DMFCs)
- Reversible fuel cells (RFCs)

Common application ranges for these fuel cells are shown in Figure 7.2.

The designation of a fuel cell type is related to the electrolyte for the reactant ions. For example, in a proton exchange membrane (PEM), a solid polymer (SPFC) such as Nafion (a registered trademark of Dupont) is used to enable conversion of hydrogen chemical energy (H_2) and an oxidizer, oxygen (air or O_2), into electrical energy. Nafion operates at temperatures between 45 and 100°C with a typical efficiency of 40%. This solid polymer membrane is the basis for the strongest interest in its use with automotive applications. The main function of such a polymer is to provide good electronic insulation and an efficient gas barrier between the two electrodes while allowing rapid proton transport and high current densities. By its consistency, the solid electrolyte does not move with respect to the electrodes. It does not diffuse or evaporate but occupies a moderate space with appreciable weight. These cells are promising for automotive applications and are expected

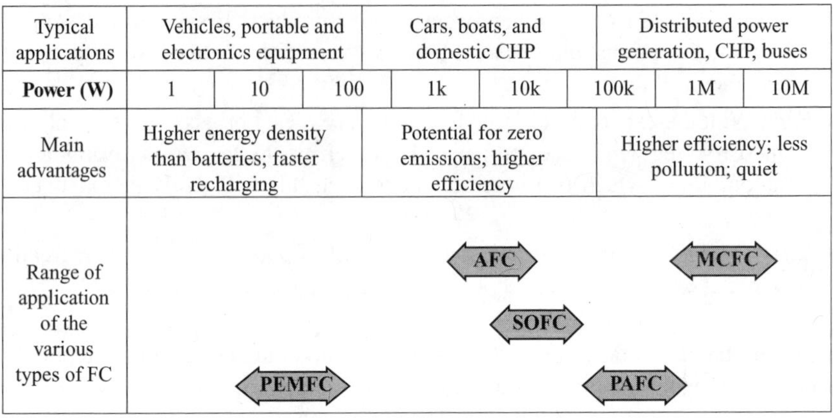

Typical applications	Vehicles, portable and electronics equipment			Cars, boats, and domestic CHP		Distributed power generation, CHP, buses		
Power (W)	1	10	100	1k	10k	100k	1M	10M
Main advantages	Higher energy density than batteries; faster recharging			Potential for zero emissions; higher efficiency		Higher efficiency; less pollution; quiet		
Range of application of the various types of FC		PEMFC		AFC	SOFC	PAFC	MCFC	

Figure 7.2 Applications of fuel cells.

to have some use in combined heat and power schemes and automotive applications.

To understand the chemical reactions and peculiarities of each fuel cell technology, some basic operation will be described next starting with the *proton exchange membrane fuel cell* (PEMFC). The chemical reaction at the anode is [3,5]

$$2H^+ + 2e^- \rightarrow H_2$$

and the cathode reaction of the fuel ionized with oxygen is

$$O_2 + 4H^+ + 4e^- \rightarrow 2H_2O$$

To make fuel cells as light and small as possible, one has to construct compact modules for easy and efficient setup. This is the case for the PEM polymeric membrane, bonded on each side by catalyzed porous electrodes. Such an electrode–electrolyte–anode or membrane electrode assembly (MEA) is made with bipolar electrode plates so that the positive side of one electrode plate is the negative side of the next plate. Therefore, the MEA is extremely thin, compact, and ready to be associated with others. Furthermore, the electrode plate pair is typically machined with flow fields for even distribution of fuel and air or oxygen to anode and cathode. It must also be provided with a compatible path for the cooling air or water at the back of the reactant flow field of the metallic electrodes. The humidity must be kept in the approximate range of 85 to 100%, as the membrane hydration assists in proton conduction through the electrodes. Thus, a small section of the MEA has to be set aside for humidification of the reactant gases. High standards for the MEA ohmic, mass transport, and kinetics characteristics are essential for its commercial application.

The *phosphoric acid fuel cell* (PAFC) has a certain similarity with the PEMFC, but the acid H_3PO_4 is the electrolyte and the cell is operated with hydrogen and air.

The power density produced by this fuel cell ranges from about 0.20 to 0.35 W/cm^2. The temperature range is about 220°C, with a typical efficiency of 40%. This cell is tolerant to poisoning up to 1% of carbon monoxide, although good water management is essential in its operation to improve reagent kinetics as dependent on temperatures varying from 150 to 220°C. The internal environment becomes very corrosive, due the high electrode potential of its operation, demanding the use of strong, corrosion-resistant materials in its construction. This type of fuel cell achieves only moderate current density. Its construction and chemical reactions are quite similar to those of the PEMFC, from which most of its features were derived. Because the temperature range of the PAFC is higher than that of the PEMFC, the PAFC is recommended for combined heat and power applications. There are many 200-kW units installed all over the world based on this technology which have a strong record of operation. The chemical reaction at the anode of a PAFC is similar to that in a PEMFC [3,5]:

$$2H^+ + 2e^- \rightarrow H_2$$

and the cathode reaction of the fuel with oxygen is

$$O_2 + 4H^+ + 4e^- \rightarrow 2H_2O$$

The *alkaline fuel cell* (AFC) was the first workable power unit used by NASA in the manned Apollo spaceship mission. Nevertheless, it was soon found that cost, reliability, ruggedness, safety, and ease of AFC operation could never compete with some of the other technologies known at that time, although it had the least expensive construction technology. It operates in temperatures between 55 and 120°C with a typical efficiency of 50%. Its major problem is that strongly alkaline electrolytes such as KOH and NaOH adsorb CO_2 and so reduce its electrolyte conductivity. This feature limits its current densities. The chemical reaction at the anode of an alkaline fuel cell is

$$2H_2 + 4OH^- \rightarrow 4H_2O + 4e^-$$

and the cathode reaction of the fuel with oxygen is

$$O_2 + 4e^- + 4OH^-$$

The electrolyte of a *molten carbonate fuel cell* (MCFC) is a mixture of alkali carbonates of potassium and lithium. A carbonate ion moves from cathode to anode, where in combination with hydrogen it forms water and carbon dioxide plus the electrical charge. The power density produced by this fuel cell ranges to about 0.10 W/cm^2, limited by ohmic losses. The high operating temperatures (around 650°C) and the high corrosivity of the molten carbonate salts demand special types of manufacturing materials, but these are justified by the noticeable improvement in reactant kinetics and reduced need for expensive noble catalysts.

The typical efficiency is above 50%. The bipolar electrodes are made from high-quality stainless steel protected by additional coatings of nickel or chrome. The high temperatures allow for internal fuel steam reforming, such as the gas methane. Its high operating temperatures recommend MCFC primarily for stationary power applications in the megawatt range. The chemical reaction at the anode of a molten carbonate fuel cell is [3,5]

$$H_2 + CO_3^{2-} \rightarrow H_2O + CO_2 + 2e^-$$

and the cathode reaction of the fuel with oxygen is

$$\tfrac{1}{2}O_2 + CO_2 + 2e^- \rightarrow {CO_3}^{2-}$$

The *solid oxide fuel cell* (SOFC) uses yttria and zirconia as oxides to operate at high temperatures (around 1000°C). Under such conditions, hydrogen and ionized oxygen are combined to form water in the anode and to liberate a pair of electrons with excellent reactant kinetics, although the cell's reversible potential is a little lower for lower-temperature cells. There is no water management problem because this fuel cell has an entirely solid-state construction and a single side tube sealing. This cell may use hydrogen or carbon monoxide as fuel with efficiencies higher than 50%. Its main drawback for high power generation is that of the materials required in its construction. Recent research developments involve three main types of reactant core design for this cell: (1) planar (similar to that of other fuel cells), which has sealing problems; (2) tubular, to improve the electronic conductivity and overcome the sealing limitations of the planar design; and (3) flattened tube, with better air guidance, to improve the packing densities of the tubular design. In any case, the tube itself forms the air electrode. The power density produced by the tubular design cells ranges from about 0.18 to 0.20 W/cm^2. The planar design cell may reach a power density of about 0.35 W/cm^2. SOFC use is restricted almost entirely to stand-alone and combined heat and power applications. The chemical reaction at the anode of a solid oxide fuel cell is [5]

$$H_2 + O^{2-} \rightarrow H_2O + 2e^-$$

and the cathode reaction of the fuel with oxygen is

$$\tfrac{1}{2}O_2 + 2e^- \rightarrow O^{2-}$$

In the *direct methanol fuel cell* (DMFC), methanol is oxidized electrochemically by water at the anode to produce carbon dioxide and positive and negative ions, in contrast to what happens in a PEMFC, where the positive ions from the hydrogen are supplied directly to the anode. To improve rejection of the carbon dioxide in a DMFC, an acid electrolyte is used; otherwise, insoluble carbonates may form in the alkaline electrolyte. Similar to what is depicted in Figure 7.1, the hydrogen ions produced at the anode permeate the polymer electrolyte in the direction of the cathode, where they react with oxygen from the air to produce water. The external circuit provides a path for the flow of electrons accumulated in the anode. The most attractive feature of the DMFC for the transportation and portable-use industries is that a fuel reformer is not required. As may be inferred from this explanation, direct methanol fuel cells have poorer performance in the anode, where finding more efficient catalysts is still a major problem. In several countries other types of alcohol are being experimented with, such as ethanol and benzol. The power density produced by direct methanol cells as reported by Newcastle University researchers in England is about 0.20 W/cm^2. Its overall efficiency does not go above 50%. The chemical reaction at the anode of a direct methanol fuel cell is [2,5]

$$CH_3OH + H_2O \rightarrow 6H^+ + CO_2 + 6e^- \qquad \text{for } E_{\text{anode}} = 0.046\,V$$

and the cathode reaction of the fuel with oxygen is

$$\tfrac{3}{2}O_2 + 6H^+ + 6e^- \rightarrow 3H_2O \qquad \text{for } E_{\text{cathode}} = 1.23\,V$$

and for the cell voltage

$$CH_3OH + \tfrac{3}{2}O_2 + H_2O \rightarrow CO_2 + 3H_2O \qquad \text{for} \quad E_{\text{cell}} = 1.18\,V$$

The *reversible fuel cell* (RFC) (also known as a *regenerative* or *unitized fuel cell*) is a special class of fuel cells that produce electricity from hydrogen and oxygen but can be reversed and powered with electricity to produce hydrogen and oxygen, as depicted in Figure 7.3. The three key concepts of the reversible or regenerative fuel cell are output power, run time, and recharge rate. The number and size of the fuel cells in the stack determine the output power, which is dependent primarily on the effective area of each electrode–membrane–electrode assembly. The run time is determined by the capacity of the hydrogen storage tank available. The recharge rate is determined by the output rate of the electrolyzer used to produce the hydrogen. An example of this is a PEM-fed uninterruptible power supply under development by Unigen as part of the U.S. Department of Energy's State Energy Programs for developing RFC modules ready for use.

Reverse fuel cells are also under development in many parts of the world. They can convert hydrogen directly from water using photovoltaic, hydro, or wind power. It has been observed that RFCs are capable of an energy density of about 450 Wh/kg, which is 10 times that of lead–acid batteries and more than twice that of current chemical batteries. One of these systems is reported by GreenVolt Power Corp. Typically, their fuel cell splits water into its components hydrogen and oxygen for use in a variety of industrial and transportation applications. The unit uses 20% less energy in a significantly smaller unit size than that of water–alkali electrolyzer devices, and requires only distilled water to produce 99.5% pure hydrogen

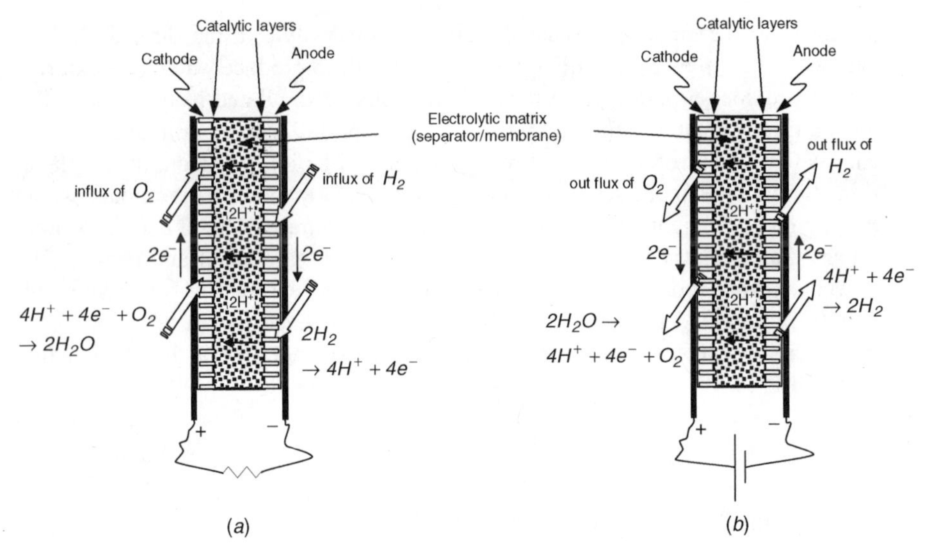

Figure 7.3 Simplified diagram of a PEM reversible fuel cell: (*a*) fuel cell mode; (*b*) electrolyzer mode.

and oxygen. When used to power hydrogen fuel cells, the water by-product created can be reused by the RFC, potentially lowering operating costs. On-site hydrogen gas generation can also improve the economical constraint of hydrogen transportation by pipelines and bottles. It is possible to fabricate ceramic microtubes by electrophoretic deposition (EPD), although further research into certain areas of this investigation is in progress.

Figure 7.3 shows the chemical reactions in a reversible fuel cell. In the fuel cell mode a PEMFC combines hydrogen and oxygen to create electricity and water. When the cell reverses its operation to act as an electrolyzer, electricity and water are combined to create oxygen and hydrogen. RFCs are expected to be useful mostly in passenger cars, solar-powered aircraft, energy storage schemes, required propulsion of satellites for orbit correction, microspacecraft, and power systems, as in load leveling in remote sources of wind turbines and solar cells.

Yet another type of fuel cell is the *zinc–air fuel cell*, which has a gas diffusion electrode, an anode separated by electrolyte, and mechanical separators. The gas diffusion electrode is a permeable membrane that allows atmospheric oxygen to pass through. After the oxygen has been converted into ions and water, the ions travel through an electrolyte and reach the zinc anode. There it reacts with the zinc, forming zinc oxide, creating an electrical potential. This electrochemical process is very similar to that of the PEM fuel cell described above, but the refueling is very different, although it has some of the characteristics of batteries. Zinc–air fuel cells are best suited for battery replacement but have another wide range of potential applications. Some developers of zinc-air fuel cells are Cinergy and Metallic Power and Electric Fuel Co.

7.4 PRACTICAL ISSUES RELATED TO FUEL CELL STACKING

It is important to emphasize that each type of fuel cell has a different internal reaction, as illustrated in Figure 7.4.

7.4.1 Low- and High-Temperature Fuel Cells

There are two broad categories of fuel cells: low-temperature or first-generation, which include AFCs and SPFCs (or PEMFCs) and the phosphoric acid type (PAFCs). The latter has been the main objective of resource investment to obtain a commercial model with compatible costs for terrestrial applications. High-temperature cells such as MCFCs and SOFCs are considered second-generation, and their commercial development is at an advanced stage, being considered for use in large-scale electric power generation [1–5].

Second-generation cells have been developed for operation at high temperatures to couple easily to reforming of the hydrogen inside the cell. Such a procedure makes the overall system very simple and increases the efficiency up to 60%, as demonstrated under laboratory conditions using external fuel treatment.

A single fuel cell can produce just a few watts. Therefore, they are stacked up to other cells, adding their effects electrically by series and/or parallel connection, very similar to ordinary batteries. A stack will then form autonomous modules of energy supply in accordance with the end purpose. There are currently very large power stacks, up to 1.5 MW.

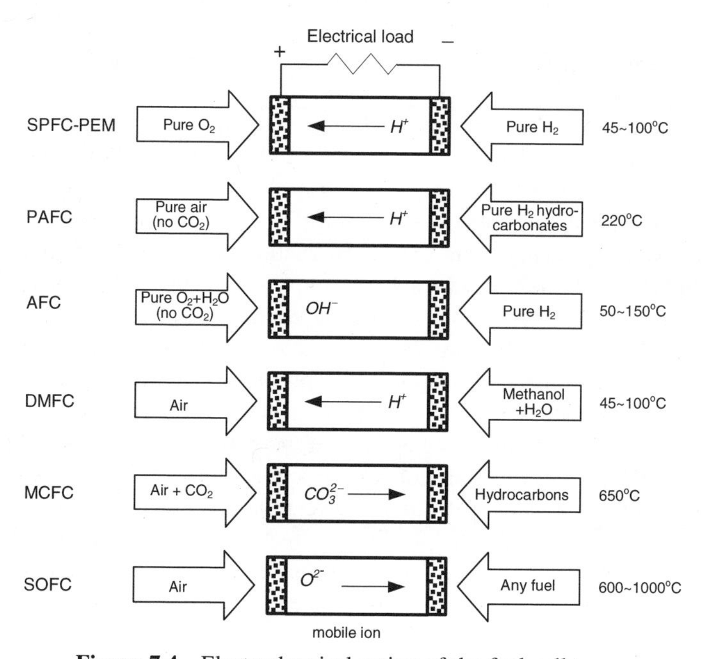

Figure 7.4 Electrochemical action of the fuel cell types.

The electric power generated by a fuel cell stack is always direct current, just as ordinary batteries. The dc power generated in the cells should be conditioned to the types of contemporary applications, as explained in Chapter 12. For all fuel cell types there is a lot of packaging research aimed at turning them into useful commercial and practical products. The following characteristics are also the subject of intense research:

- Pretreatment of fuel for cogeneration systems
- Heat recovery by preheating fuel and oxidizer
- Injection and recirculation of water or steam
- Energy administration, conditioning, conversion, and optimization
- Separation and recycling of CO_2 in MCFCs by blowing air and careful administration of the water in SPFCs to assure that the membrane stays damp for optimal conditions

7.4.2 Commercial and Manufacturing Issues

The Carnot cycle can be used to compare fuel cells with internal combustion engine (ICE) motors as generators of energy. This theoretical cycle imposes limits on the performance and construction of any energy transformation unit by high pressures and temperatures according to the upper efficiency limit $\eta_c = (T_{\text{high}} - T_{\text{low}})/T_{\text{high}}$. The dimensions of the heated machine also influence the overall efficiency directly, hopefully by about 30%. In the same way, theoretical limitations exist for fuel cells, although they may not be as narrow. The more pure the oxygen and hydrogen used, the higher the overall fuel cell efficiency rates, which are around 70%. Such rates would be justified only in an aerospace mission, or in oceans or similar environments, where the costs of hydrogen and oxygen are not as relevant as in ordinary power plants. Using methanol, for example, to obtain the fuel gas, the efficiency falls to around 44%; however, the cost reduction is dramatic. A practical alternative that is perhaps already feasible is the use of SPFCs, as in the Gemini spaceship mission and as used for many years for the production of oxygen in submarines.

One problem with fuel cells not solved completely when this book was written is related to how to obtain the largest possible contact area among electrolytes and gaseous reactants, called the *notable surface action* of the electrocatalytic conductor [1]. This conducting contact path involves the electrolyte itself, the flooded pores of the electrolyte, the platinum on the catalytic carbon particles, the polytetrafluorethylene (PTFE), and the gas through the pores.

Power systems with fuel cells for remote applications (FCPS-RA) seems to be the great opportunity for fuel cells in the current market. The power ranges between 0.1 and 5 kW would include loads such as recreation vehicles, communication stations, yachts, residences isolated from the public network, and canalizations of natural gas and oil. For these applications, the current retail prices for FCPSs are \$0.13 (per kilowatthour) (generator), \$0.29 (per kilowatthour) (battery), and \$0.50 (per kilowatthour) (photovoltaic panels) in the consuming markets. Current FCPSs in the wholesale can produce electricity for \$0.44 (per kilowatthour). In the best of hypotheses with a reduction of 50% in stack cost, an improvement of 100% in the performance, and a hydrogen generation cost of \$0.12 per gigajoule, FCPSs could supply energy at \$0.23 (per kilowatthour).

Commercial and manufacturing investment has been concentrated in three areas: (1) the development of a 5-kW reformer to be integrated with a fuel cell stack tolerant to fuel reforming; (2) commercially available fuel cells that operate with pure hydrogen and are integrated with electric vehicles, and (3) diffusion and training in the manufacture, production, and use of fuel cells.

For small- and medium-energy systems, the present state of the art seems to point toward the establishment to two already advanced technologies: proton exchange membrane and solid oxide fuel cells. Therefore, this chapter concentrates next on those two technologies.

7.5 CONSTRUCTIONAL FEATURES OF PROTON EXCHANGE MEMBRANE FUEL CELLS

Porous separation between fuel and oxidizer is achieved by the use of specialized conventional materials (e.g., tissues made of carbon or carbon paper obtained from Teflon) and is the most expensive part of the fuel cell stack. The main characteristic of these materials is that they should allow the flow of ions generated in the fuel and oxidizer reaction when going from anode to cathode, preventing the passage of electrons, so they have to circulate through an external circuit.

Figure 7.1 is a simplified diagram of a fuel cell. At the anode side of the PEM fuel cell, fuel is supplied under a certain pressure, usually enough to make it go through the flow-field channels of the electrodes and cross the electrolyte. It is assumed for the sake of discussion that the fuel is the pure gas H_2, although other gas compositions may be used, as discussed in Section 7.3. In these cases the hydrogen concentration should be determined in the mixture. The fuel spreads through the electrode until it reaches the catalytic layer of the anode, where it reacts to form protons and electrons according to the reaction

$$H_{2(g)} \rightarrow 2H^+ + 2e^- \tag{7.1}$$

Protons are transferred to the catalytic layer of the cathode through the electrolyte (a solid membrane for a PEM fuel cell). On the other side of the cell, the oxidizer flows through the channels of the plate and spreads itself through the electrode until it reaches the catalytic cathode layer. The oxidizer used in this model may be air or pure O_2. The oxygen is consumed by reacting with the H_2 protons and electrons, and water is produced with some residual heat on the surface of the catalytic particles. The electrochemical reaction in the cathode is

$$2H^+ + 2e^- + \tfrac{1}{2}O_2 \rightarrow H_2O + \text{heat} \tag{7.2}$$

Then the full FC chemical reaction becomes

$$H_2 + \frac{O_2}{2} \rightarrow H_2O + \text{heat} + \text{electrical energy} \tag{7.3}$$

Figure 7.5 is a schematic diagram of a complete electrical energy generation system using a fuel cell stack with PEMFCs to be introduced in the market. The most vital part of the system is a stack with 70 PEM fuel cells and 300 cm^2 per cell of active area, supplying 3 kW of alternating current through a dc–ac inverter. The diagram shows the stack fed with hydrogen and oxygen (air) as well as water for refrigeration and output products. Hot water and electricity are both available for the user. The overall stack output voltage is represented by V_s. Table 7.1 gives the main nominal characteristics of this system. The reformer for the cells to obtain hydrogen from a fuel with hydrocarbon and water steam is also represented. The capacity of the components in the system will depend primarily on the total output power of the stack.

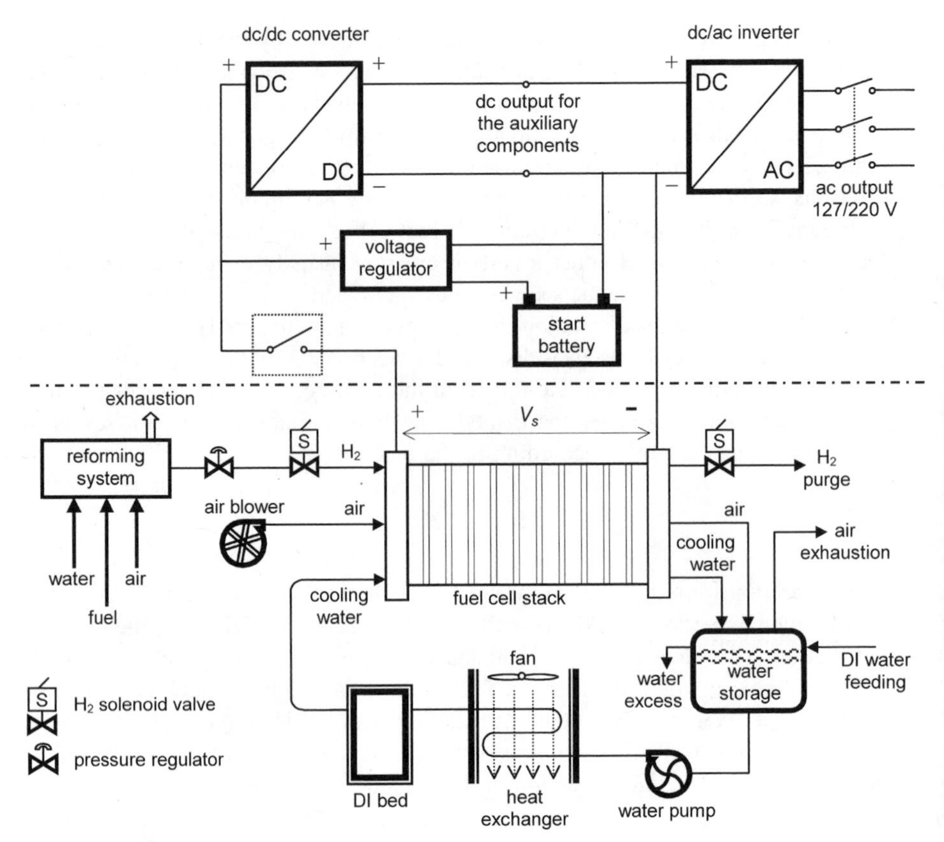

Figure 7.5 Practical outline of a small fuel cell power plant. (From Ref. [2].)

TABLE 7.1 Rated Values of a Typical Commercial Fuel Cell Stack

Description	Specification
Ac output power (kW)	3
Ac output voltage (V)	127/220
Dc voltage (V)	42–70
Temperature of the membrane (°C)	60
Normal maximum flow of the reformatted H_2 (L/min)	100
CO contents of the reformatted output (ppm)	<50

Most of the head of series consist of a low-pressure, high-volume air blower to supply the oxidizer. A water circulation system is used to cool the stack reactions and to provide humidification of the input flows of air and reformer fluids. The reformer is of the partial oxidation type, to generate fuel rich in hydrogen for the stack from the tank of natural gas or propane.

A typical system setup consists of a fuel cell stack conditioned by a dc–dc converter of 48 V to feed a blower, water pump, 4-kW power inverter, and another inverter to generate 120 V ac output voltage. Auxiliary power can typically be fed from dc–dc converters operating from the main fuel cell dc-link bus for microcontrollers, relays, solenoid valves, and so on.

A microcontroller can be used to manage the fuel cell operation, providing acquisition of operating data and to inform the user about the stack operating conditions. A startup battery must be included to bootstrap the beginning of operations, the microcontroller, and other load controls. After fuel cell starting procedures (warm-up time), the system will be fully running only with power from the fuel cell stack. The PEMFC is recommended for either stationary or vehicle power applications and there are many 5- to 250-kW units based on this technology worldwide that have had many hours of operation.

7.6 CONSTRUCTIONAL FEATURES OF SOLID OXIDE FUEL CELLS

Solid oxide fuel cells use an entirely solid oxide ion conducting ceramic of zirconia (zirconium oxide) stabilized with yttria (yttrium oxide) without any liquid-state interacting product. It works at very high temperatures, typically between 800 and 1100°C. Therefore, there is no need for electrocatalysts, the most complex item commonly associated with research in ceramics and membranes in all the other fuel cells. It may operate with hydrogen with some level of carbon monoxide, and as a result, CO_2 recycling is not necessary. As a contrast to low- and medium-temperature fuel cells, the ions crossing the electrolyte from the cathode to the anode are the oxygen and by-product water formed at the anode side. At about 800°C, zirconia allows conduction of oxygen ions starting up the energy production process. The open-circuit voltage of SOFCs is usually lower than of MCFCs but in compensation, it has lower internal resistance, thinner electrolytes, and therefore, lower losses. For these reasons, SOFCs may operate at higher current densities (around 1000 mA · cm^2).

A zirconia mixture of ceramic and metal (cermet) is widely used to construct a highly resistant and stable anode for the high SOFC temperature environment. The metal used in this mixture is nickel because of its good electrical conductivity and catalyst properties, widening the operating range of this fuel cell since the fuel-reforming process can take place at lower temperatures. On the other hand, the cathode composition is still a complicated matter because of the cost of effective conducting materials at high temperatures. Some materials presently used for this purpose are based on strontium-doped lanthanum manganite.

The most challenging issues of such a slowly maturing technology are related to high-temperature-resistant materials, combined applications with other fuel cells, heat, and power management. At this stage of development it is difficult to predict which fuel cell technology is going to be the most suitable or which is going to become a successful commercial version that takes the best possibilities of heat and power combination (CHP) and the construction of hybrid systems with various other fuel cell types. Figure 7.6 displays a possible system for SOFC cogeneration, including the electric conversion oxidizer input and fuel gases, heat exchanger, and catalytic burner [11–14].

There are other configurations considered seriously for a high-temperature combination of SOFCs with steam or gas turbines (combined cycle system) where the exhaustion gases of the fuel cell would feed the gas turbine, which would power an alternator. SOFCs of size 200 kW are being widely considered for large combined

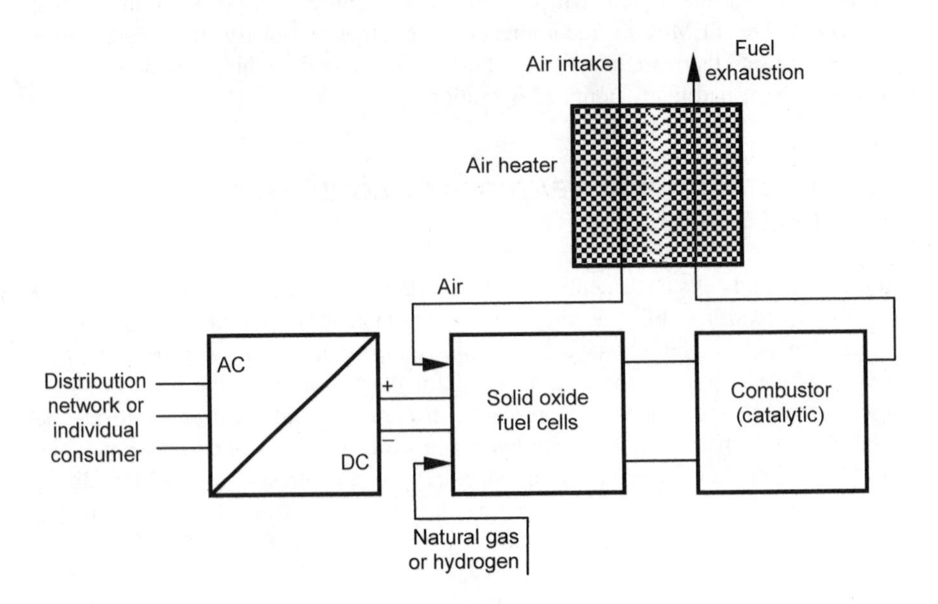

Figure 7.6 Cogeneration system using SOFCs.

heat and power generation units in shopping centers, hospitals, military headquarters, residential condominiums, public buildings, and stand-alone villages. In all these applications, the natural gas has to be desulfurized before feeding the anode, and air is admitted into the fuel cell through preheaters using exhausted anode and cathode hot gases.

7.7 WATER, AIR, AND HEAT MANAGEMENT

Optimum operation management of fuel cells is vital for maximum profits in this energy system in financial and energy use terms. The products to deal with are water, air, and heat, which are closely related to fuel cells in three aspects: (1) as a by-product of chemical reactions, (2) in the cooling circulation fluid, and (3) for heat recovery. The FC heat affects especially the speed of electrochemical reactions because it eases hydrogen extraction from hydrocarbonates and increases the overall efficiency of the fuel cell by means of heat recycling for both ambient and water heat and/or for heating boilers used as primary energy sources for turbine drivers of electrical generators and for hot water storage. As the heat is produced locally, the FC generating units may have increased their overall efficiency through the use of internally produced electricity and heat. Local production of energy decreases the transportation and distribution losses and alleviates the public network differently from what happens in concentrated power plants, usually located at very long distances from consumers [3,5].

The water resulting from chemical reactions in fuel cells has to be removed only partially since its presence increases the conductivity of the electrolyte. Conversely, if there is too much water in the electrodes bonding the electrolyte, pore flooding will cause difficulties for fuel permeation, and as a result, the concentration losses might be higher. A good relative humidity should be maintained between 85 and 100% for a reasonable operating balance. In some small fuel cells the air going through the cell may be the same for both oxygen reactions and moisture balance (stoichiometric feeding is usually at a rate higher than 2). For larger cells there are usually two independent air supplies, and the moisture may have to be complemented by an external vapor source. Use of these quantities may be expressed as [3]

$$\text{air rate} = 3.6 \times 10^{-7}\lambda \frac{P_{\text{fc}}}{V_{\text{fc}}} \quad \text{kg/s} \tag{7.4}$$

$$\text{H}_2 \text{ rate} = 1.05 \times 10^{-7} \frac{P_{\text{fc}}}{V_{\text{fc}}} \quad \text{kg/s} \tag{7.5}$$

$$\text{H}_2\text{O production rate} = 1.33 \times 10^{-4} \quad \text{kg/s} \tag{7.6}$$

The membrane relative humidity may be given in terms of the partial pressure of the water and saturated water vapor pressures:

$$\text{RH}(\%) = \frac{P_{\text{w}}}{P_{\text{sat}}}$$

The chemical reactions in a fuel cell produce heat as a result of the conversion enthalpy of a fuel into electricity. If the water product is in liquid form (without the latent water heat), the heat production may be approximated by

$$\text{heating rate} = P_{\text{fc}}\left(\frac{1.48}{V_{\text{fc}}} - 1\right) \qquad \text{J/s or W} \tag{7.7}$$

If the water product is in vapor form, the heat production may be approximated by

$$\text{heating rate} = P_{\text{fc}}\left(\frac{1.25}{V_{\text{fc}}} - 1\right) \qquad \text{J/s or W} \tag{7.8}$$

Either air heat from the cooling process or hot water from the chemical reactions may be used in heat exchangers for commercial, industrial, and residential purposes and for warming up vapor and/or air conditioning in domestic homes.

7.8 LOAD CURVE PEAK SHAVING WITH FUEL CELLS

The strategies presented in this section allow specification of the power rating for a fuel cell (FC) stack system under the approach of imposing a load curve to be as flat as possible. FC sizing can be based on the maximal load curve flatness, limiting the maximum load peak or the amount of the necessary thermal energy. Other methodologies may also be used for FC sizing, depending on operating mode and how hydrogen is going to be produced. In principle, the FC can operate at constant or variable power during a selected period, depending on the amount of hydrogen available. For each situation, a dedicated procedure should be applied [15–17].

7.8.1 Maximal Load Curve Flatness at Constant Output Power

The methodology of maximum load curve flatness at constant output power allows calculation under a set of constraints. The major consumer interest is to decrease demand during the peak period so as to reduce the cost of the energy consumed. On the other side, from a utility company point of view, there is, in addition to reduction in the maximum demand, a possibility of increasing the efficiency of the distribution network, due, for example, to a decrease in overall losses. The optimization rationale is based on limitations of load curve changes. Load can only be decreased during the load peak period (say, three hours) and can increase with the use of an electrolyzer only during nightly runs (say, a six-hour period). With this purpose, flatness of the load curve can be quantified by a form factor defined for every distribution transformer as [16]

$$\text{k}_f = \frac{P_{q\text{av}}}{P_{\text{av}}} \tag{7.9}$$

where P_{qav} is the quadratic average of the instantaneous power, defined as $P_{qav} = 1/n\sqrt{\sum_{t=1}^{n} P_t^2}, P_{av} = 1/n\sum_{t=1}^{n} P_t$, P_t is the transformer load at the time interval t, and n is the number of intervals considered.

Factor k_f is always greater than 1. If $k_f = 1$, the load curve would have an ideal flat shape. For the case of FC use,

$$P_{qav} = \frac{1}{n}\sqrt{\sum_{t=1}^{n} (P_t - P_{fc})^2} \tag{7.10}$$

where P_{fc} is the output power of the FC stack. The objective in this case is to determine P_{fc} to minimize the function $z = k_f - 1$, aiming at determination of the FC power output to make the load curve as flat as possible. The value of P_{fc} can be determined by the solution of

$$\frac{\partial z}{\partial P_{fc}(t)} = 0 \tag{7.11}$$

Current converters to feed electrolyzers are of the current source-controlled converter (CSCC) types. However, they present special characteristics of voltage and current. The rectifier current capacity depends on the amount of gas flow required. In the same way, it is possible to determine the power of the electrolyzer for production of hydrogen out of the peak period to obtain a factor k_f as close as possible to 1. The specific electrical power that can be obtained from a hydrogen-fed FC is 26.6 V_c kWh/kg, where V_c is the output voltage across the terminals of every single cell belonging to the FC stack [3,16]. This calculation uses the standard FC electrochemical model discussed in Section 7.12. The hydrogen density in the FC is 0.084 kg/m^3 (NTP), so the amount of H_2(m^3) necessary to generate the right amount of energy is determined in normal cubic meters (Nm^3) from

$$\text{volume of } H_2 = 0.44755\frac{P_{fc}h_{fc}}{V_c} \qquad \text{Nm}^3 \tag{7.12}$$

where P_{fc} is the FC rated power and h_{fc} is the number of operating hours within the peak period.

One faraday (96,489 C) produces 1 g of hydrogen (ideal case), that is, 8,105,076 As/m^3 or 2251.41 Ah/m^3. Based on Ref. [2], it can be also stated that 96,489 Ah can produce 42.84 Nm^3 of hydrogen. So 1 Nm^3 demands 2251.41 Ah. Taking into consideration that the voltage across each electrolyzer cell is V_{elec} (volts), the power demand to produce 1 Nm^3/h of H_2 would produce an ideal value of 2251.41 V_{elec} watts (neglecting losses). Now, considering the electrolyzer efficiency η_{elec}, the requisite energy to produce any amount of hydrogen is given by

$$W_{elec} = \frac{1.0077 P_{fc}\, h_{fc} V_{elec}}{V_c \eta_{elec}} \tag{7.13}$$

To incorporate the actual losses, the following equation for the electrolyzer power holds:

$$P_{e(t)} = \frac{1.0135 P_{\mathrm{fc}}\, h_{\mathrm{fc}} V_{\mathrm{elec}}}{V_{\mathrm{c}} \eta_{\mathrm{elec}} h_e} \tag{7.14}$$

where $P_{e(t)}$ is the electrolyzer power at time t, $P_{\mathrm{fc}(t)}$ is the FC power at the same instant t, and h_e is the number of operating hours of the electrolyzer. So P_{qav} can be now calculated by

$$P_{\mathrm{qav}} = \frac{1}{n}\sqrt{\sum_{t=1}^{n} \left(P_t - P_{\mathrm{fc}(t)} + P_{e(t)}\right)^2} \tag{7.15}$$

Figure 7.7*a* represents an example of the daily load curve of a small village served by an energy system with both a fuel cell and an electrolyzer operating under constant full power during both peak hours and off-peak hours. The difference between the two curves in the figure is the energy diverted to the electrolyzer from midnight to 6 p.m. and the energy received by the load from the fuel cell power plant between 6 and 9 p.m. The difference is due to the efficiency of the entire conversion loss.

There are other possible modes of peak shaving (Figure 7.7*b*) when the period of hydrogen production could be restricted to the morning hours. The fuel cell may operate at constant power or at variable power, so as to track as closely as possible consumer demand to keep the load curve at constant maximum demand.

7.8.2 Amount of Thermal Energy Necessary

The FC electrical efficiency may be determined from

$$\eta_{\mathrm{fc}} = \frac{P_{\mathrm{fc}}}{P_{\mathrm{total}}} \tag{7.16}$$

where P_{total} is the total power converted by the FC stack into electricity and heat. So P_{total} is obtained from

$$P_{\mathrm{total}} = P_{\mathrm{fc}} + Q_{\mathrm{fc}} \tag{7.17}$$

where Q_{fc} is the thermal energy produced in the stack.

Again, the thermal energy produced by the FC with power equal to P_{fc}(kW) working during h_{fc} hours may be expressed as

$$Q_{\mathrm{fc}} = P_{\mathrm{fc}}\left(\frac{1.48}{V_{\mathrm{fc}}} - 1\right) h_{\mathrm{fc}} \qquad \mathrm{kWh} \tag{7.18}$$

The stack heat losses can be separated into three categories: the rate of heat removed by the water cooling system, Q_w; the rate of heat loss in the air flowing out of the stack, Q_a; and other losses, Q_{other} (mostly stack heating and surface heat

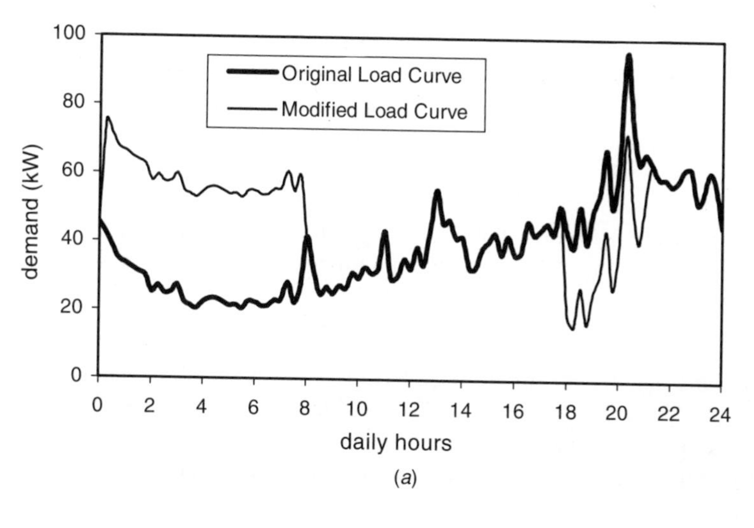

(*a*)

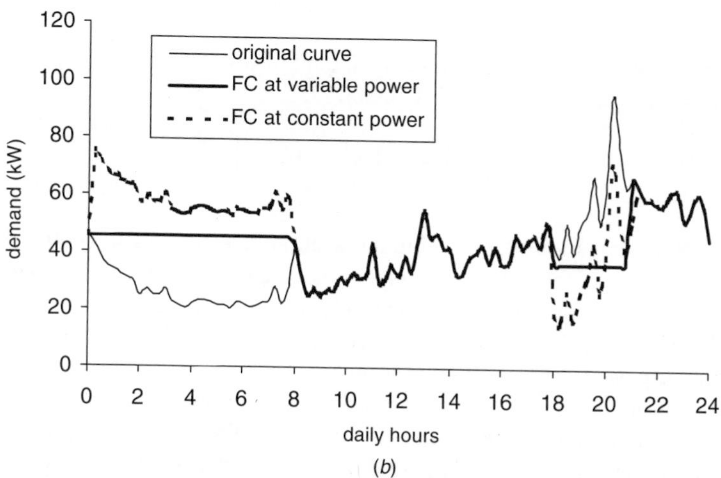

(*b*)

Figure 7.7 Typical residential load curves, including production of H_2: (*a*) hydrogen production during off-peak hours of a weekday (from 9 p.m. to 6 a.m. the next day) in a private residence; (*b*) hydrogen production during the early hours of a weekday (midnight to 8 a.m.) in a condominium.

exchange with the surroundings). The total heat loss is then given by

$$Q_{\text{fc}} = Q_w + Q_a + Q_{\text{other}} \tag{7.19}$$

It is common practice to consider the amount of heat removed by the air in a stack to be about 2% of P_{total} and the ambient losses to be 18% [16]:

$$Q_w = P_{\text{fc}}\left(\frac{1.48}{V_{\text{fc}}} - 1\right) n h_{\text{fc}} - Q_a - Q_{\text{others}} \tag{7.20}$$

An interesting case to be considered is water storage during the early hours of the day using the heat dissipated by the fuel cell between midnight and 6:00 p.m. An FC operating in this period allows enough heat dissipation in water for household applications such as those in the kitchen and bathroom in addition to air heating. The needs of a family can be accounted for on a per person basis i.e., assuming that each person uses an average of 100 L/day of hot water, so a family with n_p people will need an average flow of $100n_p$ L/day [16]. Based on these data, the necessary energy to heat up the required amount of water is calculated as

$$Q = cm\,\Delta T \tag{7.21}$$

where c is the specific heat of water 1.163×10^{-3} kWh/kg·°C), m the mass of water (liters), and ΔT is the difference (°C) or absolute temperature between the initial and final temperatures in the process.

The thermal energy needed to heat up mn_p liters of water is

$$Q = mn_pc\,\Delta T \tag{7.22}$$

The output power of an FC to heat up an amount of water (liters) during h_{fc} hours is obtained from equations (7.18), (7.19), and (7.22):

$$P_{fc} = \frac{mn_pc\,\Delta T + Q_a + Q_{other}}{(1.48 - V_{fc})h_{fc}} V_{fc} \quad \text{kW} \tag{7.23}$$

Based on the equations described in this section, it is possible to control the stack electric efficiency to obtain more thermal or electrical energy according to the appliance demand. Therefore, two operating modes may be selected for the fuel cell output power: variable power or constant power. Variable power aims at a constant load, and constant power aims at an optimal FC operating point for best operation and hydrogen consumption. In the general case, different operation regimes can be established between the storage and generation of energy. The electrolysis process can be used for these purposes in any off period or along shorter periods of time, in accord with the needs of the utility or the consumer. To evaluate operation at variable power, an iterative process has to be established. As already mentioned, due to losses in the electrolysis and in power generation with FCs, the total daily consumption of energy will increase, although now shifted to a more convenient period.

7.9 REFORMERS, ELECTROLYZER SYSTEMS, AND RELATED PRECAUTIONS

The main objective of using a reformer is to supply hydrogen locally to the fuel cell stack. It is a relatively complex system that demands automated control. The system is sensitive to several aspects, such as control failure, the presence of inflammable gases, carbon monoxide poisoning gas, and overtemperature operation. An appropriate design should foresee several redundant sensors to guarantee safe operation, monitoring the components and controlling the stack operation. Although such complex systems are usually encountered in industrial and automotive environments, a lot of research is still required for its use in efficient, safe, and reliable fuel cell systems.

There are two major fluids present in the reformer: the fuel input and the hydrogen generated. Precautions should be established for transportation, storage, distribution, and safety standards of the input fuel (methanol, propane, or natural gas). Safety requirements for hydrogen have already been adopted as related to small-scale storage and proper pipelining, but it is expected that the fuel cell industry will instill new practices with the growth of fuel cell applications.

During reforming, small amounts of rusted carbon monoxide can be produced, but as long as there is no leakage, it does not represent a danger. However, if there is any leakage before the oxidation process, it is not safe and is a potential system safety concern. Reformers usually work at a temperature of 1650°C and must be thermally insulated to maintain a bearable exterior temperature. Internally, water steam is used to cool the hydrogen. High internal temperatures can overheat and rupture the equipment. Leakage in the cooling area can produce steam jets at high temperatures with eminent danger.

Another way to obtain hydrogen is through renewable sources of energy. The process to be used for that is the electrolysis of water or direct concentration of sunlight on reactors. This process can produce almost pure hydrogen and oxygen fed only by secondary sources of energy, such as the surplus in power system installations in the early hours of the day; Earth heat, such as geothermal, deserts, and ocean temperature gradients; plus renewable sources of energy. Among renewable sources of energy with most promise for these purposes are hydroelectric power plants such as those in Hamburg and Vancouver; biomass, such as sugarcane in Brazil; solar, geothermal, and wind, as in the Middle East; and sea tidal and sea wave energy in almost every corner of the planet.

7.10 ADVANTAGES AND DISADVANTAGES OF FUEL CELLS

Fuel cells are very appropriate for power plants using cogeneration systems in circumstances where environmental issues such as a clean atmosphere, silence, and absence of vibrations are of particular concern. This is because of the absence of movable parts other than the circulation pump and gaseous fluid blowers. PAFCs or cells using fuel reduction are not very convenient for power plant purposes because they are complex and necessitate specialized engineering services. All that is just to maintain them safe and reliable, despite being smaller than those required for power plants that use rotating machines.

The modular nature of fuel cells allows their use in virtually any application that allows flexible expansion of a power plant and gradual investment to follow evolution of the loads. The energetic conversion factors are aided very favorably by flexibility in the use of fuel, high efficiency (it may get near to 70%), and low volume/power ratio. In addition, the fuel hydrogen can be homemade using water electrolysis with solar energy or reforming technology producing hydrogen from hydrocarbonate fuels.

FC systems adapt easily for energy injection applications into the utility grid because they have energy density (Wh/m^3) higher than that of standard batteries, relatively fast response to load fluctuation, high reliability, low cost of operation, and very low maintenance cost. Spilling of hydrogen will never be a major safety concern as this lighter-than-air gas flows up and away. When refueling fuel cells in vehicles, a very fast time is possible, about the same as that to fuel an ordinary gas-driven car, a very good advantage compared to recharging a battery.

The disadvantages are the typical ones in any new technology that requires time to mature and to be widely adopted. Fuel cell energy systems are unfortunately not characterized by historical data of reliability and continuous operation under harsh conditions. Another drawback of fuel cell systems is the need for expensive noble materials and susceptibilities to the contaminants present in the fuel.

Hydrogen is considered to be the most efficient FC fuel, but it is not freely available in nature. It has to be manufactured, and the full cycle of efficiency and cost must make sense. Therefore, the market for fuel cells is currently limited to special applications. On the other hand, hydrogen will not be widely available until there is an increase in demand with a corresponding economy of scale where the full cycle of efficiency sums up to a steady market. Although there are still some pessimist objections to adoption of a hydrogen economy, it is expected that sooner or later it will become a reality.

7.11 FUEL CELL EQUIVALENT CIRCUIT

Evaluation of the dynamic performance of fuel cells for studies of electrical energy generation systems is important to reduce cost and time at the design and testing stages. An electrical model may be derived from the electrochemical equations to enable determination of the open-circuit voltage and voltage drops of cells for a specified operating point [17–21]. In power generation systems, the dynamic response is of extreme importance for the control planner and system management, especially when energy is injected into the grid. So special attention has to be given to the dynamic response of FCs [22–24]. For energy injection into the grid, the generation control has to set the amount of power the FC will supply as a function of the load demand. As such, the dynamic FC response should be compatible with a fast variation in the random load curve, which is not always the case, as we discuss next [16,21,22].

From the reaction outlined in equations (7.1), (7.2) and (7.3), it is possible to obtain the electrical voltage generated in the electrochemical process in the cell if one recalls that two electrons pass through the external circuit for each water molecule produced and each molecule of hydrogen present in the process. So electrical work to move these charges is expressed as

$$G = \int 2q\, de = -2FE \tag{7.24}$$

where F is the Faraday constant (or the electron charge in every molecules) and E is the fuel cell voltage. If the system is reversible (no losses), the electrical work is equal to the *Gibbs free energy released*, which is the "energy available to do external work, neglecting any work done by changes in pressure and/or volume". All these forms of chemical energy are rather like ordinary potential energy with respect to the zero-point energy and energy variation with respect to this point. The Gibbs energy is listed in the literature [1] for the reaction of water formation from $2H_2$ and O_2 as in equation (7.3). For example, for a hydrogen cell operating at 200°C, the Gibbs energy is −220 kJ, and from equation (7.24), $E = 1.14$ V.

The activity of the reactants and products changes the Gibbs free energy of a reaction. Balmer [17] showed in 1990 that temperature and pressure do affect the reaction activity, resulting in an electromotive force given in terms of the product and/or reactant activity, called the *Nernst reversible voltage*, E_{Nernst}.

Therefore, reversible cell voltage is the cell potential obtained in an open-circuit thermodynamic balance (without load). In this section E_{Nernst} is calculated from a modified version of Nernst's equation, with two extra terms to account for changes in temperature with respect to the standard reference temperature, 25°C, and 100 kPa or 1.00 atm pressure [18–20], respectively. This is all given by

$$E_{\mathrm{Nernst}} = \frac{\Delta G}{2F} + \frac{\Delta S}{2F}(T - T_{\mathrm{ref}}) + \frac{RT}{2F}\left[\ln\left(p^*_{H_2}\right) + \frac{1}{2}\ln\left(p^*_{O_2}\right)\right] \tag{7.25}$$

where
ΔG = change in the Gibbs free energy (J/mol)
F = Faraday constant (96,487 C)
ΔS = entropy change (J/mol)
R = universal gas constant (8314 J/K·mol)
$p^*_{H_2}$, $p^*_{O_2}$ = partial pressure (atm) of the hydrogen and oxygen, respectively
T = absolute temperature of the operating cell (K)
T_{ref} = reference absolute temperature (K)

Using the foregoing standard temperature and pressure values for ΔG, ΔS, and T_{ref}, equation (7.25) can be simplified to [13–17]

$$E_{\mathrm{Nernst}} = 1.229 - 0.85.10^{-3}(T - 298.15) + 4.31 \times 10^{-5}T\left[\ln\left(p^*_{H_2}\right) + \frac{1}{2}\ln\left(p^*_{O_2}\right)\right] \tag{7.26}$$

As shown in Refs. [15–18], the electrode activation overpotential, including anode and cathode, can be calculated by

$$V_{\mathrm{act}} = -[\xi_1 + \xi_2 T + \xi_3 T\ \ln\left(c^*_{O_2}\right) + \xi_4 T\ \ln(i_{\mathrm{fc}})] \tag{7.27}$$

where the ξ_i are parametric coefficients for each cell model ($i = 1,2,3,4$) i_{fc} is the cell operating current (A), and $c^*_{O_2} = p^*_{O_2}/5.08e^{-498/T}$ is the oxygen concentration on the cathode catalytic interface (mol/cm^3). The values used in equation (7.27) are set by theoretical equations with kinetic, thermodynamic, and electrochemical foundations [15–18].

The ohmic overpotential results from the resistance to electron transfer in the collecting plates and carbon electrodes, and the resistance to proton transfer in the solid membrane. This resistance is essentially linear. In this model, a general expression for the resistance was defined to include all the important parameters of the membrane. The equivalent resistance of the membrane is then calculated by

$$R_m = \frac{\rho_M l}{A} \tag{7.28}$$

where ρ_M is the specific resistivity of the membrane for the flow of electrons ($\Omega \cdot$ cm), l the thickness of the membrane serving as cell electrolyte (cm) and A the cell active area (cm^2).

The Nafion membrane type considered in this book is used widely in PEMFC technology. Dupont uses the product designations shown in Table 7.2 to denote Nafion membrane thickness. The following numerical expression for the resistivity of Nafion membranes is used:

$$\rho_M = \frac{181.6\left[1 + 0.03(i_{\text{fc}}/A) + 0.062(T/303)^2(i_{\text{fc}}/A)^{2.5}\right]}{[\psi - 0.634 - 3(i_{\text{fc}}/A)]e^{4.18(T-303)/T}} \tag{7.29}$$

where $181.6/(\psi - 0.634)$ is the specific resistivity ($\Omega \cdot$ cm) at no load current and 30°C, T is the absolute temperature of the cell (K), ψ is an adjustable parameter with a possible maximum value of 23; and $e^{4.18(T-303)/T}$ is a temperature factor correction if the cell is not at 30°C. The parameter ψ is influenced by the membrane preparation procedure and is a function of the relative humidity and stoichiometric rate of the anode gas. It can have a value on the order of 14 under the ideal conditions of 100% relative humidity, and values on the order of 22 and 23 have been reported under oversaturated conditions.

Equation (7.28) may be used to obtain the ohmic overpotential of the membrane resistance:

$$V_{\text{ohmic}} = i_{\text{fc}}(R_M + R_C) \tag{7.30}$$

TABLE 7.2 Dupont Designations for Nafion Membrane Thickness

Dupont Designation	Thickness l
Nafion 117	7 mils (178μm)
Nafion 115	5 mils (127μm)
Nafion 112	2 mils (51μm)

where R_C represents the resistance of the electrodes and contacts to the ion transfer through the membrane (electrolyte), usually considered constant. The concentration or mass transport affects the hydrogen and oxygen concentrations. This, in turn, causes a decrease in the partial pressures of these gases. Reduction in the pressure of oxygen and hydrogen depends on the electrical current and physical characteristics of the system. To determine an equation for this voltage drop, it is defined as the maximum current density, $J_{\max}$, under which the fuel is being used at the same maximum supply rate. The current density cannot surpass this limit because the fuel cannot be supplied at a greater rate. Typical values for $J_{\max}$ are in the range 1000 to 1500 mA/cm^2. Thus, the voltage drop due to mass transport is

$$V_{\text{con}} = -B \ln\left(1 - \frac{J}{J_{\max}}\right) \tag{7.31}$$

where B is a constant depending on the cell and its operating state (V) and J is the actual current density of the cell electrode (A/cm^2).

Before setting up an equivalent circuit to represent the cell dynamics, it is interesting to examine in greater detail the phenomenon known as the *charge double layer*. Such a phenomenon normally exists on every contact between two different materials, due to a charge accumulation on the opposite surfaces or charge transfer from one to the other. The charge layer on both electrode–electrolyte interfaces (or close to the interface) is the storage of electrical charges and energy; in this way it behaves as an electrical capacitor. If the current changes, there will be some elapsed time for the load (and its associated voltage) to decay (if the current decreases) or to increase (if the current increases). Such a delay affects the activation and concentration potentials. It is important to point out that the ohmic overpotential is not affected, since this has a linear relationship with the cell current through Ohm's law. Thus, a change in the current causes an immediate change in the ohmic voltage drop. In this way it can be considered that a delay of first order exists due to the activation and concentration voltages only. The associated time delay $\tau(s)$ is the product [18–21]

$$\tau = CR_a \tag{7.32}$$

where C is the equivalent capacitance of the system (farads) and R_a is the equivalent variable resistance to the activation and concentration losses (ohms).

The value of the capacitance is some few farads. The resistance R_a is determined from steady-state evaluation of the cell current and the activation and concentration voltages. That is, after the end of the transient, there will be stable final values of voltage and current, which are used to calculate the resistance through [22,23]

$$R_a = \frac{V_{\text{act}} + V_{\text{con}}}{i_{\text{fc}}} \tag{7.33}$$

In broad terms, the capacitive effect assures the good dynamic performance of the cell, since the voltage moves smoothly to a new value in response to a change in the load current. The electrical output energy of the cell is linked to a certain load, such as the load represented in Figure 7.5. There is no restriction with respect to load type as long as the power supplied by the stack is capable of feeding it, or if it does not represent starting motors and fast transient response loads. For example, in systems for injection of energy into a distribution network, the load can be a dc–dc boost converter, followed by a dc–ac inverter, connected to the public network through a transformer. In stand-alone systems, it can be a pure resistive load (heating) or a resistive–inductive load (motor) [24–27]. In any case, the average current density $J(\mathrm{A/cm^2})$ through the cell cross-sectional area is defined as

$$J = \frac{i_{\mathrm{fc}}}{A} \tag{7.34}$$

The instantaneous electrical power supplied by the cell to the load can be determined by

$$P_{\mathrm{fc}} = V_{\mathrm{fc}} i_{\mathrm{fc}} \tag{7.35}$$

where V_{fc} is the cell output voltage for each operating condition (volts) and P_{fc} is the corresponding power (watts). In Table 7.3 are listed values of the parameters discussed in this section for a typical Ballard Mark V fuel cell.

The equations derived above in this section allow the construction of an equivalent circuit model as shown in Figure 7.8 to represent FC dynamic behavior as given in many references [2,13,14,16–18]. Equation (7.26) represents the active source E_{Nernst}. The membrane and contact ohmic losses are represented by $R_r = R_m + R_c$ [equation (7.33)]. The time delay given in equation (7.32) includes the capacitor, C, to represent the double-charge-layer effects on the output voltage and current. This capacitance is very large (a few farads), since it is directly proportional to the electrode area and the manufacturing material, and it is inversely proportional to the electrode thickness, which is numerically very small (a few nanometers).

TABLE 7.3 Typical Parameters of a Ballard Mark V Fuel Cell

Parameter	Value	Parameter	Value
T(K)	343.15	ξ_1	−0.948
A (cm^2)	50.6	ξ_2	$0.00286 + 0.0002 \ln A + (4.3 \times 10^{-5}) \ln C^*_{H_2}$
l (μm)	178	ξ_3	7.6×10^{-5}
$c^*_{O_2}$ (mol/cm^3)	1.10^{-4}	ξ_4	$-1.93\ 10^{-4}$
$c^*_{H_2}$ (mol/cm^3)	1.10^{-4}	ψ	23.0
$p^*_{O_2}$ (atm)	1	B	0.016 V
$p^*_{H_2}$ (atm)	1	$J_{\max}$	1500 mA/cm^2

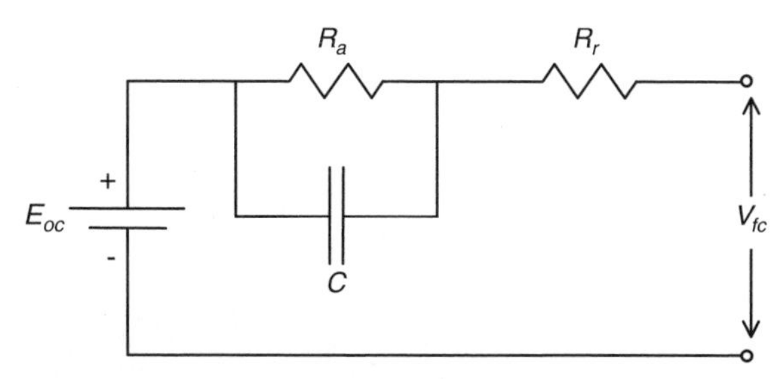

Figure 7.8 Equivalent circuit of the dynamic behavior of a fuel cell.

The activation and concentration losses (mass transport) cause a voltage drop across the resistor represented by R_a as indicated by equation (7.30). The fuel crossover and internal currents are essentially small, equivalent, and less important than other losses in terms of the cell operating efficiency. However, it has a very notable effect on the open-circuit voltage of low-temperature cells. Therefore, from Figure 7.8, the output voltage of a single cell can be defined as the result of the expression

$$V_{\mathrm{fc}} = E_{\mathrm{Nernst}} - V_{\mathrm{act}} - V_{\mathrm{con}} - V_{\mathrm{ohmic}} \tag{7.36}$$

Typical values for the components used in the circuit model of Figure 7.8 may vary widely with the size, construction, and type of fuel cell, but as an example, the following parameters could be used: $E_{\mathrm{Nernst}} = 1.48V$ and $C = 3F$; R_a is a function of i_{fc} as given in equations (7.27) and (7.33).

The first term appearing in equation (7.36) represents the open-circuit voltage of the FC (operating voltage without load); the last three terms represent voltage drops to the net voltage of the cell, V_{fc}, at a certain operating current. The term E_{Nernst} is the thermodynamic potential of the cell and represents its reversible voltage; V_{act} is the voltage drop due to the anode and cathode activation over potential, a measure of the voltage drop associated with the electrodes; V_{con} represents the voltage drop resulting from the concentration or mass transportation of oxygen and hydrogen (concentration over potential); and V_{ohmic} is the ohmic voltage drop (ohmic over potential), a measure of the ohmic losses associated with the conduction of protons through the solid electrolyte and internal electronic resistances [8]. Many other models can be found in the literature. [16–22].

Figure 7.9 shows the performance curves for a typical membrane used in fuel cells, which may be superimposed on the theoretical curves described by equations (7.35) and (7.36), respectively, for P_{fc} and V_{fc}. The efficiency η is defined as the relationship between the electric output power and the energy corresponding to the fuel input [3]. When the water product is in liquid form, it is usually given as

$$\eta = \mu_f \frac{V_c}{1.48} \tag{7.37}$$

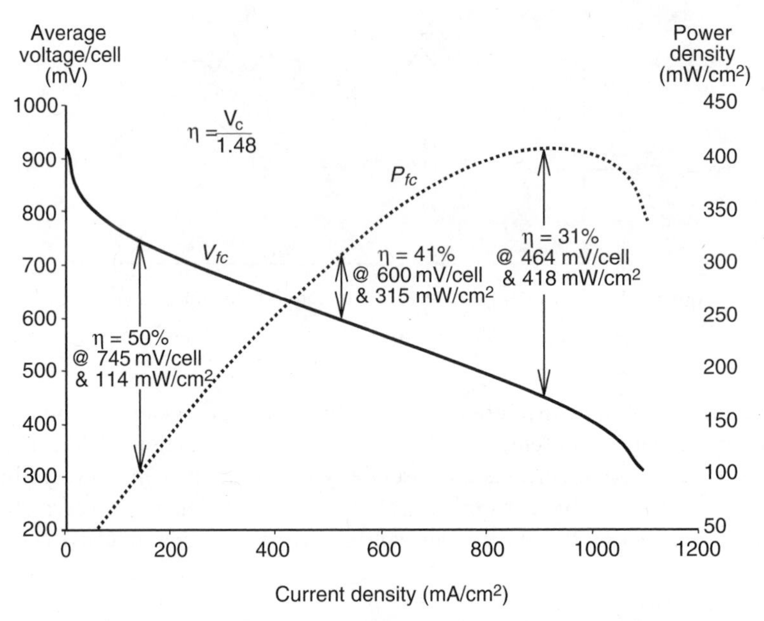

Figure 7.9 Typical curves of fuel cells.

On the other hand, when the water product is in vapor form, the efficiency is generally given as

$$\eta = \mu_f \frac{V_c}{1.25} \tag{7.38}$$

where μ_f is the fuel utilization factor, generally about 95%, and 1.48 and 1.25 V are the maximum voltages that can be obtained using the highest and lowest values, respectively, of the cell enthalpy.

7.12 PRACTICAL DETERMINATION OF THE EQUIVALENT MODEL PARAMETERS

Several techniques have been used to obtain equivalent circuit parameters. A possible method is the electrical impedance spectroscopy in which a variable-frequency alternating current is applied through the cell, causing a voltage drop across its terminals. The cell equivalent impedance is derived from the relationship between this voltage and the applied alternating current. The frequencies used in these tests may be as low as 10 mHz.

Another very simple method used to obtain equivalent model parameters is the current interruption technique. In this method, the concentration or mass transport overvoltage has to be neglected. Assume that a load is connected across the FC terminals at V_L (volts) and I_L (amperes) represented in Figure 7.8. An

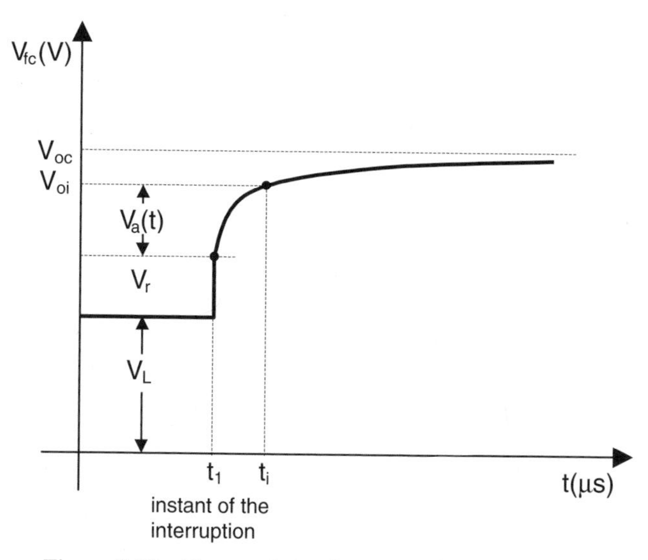

Figure 7.10 Characteristic of a current interruption test.

interruption in the FC load current will cause a sudden voltage increase V_r across its terminals. This is measured through a data acquisition system or a storage oscilloscope to show that there is a sudden jump of voltage followed by a capacitor charge–like response, as illustrated in Figure 7.10. Therefore, there is a need not only for instantaneous switching but also for fast sampling of the voltage increase to enable a clear separation of the activation overvoltage and the overvoltage due to the ohmic losses. This can be interpreted as if at the exact instant of the current interruption the voltage across the charge-double-layer capacitance cannot immediately change [21–24]. After some time the FC output voltage would tend to $E_{\text{oc}} = V_L + V_r + V_a$. A small data extrapolation may be necessary to intercept the vertical line of the exact instant when there was the current interruption. As zero current is assumed to occur after the load current, there will only be capacitor discharge through R_a. The following expressions could be used to express these changes:

$$R_r = \frac{V_r}{I_L} \tag{7.39}$$

$$R_a + R_r = \frac{E_{\text{oc}} - V_L}{I_L} \tag{7.40}$$

Notice that for a given electrode area, R_a is dependent on the load current, temperature, and specific resistivity. So the voltage across the double-layer capacitance $V_{t1} = I_L R_a = E_{\text{oc}} - V_L - V_r$ at the very instant the load current was interrupted may be considered as the discharge of a capacitor through the variable resistance

R_a. Therefore, the activation voltage drop may be expressed as

$$v_a(t) = V_{t1}\left[1 - \exp\left(\frac{t - t_1}{R_a C}\right)\right] \quad \text{for } t > t_1 \tag{7.41}$$

or for a generic instant t_i,

$$\tau_i = R_{ai}C = \frac{t_i - t_1}{\ln[V_{t1}/(V_{t1} - v_{ai})]} \tag{7.42}$$

The tangent slope expression at the current interruption instant is complex but may be experimented by

$$\left.\frac{dv_a}{dt}\right|_{t=t1} = \frac{V_{t1}}{R_a C} \tag{7.43}$$

The instantaneous terminal voltage of the fuel cell at any generic instant t_i is given by

$$E_{oi} = V_L + V_r + v_{ai} \tag{7.44}$$

Figure 7.8 is the equivalent dynamic model of a fuel cell. Its best representation depends on the precise detection of the voltage jump V_r at the exact instant of the current interruption, which demands a high-speed data acquisition system or storage oscilloscope. The foregoing parameters can be calculated through the following procedure:

1. Perform a current interruption test and plot the current interruption characteristic as in Figure 7.10.
2. On this plot, measure V_{oc},V_r, V_L, and I_L.
3. $R_r = V_r/I_L$.
4. $R_{a1} = [(V_{oc} - V_r)/I_L] - R_r$.
5. $V_{t1} = I_L R_{a1}$.
6. From equation (7.44) comes $v_{ai} = V_{oi} - (V_L + V_r)$.
7. Assuming a large, constant values of C and R_a, for a time period very close to the current interruption instant comes

$$\frac{dv_a}{dt} = \frac{V_{t1}}{R_a C} \simeq \left.\frac{\Delta V}{\Delta t}\right|_{t \to t1} \tag{7.45}$$

and from equation (7.43) comes

$$\begin{aligned}\tau_i = R_{ai}C &= \frac{t_i - t_1}{\ln[V_{t1}(V_{t1} - v_{ai})]} \quad \text{and} \quad \tau_{i+1} = R_{ai+1}C \\ &= \frac{t_{i+1} - t_1}{\ln[V_{t1}/(V_{t1} - v_{ai+1})]} \quad \text{for } i > 1\end{aligned} \tag{7.46}$$

Dividing any two successive values given by the recurrence formula in equation (7.43) in an orderly way beginning with $i = 1$ yields

$$R_{ai+1} = R_{ai}\frac{\tau_{i+1}}{\tau_i}$$

7.12.1 Example of Determination of FC Parameters

Assume a 32-cell PEM stack with a cell area $A = 100\,\text{cm}^2$ and the characteristic given in Figure 7.8 for which the current interruption characteristic as plotted in Figure 7.11 to obtain the stack parameters.

Step 1. Using a digital storage oscilloscope, plot the current interruption characteristic of the cell as in Figure 7.10.

Step 2. Measure V_{oc}, V_r, V_L, and I_L:

$$E_{oc} = 29.440V \qquad V_L = 16.269V$$
$$V_r = 6.004V \qquad I_L = 100\ \text{cm}^2 \times 520\ \text{mA/cm}^2 = 52\ \text{A}$$

Step 3. $R_r = V_r/I_L = 6.004/52 = 0.115\ \Omega$.

Step 4.

$$R_{a1} = \frac{E_{oc} - V_r}{I_L} - R_r = \frac{29.44 - 6.004}{52} - 0.196 = 0.336\ \Omega$$

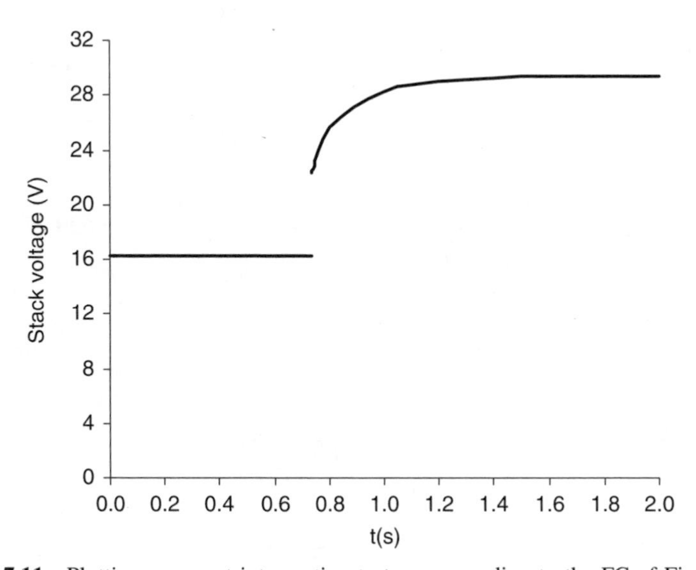

Figure 7.11 Plotting a current interruption test corresponding to the FC of Figure 7.9.

Step 5. The current interruption test begun at $t_1 = 0.7334\ \mu s$, when it was calculated that

$$V_{t1} = I_L R_{a1} = 17.472\ V$$

Step 6. From the digital oscilloscope data, at $t_2 = 0.8\ \mu s$ and $t_3 = 1.0\ \mu s$ from the beginning of the test, the respective values of the output voltage were $V_{o2} = 25.826\ V$ and $V_{o2} = 28.231\ V$. Therefore, the activation voltages were

$$v_{a2} = V_{o2} - (V_L + V_r) = 25.826 - (16.269 + 6.004) = 3.553\ V$$
$$v_{a3} = V_{o3} - (V_L + V_r) = 28.231 - (16.269 + 6.004) = 5.958\ V$$

Step 7. Assuming a constant and large value of C for a time instant very close to the current interruption instant, we obtain

$$\frac{dv_a}{dt} = \frac{V_{t1}}{R_a C} \cong \left.\frac{\Delta V}{\Delta t}\right|_{t \to t1}$$

or in terms of the capacitance,

$$C = \frac{V_{t1}}{R_a \left.\dfrac{\Delta V_o}{\Delta t}\right|_{\Delta t \to 0}} = \frac{17.472}{0.336 \dfrac{17.975 - 17.472}{(7.362 - 7.334) \times 10^{-6}}} = 3232\ \mu F$$

Combining this value with that from step 4 yields

$$\tau_1 = R_{a1} C = 0.336(3232 \times 10^{-6}) = 1.086\,\text{ms}$$

and from equation (7.43) comes

$$\tau_2 = R_{a2} C = \frac{t_2 - t_1}{\ln\left(\dfrac{V_{t1}}{V_{t1} - v_{a2}}\right)} = \frac{0.8000 - 0.7334}{\ln\left(\dfrac{17.472}{17.472 - 3.553}\right)} = 83.600\,\text{ms}$$

$$\tau_3 = R_{a3} C = \frac{t_3 - t_1}{\ln\left(\dfrac{V_{t1}}{V_{t1} - v_{a3}}\right)} = \frac{1.0000 - 0.7334}{\ln\left(\dfrac{17.472}{17.472 - 5.958}\right)} = 639.273\,\text{ms}$$

Dividing these two successive values in an orderly way for $i = 1, 2, 3$ gives us

$$R_{a1} = 0.336\,\Omega$$
$$R_{a2} = R_{a1}\frac{\tau_2}{\tau_1} = 0.336\left(\frac{83.600}{1.086}\right) = 76.98\,\Omega$$
$$R_{a3} = R_{a2}\frac{\tau_3}{\tau_2} = 76.98\left(\frac{639.273}{83.600}\right) = 588,54\,\Omega$$

and so on.

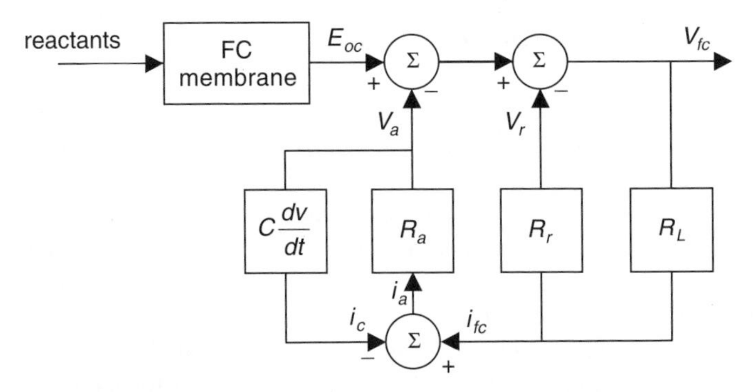

Figure 7.12 Flow diagram calculation of FC output voltage.

Several computer software products can be used for solving or simulating electrical circuits, such as MatLab/SIMULINK, PSim, or PSpice, with a computation flow as in the graph illustrated in Figure 7.12. In this calculation the internal voltage drops are calculated from the fuel cell internal parameters measured in laboratory tests. Either solutions suggested in Figure 7.13 could be used to obtain the voltage across the equivalent capacitance of the charge double layer [18,19]. The solution in Figure 7.12*a* is preferred since it is very easy to calculate the initial value of i_a assumed soon after any constant-state condition prior to the desired calculation interval:

$$i_{ao} = I_L = \frac{V_{t1}}{R_{ao}} = \frac{V_{\mathrm{oc}} - V_L - V_r}{R_{ao}}$$

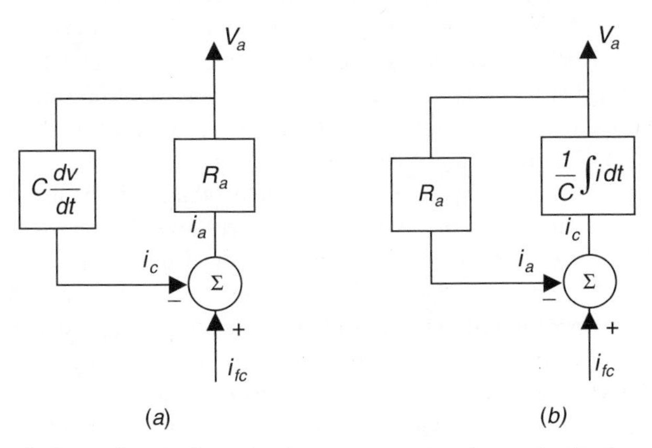

Figure 7.13 Options of equivalent circuits to represent a charge double layer: (*a*) solution by derivative; (*b*) solution by integration.

7.13 ASPECTS OF HYDROGEN AS FUEL

Hydrogen is the lightest and most buoyant element, so if it is released into an open space, it disperses quickly, reducing dramatically the chance of ignition. In general, for every 96 parts of air, at least 4 parts must be hydrogen before there is a threat of combustion. This is actually quite a high concentration relative to other commonly used fuels. For this reason it is good to know the conditions under which this happens based on such properties as: ignition energy, ease of flotation, diffusivity, flammability limits in the air, and combustion energy.

The combustion energy of the hydrogen is very low, which makes it even easier to catch fire, being in the ratio 1:10 for gasoline and 1:15 for natural gas or propane. All these gases are in fact of very low ignition energy, such that the probability of ignition of a mixture of any of them is relatively high even for a weak ignition source. Comparable examples include mixtures of 4 to 75% of hydrogen in the air, 5 to 16% of natural gas in the air, and 1.4 to 7.6% of gasoline steam in the air.

Hydrogen possesses 2.4 times more combustion energy stored per unit of mass than does natural gas or gasoline. In volumetric terms hydrogen has much less energy: It has 25% of the explosion energy of natural gas and 0.3% of the liquid gasoline per unit of volume under normal conditions of temperature and pressure. The amount of energy stored in small systems of hydrogen is usually less than that corresponding to 4 L of gasoline.

Advantages of hydrogen include its high diffusivity and flotation capacity. Such properties help to avoid fuel mixtures, and when this happens, they last for shorter times. Hydrogen is four times more diffusive than natural gas and eight times more diffusive than gasoline.

The first strategy used to prevent hydrogen from starting a fire is based on reducting the possibility of forming a combustible mixture, which can be accomplished using very well sealed canalization to avoid leakage. When one of these happens, the hydrogen will be dispersed quickly unless it is contained. The second strategy would be an ambient well ventilated to reduce or, in some cases, to eliminate the area in which a combustible mixture may develop. In the case of leaks, this reduces the exposure time to the possibility of developing an eventual combustible mixture. The third strategy is the minimization of near-ignition sources such an as static discharges, open fires, hot surfaces (temperatures higher than 585°C), and other equipment able to produce sparks.

The flammability of the hydrogen is the greatest danger in its use. However, the problem may be limited by storage of only small volumes. The most dangerous and susceptible parts refer to canalization for the gas, electrical equipment, the fuel cell itself, the control system, and the system reformer. A control system should be used to establish safe operation of the power plant. If it extrapolates the safety limits, the control system should turn off the power plant, switching alarms on and possibly informing operators regarding details of the dangerous conditions. The control system should also minimize dangerous situations due to single, multiple, or simultaneous failure of components.

The efficiency of a fuel cell is a function of the operating voltage of the entire stack. Higher voltages produce higher efficiency and therefore less consumption of hydrogen per kilowatthour, but less output power is supplied. For a better trade-off, the choices for a given output power are between (1) operating at higher voltages, increasing the number of cells in the stack and thus the capital costs or (2) operating at higher densities of current with fewer cells, higher fuel costs, and a shorter useful life.

Fuel cell and hydrogen costs, together with stack efficiency, will determine the final costs ($/kWh). The current technology, shown in Figure 7.9, displays the performance of an individual cell belonging to a PEM stack with 70 cells, providing of 33% efficiency in generating electric power at a cost of $0.23/kWh based on a hydrogen cost of $12/GJ. A higher fuel cost would require a stack with more cells operating at higher efficiency to minimize the total cost.

7.14 FUTURE PERSPECTIVES

Since fuel cell principles were discovered more than one and half centuries ago by a layperson and continue to be studied today in sophisticated laboratories, fuel cell technology is about to become mature and to change the way our society handles the production, storage, and delivery of energy. Enormous investments are being committed to the development of a reliable and workable energy system able to replace the fossil fuel model in current use. The benefits of such a breakthrough would be to decrease pollution in urban areas, reduce greenhouse gas emissions, and increase the energy independence of oil-consuming countries. But hydrogen cannot be compared directly with fossil energies, because it is only an energy vector, not an energy source. As such, it simply makes it possible to transmit a given quantity of energy from the place of production to the place of consumption.

As a result of current social and economic conflicts such as the continuing profitability of fossil fuel providers, it is not perhaps lack of technological knowledge that causes the greatest difficulty but selection of the right moment to begin the transforming process. Automobile manufacturers have been very aggressive and investing massively in developing not only fuel cell systems but all the required peripherals for easy integration to household use.

The most important barriers to development of fuel cell applications, besides cost, are the absence of production infrastructure and distribution networks, as well as the difficulties encountered in developing hydrogen storage technologies. Therefore, use of hydrogen in the transport sector should remain relatively limited in the short run. On the other hand, the production of electricity (stationary fuel cells) and the storage of energy for mobile equipment (cell phones, laptops, etc.) are expected to be commercially available at competitive costs in the very near future.

The expected worldwide shift in the political power balance, new commercial developments, environmental concerns, and energetic solutions are being consolidated. A new equilibrium point will eventually be found for a future era that will provide the same level of comfort, lifestyle, and power strategies or better than those of our present society.

REFERENCES

[1] M. R. Fry, Environmental impacts of electricity generation: fuel cells, *IEE Proceedings*, Vol. 140, No. 1, pp. 40–46, January 1993.

[2] J. L. Pepper, Fuel cells: clean energy for sustainable development, *Reviews of Modern Electricity*, pp. 224–236, April 1999.

[3] J. E. Larminie and A. Dicks, *Fuel Cell Systems Explained*, Wiley, Chichester, West Sussex, England, 2000.

[4] Teledyne Brown Engineering, *Energy Systems Using Fuel Cells for Remote Applications*, Final Report, Teledyne Brown Engineering Energy Systems, Schatz Energy Research Center Humbolt State University, Arcata, CA, and Palm Desert City Canada, February 1998.

[5] G. Hoogers, *Fuel Cell Technology Handbook*, CRC Press, Boca Raton, FL, 2003.

[6] Advanced Alternative Energy Corp., Business opportunities, http://www.aaecorp.com, June 1999.

[7] TVE International, Intermediate Technology Development Group, Schumacher Centre for Technology and Development, Information Service Unit, Rugby, Warwickshire, England, http://www.oneworld.org/itdg, or http://www.irdg.org.pe, 1998.

[8] W. L. Hughes, *Energy in Rural Development: Renewable Resources and Alternative Technologies for Developing Countries*, Advisory Committee on Technology Innovation, Board on Science and Technology for International Development, Commission on International Relations, National Academy of Sciences, Washington, DC, 1976.

[9] Carl Von Duisberg Gesellschaft, *Sustainable Germany: The Contribution to Sustainable Global Development*, Wupertal Institute for Climate, Environment, Energy in the North Rhine-Westphalian Science Centers, Bünde, Germany, 1995.

[10] R. Hill and A. E. Baumann, Environmental costs of photovoltaic energy, *IEE Proceedings*, Vol. 140, No. 1, pp. 76–80, January 1993.

[11] R. F. Mann, J. C. Amphlett, M. A. I. Hooper, H. M. Jensen, B. A. Peppley, and P. R. Roberge, Development and application of a generalized steady-state electrochemical model for a PEM fuel cell, *Journal of Power Sources*, Vol. 86, pp. 173–180, 2000.

[12] J. J. Baschuck and X. Li, Modeling of polymer electrolyte membrane fuel cells with variable degrees of water flooding, *Journal of Power Sources*, Vol. 86, pp. 181–196, 2000.

[13] J. C. Amphlett, R. F. Mann, B. A. Peppley, P. R. Roberge, and A. Rodrigues, A model predicting transient responses of proton exchange membrane fuel cells, *Journal of Power Sources*, Vol. 61, pp. 183–188, 1996.

[14] G. R. Ault and J. R. McDonald, An integrated SOFC plant dynamic model for power systems simulation, *Journal of Power Sources*, Vol. 86, pp. 495–500, 2000.

[15] D. Chu and R. Jiang, Performance of polymer electrolyte membrane fuel cell (PEMFC) stacks, part I: Evaluation and simulation of an air-breathing PEMFC stack, *Journal of Power Sources*, Vol. 83, pp. 128–133, 1999.

[16] L. N. Canha, V. A. Popov, A. R. Abaide, F. A. Farret, A. L. König, D. P. Bernardon, and L. Comassetto, Multicriterial analysis for optimal location of distributed energy sources considering the power system, presented at the 9th Symposium of Specialists in Electric, Operational and Expansion Planning, Rio de Janeiro, Brazil, May 2004.

[17] R. Balmer, *Thermodynamics*, West Publishing, st. Paul, MN, 1990.

[18] R. M. Nelms, D. R. Cahela, and B. J. Tatarchuk, Modeling double-layer capacitor behavior using ladder circuits, *IEEE Transactions on Aerospace and Electronic Systems*, Vol. 39, No. 2, April 2003.

[19] J. M. Corrêa, F. A. Farret, L. N. Canha, and M. G. Simões, An electrochemical-based fuel-cell model suitable for electrical engineering automation approach, *IEEE Transactions on Power Delivery*, Vol. 17, No. 2, pp. 467–476, April 2002.

[20] Y. Kim and S. Kim, An electrical modeling and fuzzy logic control of a fuel cell generation system, *IEEE Transactions on Energy Conversion*, Vol. 14, No. 2, pp. 239–244, June 1999.

[21] C. J. Hatziadoniu, A. A. Lobo, F. Pourboghrat, M. Daneshdoost, J. T. Pukrushpan, A. G. Stefanopoulou, and Huei Peng, A simplified dynamic model of grid-connected fuel-cell generators, *IEEE Transactions on Industrial Electronics*, Vol. 51, No. 5, pp. 1103–1109, October 2004.

[22] J.T. Pukrushpan, A.G. Stefanopoulou, and H. Peng, Modeling and control for PEM fuel cell stack system, in *Proceedings of the American Control Conference*, Anchorage, AK, pp. 3117–3122, May 2002.

[23] C. Wang and M. H. Nehrir, PSpice circuit model for PEM fuel cells, *IEEE Transactions on Power Electronics*, Vol. 19, No. 5, September 2004.

[24] W. Friede, S. Raël, and B. Davat, Mathematical model and characterization of the transient behavior of a PEM fuel cell, *IEEE Transactions on Power Electronics*, Vol. 19, No. 5, pp 1234–1241, September 2004.

[25] J. C. Amphlett, R. F. Mann, B. A. Peppley, P. R. Roberge, and A. Rodrigues, *A Practical PEM Fuel Cell Model for Simulating Vehicle Power*, Royal Military College of Canada, Kingston, Ontario, Canada, pp. 221, 226.

[26] F. A. Farret, J. R. Gomes, and A. S. Padilha, Comparison of the hill climbing methods used in micro power plants, *in Proceedings of INDUSCON 2000*, pp. 756–760.

[27] Dachuan Yu and S. Yuvarajan, A novel circuit model for pem fuel cells, *Proceedings of the Nineteenth Annual IEEE Conference and Exposition on Applied Power Electronics*, APEC2004, Vol. 1, pp 362–366.

CHAPTER 8

BIOMASS-POWERED MICROPLANTS

8.1 INTRODUCTION

Biomass energy (or *bioenergy*) is energy derived from organic matter. This energy has the potential to greatly reduce worldwide greenhouse gas emissions. Organic components from municipal and industrial waste, plants, agricultural and forestry residues, home waste, and landfills can be used very efficiently in our society [1–3]. Table 8.1 illustrates the biomass potential of several sources. Dedicated energy plantations are spread throughout the world. For example, there is a program in China targeting 13.5 million hectares of fuel wood by 2010. Three million hectares of eucalyptus is used for charcoal in Brazil, 16,000 ha of willow plantations is used for generation of heat and power in Sweden, and in the United States 50,000 ha of agricultural land has been converted to woody plantations and may rise to as much as 4 million hectares (10 million acres) by 2020. The U.S. Department of Energy road map for biomass is illustrated in Figure 8.1.

Biomass generates about the same amount of carbon dioxide as do fossil fuels (when burned), but from a chemical balance point of view, every time a new plant grows, carbon dioxide is actually removed from the atmosphere. The net emission of carbon dioxide will be zero as long as plants continue to be replenished for biomass energy purposes. If the biomass is converted through gasification or pyrolysis, the net balance can even result in removal of carbon dioxide. Energy crops such as

Integration of Alternative Sources of Energy, by Felix A. Farret and M. Godoy Simões

TABLE 8.1 U.S. Potential in Biomass

Classification	Biomass Type	Amount
Municipal solid waste/landfills	Quantity of raw material	167 million tonnes
	Direct use from combustion	217,722 TJ
	Electricity generation capacity	2,862,000 kW
	Electricity generating	71,405 TJ
	Total energy production	289,127 TJ
Forestry/wood processing	Electricity generating capacity	6,726,000 kW
	Electricity generation	124,712 TJ
	Direct use from combustion	2,306,026 TJ
	Total energy production	2,430,738 TJ
Agricultural residues		
Corn	Quantity of raw material	13.5 million tonnes
	Ethanol fuel production capacity	152,376 TJ/yr
	Yield of ethanol	8. 8 GJ/tone
	Ethanol fuel production	118,010 TJ
Soy bean and waste food oils	Biodiesel production capacity	6,708 TJ/yr
	Yield of biodiesel	40 GJ/tone
	Biodiesel production	671 TJ
Wood pellets	Quantity of raw material available	0.582 million tonnes
	Direct use from combustion	8,872 TJ
Other biomass	Electricity generating capacity	10,602,000 kW
	Electricity generation	11,328 TJ
	Direct use from combustion	102,084 TJ
	Total energy production	113,412 TJ

Source: www.worldenergy.org

fast-growing trees and grasses are called *biomass feedstocks*. The use of biomass feedstocks can help increase profits for the agricultural industry.

Combustion deriving from biomass, such as burning wood, has been used from prehistoric times to the present. However, it is not very efficient. Converting solid biomass to a gaseous or liquid fuel by heating it with limited oxygen prior to combustion greatly increases the heating value and overall efficiency, making it possible to convert the biomass to other valuable chemicals or materials. The gasification of biomass is a developing energy technology among various systems

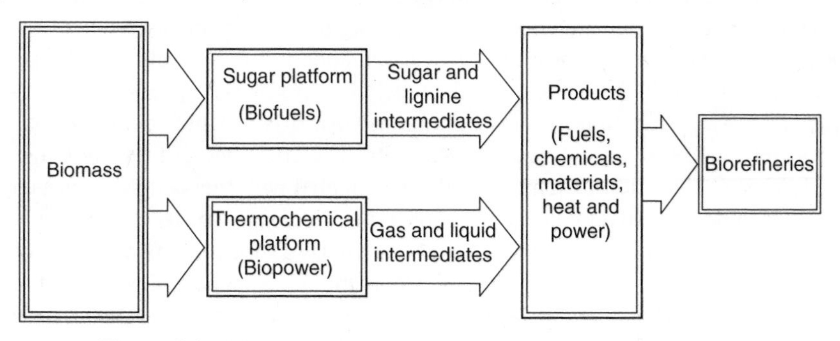

Figure 8.1 Biomass energy conversion. (From Refs. [1] and [2].)

for the energetic utilization of biomass. Biomass gasification has the following principal advantages over conventional combustion technologies:

1. The combined heat and power generation (via biomass gasification techniques connected to gas-fired engines or gas turbines) can achieve significantly higher electrical efficiencies, between 22 and 37%, than those of biomass combustion technologies with steam generation and steam turbine, 15 to 18%. If the gas produced is used in fuel cells for power generation, an even higher overall electrical efficiency can be attained, in the range of 25 to 50%, even in small-scale biomass gasification plants and under partial load operation.
2. Due to the improved electrical efficiency of the energy conversion via gasification, the potential reduction in CO_2 is greater than with combustion. The formation of NO_x compounds can also be largely prevented, and the removal of pollutants is easier for various substances. The NO_x advantage, however, may be partly lost if the gas is subsequently used in gas-fired engines or gas turbines. Significantly lower emissions of NO_x, CO, and hydrocarbons can be expected when the gas produced is used in fuel cells rather than in gas-fired engines or gas turbines.

Researchers at the U.S. Department of Energy Biomass Program are leading a national effort to develop thermochemical technologies to tap the enormous energy potential of lignocellulosic biomass more efficiently [4,5]. In addition to gasification, pyrolysis, and other thermal processing, program research focuses on cleaning up and conditioning the converted fuel. This is a key step in the effective commercial use of thermochemical platform chemicals.

Biomass pyrolysis refers to a process where biomass is exposed to high temperatures in the absence of air, causing the biomass to decompose. The end product of pyrolysis is a mixture of solids (char), liquids (oxygenated oils), and gases (methane, carbon monoxide, and carbon dioxide). Flash pyrolysis gives high oil yields, but because of the technical effort needed to process pyrolytic oils, this energy-generating system does not seem very promising at the present stage of development. However, pyrolysis as a first stage in a two-stage gasification plant for straw and other agricultural feedstocks that pose technical difficulties in gasification does deserve consideration.

Biomass gasification technologies have been a subject of commercial interest for several decades. Gasification operates by heating biomass in an environment where the solid biomass breaks down to form a flammable, low-caloric gas. The biogas is then cleaned and filtered to remove chemical compounds. The gas is used in more efficient power generation systems called *combined cycles*, which combine gas turbines and steam turbines to produce electricity. The efficiency of these systems can reach 60%. Gasification systems may also be coupled with fuel cell systems using a reformer to produce hydrogen and then convert hydrogen gas to electricity (and heat) using an electrochemical process. Interest in biomass gasification increased substantially in the 1970s because of uncertainties in petroleum supplies. Most development activities were concentrated on small-scale systems. Low-energy gasifiers are now commercially available, and dozens of small-scale facilities are in operation.

During the 1980s, government and private industry sponsored R&D for gasifier systems primarily to gain a better understanding of reaction chemistry and economies-of-scale issues. In the 1990s, combined heat and power was identified as a potential near-term opportunity for biomass gasification because of incentives and favorable power market drivers. R&D concentrated on integrated gasification combined cycle (IGCC) and gasification cofiring demonstrations, which culminated in a number of commercial-scale systems. In U.S. projects, most processes are very recalcitrant feeds such as bagasse and alfalfa.

The use of biogas for energy generation represents a high-efficiency alternative source to replace fossil fuels. The annual world bioenergy potential is about 2900 EJ, although only about 10% of that could be considered available on a sustainable basis and at competitive prices. In urban areas, another important benefit is the use of the enormous amounts of organic garbage or liquid effluents such as industrial residues, sewers, and trashcans. Transformation of these items into industrial or automotive fuels can avoid environmental damage. Figure 8.2 depicts the main processes of biomass conversion into biofuel [6–9].

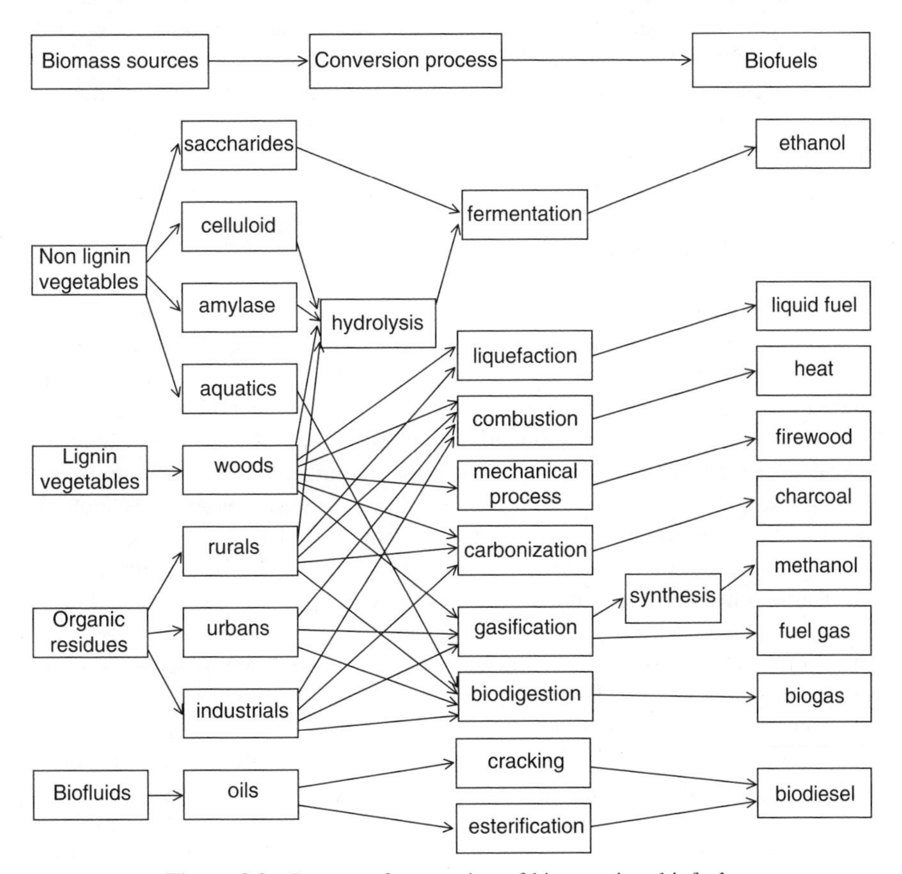

Figure 8.2 Process of conversion of biomass into biofuels.

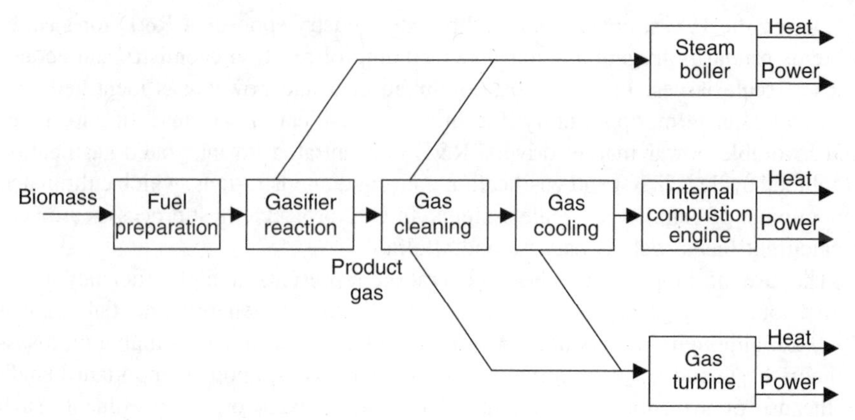

Figure 8.3 Gasification paths to heat and power. (From Ref. [9].)

The use of biomass and biodigesters introduces several advantages for rural applications, where leftover cultural and animal residues can be used to obtain biofertilizer (i.e., the organic material processed in biodigesters can be used as fertilizer). Furthermore, biomass and biodigesters can be used to provide necessary energy for illumination, heating, and to drive motors. The market potential and future prospects of small-scale electricity generation from RES are covered in an OPET report [9] and summarized in Figure 8.3.

8.2 FUEL FROM BIOMASS

Biomass seems to be the only renewable alternative for liquid transportation fuel. Biomass use strengthens rural economies, decreases U.S. dependence on imported oil, avoids use of methyl tertiary butyl ether (MTBE) or other highly toxic fuel additives, reduces air and water pollution, and reduces greenhouse gas emissions. Fuel from biomass is related to ethanol, biodiesel, biomass power, and industrial process energy. To expand the role of biomass in our modern society, the Department of Energy's Office of the Biomass Program is fostering biomass technologies with a balanced portfolio of research and development [10].

It is important to consider the use of biomass, in that most of the electricity, heat, and steam produced by industry is consumed on site. However, some manufacturers sell excess power to the grid. Wider use of biomass resources will directly benefit many companies whose growth generates more residues, such as wood or animal wastes, than they can use internally. New markets for these excess materials will also support business expansion.

The pulp and paper industry is a very important stakeholder in the forest industry, because it is estimated that 85% of wood waste is used for energy in the United States, with a tremendous potential for market growth for other woody waste products and biomass renewable sources of fuel. Gasification technologies using biomass pulp by-products has aided the pulp industry, and the paper industry has been improving chemical recovery and generating process steam and electricity at higher efficiencies and with lower capital costs than have conventional technologies. Pulp and paper industry by-products that can be gasified include hogged wood, bark, and spent black liquor.

For biomass gasification, organic materials are heated with less oxygen than that needed for efficient combustion. Since combustion is a function of the mixture of oxygen with hydrocarbon fuel, there is a stoichiometric amount of oxygen and other conditions to determine if biomass gasifies or pyrolyzes to a mixture of carbon monoxide and hydrogen, known as *synthesis gas* or *syngas*. Gaseous fuels mix with oxygen more easily than do liquid fuels, which in turn mix more easily than solid fuels. Therefore, inherently, syngas, burns more efficiently and cleanly than the solid biomass from which it is made.

Biomass gasification can improve the efficiency of large-scale biomass power facilities. Forest industry residues and specialized facilities such as black liquor recovery boilers used in pulp and paper industry can be major sources of biomass power. Syngas can also be burned in gas turbines, a more efficient electrical generation technology than steam boilers, for whose use solid biomass and fossil fuels are limited.

Most electrical generation systems are relatively inefficient, because half to two-thirds of the energy is lost as waste heat. If such heat is used for an industrial process, space heating, or another purpose, the efficiency can be greatly increased. Small, modular biopower systems are more easily used for such cogeneration than are most large-scale electrical generation.

Syngas mixes more readily with oxygen for combustion. It also mixes more readily with chemical catalysts than solid fuels do, thus enhancing its ability to be converted to other valuable fuels, chemicals, and materials. The *Fischer–Tropsch* process converts syngas to liquid fuels, which aids in transporting this fuel [11]. The process is named after two German coal researchers, who discovered the method in 1923. A variety of other catalytic processes can turn syngas into a myriad of chemicals or other potential fuels or products, and it is used for the synthesis of hydrocarbons and other aliphatic compounds. Synthesis gas reacts in the presence of an iron or cobalt catalyst; much heat is involved, and such products as methane, synthetic gasoline and waxes, and alcohols are made, with water or carbon dioxide produced as a by-product. In other words, the water-gas shift process converts syngas to more concentrated hydrogen for fuel cells. This important source of the hydrogen–carbon monoxide gas mixture is the gasification of coal.

Biomass gasification involves thermal conversion to simple chemical building blocks that can be transformed into fuels, products, power, and hydrogen. Components include feed preparation, the biomass gasifier, and a gas treatment and cleaning train. Syngas initially contains particulates and other contaminants and must be cleaned and conditioned prior to use in fuels or chemical or power conversion systems (e.g., catalyst beds, fuel cells). The gasification process readily converts all major components of biomass, including lignin, which is resistant to biological conversion, to intermediate building blocks. Utilization of the lignin, which is typically

25 to 30% of the biomass, is essential to achieve high efficiencies in the biorefinery. The gasification process can convert most biomass feedstocks or residues to a clean synthesis gas. Once such a gas is obtained, it is possible to access and leverage the process technology developed in the petroleum and chemical industries to produce a wide range of liquid fuels and chemicals.

There are several widely used process designs for biomass gasification: (1) staged reformation with a fluidized-bed gasifier, (2) staged reformation with a screw auger gasifier, (3) entrained flow reformation, or (4) partial oxidation. In staged steam reformation with a fluidized-bed reactor, the biomass is first pyrolyzed in the absence of oxygen. Then the pyrolysis vapors are reformed to synthesis gas with steam, providing added hydrogen as well as the proper amount of oxygen and process heat that comes from burning the char. With a screw auger reactor, moisture (and oxygen) is introduced at the pyrolysis stage, and process heat comes from burning some of the gas produced in the latter. In entrained flow reformation, external steam and air are introduced in a single-stage gasification reactor. Partial oxidation gasification uses pure oxygen with no steam, to provide the proper amount of oxygen. Using air instead of oxygen, as in small modular uses, yields produce gas (including nitrogen oxides) rather than synthesis gas.

Biomass gasification is also important for providing a fuel source for electricity and heat generation for the integrated biorefinery. Virtually all other conversion processes, whether physical or biological, produce residue that cannot be converted to primary products. To avoid a waste stream from the refinery, and to maximize the overall efficiency, these residues can be used for combined heat and power production (CHP). In existing facilities, these residues are combusted to produce steam for power generation. Gasification offers the potential to utilize higher-efficiency power generation technologies, such as combined cycle gas turbines or fuel cells. Gas turbine systems offer potential electrical conversion efficiencies approximately double those of steam-cycle processes, with fuel cells being nearly three times as efficient.

A workable gasification process requires development of some technology: for example, feed processing and handling, gasification performance improvement, syngas cleanup and conditioning, development of sensors, analytical instruments and controls, process integration, and materials used for the systems.

8.3 BIOGAS

Biogas, also termed methane or gobar gas, comprises a mixture of gases. It is a fuel of high caloric value resulting from anaerobic fermentation of organic matter called *biomass*. Composition of this gas varies with the type of organic material used. Its basic composition is listed in Table 8.2.

Methane is a colorless, odorless gas with a wide distribution in nature. It is the principal component of natural gas, a mixture containing about 75% methane (CH_4), 15% ethane (C_2H_6), and 5% other hydrocarbons, such as propane (C_3H_8) and butane (C_4H_{10}). The *firedamp* of coal mines is chiefly methane [6,7], and methene is the main component of biogas. It is colorless and odorless, highly flammable, and

TABLE 8.2 Probable Composition of Some Combustible Gases

Methane (CH_4)	60 to 70%
Carbonic gas (CO_2)	30 to 40%
Nitrogen (N)	Traces
Hydrogen (H)	Traces
Gas sulfidric (H_2S)	Traces

in combustion presents a lilac-blue flame and small red stains. It does not leave soot and produces minimal pollution. The caloric power of biogas depends on the amount of methane in its composition, which could reach 5000 to 6000 kcal/m^3. Biogas can be used for stove heating, campaniles, water heaters, torches, motors, and other equipment.

8.4 BIOMASS FOR BIOGAS

Biomass can be considered as all materials that have the property of being decomposed by biological effects, that is, by the action of bacteria. Biomass can be decomposed by methanegenic bacteria to produce biogas, in a process depending on factors such as temperature, pH, carbon/nitrogen ratio, and the quality of each. Usable and accessible organic matter includes animal residue, agricultural residue, water hyacinth (*Eichornia crassipes*), industrial residue, urban garbage, and marine algae.

There exists a potential biomass use for animal feces produced in cattle farming, which is easily mixable with water. In building a biodigester, the first step is to determine its capacity, observing the amount of biomass available locally. For this calculation, one simply multiplies the weight of a live animal by 0.019 to find the approximate daily production of manure [11–13].

Agricultural residue produces an average of seven times more biogas than animal waste, but is not very practical with respect to harnessing, provisioning, and discharging in biodigesters. For this biomass, intermittent biodigesters are most appropriate. Agricultural remains are not mixable with water because they may be lighter; they emerge only after initial decomposition, when the production of gas begins. If triturated, they can be used in any type of biodigester, although that represents greater energy and labor. Provisioning is made once every 15 or 20 days, and the fermentation period and production of biogas take from 60 to 120 days. Residues containing agricultural chlorinated pesticides or herbicides are not used because chlorine does not allow development of methanegenic bacterium. Table 8.3 lists the approximate volume of biogas produced by agricultural waste.

Under favorable conditions, water hyacinth produces up to 600 kg of dry matter per hectare a day, offering excellent biogas production. According to research data, 350 to 410 L of biogas per kilogram of dry water hyacinth was captured. Each

TABLE 8.3 Biogas Volume Produced by Organic Residues

Biomass	Production of Biogas (m^3/ton)	Methane (%)
Sunflower leaves	300	58
Rice straws	300	Variable
Wheat straws	300	Variable
Bean straws	380	59
Soy straws	300	57
Linen stem	359	59
Grapevine leaves	270	Variable
Potatoes leaves	270	Variable
Dry leaves of trees	245	58

square meter of plantation can produce, on average, 18 L/day, or 30,000 m^3/ha. The amount of methane reaches 80% in volume, greater than that of other types of biomass.

Use of industrial residues is minimal. Except for the industrialization of fruits, meats, cereals, and alcohol, such biomasss can be used when the residue has been dumped in rivers or elsewhere outdoors and is polluting the environment. Urban garbage can be transformed into sources of energy if processed in biodigesters. Biodigestion could be considered a powerful instrument for the disposition of garbage if associated with noble products such as electricity.

Marine algae offer a strong option as a prime matter in the production of biogas. Scientific studies show that the biogas produced from marine algae is of good quality, characterized by the lack of an odor of sulfur, and providing a clear to bluish flame. On average, this biogas produces 300 L/kg of dry raw material and is 60 to 70% gas methane. Its caloric value reaches about 6600 to 7200 calories.

8.5 BIOLOGICAL FORMATION OF BIOGAS

Animal manure is a biomass found in considerable quantities on rural properties [4,5]. Anaerobic compost is formed from a mixture of organic manure with animal–water residues in three stages. In phase 1 (the solid stage), substances such as carbohydrates, lipids, and proteins are attacked by ordinary fermentative bacteria for the production of fatty acids, glucose, and amino acids. In phase 2 (the liquid stage), the substances formed previously are attacked by the propionic-bacteria, acetogenic bacteria, and acidogenic bacteria, forming organic acids, mainly propionic and acetic, forming carbon dioxide, acetates, and H_2. In phase 3 (the gaseous stage), the methanegenic bacteria act on the organic acids to produce primarily methane CH_4 and carbon dioxide CO_2 (biogas). The third stage is the most important, because bacteria demand special care; they are responsible for limiting the speed of a chain of reactions. This is due primarily to the formation of microbulbs of methane and carbon dioxide around the methanegenic bacteria, which isolates them from the mixture in digestion. Furthermore, they require temperature and acidity suited for their reproduction. Methanegenic fermentation is a highly sensitive biological process involving many microorganisms, such as the thermofilic and mesofilic methanegenic bacteria, which is reproduced between 45 and 50°C and 20 and 45°C, respectively.

8.6 FACTORS AFFECTING BIODIGESTION

Fermentation is more intense when the temperature of the material is between 30 and 35°C. At these conditions, the biogas production per kilogram of raw material is higher and faster. Therefore, the biodigester is usually buried because underground temperatures are higher and more constant. Additionally, a buried biodigester is easier to handle. Ideal temperatures for the material used to produce methanegenic bacteria are as follows: psicrofilics: highest growth at 20°C; mesofilics: highest growth at 35°C; and thermofilics: highest growth at 55°C. In general, the biomass inside a digester should be at a temperature of approximately 35°C. Table 8.4 lists the amount of waste per semistabled animal and the corresponding production of gas.

The biodigestion acidity measured by the hydrogenionic potential (pH) should remain between 6 and 8. If the pH is below those values, the environment becomes better adapted to the acid bacteria that will prevent the growth of methanegenic bacteria.

For any biodigester system, it is imperative that agitation of the mixture achieves a uniform distribution of temperature, thus improving distribution of the intermediary and final products in the biodigester and reducing top mud layer growth [4,5].

The concentration of nutrients in a mixture is another important factor. Nutrients for development of bacteria include carbon, nitrogen, nitrate, phosphorus, sulfur, and sulfates. Animal urine (urea) or chemical fertilizers such as ammonia sulfate can be used to cause fermentation. For such, the following points should be observed:

1. The carbon/nitrogen ratio: Nitrogen is a nutrient that contributes to the formation and multiplication of bacteria, but in excess it induces ammonia

TABLE 8.4 Production of Gas from Manure from Semistabled Animals

Manure Animal	Production (kg/day)	m^3 of Gas/kg of Manure	m^3 of Gas/ Animal/Day
Birds	0.09	0.055	0.0049
Bovines	10.00	0.040	0.4000
Equines	6.50	0.048	0.3100
Oviparous	0.77	0.070	0.0500
Suidae	2.25	0.064	0.1400

Source: Refs. [4] and [7].

TABLE 8.5 Average Percent of Dry Matter in Selected Animal Manure

Producing Animal	Dry Matter in the Manure (%)
Bovine	16.5
Equines	24.2
Oviparous	34.5
Capra	34.8
Suidae	19.0
Birds	18.0

Source: Refs. [4] and [7].

formation, which can increase and even stop production of biogas. Carbon is responsible for the production of energy. As most animal solid waste has a low carbon/nitrogen ratio, those elements should be corrected with vegetable residues to reach the ideal point.

2. The concentration of solids should be between 7 and 9%, for the following reasons:
   - To avoid mud formation on the top of the digester
   - To ease movement of the material in the digester
   - To obtain good fermentation
   - To allow good digestion

Tables 8.5 and 8.6 list the average dry material content, used to establish the correct amount of water in the mixture [4,7].

The retention period of the substratum in the biodigester is the time that elapses between the input and output of the biomass in the digester, it is the digestion period, and it varies as a function of the pH, volume of material used, agitation inside the biodigester, and internal digester temperature. Generally, this period

TABLE 8.6 Manure/Water Ratio for 7, 8, and 9% Solid Contents

Producing Animal	Solids in the Biomass Preparation		
	7%	8%	9%
Bovine	0.74 : 1	0.94 : 1	1.20 : 1
Equines	0.41 : 1	0.49 : 1	0.59 : 1
Oviparous	0.26 : 1	0.30 : 1	0.35 : 1
Capra	0.25 : 1	0.30 : 1	0.35 : 1
Suidae	0.58 : 1	0.73 : 1	0.90 : 1
Birds	0.64 : 1	0.80 : 1	1.00 : 1

Source: Refs. [4] and [7].

ranges from 20 to 50 days, lengthening the digestion time when the biofertilizer is not smelly at the output, detected by the presence of flies.

Digestion can produce some toxic substances. Thus, as the biogas is not a decomposition product of sterilized organic residues, it should be mixed with non-chlorinated water. At the same time, the presence of disinfectants and excess antibiotics and/or copper in animals should be watched for, as these are fatal to methanegenic bacteria and thus prevent the production of gas.

8.7 CHARACTERISTICS OF BIODIGESTERS

Several biodigester models are available. For this study, a typical biodigester of average capacity of the Indian type will be referenced; this type has been used to reference most facilities worldwide [4,6]. In the Indian model, the system of biogas production consists primarily of a digester and a gasometer (see Figure 8.4). The digester is a reservoir built of bricks or concrete below ground level. A wall divides the biodigester into two semicylindrical parts for the purpose of retaining and providing circulation for the biomass loaded in a biofertilization process. The biodigester is loaded through the charge box, which also serves as a prefermenter. The charge box communicates with the digester through a pipe going down to the bottom. The output of the biofertilizer is through another pipe at a level that assures that the amount of biomass entering the biodigesters is the same as that leaving it in biofertilizer form. It should also have a discharge box, tank, or dam to pump and/or deliver the biofertilizer directly to the consumer.

The gasometer is responsible for storing the biogas produced and to exert a pressure equivalent to its weight to supply it with almost constant pressure. Sometimes, counterweights are placed in the gasometer to supply gas at higher pressure

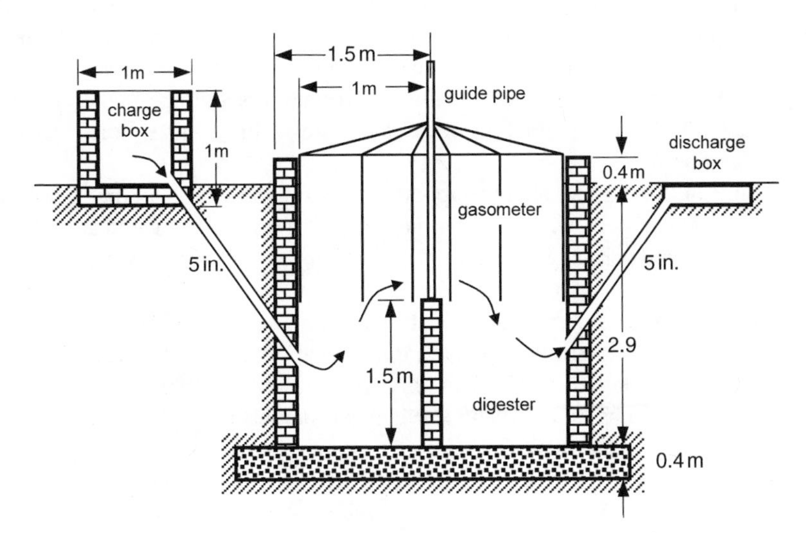

Figure 8.4 Production of biogas in digesters (From Ref. [4].)

at the output; they move up or down in accordance with the volume of the stored biogas. The counterweights are often bags of sand or cement blocks. The gasometer is generally made of steel foil 14, which should be welded in a metallic structure made of $\frac{3}{4}$-inch carbon steel. It is usually cylindrical with an arched cover (in cone form) to avoid deposition of sludge and water in its exterior parts. The gasometer is guided in its up or down movement by a galvanized guide tube 2.5 to 3.0 inches in diameter.

The biogas output is a device on the top of the gasometer through which the biogas leaves the interior of the gasometer for the consumption point. It should be a flexible hose, to ease gasometer movement. The gas is transported in a 1-inch-diameter tube.

In the dimensions shown in Figure 8.4, the biodigester produces 6 m^3 of gas per day, corresponding to 8568 kWh (1 m^3/day = 1428 kWh). The tank will be charged with 240 L of biomass a day, of which 120 L is water and 120 L is bovine manure. Twelve or thirteen adult semistabled bovine animals are necessary for this level of production.

8.8 CONSTRUCTION OF BIODIGESTER [4,7]

The biodigester capacity to be installed on a rural property is determined by the number of people and their minimum energy needs. The biodigester should be placed between the point of consumption (kitchen, refectory, and drying deposits) of the gas and the site of manure collection, at a point 20 to 25 m from the consumption point. Sloped grounds, to ease withdrawal and distribution of the biofertilizer, are preferable.

The digester well should be cylindrical, with a diameter a little larger than the external diameter of the digester and a depth 20 to 30 cm smaller than its useful height. During excavation of the well, two rifts should be constructed in opposite directions, with an approximate slope of 45° for placement of the charge and discharge pipes. If the land is flat, a shallower excavation should be made, using the soil extracted to form a mound around the digester, which will simplify charging and discharging.

The floor of the digester should support the entire hydraulic load that it holds, in one of ways: (1) in wells with a rock or very firm dirt bottom on which a brick floor is set on fine sand, or (2) in wells without a firm bases, in which case the floor should be of concrete 10 to 15 cm thick.

The external and dividing walls of the digester can be built of concrete or of ordinary bricks settled firmly with no gaps through the walls, to avoid undesired leaks or cracks in the digester. The dividing walls have the double purpose of driving the mass flow to be digested and serving as a support for the gasometer guides. The latter function requires very resistant supports to withstand the gasometer movement along the guides; building double walls or concrete dividing walls allowing apertures of 40 to 50 cm where the pipe guides of the gasometer will be fixed is recommended.

The digester charges and discharges through pipes, allowing mass circulation during the digesting process. The pipes can be made of 100-mm-diameter PVC and placed in rifts in the digester well, fixed on the bottom and fastened on the top of the wall.

The charge tank can be used either as the prefermentation tank or for mixing the material. Therefore, its volume should be a little larger than the daily charge and a little bit above the liquid level of the digester. The discharge tank is constructed primarily to protect the output flow.

The cladding used for the digester is of fundamental importance to its construction and operation. The cladding should be made of a 1 : 6 ratio of cement to fine sand and should be at least 10 mm thick. Finally, to guarantee the accuracy of the diameter measurements the use of a standard scale fixed on the guide pipes of the plant so as to allow free rotation around it is recommended.

8.8.1 Sizing a Biodigester

Figure 8.4 is a schematic diagram of a typical biodigester with a production capacity of 6 m^3/day and a retention time of 50 days. For the daily production of gas, the following is required:

Amount of biomass: 240 L/day

Retention time: 50 days

Total useful volume: (240 L/day) (50 days) (1 m^3/103 L) = 12 m^3

Diameter of the chamber: 2.5 m

Total height: 2.85 m

Useful height: 2.45 m

Total volume: 14 m^3

Volume of charge: 12 m^3

Volume of the gasometer: 6.3 m^3

External diameter of the gasometer: 2.40 m

Plant height of the gasometer: 1.85 m

Lateral height of the gasometer: 1.40 m

Height of the dividing wall: 1.52 m

The charge tank dimensions to carry 300 L of biomass: 1 m × 1 m × 0.8 m.

8.9 GENERATION OF ELECTRICITY USING BIOGAS

For installation of a biogas microplant as an alternative source of electric energy for small power generators, it is necessary to use gas motors coupled with turbines and electrical generators, as illustrated in Figure 8.5. Many manufactured generator–turbines work with gas. Because the majority of the smaller commercial units are in the high-power range (1.6 to 216 MW) [8,9], some adaptations are required. For micropower plants, alcohol and gasoline motors can be made to operate

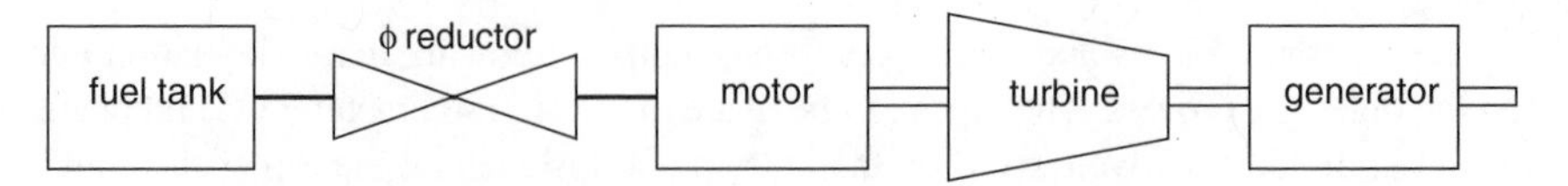

Figure 8.5 Generating group of electric energy using biogas.

with methane without affecting their operational integrity. This adaptation is made by installing a cylinder of biogas in place of using conventional fuel. For gas flow regulation, a reducer is placed close to the motor. As the biogas flow operates at low pressures, it demands a larger aperture in the injector (with a diameter of 1.5 to 2 mm). The adaptation of equipment to work with PLG gases as biogas involves simply replacement of the injector or an increase in the diameter of biogas flow to the motor. Table 8.7 shows the average consumption of biogas for motors of several power levels.

Observing that a biogas flow of $1\ \text{m}^3/\text{h} = 1.428\text{kWh}$, it is sufficient to use a simple rule of thumb to determine the power consumption in kilowatts of each motor, as in the following example of a 9-hp motor (see Table 8.7):

$$
\begin{aligned}
x &= \frac{(1.428\ \text{kWh})\ (16\ \text{m}^3/\text{h})}{1\ \text{m}^3/\text{h}} \\
&= 4.5\ \text{kWh}
\end{aligned}
$$

For a better understanding of these quantities, the equivalent energy of a cubic meter of biogas to obtain 6148.98 kcal of heat in accordance with Table 8.8 can

TABLE 8.7 Average Consumption of Some Biogas Motors

Motor Power (hp)	Average Consumption (m^3/h)
1.0	0.45
2.0	0.92
5.5	2.24
9.0	3.16

TABLE 8.8 Equivalent Energy per Cubic Meter of Biogas to Obtain 6148.98 kcal of Heat

Fuel	Equivalent Amount
Gasoline	0.98 L
Alcohol	1.34 L
Crude oil	0.72 L
Natural gas	1.50 m^3
Coal	1.51 m^3
Electricity	2.21 kWh

Source: Adapted from Ref. [4].

TABLE 8.9 Daily Domestic Amounts of Gas Use by a Family of Five or Six People

Use Purpose	Biogas
Kitchen	1.960 m^3
Bathtub	0.588 m^3
Shower	0.336 m^3
Refrigeration of victuals	2.800 m^3
Illumination	0.140 m^3
Gasoline (fuel)	0.5 $m^3/hp \cdot h$

Source: Adapted from Ref. [4].

be considered. Table 8.9 shows the daily amounts of biogas used for domestic purposes by a family of five or six people.

Figure 8.6 shows a renewable fuel gas generator where a gas production module converts biomass to a clean gas for power and heat, biofuels research, and hydrogen generation. Depending on a number of variables, the fuel cost for the product gas measured in dollars per million Btu is at least 50% less than that of nonrenewable natural gas and propane. Figure 8.6 shows integration with a gas turbine generator, but it can typically be converted to operate with microturbines, thermoelectric power generators, and fuel cells.

Another application related to biomass is the utilization of digester gas derived from anaerobic sludge digestion in wastewater treatment plants. Many wastewater treatment plants in the United States employ anaerobic sludge digestion as a component of their treatment process. Microturbines are being used to transform the

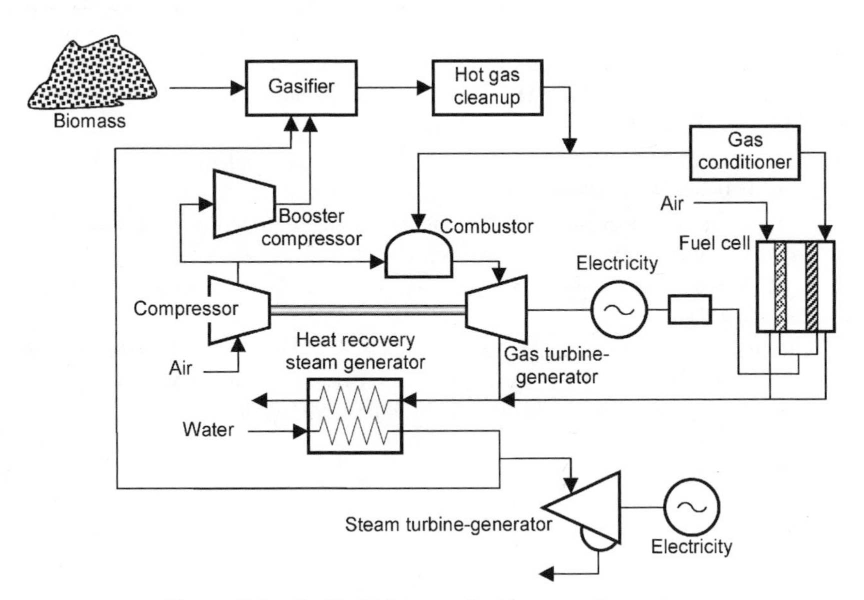

Figure 8.6 Gasified biomass electric generator system.

energy contained in the methane-based biogas produced as a by-product of the sludge digestion process into useful forms of electricity and thermal energy rather than disposing of the biogas as a waste product. There are a variety of technical and financial factors that must be considered and various barriers to the adoption of the microturbine technology that must be overcome before a microturbine installation can be deemed practical for a given site.

REFERENCES

[1] U.S. Environmental Protection Agency, *Municipal Solid Waste Handbook*, Database Version 3.0, USEPA, Washington, DC, March 1996.

[2] U.S. Environmental Protection Agency, *Characterization of Municipal Solid Waste in the United States: 1995 Update*, EPA/530-S-96-001, USEPA, Washington, DC, March 1996.

[3] Governmental Advisory Associates, *Municipal Waste Combustion in the United States: 1996–97 Yearbook, Directory, and Guide*, GAA, Westport, CT, 1997.

[4] http://www.worldenergy.org.

[5] U.S. Department of Energy, *Energy Efficiency and Renewable Energy*, Office of the Biomass Program, Technical Plan Summary, www.bioproducts-bioenergy.gov and http://www.eere.energy.gov/informationcenter.

[6] C. A. Cicchi, Co-generation with a gas turbine, *Força Energética*, Vol. 2, No. 4, pp. 37–38, July 1993.

[7] Gas powered car, *Força Energética*, Vol. 1, No. 1, p. 39, May–June 1992.

[8] J. O. C. Nogueira, Analysis of equivalent costs between the energy supplied by biodigesters and an electric network in the rural area of a district of Santa Maria, M.E.P. thesis, Federal University of Santa Maria, Santa Maria, Brazil, September 1983.

[9] T. Lensu and A. Alakangas, *Small Scale Electricity Generation from RES: A Glance at Selected Technologies, Their Market Potential and Future Prospects*, OPET Report 13, VTT Processes, Jyväskylä, Finland, May 2004.

[10] U.S. Department of Energy, Biomass summary in *Multi-year Technical Plan*, mytp_040804, DOE, Washington, DC.

[11] Syntroleum, in cooperation with Dr. Anthony Stranges, Professor of History at Texas A&M University, http://www.fischer-tropsch.org/.

[12] L. M. Auerbach, Home power unit: methane generator, in *Alternative Energy Systems*, 2nd ed., 1974.

[13] TVE International, Intermediate Technology Development Group, Schumacher Centre for Technology and Development, Information Service Unit, Rugby, Warwickshire, England, http://www.oneworld.org/itdg, or http://www.irdg.org.pe, 2004.

CHAPTER 9

MICROTURBINES

9.1 INTRODUCTION

Microturbines were developed by industry through improvements in auxiliary power units originally designed for aircraft and helicopters and customized for customer-site electric user applications. Utility gas turbine generators ranging from 500 kW to 250 MW are too large for distributed power applications. Therefore, microturbines from 30 to 400 kW were developed for small-scale distributed power either for electrical power generation alone, in distributed electrical power generation, or in combined cooling or heat and power (CCHP) systems.

Microturbines can burn a variety of fuels, including natural gas, gasoline, diesel, kerosene, naphtha, alcohol, propane, methane, and digester gas. The majority of commercial devices presently available use natural gas as the primary fuel. Modern microturbines have evolved dramatically, with advanced components such as inverters, heat exchangers (recuperators), power electronics, communications, and control. Microturbines have several inherent advantages over conventional fossil-fueled power systems:

- Microturbines are very low in weight per horsepower, resulting in light generator sets.

Integration of Alternative Sources of Energy, by Felix A. Farret and M. Godoy Simões

- Microturbines demonstrate pure rotary motion as opposed to stroking, resulting in less vibration, low noise compared to diesel generators, high mechanical performance, and very high reliability.
- A liquid cooling system is not required.
- Some microturbines run on air bearings with very low maintenance.
- Microturbines exhibit a very fast response to load variation, since they do not need to build up pressure as in steam turbines or have high momentum as in reciprocating engines.
- Microturbines operate on a variety of fuels.
- Combustion usually runs on an excess of air, resulting in very low emissions.
- Even low-power microturbines can provide recoverable heat for water and space heating, and the larger units can be used for industrial purposes or in a combined cycle with other turbines.

The U.S. Department of Energy (DOE), industry, and the National Fuel Cell Research Center have been analyzing extensively a hybrid power system that combines a gas turbine (GT) with a high-temperature fuel cell (HTFC). These efforts have demonstrated that such a combination is capable of providing remarkably high efficiencies [1].

Reciprocating internal combustion engines have some advantages over microturbines, such as efficiencies over 38%, whereas microturbines have efficiencies ranging from 28 to 32%. In addition, microturbines are very sensitive to ambient air temperature, pressure, and humidity, thus requiring derating for those ambient variables. A major drawback of microturbines is the requirement for better skilled technicians to repair and do maintenance.

Microturbines are ideally suited for distributed generation applications [2] because of their flexibility in connection methods, ability to be stacked in parallel to serve larger loads, ability to provide stable and reliable power, and low emissions. The most typical applications include:

- Peak shaving and base load power (grid parallel)
- Combined heat and power
- Stand-alone power
- Backup/standby power
- Ride-through connection
- Primary power with grid as backup
- Microgrid

Target customers include financial services, data processing, telecommunications, restaurant, lodging, retail, office building, and other commercial sectors. Microturbines are currently operating in resource recovery operations at oil and gas production fields, wellheads, coal mines, and landfill operations, where by-product gases serve as essentially free fuel. Unattended operation is important since these locations may be remote from the grid, and even when served by the grid may experience costly downtime when electric service is lost due to weather, fire, or animals. In CHP applications, the waste heat from microturbines can be used to produce hot water, to heat building space, to drive absorption cooling or desiccant dehumidification equipment, and to supply other thermal energy needs in a building or industrial process.

9.2 PRINCIPLES OF OPERATION

Figure 9.1 illustrates a drawing of the mechanical features of a small gas turbine engine. From left to right there is a compressor, a combustor, a turbine, and a generator, all sharing the same shaft. In general terms, an auxiliary machine starts the entire process initially by spinning the turbine shaft to introduce air compressed by the blades indicated on the left side of the shaft in Figure 9.1. The air is then mixed with fuel gas and passed into the combustion chamber of the turbine. This mixture is compressed to a point of continuous combustion, which drives the set of blades indicated on the right side of the shaft, increasing the speed of the shaft, which in turn admits more incoming air and fuel gas. A high-speed generator is connected on the shaft, which converts this mechanical energy into electricity. The operation is based on the Brayton cycle presented in Chapter 2. At the end of the process, the hot residual gases from combustion are expelled.

Figure 9.1 shows air coming in through the inlet. The shape of the inlet must be designed carefully to prevent particulates and foreign objects from being drawn into the engine. The air passes through a fanlike assembly of rotating impellers fitted with curved vanes at the intake area, which guides the air in, pulling it at higher speed and pressure. There are several such turbine blades in succession, each row taking the air shoved by the preceding wheel and providing an even greater push back into the combustion area, gradually accumulating significant differential pressure. The pressure of the air is raised about 10 times the atmospheric pressure, and eventually enters the combustion chamber.

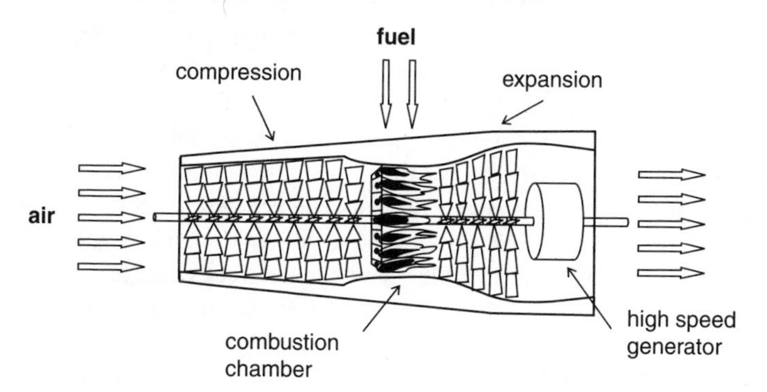

Figure 9.1 Principles of gas microturbine power conversion.

The combustion assembly might include a single or multiple combustors. The combustion chamber is a reacting system (i.e., a chemical reaction takes place). In the combustion chamber, heat is added to the gas at constant pressure. The density decreases and the specific volume and temperature are increased. Entropy is also increased, since combustion is not a reversible process. In the turbine the situation is the opposite of the compressor, where the pressure decreases and the temperature decreases (for ideal expansion the entropy is constant). The compressor and other components are considered nonreacting. The combustion chamber takes the internal energy of the fuel, mixes the burning fuel with air at constant pressure, and increases the temperature. Usually, natural gas is used as fuel source. The best natural gas input should be at 80 psig or higher pressure. As the gas pipeline gets closer to consumers, a regulator station reduces the pressure in the pipeline to about 60 psi or less. Therefore, precompression of natural gas can be required for microturbines.

The combustion chamber is of a lean premix emission type, achieving low emissions of NO_x, CO, and unburned hydrocarbons in the exhaust gases. The hot high-speed mix of air with burned fuel expands rapidly as it enters the expansion turbine stage. The expansion turbine is the power-producing element of the overall set. It consists of turbine wheels, each of greater diameter than the preceding wheel. Work is extracted from the hot gas by letting it expand. The turbine has a carefully designed cross section, wheel diameters, and vane angles, which force the gas to slow, cool, and expand at an optimal rate to collect the most energy out of the gas as it passes from one blade to another. Eventually, less than one-third of the power extracted from the hot gases is available to drive the generator, while the other two-thirds is used to drive the compressor and auxiliary equipment such as a fuel booster, ventilation fan, oil pump, water cooling pump, oil separator, and buffer air pump.

The generator is typically a high-speed permanent-magnet machine. The air leaves the turbine through the exhaust outlet. Although the operation seems simple, it must be designed and tailored to the turbine to have the correct speed and pressure combination to avoid back pressure and optimize overall performance.

The exhaust gases can be used in process heating applications or can be used indirectly with a heat exchanger to produce hot water. Typically, microturbines for distributed generation do not operate at very high temperatures, so no steam is produced, although large gas turbines can produce steam and may be coupled and combined with other cycles.

The shaft design used by some manufacturers is supported by airfoil bearings requiring no lubrication. If heat recovery is implemented, a careful study must be done to minimize piping and loss of heat. The heat exchangers (or recuperators) can heat water or glycol but may not be able to generate steam. A good combination is to use microturbines to heat boiler feedwater, effectively reducing boiler fuel needs. However, recuperators also cool the exhaust gas and hence limit the residual thermal energy available for use. Some microturbine manufacturers include a recuperator bypass valve to reduce the electrical efficiency but to increase the overall system efficiency by increasing the recoverable heat available. This option also provides increased flexibility to balance the electricity and heat production demands on site.

9.3 MICROTURBINE FUEL

All gas turbines use the same energy source (i.e., an expanding, high-pressure gas produced by the combustion of mixed fossil fuel and air). This mixture has the same physical properties regardless of the size of the gas turbine. Therefore, the tips of the turbine blades must move at a speed appropriate to capture the energy from this expanding gas. This means that large utility turbines with 2.5 m wheels will spin at 1800 or 3600 rpm, while smaller turbine with wheels only 0.15 m in diameter will have a much higher rotational speed, up to 100,000 rpm. Therefore, large gas turbines use the same philosophy as that of central-station steam turbines or diesel technology driven at constant speed. On the other hand, microturbines have special generators running at very high speed which must be able to supply stand-alone loads. Therefore, a variable-speed generator is typically used as an intermediate power electronics stage to convert the variable voltage and frequency to a fixed grid frequency (i.e., the output of the high-speed generator is rectified and coupled to a dc–ac power converter).

Most gas turbines are equipped to burn natural gas. A typical range of heating values of gaseous fuels acceptable for gas turbines is 900 to 1100 Btu per standard cubic foot (SCF), which covers the range of pipeline-quality natural gas. Clean liquid fuels are also suitable for use in gas turbines. Larger gas turbines may have special combustors to handle cleaned gasified solid and liquid fuels. Fuels are permitted to have low levels of specified contaminants (typically less than 10 ppm total alkalis and single-digit ppm amounts of sulfur).

Liquid fuels require their own pumps, flow control, nozzles, and mixing systems. Many gas turbines are available with either gas- or liquid-firing capability. In general, such gas turbines can be converted from one fuel to another. Several gas turbines are equipped for dual-fuel firing and can switch fuels with minimal or no interruption. Lean-burn/dry low NO_x combustors can generate NO_x emissions levels as low as 9 ppm (at 15% O_2), and those with liquid fuel combustors have NO_x emissions as low as approximately 25 ppm (at 15% O_2). Although there is no substantial difference in general performance with either fuel, the different heat levels of combustion result in slightly higher mass flows through the expansion turbine when liquid fuels are used, and thus a very small increase in power and efficiency performance is obtained. Fuel pump power requirements for liquid fuels are less than those of fuel gas booster compressors, further increasing net performance with liquid fuels.

Quite often, a fuel gas booster compressor is required to ensure adequate fuel pressure for gas turbine flow control and combustion systems. Gas turbine combustors operate at pressure levels of 75 to 350 psi. Although the pressure of pipeline natural gas is higher in interstate transmission lines, the pressure is typically reduced at the city gate metering station before it flows into the local distribution piping system. The cost of the fuel gas booster compressors adds to the installation capital cost. High reliability will often require redundant booster compressors. Liquid-fueled gas turbines use pumps to deliver the fuel to the combustors.

Gas turbines are among the cleanest fossil-fueled power-generation equipment commercially available. The primary pollutants from gas turbines are oxides of nitrogen (NO_x), carbon monoxide (CO), and volatile organic compounds (VOCs). Other pollutants, such as oxides of sulfur (SO_x) and particulate matter, depend primarily on the fuel used. Emissions of sulfur compounds, primarily SO_2, reflect the sulfur content of the fuel. Gas turbines operating on natural gas or distillate oil that has been desulfurized in a refinery emit insignificant levels of SO_x. In general, SO_x emissions are significant only if heavy oils are fired in the turbine, and SO_x control is a fuel purchasing issue rather than a turbine technology issue. Particulate matter is a marginally significant pollutant for gas turbines using liquid fuels. Ash and metallic additives in the fuel may contribute to particulate matter in the exhaust.

It is important to note that the gas turbine operating load has a significant effect on the emission levels of the primary pollutants (NO_x, CO, and VOCs). Gas turbines are typically operated at high loads and therefore are designed to achieve maximum efficiency and optimum combustion at these high loads.

9.4 CONTROL OF MICROTURBINES

Modeling of microturbines is a very complex matter, and although there are several modeling efforts [3–7], there is no complete electrical engineering–based model available for feedback control and automation development purposes. To support the modeling and control approach taken in this book, several assumptions and simplifications had to be made. For example, constant physical and chemical properties are assumed, such as specific heat at constant pressure and volume and specific gas constant. Heat loss is neglected and there is no internal energy storage. Through a thermodynamic evaluation, a microturbine can be considered a topping cycle (i.e., electricity or mechanical power is produced first and heat is recovered to meet the thermal loads). Topping cycle systems are generally used when there is no requirement for high temperature processes. Modern commercial microturbines operate on the nonideal Brayton open cycle with heat recovery.

9.4.1 Mechanical-Side Structure

The control structure of a microturbine will depend on whether the unit has a single- or two-shaft structure. The physical arrangement of a microturbine can be used for classification: for example, single-shaft or two-shaft, simple cycle or recovered, intercooled or reheated. Unrecovered turbines have the compressed air mixed with fuel and burned under constant-pressure conditions. The resulting hot gas is allowed to expand through a turbine to perform work. Simple-cycle microturbines have lower efficiencies, lower capital costs, higher reliability, and more heat available for cogeneration applications than recuperated units. Recuperated units use a sheet-metal heat exchanger that recovers some of the heat from an exhaust stream and transfers it to the incoming airstream, boosting the temperature of the airstream supplied to the combustor. Further exhaust heat recovery can be achieved in a cogeneration configuration. Advanced materials such as ceramics and thermal barrier coatings are some of the key enabling technologies used to improve microturbines further. Efficiency gains can be achieved with materials like ceramics, which allow a significant increase in engine operating temperature.

A lot of knowledge of large steam turbines and large synchronous generators built a very solid understanding of the steady state and dynamic behavior for such systems. The basic operation has two modes: steady state and load transient. At steady state, the power of the steam rate into the turbine is equal to the electrical power removed from the generator. The speed of the generator and turbine is considered to be synchronous, implying that the output electrical sinusoidal voltage is in phase with the grid. This operation requires good speed control of the turbine. During a load transient, the change in power is taken from the rotor speed of the large turbine–generator set. Because these devices are enormous, there is considerable stored energy in the rotating masses. The speed control of the turbines sees this speed change and corrects the rate at which the steam is supplied to the turbine, correcting the speed until the set point is achieved. In this manner, the turbine generator set is capable of nearly instantaneous load tracking.

Unfortunately, a similar base of knowledge is not yet fully developed for microturbines and the correspondent generators. However, the following principles are important in establishing the control approach for a microturbine:

1. At steady state, the power of the natural gas combustion and air into the turbine is ideally equal to the electrical power removed from the generator. The speed of the generator and turbine is not critical, as the output sinusoids from the generator are rectified. The dc-link voltage needs to be supported to ensure that the power requirements are conserved. This operation requires good speed control of the turbine.
2. During a load transient, the change in power is taken from the rotor speed of the microturbine. However, because these devices are small, there is very little stored energy in the rotating masses and the rotor speed changes very quickly. The speed control of the microturbine sees this speed change and corrects the rate at which the fuel is supplied to the microturbine, correcting the speed until the set point is achieved. The speed of a microturbine needs to be changed quickly to ensure that the generator does not stall. In this manner, the turbine generator set is capable of load tracking.

The shaft construction defines many important characteristics that will eventually influence the control system. There are essentially two types of shaft construction. One is a high-speed single-shaft design with the compressor and turbine mounted on the same shaft as the alternator rotates at speeds of 90,000 to 120,000 rpm. The split-shaft design has a power wheel on a separate shaft and transfers its output to a conventional generator via a gear reducer [8,9], as indicated

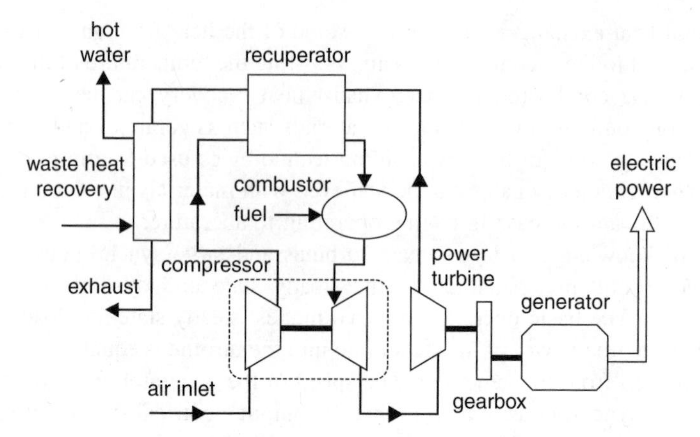

Figure 9.2 Split-shaft microturbine system.

in Figure 9.2. This system philosophy employs the same technology as that for two-pole generator sets running at 3600 rpm or may use an induction generator with a static VAR compensator to assist as a source of reactive power. This is a proven robust scheme, and although manufacturers claim that it does not need power electronics, it requires synchronizing equipment and relays for connection to the electric grid. The gear reducer requires maintenance along with the lubrication system. There are two turbines, one a gasifier turbine driving a compressor and another a free power turbine driving a generator at a rotating speed of 3600 rpm. Therefore, there is one combustor and one gasifier compressor. In conventional power plants, a two-pole permanent magnet generator is driven via a gearbox, running at constant speed, since it is synchronized to the electric network.

At the intake, gas is controlled through a nozzle, where its velocity is increased and the pressure decreases. The high-speed gas exerts a force on the turbine blades, and the geometry of the blades causes the turbine to rotate, thus producing mechanical work. After the turbine rotor, the gas passes through the diffuser of the turbine, where the velocity is again decreased and the pressure is increased, but not as much as after the compressor. Most of the same steam power–based technology is used to control a split-shaft gas turbine system [8,9].

A commercial single-shaft microturbine system is portrayed in Figure 9.3. It is an integrated design where the rotor spins on a thin film of pressured air rather than oil inside its bearings. The high-speed generator is integrated to the shaft as well as the recuperator. The compressor and turbine blades are designed to shape the airflow to go through the combustion chamber and exhaust the hot gas at optimal pressure.

9.4.2 Electrical-Side Structure

One advantage of a high-speed generator is that the size of the machine decreases almost in direct proportion to the increase in speed, leading to a very small unit that

Figure 9.3 Capstone single-shaft microturbine. (Courtesy of National Renewable Energy Laboratory.)

can be integrated with the gas turbine. In the split-shaft design, a starter is needed to bring the turbine up to operational speed, whereas in the single-shaft design, the generator may operate in motoring mode through a bidirectional power electronic system for startup. The generator is a rare-earth permanent-magnet alternator. The stator core is made with thin laminations of low-loss electrical steel and coils are made of Litz wire to achieve good high-frequency characteristics. A four-pole design contributes to short end windings that make possible a short distance between bearings.

Water cooling might be used to keep the temperature low in the windings. Vacuum-pressurized impregnation and insulation thicker than standard ensures a long lifetime for the windings. An important issue is to ensure that the rotor never reaches a temperature that would demagnetize the magnets. This is done by reducing the rotor losses through providing efficient cooling in the airgap.

Power electronic inverters are also integrated to the generator system, and high switching transistors with customized monitoring system control the operation of the generator. As opposed to the split-shaft design, the use of an inverter provides several advantages, such as elimination of a gearbox and start-up features. The power electronics provide most of the protection and interconnection functionalities in addition to power factor correction and control flexibilities impossible with the split-shaft concept.

Figure 9.4*a* shows a bidirectional power electronics converter that allows motoring and regeneration control of the high-frequency permanent-magnet generator

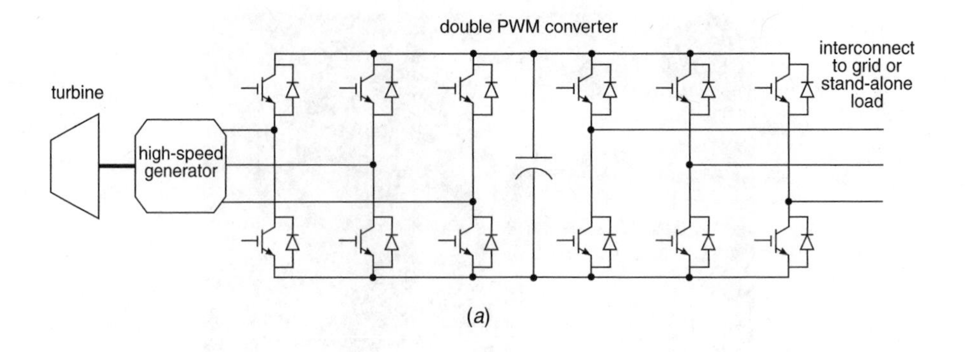

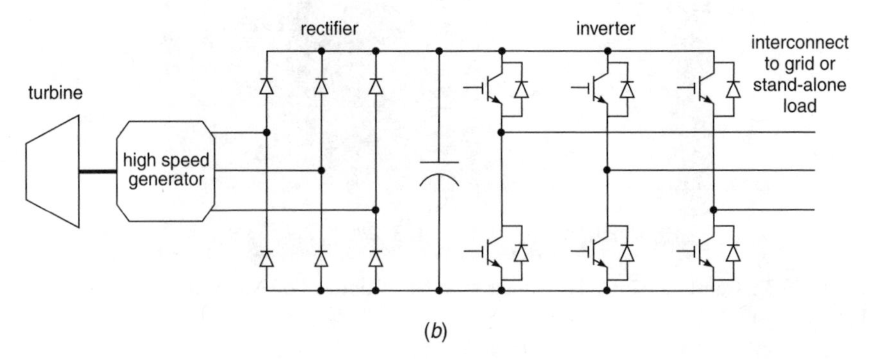

Figure 9.4 Power electronic topologies for permanent magnet generators: (*a*) bidirectional; (*b*) unidirectional.

[10–12]. Figure 9.4*b* shows a more simplified scheme where the generator voltage is rectified and delivered to an inverter tied to the grid. The scheme provided by Figure 9.4*a* is more versatile; it requires very fast devices and very fast signal processing. This is the current challenge for power electronics engineers. On the other hand, the scheme in Figure 9.4*b* requires standard power electronic devices at the expense of an auxiliary mechanical startup on the shaft for bringing the turbine to the optimal generation speed range. Most commercial turbines are constructed with a front-end rectifier such as Figure 9.4*b* for lower cost and less complexity.

The overall single-shaft microturbine system is very intricate, as depicted by Figure 9.5. The main input to the regulator is an electric power reference (i.e., the amount of electric power the generator should produce). The main output of a regulator is the fuel rate (i.e., how much fuel will be injected in the combustion chamber). There is a maximum value, with the fuel valves fully open, and a minimum value, when the valves are open just to keep the flame alive.

9.4.3 Control-Side Structure

There are measurements in the feedback loop for the speed, the power currently produced in the generator, and the outlet temperature at the turbine. When the

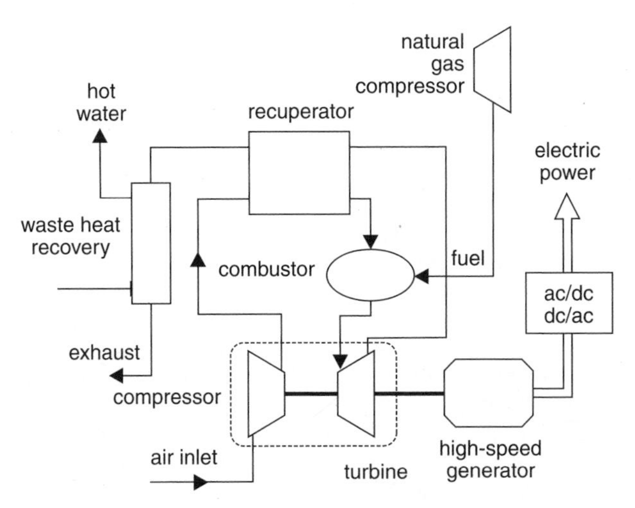

Figure 9.5 Single-shaft microturbine overall system.

microturbine is connected to the grid, the gas turbine provides heat and power at a constant level to a local load. If the power demand in the local power grid increases to a higher value than the microturbine maximum output, the extra power is taken from the outer power grid. The microturbine usually runs at its optimum, full load. Since the control system does not need to meet any sudden changes, all changes can be done slowly and the thermal fluctuations inside the gas turbine are minimized.

The life span of a gas turbine is closely related to the size and speed of the temperature changes it experiences. These changes can be slowed by including a rate-limiter function. The total efficiency is optimal at a certain temperature and speed, which makes the microturbine more economical if always run at the optimum point. The purpose of the control system is to produce the power demanded while maintaining optimal temperature and speed characteristics.

In stand-alone mode the control system must maintain a constant voltage and frequency insensitive to variations in the load power. Therefore, a very fast control loop is required for stand-alone systems. The faster control is achieved at the expense of higher thermal variations in the machine, thus reducing its life span. Figure 9.6 depicts a simulation block diagram of a microturbine system from speed control to mechanical power delivery [13]. There are several time constants and coefficients to be determined, but it is not easy to obtain this information from a manufacturer's data sheet. System identification procedures [14–16] must be conducted to define:

- Fuel governor response
- Temperature rise time
- Valve positioner
- Combustion chamber response

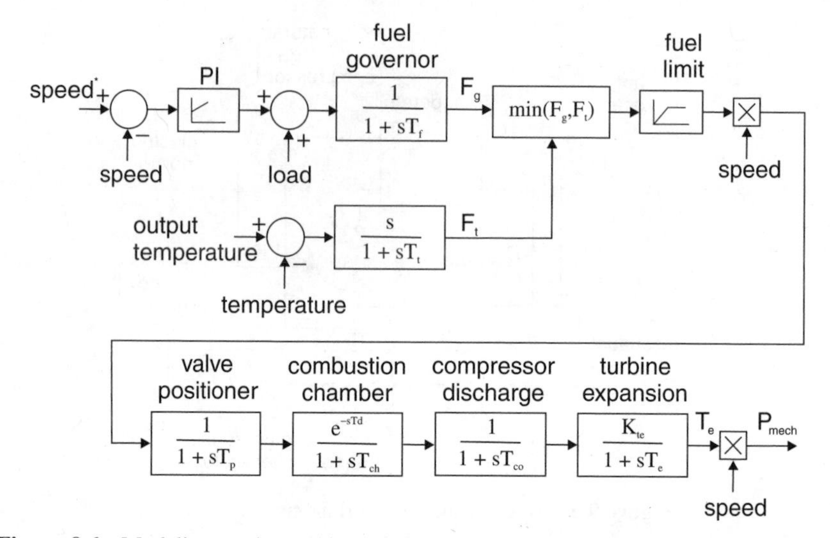

Figure 9.6 Modeling a microturbine system from speed to mechanical power delivery.

- Time delay of the combustion chamber
- Compressor discharge
- Turbine expansion

The control system consists of a normal speed regulator that tries to maintain a predefined reference speed (e.g., 65,000 rpm). Such speed may not be optimal for lower power loads, but the regulator tries to maintain the speed by controlling the fuel input to the turbine. If the load is increased suddenly, it is very difficult to increase the speed without reducing the power load for a moment, and a very complex control must consider all those variables. The upper performance limit here is the limitation of the fuel injectors. The maximum amount of fuel that can be injected is based on temperature limitations.

The full system of Figure 9.6 can be simplified and a first-order dominant pole can be assumed to represent the torque response to the fuel governor. In this case, Figure 9.7 represents a possible control system diagram that integrates a microturbine generator with a front-end rectifier with control of the dc-link voltage [17,18].

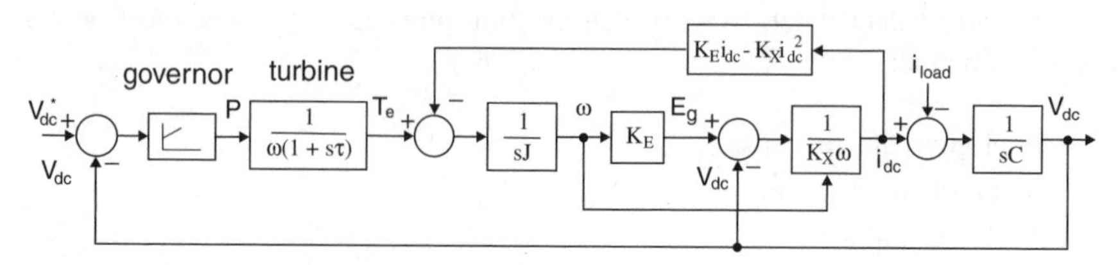

Figure 9.7 Combined electromechanical system model for a microturbine system.

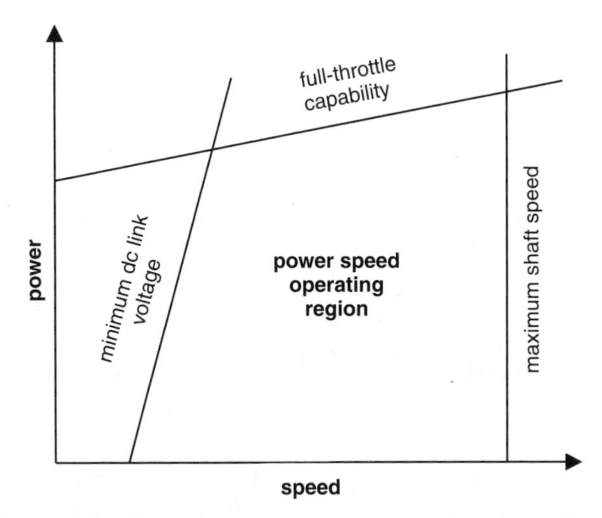

Figure 9.8 Constraints on power and speed of a microturbine.

Some parameter and system identification procedures are still required to determine the proportional integral parameters for the governor and turbine time constant τ, system inertia J, and apparent turbine impedance to the dc link system by constants K_e and K_x. The ac load I_{load} will eventually draw power from the dc link. Therefore, this model considers an equivalent current at the dc link, which is reflected by the ac load.

The power and speed operating regions are constrained by three variables, as indicated in Figure 9.8. A minimum dc-link voltage is required for maintaining a linear pulse width modulation. Therefore, such a constraint imposes a minimum shaft speed. Maximum power is determined by the fuel injection rate to the combustors, limiting the maximum power over the operating range. Bearings, vibrations, and overall mechanical design constraints will limit the maximum speed on the right side of the power–speed envelope of Figure 9.8.

Inside the power and speed-operating region of Figure 9.8, a mapping of power to speed can be defined either by system identification or by nonlinear methods. If a family of curves defines such a power–speed mapping, a very robust direct self-control system can be implemented as indicated by Figure 9.9. For the sake of simplicity, the dc bus capacitance response is neglected in Figure 9.9. As dc-link voltage has a trend to increase or decrease, the power estimation map will define the estimated output power and will generate the speed reference. The shaft speed will accommodate the microturbine within the power and speed-operating region by a self-tracking procedure [19]. Figure 9.10 depicts the signal flow of all the interconnecting modules (i.e., the system controller controls the fuel injector by direct self-control after reading the dc bus output voltage, output voltage, current, and shaft speed) [20–22]. The output inverter is controlled by sinusoidal pulse-width modulation (SPWM) to deliver clean output voltage.

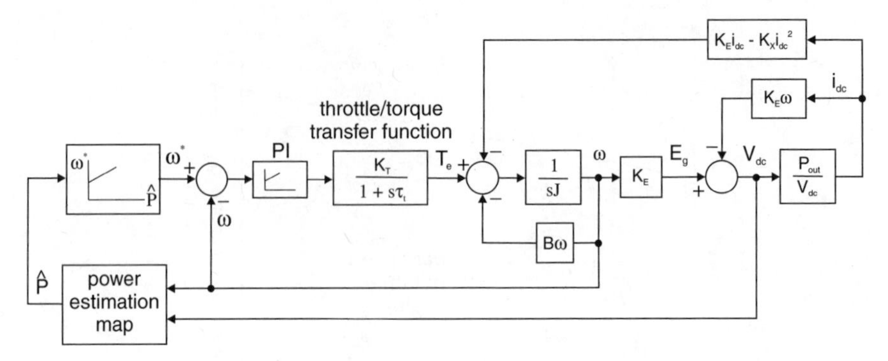

Figure 9.9 Direct self-control approach of a microturbine with power–speed mapping.

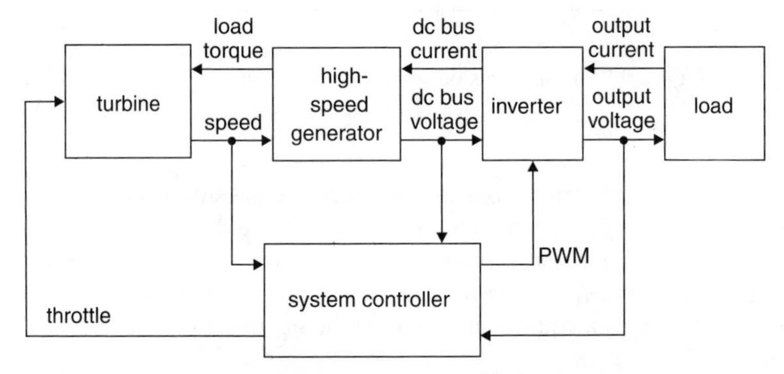

Figure 9.10 System control of a microturbine and output inverter.

9.5 EFFICIENCY AND POWER OF MICROTURBINES

The work produced by the turbine can be calculated by an equation using the inlet and outlet pressures (p_1 and p_2) and input temperature (T_1) at isentropic conditions. Those conditions are assuming ideal conditions (the process can be reversed without losing energy) and no heat flow in or out of the system:

$$W = \frac{\kappa}{\kappa - 1} RT_1 \left[\left(\frac{p_2}{p_1} \right)^{(\kappa-1)/\kappa} - 1 \right] \tag{9.1}$$

where κ is the ratio of specific heat at constant pressure and constant volume, and R is the gas constant for the gas specified, in this case, air.

In addition to simplifying ideal isentropic conditions, friction, turbulent eddies behind the vanes, and the efficiency of the compressor all decrease the output power. Therefore, the thermodynamic variable called the isentropic efficiency, η_{is}, defined as the ratio of the enthalpy change at isentropic conditions and the actual enthalpy change, is used to compensate for the actual conditions. The value of the isentropic efficiency can be determined through experiments and tabulated for variable temperature and pressure ratio. The work multiplied by the mass flow rate $\dot{m}$, the isentropic efficiency, and the mechanical efficiency due to the bearing and windage losses, η_{mec}, give the output power, as indicated by:

$$P_{turbine} = \dot{m}\eta_{is}\eta_{mec}\frac{\kappa}{\kappa - 1}RT_1\left[\left(\frac{p_2}{p_1}\right)^{(\kappa-1)/\kappa} - 1\right] \quad (9.2)$$

Formulation of equations (9.1) and (9.2) embeds the heating value of the natural gas in the outlet pressure increase in the air. However, for economical evaluation, it is very important to define the turbine efficiency from fuel power to the output power. The natural gas input power in Btu/hr or kJ/s can be evaluated by

$$P_{fuel} = \mathrm{HV}_s Q \frac{P}{P_s}\frac{T_s}{T} \quad (9.3)$$

where HV_s is the heating value of the natural gas at standard temperature and pressure (energy/volume), Q the natural gas volume flow rate, P the natural gas pressure, P_s the standard pressure, T the natural gas temperature, and T_s the standard temperature. For example, for natural gas at $T_s = 20°\mathrm{C}$ and $P_s = 14.7$ psi, the heating value is $HV_s = 1042\ \mathrm{Btu/ft^3}$.

As depicted in the power equations above, one can consider a microturbine as a mass flow machine where changes in the air density affect the performance significantly. The industry has set the inlet air temperature at 15°C for determining performance. Therefore, if a microturbine is rated at 60 kWh output, it is for 15°C inlet air temperature. Manufacturers provide information on the derating of power and efficiency for variable inlet air temperature. An example is given in Figure 9.11.

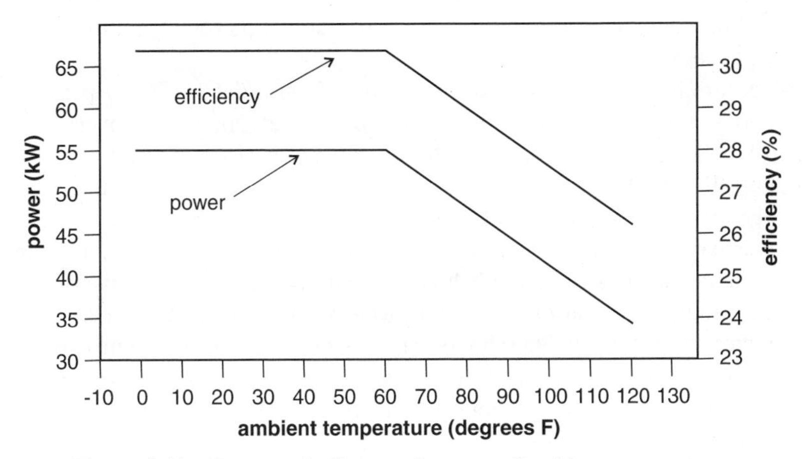

Figure 9.11 Power and efficiency in terms of ambient temperature.

The elevation is also very relevant technical information, and a manufacturer data sheet is usually available to provide the relative sensitivity with altitude. A rule of thumb is that for every 300 m above sea level, the net power should be derated 3%. However, a combination of altitude and inlet temperature must be used to determine output power [23]. Since exhaust gases are usually ducted for either heat recovery or for use as a drying furnace, it is important to calculate the decay in power and efficiency, as this backpressure affects microturbine performance. Losses as a function of the duct diameter, duct length, number of bends in the duct, and type of material that determines the roughness and friction factors must be accounted for as the total head loss and must be limited to the level specified by the microturbine manufacturer.

9.6 SITE ASSESSMENT FOR INSTALLATION OF MICROTURBINES

Microturbines can be used for standby power, power quality and reliability, peak shaving, and cogeneration applications. In addition, because microturbines are being developed to utilize a variety of fuels, they are being used for resource recovery such as landfill gas applications. Siting and sizing a microturbine require a comprehensive economical evaluation and a study of safety, building, and electrical codes. Microturbines are well suited for small commercial building establishments such as restaurants, hotels/motels, small offices, retail stores, and many others.

Microtubines provide an aesthetic power source, improving sightlines and views with off-the-grid systems that eliminate the need for overhead power lines and enable cost savings by reducing the peak demand at a facility. As a result, it lowers demand charges and provides better power reliability and quality, especially for those in areas where brownouts and surges are common or utility power is less dependable. Microturbines can provide power to remote applications where traditional transmission/distribution lines are not an option, such as construction sites and offshore facilities and possess combined heat and power capabilities for sensitive applications.

Microturbine capital costs currently range from $700 to $1100 per kilowatt, including hardware, software, and initial training. Adding heat recovery increases the cost by $75 to $350 per kilowatt. Installation costs vary significantly by location but generally add 30 to 50% to the total installed cost. Microturbine manufacturers are targeting a future cost below $650 per kilowatt. This appears to be feasible if the market expands and sales volumes increase. With fewer moving parts, microturbine vendors hope the units can provide higher reliability than that for conventional reciprocating engine generator technologies. Most manufacturers are targeting maintenance intervals of 5000 to 8000 hours. Maintenance costs for microturbine units are still based on forecasts with minimal real-life situations. Estimates range from $0.005 to $0.016 per kilowatt hour, which would be comparable to that for small reciprocating engine systems.

Placement of microturbines is critical to avoid piping and wiring costs. Therefore, ideal locations are near already existing emergency diesel generators and boiler rooms with available connectivity to natural gas, electricity, and hot water. Often, outdoor installations are more cost-effective because audible noise is typically on the level of 70 dBA at 100 m and it is probably convenient to find an outdoor place in the building with easy connections to exhaust and fresh air ducting and close to utility connections.

Typical microturbine power and heat output efficiencies for power generation are at around 30% at the moment. The electrical efficiency falls to about half of that when the recuperator bypass is engaged. In this case, the overall thermal efficiency may rise to about 80%. The exhaust gas temperature is typically about 260°C while using the recuperator and 870°C with the recuperator bypassed. Not all of this heat can be transformed effectively into useful energy. For example, the Capstone microturbine can produce hot water at about 90°C when joined with a CHP heat recovery unit.

REFERENCES

[1] S. Samuelsen, Fuel cell/gas turbine hybrid systems, NFCRC report, http://www.asme.org/igti/resources/articles/turbo-fuel-cell_report_feb04.pdf.

[2] W. G. Scott, Microturbine generators for distribution systems, *IEEE Industry Applications*, Vol. 4, No. 3, pp. 57–62, May–June 1998.

[3] P. Ailer, Modeling and nonlinear control of a low-power gas turbine, M.Sc. thesis, Department of Aircraft and Ships, Budapest University of Technology and Economics, Budapest, Hungary, 2002.

[4] S. Haugwitz, Modeling of microturbine systems, M.Sc. thesis, Department of Automatic Control, Lund Institute of Technology, Lund, Sweden, 2002.

[5] A. A. Pérez Gómez, Modeling of a gas turbine with Modelica, M.Sc. thesis, Department of Automatic Control, Lund Institute of Technology, Lund, Sweden, May 2001.

[6] Y. Zhu and K. Tomsovic, Development of models for analyzing the load following performance of microturbines and fuel cells, *Journal of Electric Power Systems Research*, Vol. 62, No. 1, pp. 1–11, May 2002.

[7] W. I. Rowen, Simplified mathematical representations of heavy duty gas turbines, *ASME Journal of Engineering for Power*, Vol. 105, pp. 865–869, October 1983.

[8] L. N. Hannett, G. Jee, and B. Fardanesh, A governor/turbine model for a twin-shaft combustion turbine, *IEEE Transactions on Power Systems*, Vol. 10, No. 1, pp. 133–140, 1995.

[9] A. Al-Hinai, K. Schoder, and A. Feliachi, Control of grid-connected split-shaft microturbine distributed generator, in *Proceedings of the 35th Southeastern Symposium on System Theory*, March 16–18, 2003.

[10] R. H. Staunton and B. Ozpineci, *Microturbine Power Conversion Technology Review*, Report ORNL/TM-2003/74, Oak Ridge National Laboratory, Oak Ridge, TN, April 2003.

[11] E. Mollerstedt and A. Stothert, A. model of a microturbine line side converter, in *Proceedings of the International Power System Technology Conference*, December 4–7, 2000, Vol. 2, pp. 909–914.

[12] O. Aglen, Back-to-back tests of a high-speed generator, in *Proceedings of the IEEE International Electric Machines and Drives Conference, IEMDC'03*, June 1–4, 2003, Vol. 2, pp. 1084–1090.

[13] J. Padulles, G. W. Ault, and J.R. McDonald, An integrated SOFC plant dynamic model for power systems simulation, *Journal of Power Sources*, Vol. 86, pp. 495–500, 2000.

[14] Y. Tanaka, Alpha parameter method and its application to a gas turbine, in *Proceedings of the 35th Conference on Decision and Control*, Kobe, Japan, December 1995, pp. 2792–2793.

[15] P. D. Fairchild, S. D. Labinov, A. Zaltash, and B. D. Rizy, Experimental and theoretical study of microturbine-based BCHP system, presented at the 2001 ASME International Congress and Exposition, New York, November 11–16, 2001.

[16] A. Cano and F. Jurado, Modeling microturbines on the distribution system using identification algorithms, in *Proceedings of the IEE Conference on Emerging Technologies and Factory Automation, ETFA'03*, September 16–19, 2003, Vol. 2, pp. 717–723.

[17] R. Lasseter, Dynamic models for microturbines and fuel cells, in *Proceedings of the IEEE Power Engineering Society Summer Meeting*, July 15–19, 2001, Vol. 2, pp. 761–766.

[18] M. Etezadi-Amoli and K. Choma, Electrical performance characteristics of a new microturbine generator, in *Proceedings of the IEEE Power Engineering Society Winter Meeting*, January 28–February 1, 2001, Vol. 2, pp. 736–740.

[19] A. M. Azmy and I. Erlich, Dynamic simulation of fuel cells and microturbines integrated with a multimachine network, in *Proceedings of the IEEE Power Tech Conference*, Bologna, Italy, June 23–26, 2003, Vol. 2, pp.

[20] A. Al-Hinai and A. Feliachi, Dynamic model of a microturbine used as a distributed generator, *Proceedings of the 34th Southeastern Symposium on System Theory*, March 18–19, pp. 209–213.

[21] Q. Zhang and P. L. So, Dynamic modeling of a combined cycle plant for power system stability studies, in *Proceedings of the IEEE Power Engineering Society 2000 Winter Meeting*, Singapore, January 23–27, 2000.

[22] G. Venkataramanan, M. S. Illindala, C. Houle, and R. H. Lasseter, *Hardware Development of a Laboratory-Scale Microgrid Phase 1-Single Inverter in Island Mode Operation*, Report NREL/SR-560–32527, National Renewable Energy Laboratory, Golden, CO, November 2002.

[23] B. F. Kolanowski, *Guide to Microturbines*, Fairmont Press, Lilburn, GA, and Marcel Dekker, New York, 2004.

CHAPTER 10

INDUCTION GENERATORS

10.1 INTRODUCTION

Rotating generators for renewable energies can technically use any type of electrical winding and construction, such as dc or ac, where the ac can be either synchronous or asynchronous (induction) types. Today, dc machines can only be justified for very small power plants because they are bulky, require maintenance, and are relatively inefficient. There are some other contemporary permanent magnet–based design machines, but they are also used for low power ranges. Typically, small renewable energy power plants rely mostly on induction machines, because they are widely commercially available and very inexpensive. It is also very easy to operate them in parallel with large power systems, because the utility grid controls voltage and frequency while static and reactive compensating capacitors can be used for correction of the power factor and harmonic reduction [1,2]. Since induction generators are important for applications in small hydropower and wind systems, in this chapter we discuss briefly details that will enable the reader to understand how to use this electrical machine.

Although the induction generator is mostly suitable for hydro and wind power plants, it can be used efficiently in prime movers driven by diesel, biogas, natural gas, gasoline, and alcohol motors. Induction generators have outstanding operation

Integration of Alternative Sources of Energy, by Felix A. Farret and M. Godoy Simões

as either motors or generators; they have very robust construction features, providing natural protection against short circuits, and have the lowest cost among generators. Abrupt speed changes due to load or primary source changes, usually expected in small power plants, are easily absorbed by its solid rotor, and any current surge is damped by the magnetization path of its iron core without fear of demagnetization, as opposed to permanent magnet–based generators. In this chapter we consider an analysis of the induction generator in both stand-alone and grid-connected mode [3–5].

In this book, the term *asynchronous generator* is applied beyond its standard meaning of induction generator. As described in Chapters 12 and 13, we include under the asynchronous group other types of generators without synchronism with the main source frequency. That is the case for photovoltaic-, fuel cell-, and battery-powered systems. This distinction is important for alternative sources of energy, because lack of synchronism is typically the major characteristic of any of these sources. The lack of synchronism may be responsible for frequency oscillations and instabilities in the power systems to which they are connected. Main utility providers have policies to constrain the amount of injected asynchronous energy up to 15% of their total capacity. Therefore in the future, widespread and deep penetration of distributed generation will only be possible with sophisticated controls and interaction with the utility grid.

10.2 PRINCIPLES OF OPERATION

The induction generator has the very same construction as induction motors with some possible improvements in efficiency [1–4]. There is an important operating difference; the rotor speed is advanced with respect to the stator magnetic field rotation. In quantitative terms, if n_s is the synchronous mechanical speed (rpm) at the synchronous frequency f_s (Hz), the resulting output power comes from the induced voltage proportional to the relative speed difference between the electrical synchronous rotation and the mechanical rotation within a speed-slip factor range given by

$$s = \frac{n_s - n_r}{n_s} \tag{10.1}$$

where s is the slip factor and n_r is the rotor speed (rpm).

Calling the electrical stator frequency f_s, the rotor frequency f_r can be related to the electric frequency through the expression

$$f_r = \frac{p}{120}(n_s - n_r)\frac{n_s}{n_s} = sf_s \tag{10.2}$$

When there is no relative movement of rotation between rotor and stator $\omega_r = \omega_s$, the rotor frequency is dc. For other angular speeds, in quantitative terms, the voltage induced on the rotor is directly proportional to the slip factor s, and from there, transferred to the stator windings [1,5]. From this definition the voltage induced E_r on the rotor is established for any speed with respect to the blocked rotor voltage E_{r0}; that is,

$$E_2 = E_r = sE_{r0}$$

using the transformer turn ration a_{rms} stator voltage comes from

$$E_1 = a_{\text{rms}} \frac{E_r}{s} \tag{10.3}$$

Similarly, the rotor current can be established as

$$I_2 = \frac{I_r}{a_{\text{rms}}}$$

and

$$Z_r = \frac{E_r}{I_r} = \frac{sE_{\text{r0}}}{I_r} = R_r + jsX_{\text{r0}} \tag{10.4}$$

From this definition, the rotor impedance for any rotor speed can be established from the blocked rotor test as

$$Z_{\text{r0}} = \frac{E_{\text{r0}}}{I_r} = \frac{R_r}{s} + jX_{\text{r0}} = \frac{Z_r}{s} \tag{10.5}$$

where X_{r0} is the blocked rotor reactance and R_r is the rotor winding resistance. Then the rotor impedance reflected on the primary can be obtained from equations (10.3), (10.4), and (10.5) as

$$Z_2 = a_{\text{rms}}^2 \left(\frac{R_r}{s} + jX_{r0} \right) \tag{10.6}$$

In terms of rotor values referred to stator values comes

$$Z_2 = \frac{E_1}{I_2} = \frac{R_2}{s} + jX_2 \tag{10.7}$$

and the rotor current is

$$I_2 = \frac{E_1 s}{R_2 + jX_2 s} \tag{10.8}$$

Voltage induction in induction motors is similar to an electric transformer except that the transformer secondary windings (in the induction motor) are rotating. Because there is an airgap in electric machines, the magnetic coupling between the primary and secondary windings makes the *B–H* characteristic slope under the magnetomotive force (mmf) curve of these machines much less accentuated than that of a good-quality transformer. Therefore, saturation of the induction generator is decreased significantly, and as a consequence, the necessary magnetizing current is a lot higher with respect to a transformer (X_m is much lower).

The apparent rotor power (in complex variable) can also be established from equations (10.5) and (10.6) as

$$S_2 = E_1 I_2 = I_r^2 \left(\frac{R_r}{s} + jX_{r0} \right)$$

or

$$S_2 = \frac{E_1^2 s}{R_2 + jX_2 s} \tag{10.9}$$

From these expressions it is clear that the rotor power factor of the induction generator depends on the slip factor and other parameters rather than the load. The quadrature leading current component (in respect to voltage) is almost constant for all voltages across the output terminals and fixed frequencies. So the reactive power absorbed by this circuit must be supplied by an external source, which can be a utility grid with a synchronous generator, a capacitor bank, or an electronic power electronic compensator.

10.3 REPRESENTATION OF STEADY-STATE OPERATION

In Section 10.2 we suggested a formulation that leads to an equivalent circuit of the induction machine similar to a transformer model as illustrated in Figure 10.1. This model is also known as a *per-phase equivalent model*, valid only for perfect sinusoidal excitation and balanced conditions [3–6]. For a more representative approach, every element of the induction generator model has to be excited by the grid voltage, considering any distortions and harmonics. However, the per-phase

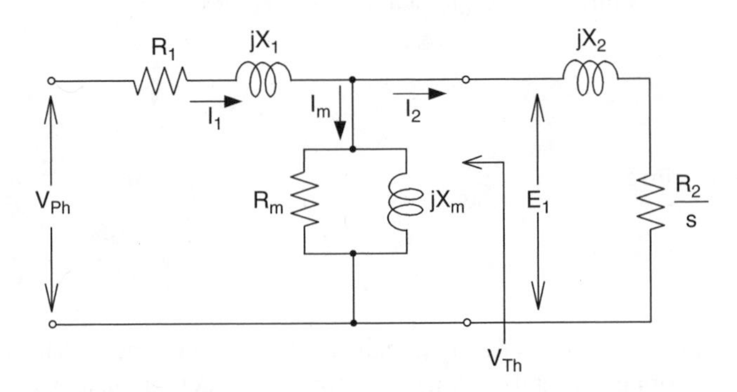

Figure 10.1 Equivalent model of the induction generator.

equivalent model in Figure 10.1 is usually considered for analysis, where R_1 and X_1 are the resistance and leakage reactance of the stator, R_2 and X_2 are the resistance and reactance of the rotor, and R_m and X_m are the equivalent core loss resistance and magnetizing reactance.

The only equivalent resistance in the rotor model R_2/s depends on the rotor speed (i.e., on the slip factor). This is a very important difference with respect to a transformer model, because rotor voltage is subjected to a variable frequency, causing E_r, R_r, and X_r to be variable. In addition, it must be emphasized that the circulating current in the rotor will also depend on the impedance, whose resistance and inductance change slightly because of skin effect. But the inductance is indeed affected in a more intricate way. For very large slip, the rotor impedance and skin effect will impose nonlinear torque and airgap power relationships. For small slip factors $(s \rightarrow 0)$, the rotor impedance becomes predominantly resistive and the rotor current can be considered primarily to vary linearly with s.

For voltage excitation and below the synchronous speed, the slip factor is positive and the current is delayed with respect to the voltage. For current excitation (e.g., with vector-controlled drives) the slip can be either positive or negative. Above the synchronous speed, the rotor moves faster than the magnetic field; and for voltage excitation, the slip factor s becomes negative and the current I_2 leads the E_2 voltage. At first glance, this fact seems contradictory since the rotor circuit is essentially inductive. But the reason for such phase inversion is the reversal of the relative move of the rotating magnetic field with respect to the rotor with changes in s when both real and imaginary parts of the current will be reversed.

The equivalent per-phase impedance seen by the terminals represented in Figure 10.1 is [4,5]

$$Z = R_1 + jX_1 + \frac{1}{\dfrac{1}{jX_m} + \dfrac{1}{R_m} + \dfrac{1}{(R_2/s) + jX_2}} \tag{10.10}$$

To simplify the analysis and improve the formulation of the performance calculation, it is convenient to use a Thévenin circuit. Thus, the impedances of the induction generator can be taken with the following notation:

$$Z_1 = R_1 + jX_1 \qquad Z_2 = \frac{R_2}{s} + jX_2 \qquad Z_m = \frac{1}{1/R_m + 1/jX_m} \tag{10.11}$$

10.4 POWER AND LOSSES GENERATED

The power balance in an induction generator can be expressed as

$$P_{\text{out}} = P_{\text{in}} - P_{\text{losses}} \tag{10.12}$$

The inherent losses can be expressed by their individual contribution as

$$P_{\text{losses}} = P_{\text{copper}} + P_{\text{iron}} + P_{\text{f\&w}} + P_{\text{stray}} \tag{10.13}$$

Copper losses in the stator are straightforward and obtained from $P_{\text{stCu}} = I_1^2 R_1$. Iron losses are caused by hysteresis current (magnetizing) and Foucault current (current induced in the iron). It should be emphasized that stator and rotor iron losses are usually grouped together, as calculated by experimental procedures. It is very difficult to separate their contributions. Therefore, the loss resistance R_m in the equivalent circuit of the induction generator practically represents all these losses combined with the mechanical losses. In these losses are included the friction losses and those resulting from the rotor movement against the air and spurious losses. The spurious losses can be reunited, among other additional losses, in high-frequency losses in the stator and rotor dents, mainly through the parasite currents produced by fast flux pulsation when dents and grooves move from their relative positions [4,5]. Subtracting these losses from input power leaves the airgap power transferred from stator to the rotor. Then in such power received by the rotor, one must subtract the rotor copper losses, $P_{\text{rCu}} = I_2^2 R_2$, as well. This final value remains for transformation from mechanical into electrical power (in motoring mode). Finally, the output power is the *net power in the shaft*, also called *shaft useful power* for motoring mode. As a generator a reverse similar reasoning is applied, i.e., power is input to the shaft and converted to the machine terminals.

For the generating mode, the reasoning must be reversed (i.e., starting from rotor and moving the power conversion toward the machine terminals). The increase in shaft rotation also increases losses by friction, windage, and spurious effects. In compensation, the losses in the core are reduced to the synchronous speed. These, called *rotating losses*, are approximately constant, since some of them increase and others decrease with rotation. To quantify losses, a current I_1 will produce losses in the stator winding of the three phases by

$$P_{\text{stCu}} = 3I_1^2 R_1 \tag{10.14}$$

The iron core losses are given by

$$P_{\text{iron}} = \frac{3E_1^2}{R_m} \tag{10.15}$$

The only possible power dissipation (active power) in the secondary part of the circuit of Figure 10.1 is through the rotor resistance; therefore,

$$P_g = 3I_2^2 \frac{R_2}{s} \tag{10.16}$$

However, from the rotor equivalent circuit shown in Figure 10.1, the resistive losses are given by

$$P_{r\text{Cu}} = 3I_r^2 R_r \tag{10.17}$$

In any ideal lossless transformer, the rotor power would remain unchanged when referred to the stator and, therefore, from equations (10.16) and (10.17),

$$P_{r\mathrm{Cu}} = 3I_2^2R_2 = sP_g \tag{10.18}$$

The losses following the electrically converted power are in the mechanical parts, in the copper, in the core, and in other spurious losses. They can be all expressed as

$$P_{\mathrm{losses}} = P_{\mathrm{stCu}} + P_{r\mathrm{Cu}} + P_{\mathrm{iron}} + P_{\mathrm{f\&w}} + P_{\mathrm{stray}}$$

With equations (10.14) and (10.18), the copper losses in the stator and rotor can be regrouped as

$$P_{\mathrm{copper}} = 3I_1^2(R_1 + R_2) = \frac{3V_{\mathrm{ph}}^2(R_1 + R_2)}{(R_1 + R_2/s)^2 + F^2(X_1 + X_2)^2} \tag{10.19}$$

The mechanical power converted into electricity, or the power developed in the shaft for a negative s, is the difference between the power going through the airgap and that dissipated in the rotor [4–6]. Then, from equations (10.16) and (10.18) comes

$$P_{\mathrm{mech}} = 3I_2^2\frac{R_2}{s} - 3I_2^2R_2$$

after simplification it becomes

$$P_{\mathrm{mech}} = 3I_2^2R_2\frac{1-s}{s} = (1-s)P_g \tag{10.20}$$

With the definitions given in equation (10.13), it is possible to establish a routine to calculate the relationships given in the flowchart of Figure 10.2, where voltage, current, and power are represented in the equivalent circuit of Figure 10.1. Notice that if $Z_m \gg Z_1$ and $Z_m \gg Z_2$, as is usually the case, the stator Thévenin voltages become $V_{\mathrm{ph}} \approx V_{\mathrm{Th}}$ and $Z_1 \approx Z_{\mathrm{Th}}$. Conversely, assuming that $R_m \to \infty$, Z_1 and V_{ph} will have to be replaced by their Thévenin equivalent.

It is possible to take into account how the frequency affects the parameters of the induction generator. These differences are included in the per unit (p.u.) frequency related to the frequency used in the parameter measurement [usually, $\omega_{\mathrm{meas}} = 2\pi(60)$], defined as

$$F = \frac{f}{f_{\mathrm{meas}}} = \frac{\omega}{\omega_{\mathrm{meas}}} = \frac{p}{120}\frac{n_r}{60} \tag{10.21}$$

where f, ω = actual operating frequencies (Hz and rad/s, respectively)
$f_{\mathrm{meas}}, \omega_{\mathrm{meas}}$ = frequencies of the parameter measurement (Hz and rad/s, respectively)
p = number of poles of the machine
n_r = rotor speed (rotations per minute)

Corrections using F on the parameter values are extremely important when using the induction generator model in frequencies other than that used to obtain the machine parameters in laboratory tests.

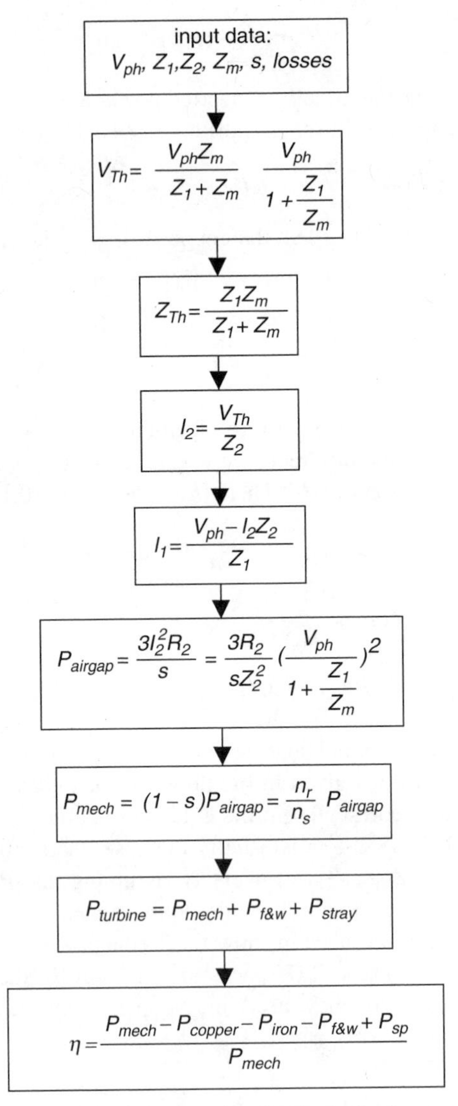

Figure 10.2 Calculation routine for the induction generator.

10.5 SELF-EXCITED INDUCTION GENERATOR

For its operation, the induction generator needs a reasonable amount of reactive power which must be fed externally to establish the magnetic field necessary to convert the mechanical power from its shaft into electrical power. In interconnected applications, the synchronous network supplies such reactive power. In stand-alone applications, the reactive power must be supplied by the load itself, by a bank of capacitors connected across its terminals, or by an electronic inverter. For this

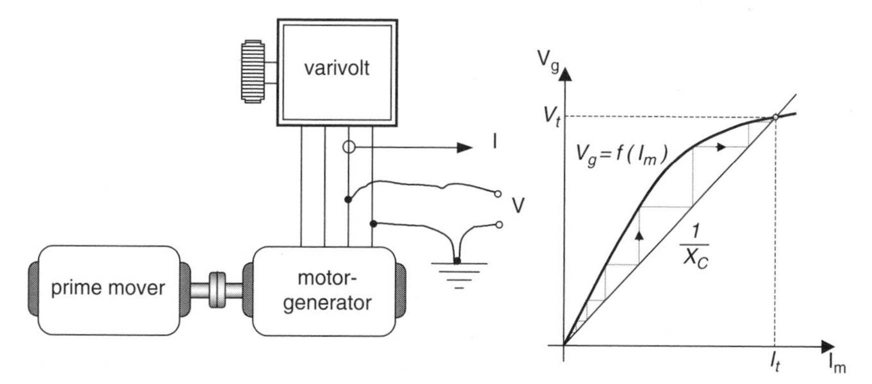

Figure 10.3 Self-excitation process of the induction generator.

reason, the external reactive source must remain permanently connected to the stator windings responsible for the output voltage control. When capacitors are connected to induction generators, the system is usually called a SEIG (a self-excited induction generator). Economically, self-excitation of the induction generator is usually recommended only for small power plants [5–11].

The self-excitation process occurs when an external capacitance is associated with any ordinary induction motor. When the shaft is rotated externally, such movement interacts a residual magnetic field and induces a voltage across the external capacitor, resulting in a current in the parallel circuit, which, in turn, reinforces the magnetic field and the system builds up an increasing excitation. This process is illustrated in Figure 10.3. Notice that in practical schemes it is advisable to connect each excitation bank of capacitors directly across each motor winding phase in either Δ–Δ or Y–Y connection [5,6,12–14].

Each induction machine will require a certain capacitance bank that will depend on the primary energy, the magnetization curve of the core, and the instantaneous load. So interaction among the operational state of the primary source of energy, the induction generator, the self-excitation parameters, and the load will define the overall performance of the power plant [15–21]. The performance is greatly affected by the variance of many of the parameters related to the availability of primary energy and load variations.

The equivalent circuit given in Figure 10.1 is conveniently transformed in the circuit of Figure 10.4 to include the self-excitation capacitor and the load in p.u. frequency values F as given in equation (10.21), representing a more generic form of the stand-alone generator [2,3]. From the definition of secondary resistance (rotor resistance) shown in Figure 10.1 and from equation (10.6), the following modification can be used to correct R_2/s to take into account changes in the stator and rotor p.u. frequencies:

$$\frac{R_2}{Fs} = \frac{R_2}{F(1 - n_r/n_s)} = \frac{R_2}{F - v}$$

where n_r is the rotor speed in p.u. referred to the test speed used in the rotor.

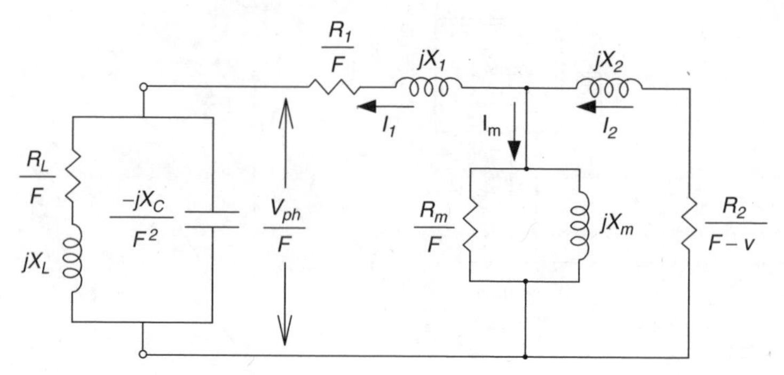

Figure 10.4 Equivalent circuit of the self-excited induction generator. (From Ref. [20].)

Although most transient modeling studies consider magnetizing reactance under magnetic saturation, the other parameters are typically considered constants [22–25]. However, the use of uncorrected parameters may lead to misrepresentation of overall machine performance.

Unfortunately, when working in stand-alone mode, a self-excited induction generator may collapse because of high load or short-circuit current through its terminals, resulting in complete loss of residual magnetism. Although such a feature is an intrinsic protection feature, it may affect the restarting of the overall system. Four methods for its remagnetization are possible: (1) maintaining a spare capacitor always charged and, when necessary, discharging it across one of the generator phases; (2) using a charged battery; (3) using a rectifier fed by the distribution network; and (4) keeping it running without load until the minute residual magnetism available is able to build itself up again.

10.6 MAGNETIZING CURVES AND SELF-EXCITATION

The equivalent impedance seen by the voltage across the terminals of the stator reactance can be evaluated with Figure 10.1 at the synchronous speed ($F = 1$) as

$$Z = R_1 + jX_1 + \frac{1}{1/jX_m + 1/(R_2/s) + jX_2} \tag{10.22}$$

The impedance given by equation (10.22) is useful in determination of load limits, in both amplitude and amount of reactive power required for the induction generator when interconnected to the utility grid [25–27].

It is often more convenient initially to assume the electric frequency supplied to the load and then obtain the mechanical rotation. Therefore, the equation used for calculation of these variables should be altered conveniently to accommodate the electrical frequency as [3]

$$\omega_r = \frac{p}{2}\left(2\pi\frac{n_r}{60}\right) = \frac{p}{2}\omega_r' \tag{10.23}$$

where $\omega_r' = 2\pi(n_r/60)$ is the angular mechanical speed of the rotor (rad/s).

The required capacitance connected across the induction generator terminals can be determined by the intersection of the magnetizing curve $V_g = f(I_m)$ of the machine and the straight line represented by the capacitive reactance given by $1/\omega C$ of the self-exciting bank, as depicted in Figure 10.3. This intersection point gives the terminal voltage V_t and current I_t of the induction generator at no-load conditions. At this point, the magnetizing current is lagging with respect to the terminal voltage (about 90°), depending on the losses of the motor at no-load conditions, while the current through the capacitor is advanced by approximately 90°. This means that the intersection of these two lines is the point on which the necessary reactive power of the generator is supplied only by the capacitors and can be understood as the resonant conditions for the reactive power flow. Location of this point should be in between the rated current of the machine and the end of the straightest portion of the airgap line.

The magnetizing characteristic shown in Figure 10.3 can be determined with the help of a secondary driving machine coupled to the shaft of the induction motor–generator, compensating for any shaft mechanical loading during the experimental procedure for obtaining the magnetizing curve. With a constant rotation, say at 1800 rpm for a 60-Hz four-pole machine, an ac voltage is imposed across the terminals of the motor–generator by a varivolt that induces the nominal current. From this point, several decreasing steps of voltage have to be recorded with the corresponding current until the zero voltage. A constant rotation must always be guaranteed during the tests, because any mismatch will change the frequency and terminal voltage of the rotor. The mutual inductance is essentially nonlinear, and for best results in numerical computations it should be determined for each value of the instantaneous magnetizing current, I_m [23–25].

From the discussion above it can be observed that the induction generator accepts constant and variable loads; it starts either loaded or without load, it is capable of continuous or intermittent operation, and it has a natural protection against short circuits and over currents through its terminals. When the load current goes above certain limits, the residual magnetism falls to zero and the machine is deexcited.

10.7 MATHEMATICAL DESCRIPTION OF THE SELF-EXCITATION PROCESS

A mathematical description of the self-excitation process is based on the Doxey model, which is derived from the steady-state model presented in Figure 10.4. Since the excitation impedance is much larger than the winding impedance, it is usual to represent the induction generator model with a separate resistance for losses; an RL branch is included to represent a typical load connected across the generator's terminals [2,5,24,25].

By observing Figure 10.4, we see that the inductive load will decrease the effective value of self-exciting capacitance because some of the reactive power will be deviated to the load branch. Equation (10.24) quantifies the effective capacitance due to such an effect:

$$C_{\text{eff}} = C - \frac{L}{R_L^2 + (\omega_s L)^2} \tag{10.24}$$

Equation (10.24) leads to the conclusion that for small L, the inductive reactance of the load does not have much influence on C_{eff}. However, for high values of L, this inductance can be decisive to the output voltage because it approximates the straight line of the capacitive reactance slope to the almost linear portion of the excitation curve slope (airgap line), drastically reducing the terminal voltage until total collapse. The beneficial side is that for too small load resistance, the discharge of the self-excitation capacitor is very quick and establishes a natural protection against high currents and shortcircuits. On the other hand, high values of the self-excitation capacitance are limited by the iron saturation. For heavy saturation, the intersection point on the straight line of the capacitive reactance matches very high current values for the generator, causing the winding to carry high circulation current and possible permanent damage in the insulation and in the magnetic properties of the iron.

By applying the basic principle of power balance, Figure 10.4 helps to obtain equations (10.25) and (10.26) equating the active and reactive power:

$$I_2^2 R_2 \frac{1-s}{s} + I_2^2 (R_1 + R_2) + \frac{V_f^2}{R_m} + \frac{V_f^2}{R_{Lp}} = \sum P = 0 \tag{10.25}$$

$$\frac{V_f^2}{X_p} + \frac{V_f^2}{X_m} - \frac{V_f^2}{X_c} + \frac{V_f^2}{X_{Lp}} = \sum Q = 0 \tag{10.26}$$

where $X_m = \omega_s L_m$ and R_{Lp} and X_{Lp} are the parallel equivalent load resistance and reactance connected across the generator's terminals.

To evaluate the performance of induction generators, one should take into account that the only power entering the circuit is that from the primary machine under the form of active power given by equation (10.25). The reactive power should also be balanced as in equation (10.26). The first term of equation (10.25) is the energy supplied to the generator from the mechanical shaft. The sign of the first term is inverted when $s1$ or $s0$, depicting the generating mode of the induction machine as a generator or brake, respectively [5,22,25]. Dividing equation (10.25) by I_2^2, after simplification one gets

$$\frac{R_2}{s} + R_1 + \frac{V_f^2}{I_2^2}\left(\frac{1}{R_m} + \frac{1}{R_{Lp}}\right) = 0 \tag{10.27}$$

From Figure 10.4 we obtain

$$\frac{V_{ph}^2}{F^2 I_2^2} = \left(\frac{R_2}{Fs} + \frac{R_1}{F}\right)^2 + (X_1 + X_2)^2 \tag{10.28}$$

and with equation (10.27) we have

$$\left(\frac{R_2}{s} + R_1\right) R_{mL} + \left(\frac{R_2}{s} + R_1\right)^2 + F^2 (X_1 + X_2)^2 = 0 \tag{10.29}$$

Equation (10.29) is of second order and the slip can be calculated with equation (10.30) (assuming constant parameters):

$$s = \frac{2R_2}{-2R_1 - R_{mL} \pm \sqrt{R_{mL}^2 - 4F^2 (X_1 + X_2)^2}} \tag{10.30}$$

The roots of the denominator in equation (10.30) will always be negative to satisfy the practical condition that if $X_1 + X_2$ is very small, the load R_{mL} would have almost no influence on s, a constraint not true. So it may be convenient to approximate equation (10.30) by the following expression:

$$s \cong -\frac{R_2}{R_1 + R_{mL}} = -\frac{R_2 (R_m + R_{Lp})}{R_1 R_m + R_1 R_{Lp} + R_m R_{Lp}} \tag{10.31}$$

An iterative method may be used when using equation (10.30) instead of equation (10.31), because the computation of F depends on ω_s, which, in turn, depends on s, which again is dependent on F. The efficiency can be estimated by

$$\eta = \frac{P_{out}}{P_{in}} = \frac{P_{in} - P_{losses}}{P_{in}} \tag{10.32}$$

The input mechanical power must subtract the copper losses $(R_1 + R_2)$ and the loss represented by the resistance R_m. Using equation (10.12) with equations (10.25) and (10.28) for the three-phase case without the portion corresponding to the load power $(R_{Lp} \to \infty)$ yields

$$P_{out} = P_{in} - P_{losses} = 3I_2^2 R_2 \frac{1-s}{s} + 3I_2^2 (R_1 + R_2) + \frac{3V_{ph}^2}{R_m}$$

where $P_{in} = 3I_2^2 R_2[(1-s)/s]$. Therefore, using these values in equation (10.32), after simplification we get

$$\eta = \frac{I_2^2 R_2[(1-s)/s] + I_2^2 (R_1 + R_2) + V_{ph}^2/R_m}{I_2^2 R_2[(1-s)/s]} \tag{10.33}$$

Dividing the numerator and the denominator of equation (10.33) by I_2^2 and using equation (10.28) once more results in

$$\eta = \frac{[(R_2/s) + R_1] + 1/R_m[((R_2/s) + R_1)^2 + F^2(X_1 + X_2)^2]}{R_2[(1 - s)/s]} \tag{10.34}$$

10.8 INTERCONNECTED AND STAND-ALONE OPERATION

When interconnected directly with the distribution network, an induction machine must change its speed (increased up to or above the synchronous speed). The absorbed mechanical power at the synchronous rotation is enough to withstand the mechanical friction and resistance of the air. In case the speed is increased just above the synchronous speed, a regenerative action happens, yet without supplying energy to the distribution network. This will happen only when the demagnetizing effect current of the rotor is balanced by a stator component capable of supplying its own iron losses and, above that, supplying power to the external load.

When considering interconnection with or disconnection from the electric distribution network, it should be observed that there is a rotation interval above the synchronous speed in which the efficiency is very low. This effect is caused by the fixed losses related to the low level of generated power and torque in these low speeds (see Figure 10.5) [27–29].

Another relevant aspect is the maximum rotation. At this point, disconnection from the distribution network should occur so that a control system will act as a brake for the turbine under speed-controlled operation. The disconnection should be performed for electric safety of the generator in the case of a flaw in the control brake and for the safety of the local electric power company maintenance teams, when the generator should be disconnected from the distribution network.

Connection of induction generators to the distribution network is quite a simple process as long as the interconnection and protection guidelines are followed by the local utilities. Technically, the rotor is turned in the same direction as that in which the magnetic field is rotating, as close as possible to the synchronous speed to avoid unnecessary speed and voltage clashes. A phenomenon similar to the connection of motors or transformers to the distribution network will result in a transient exchange of active and reactive power between generator and load. The load considered here can be any ordinary electrical load or a power inverter for interconnection with the distribution network or any ac load.

The active power supplied to the load by an induction generator, similar to what happens with synchronous generators, can be controlled by speed change, which is related to controlling the mechanical primary power. For the case of stand-alone operation, the magnetizing current is to be obtained from the self-excitation process as discussed earlier in the chapter. The generator supplies a capacitive current, because the current I_l is leading V_{ph}, as depicted in Figure 10.1. This is because the mechanical energy of rotation can influence only the active component of the current, with no effect on the reactive component, and as a consequence, the cage rotor

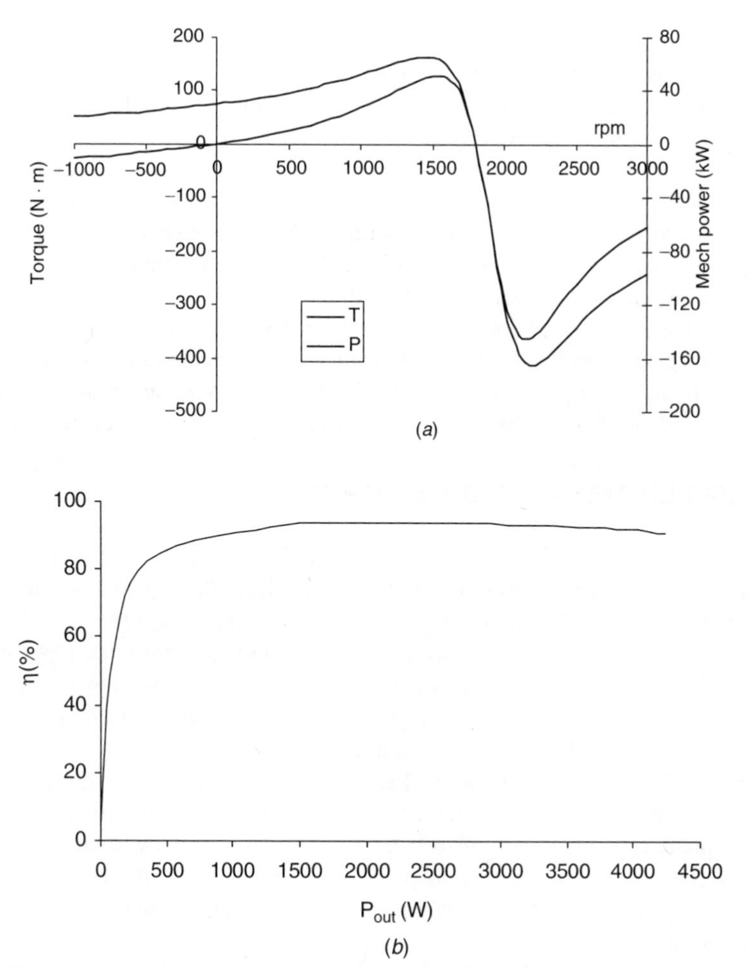

Figure 10.5 Typical performance of a four-pole 50-kW induction generator: (*a*) speed versus torque and output power; (*b*) generator efficiency characteristic.

does not supply reactive current. As a component of the magnetizing current is necessary in the main field direction, $I_{mx(B)}$, to maintain this flux, another reactive current to maintain the dispersion flux also becomes necessary. A leading current with respect to the stator current should supply such currents.

For an induction generator in parallel with a synchronous machine, the excitation depends on the relative speed between them, so the short-circuit current supplied depends on the voltage drop produced across the terminals of the synchronous generator. However, a very intense short circuit transient current arises of extremely short duration. If the voltage across the terminals goes to zero, the steady-state short-circuit current is zero. A small current is supplied in the case of a partial short circuit because the maximum power the induction generator can supply with fixed slip factor and frequency is proportional to the square of the voltage across its terminals [see equation (10.19)]. The incapacity of sustaining short circuits reduces possible damage caused by electrical and mechanical stresses. As a result, it allows a lower short-circuit power and lower rating protection circuit at lower cost than for synchronous generators alone.

The induction generator dampens oscillations as long it does not have to work at synchronous speed. All the load variation is followed by a speed variation and small phase displacement, much the same as with a synchronous generator. The mechanical speed variations of the primary machine driving the generator are so small that they produce only minor variations of load.

Some steady-state conditions can be used for design and performance prediction of induction generators in operation with wind, diesel, or hydro turbines. Aspects such as high slip factor, torque curve, winding sized to support higher saturation current, and increased number of poles can enhance the overall performance of the system.

10.9 SPEED AND VOLTAGE CONTROL

Alternative sources of energy require several levels of control, such as system- and equipment-based controls, as illustrated in Figure 10.6. Some variables are used in accordance with the overall strategy, such as active and reactive power levels or shaft rotation. Such a system is usually a multiple-input/multiple-output control system where the output voltage is associated with the reactive power, frequency is associated with the mechanical output power, and variables are monitored continuously and fed back to follow reference signals [2,29].

When designing the control of an alternative source of energy for electricity generation, it is necessary to consider whether the source is of the rotating or static type. Typical rotating types are related to the wind- and hydroelectric energy, whereas static types are related to photovoltaic and fuel cells. For the rotating type, three options of generators can be used: dc generator, synchronous generator, or asynchronous generator.

In principle, all generators operate internally as ac, but the constructional features of generator will make the output become dc or ac. The ordinary type of dc generator is also known as a commutating machine, because an internal mechanism

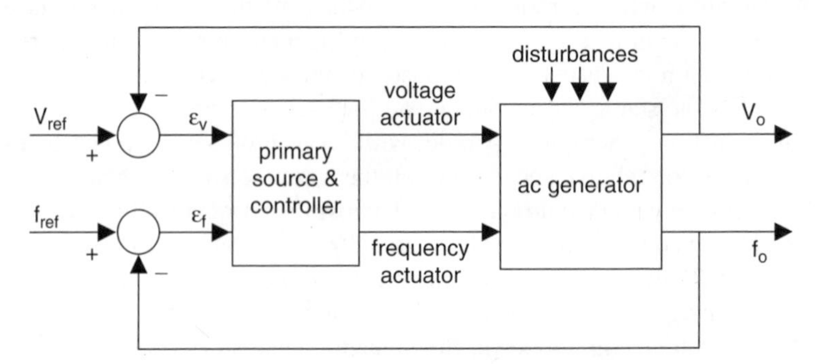

Figure 10.6 Basic control diagram for rotating generators.

is used such that the machine output becomes a direct current while internally the flux alternates. The brushless dc generator, which is in fact an ac generator, is so called for two reasons: (1) because the alternating current must be variable frequency and so derived from a permanent magnet or dc supply; and (2) its speed–torque characteristics are very similar to those of an ordinary dc generator (with brushes). The rotor consists of strong permanent magnets passing between the winding poles of the stator to induce an alternating current. A major advantage is that there is no need for induced current as in the induction generator, making dc generators more efficient and with slightly greater specific power. A problem with this generator is that the torque is very unsteady, causing some stress on the turbine and the mechanical drivetrain. This characteristic may be improved by distributing three or more harnessing coils around the stator, which, in turn, increases the frequency of the output voltage for the same rotation (increased number of poles). Such a feature is particularly interesting for wind energy because it may eliminate the gearbox. The rotating magnet will generate a back electromotive force in the coil that it is approaching. Dc generation has been massively replaced by ac generation because most modern ac power sources are lighter in construction, are mass produced, have very sophisticated power electronic controllers and are very easy and flexible to operate. A complete discussion of other types of generators is not included here; the reader is encouraged to refer to an electric machines text.

Two main variables are important to control ac generators: the output voltage level and the power frequency. Synchronous generators have two control inputs (field current i_f and motor torque T_m) and four control outputs [output active power, output reactive power, voltage magnitude (represented by P_o, Q_o, and $|V|$, respectively) and frequency f]. All these variables are related to each other under a very complex interaction; torque variation causes changes in all four outputs, and there is no stronger relationship between τ_m and Q_o. The degree of cross-coupling will depend on the structure and system size. Renewable energy–based systems are usually small, with a reduced shaft inertia easily subject to oscillations, except for static sources [26–29].

One point in common between induction generators and other asynchronous sources of energy, such as photovoltaic and fuel cells, is their asynchronism with respect to the utility grid voltage. Therefore, a similar control philosophy may be extended to them. In the case of induction generators, the control variables are flux, torque, and impressed frequency voltage terminal. For photovoltaic and fuel cells, a dc link is a loop control associated with machine speed, and all the other variables (torque, flux, and frequency) also correspond. Induction generators may control the input power by mechanical or electromechanical means, or electrically by load matching of the electrical terminals. PV arrays can have mechanical controls for positioning the array. Fuel cells may have hydrogen feedline control, and both can have electrical load matching through power electronic inverters.

10.9.1 Frequency, Speed, and Voltage Controls

Mechanical controls are frequently applied in controlling the input shaft power of induction generators; for example, centrifugal weights might be used to close or open the admission of water for hydro turbines or for exerting pitch blade control in wind turbines. Some sophisticated mechanical controls include flywheels (to store transient energy), anemometers, or flow meters (to determine the wind intensity and water flow, respectively), in addition to wind tails and wind vanes (for detection of the direction of the winds), used for control and to limit the turbine rotor speed. Despite being robust, they are slow, bulky, and rough. However, they are still used in practice, as they are considered proven, reliable, and efficient solutions. Electromechanical controls are more accurate and relatively lighter and faster than mechanical controls. They use governors, actuators, servomechanisms, electric motors, solenoids, and relays to control the primary energy. Some modern self-excited induction generators use internal compensatory windings to keep the output voltage almost constant.

Electroelectronic controls use electronic sensors and power converters to regularize energy, speed, and power. They can be quite accurate in adjusting voltage, speed, and frequency, are lightweight, and are easily suitable for remote control and telemetry, as discussed in the following section.

10.9.2 Load Control Versus Source Control for Induction Generators

Power electronics technology has enabled the use of alternative energy sources to many types of electrical machines: permanent magnet–based, reluctance, and particularly, the induction generator. From the ordinary closure of ac and dc switches to the very fast and precise electronic switching techniques used in pulse-width-modulation (PWM) and pulse-density-modulation (PDM) control it is possible to adjust the machine to obtain any desired operating control. To obtain the best features for the system required, it is important for the hardware designer to be aware of what can be done in terms of control.

A very important machine, typically used for high-power applications, is the doubly fed induction generator (DFIG), known in the past as a Scherbius variable-speed driver. The DFIG is a wound rotor machine with the rotor circuit connected to an external variable voltage and frequency source via slip rings, and the stator connected to the grid network as illustrated in Figure 10.7. It is also possible to alter the rotor reactance by effectively modulating some inductors in series with the original rotor reactance. Adjusting the frequency of the external rotor source of current controls the speed of the doubly fed induction generator, which is usually limited to a 2 : 1 range [5,29].

Doubly fed machines have not been very popular, due to the maintenance required for the slip rings. More recently, with the development of new materials, powerful digital controllers, and power electronics, the doubly fed induction generator has became a solution in power generation for ratings of up to several hundreds ratings of kilowatts. Power converters (whose cost for large units becomes irrelevant compared to the entire system) usually create a need for a variable-frequency source for the rotor.

As we said above, the control of induction generators can be exerted through either the stator or rotor variables. The controllable stator variables are the number

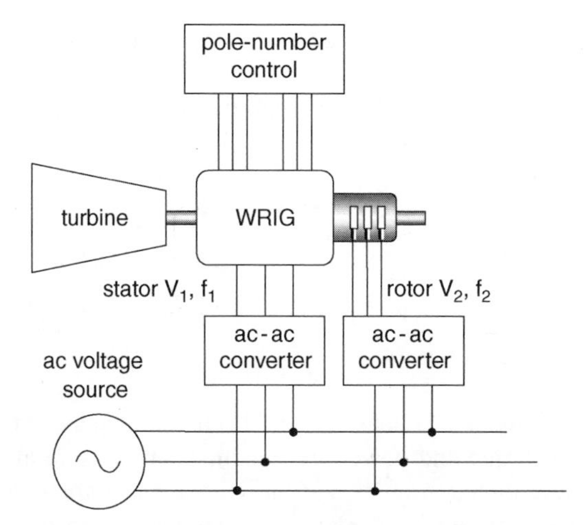

Figure 10.7 Experimental setup for overall stator and rotor parameter control.

of poles, the voltage, and the frequency. The rotor variables for squirrel-cage rotors can be design resistance, design reactance, and speed. The doubly fed induction generator is affected by the second power of the grid voltage, and the controllable variables are current, voltage, frequency, and voltage phase shift with respect to the stator voltage angle. An experimental setup to implement and monitor changes in the stator and rotor characteristics is depicted in Figure 10.7. Obviously, in most applications, this setup can be simplified. If, for example, there is no interest in stator frequency or voltage changes, the stator power converter is in practice not required.

Figure 10.8 helps to understand the DFIG. For a voltage source $V_s\angle 0°$, a voltage $V_r/s\angle\phi$ is introduced in the rotor circuit to control either rotor voltage or frequency through a PWM control, for example. It then introduces an apparent and

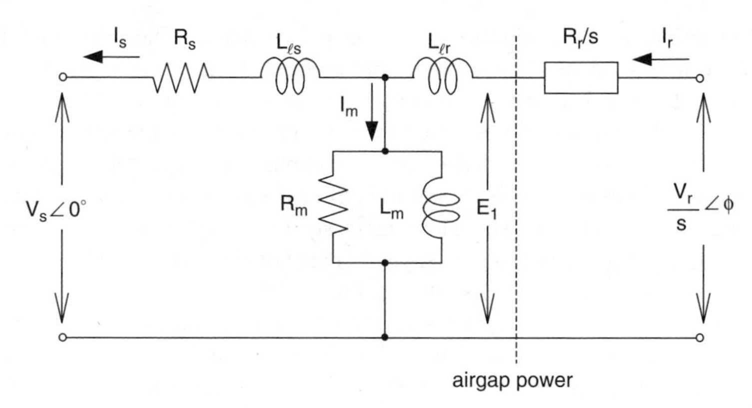

Figure 10.8 Equivalent model of a WRIG.

effective change in the rotor reactance by manipulation of the rotor voltage or current source.

From the equivalent model in Figure 10.8, the rotor current and torque may be obtained as

$$\bar{I}_r = \frac{V_s\angle 0^\circ - (V_r/s)\angle\phi}{\sqrt{(R_s+R_r/s)^2+\omega_e^2(L_{ls}+L_{lr})^2}\angle\tan^{-1}(\omega_e(L_{ls}+L_{lr})/[R_s+R_r/s])} \tag{10.35}$$

$$T_e = 3\left(\frac{P}{2}\right)\frac{R_r}{s\omega_e} - \frac{V_s^2 - 2(V_sV_r/s)\cos\phi + (V_r/s)^2}{(R_s+R_r/s)+\omega_e^2(L_{ls}+L_{lr})^2} \tag{10.36}$$

As predicted by the equivalent model, from an increase in stator voltage, increases in output torque and power are obtained. There is an almost direct proportion among these three variables and the increase in rotor speed, as shown in Figure 10.9*a*. Similar proportional results are obtained when there is an increase in the rotor external source voltage for the power injected through the rotor as shown in Figure 10.9*b*. Notice that in this torque versus speed characteristic, there is a shift to the left (lower rotor speeds) of the pullout torque caused by changes in the rotor parameters. The pullout torque is defined as the maximum torque to which generator is subject. This is a significant parameter in the sense that the maximum power transfer from the induction generator to the load happens whenever the source impedance is equal to the load impedance:

$$\frac{R_r}{s} = \sqrt{R_s^2 + (X_s + X_r)^2}$$

which replaced in the torque equation (10.36) for $V_r = 0$ gives

$$T_e = \frac{3P}{2\omega_e}\frac{V_s^2\sqrt{R_s^2+(X_s+X_r)^2}}{\left[R_s+\sqrt{R_s^2+(X_s+X_r)^2}\right]^2+\omega_e^2(L_{ls}+L_{lr})^2} \tag{10.37}$$

Wider variations are obtained through higher rotor voltages. The shift to the left in the pullout torque becomes much more pronounced if the phase angle ϕ between the stator voltage and rotor voltage is varied as shown in Figure 10.9*c*.

The slope of the torque curve close to the synchronous speed determines the variation range of the rotor speed during load operation. The steeper the torque is when crossing the synchronous speed point, the stiffer the system becomes and the narrower the speed regulation. This slope increases to the left as the rotor resistance increases in the torque characteristics, yet there is not much effect on the magnitude of the pullout torque, as demonstrated in Figure 10.9*d*. One problem here is that the heat so generated in the rotor needs to be dissipated to the outside, so limiting the slip factor in large machines. The stall control used in these large machines has been replaced by a fast pitch control to compensate for the rotor speed stiffness caused by the small slip factor.

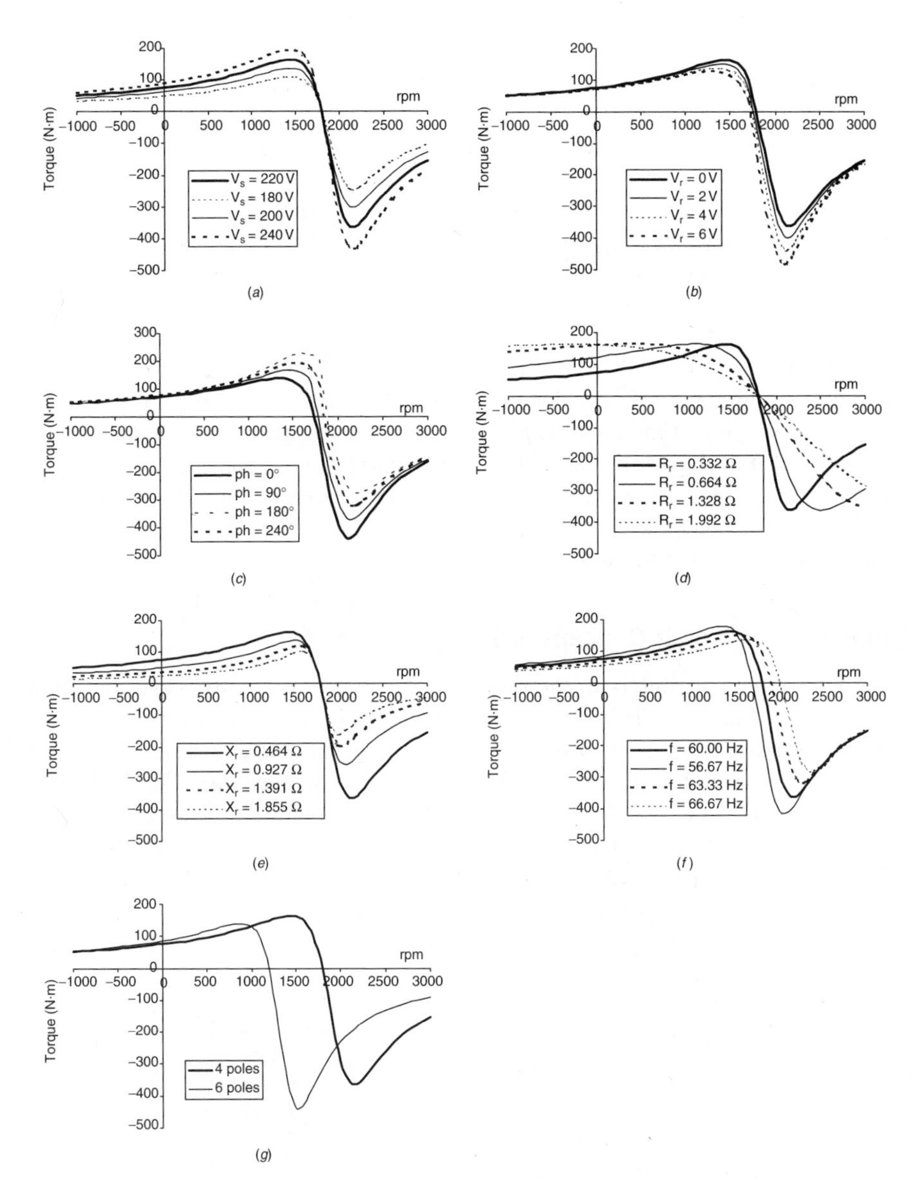

Figure 10.9 Control variables in the torque–speed characteristics of the induction generator: (*a*) stator voltage; (*b*) rotor voltage; (*c*) stator–rotor voltage phase shift; (*d*) rotor resistance; (*e*) rotor reactance; (*f*) stator voltage; (*g*) number of stator poles.

Rotor reactance can be changed through rotor design, rotor frequency control, or external insertion of modulated values of inductance. The quadratic effects of the rotor reactance on the torque versus speed characteristics are noticeable in Figure 10.9*e*. Different rotor angular frequencies can only be observed by the fact that the resistance does depend on the slip factor. It is interesting to point out here that these changes in X_r have very little effect on the rotor losses, which is not the case with the actual rotor resistance changes. It is also important to note that mismatching the rotor frequency too much may cause interaction between rotor frequency and stator frequency, which can lead to undesirable resonant and anti-resonant effects.

A very effective way of changing the synchronous speed of the induction generator is through stator frequency changes. These changes are exactly proportional figures related to the load frequency and the line crossing over the rotor speed, as illustrated in Figure 10.9*f*. One way to change the grid speed ranges of the induction generator is through machine pole changing. The phase circuitry of the stator is designed so that a different configuration of poles (pole number) can produce quite different synchronous speeds. None of the other variables are comparable in their effects on the pullout torque, as is a change in the pole number (see Figure 10.9*g*).

Notice that in all parts of Figure 10.9 except part (*d*), the changes proposed in the stator and rotor parameters have a pronounced influence not only on the crossover of the torque characteristic of the rotor speed axis, but also on the magnitude of the corresponding pullout torques.

10.9.3 The Danish Concept [5,30]

The *Danish concept* refers to the directly grid-connected squirrel-cage induction generator. It became very successful in commercial wind turbines for its simple and inexpensive design. Most of these systems use one large generator with one small induction generator connected alternately in parallel (see Figure 10.10). The large generator initially operates as a motor fed by the grid to finally put it at a speed slightly above the synchronous speed. If the wind speed is just above the cut-in speed of the turbine, the small generator is connected to the grid. As the wind speed increases, the small generator is switched off and the large generator is switched on. The wind speed difference between the two generators may be quite

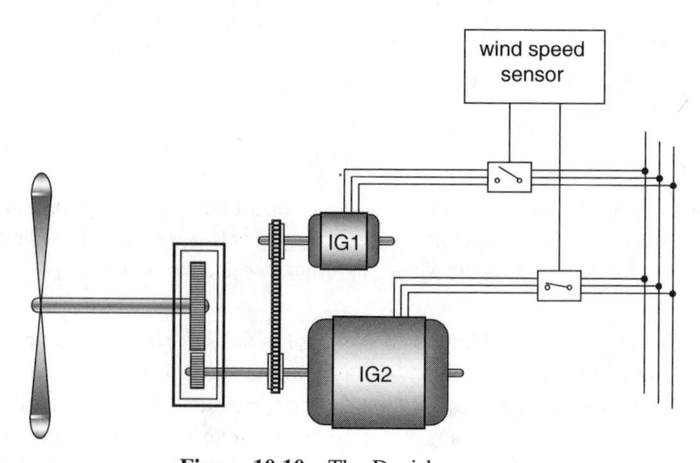

Figure 10.10 The Danish concept.

large to accommodate small and large rated electrical loads, from two to three times in magnitude. The explanation for this is in Chapter 4, where it is shown that the best results are obtained for a wind turbine when the power coefficient C_p is kept at its maximum value. For high wind intensities and small loads, a large electrical generator tends to overspeed, and vice versa with small generators tending to underspeed. This is the main reason that small generators are convenient to use alternately in parallel with large generators, as in the Danish concept.

Although the Danish concept results in a bulkier configuration, it is advantageous for the relatively low cost of the control mechanism and better suitability to the rather pronounced natural variations in wind speed throughout the year. Its disadvantages refer to the heavy mechanical stress on the gearbox because of sudden interchanges of generators and the need for larger slip factor designs for the large generator that allow them to work smoothly. To date, this concept is suitable only for relatively small wind power units; larger units can use other techniques, such as pole changing. Power electronics may become a very powerful tool to resolve the speed stiffness of the induction generator for existing machines or for new designs, because it allows changes in the rotor resistance through an external control, as suggested in Figure 10.9*d*, as well as for pole changing through a stator or externally driven rotor circuit, as suggested in Figure 10.7.

10.9.4 Variable-Speed Grid Connection

Slip factor control is necessary to withstand heavy changes in the generator load or in the primary energy, say the wind intensity. This control may typically be exerted through changes in the rotor resistance R_2 to allow more room for variations in the rotor speed of large generators (see Figure 10.9*d*). Costs for the ac–ac power converters in these systems are relatively small and they act to decouple, as much as possible, the rotor speed and the grid frequency (changes in the slip factor), keeping the wind turbine loads and power fluctuations within limits. Very large turbines must use as much ordinary speed control (pitch control) as possible to avoid power dissipation in the rotor circuit of the turbine. The rotor resistance is increased only during transient peaks of load or wind intensity. In the usual operation, programming of the experimental circuit depicted in Figure 10.7 involves the stator and rotor converters, both supervised by a control system that monitors the wind speed. For low-power turbines, the decoupling between rotor speed and the ac system is usually made with cascaded back-to-back converters with an intermediate dc–dc converter in between them.

Action of this system is somehow like the Danish concept except by the fact that the smaller generator is replaced by an electronic power circuit, which is known as an over-synchronous static Kraemer system. For lower wind speeds, the generator operates as a squirrel-cage rotor and the power electronics circuitry equivalent to the small generator is switched off (see Figure 10.10). For the design of normal operating ranges of wind speed, the wind turbine operates at the maximum power coefficient C_p (optimum power), as detailed in Chapter 4. For very high wind speeds, there is speed limiting by pitch control until the unit is cut off.

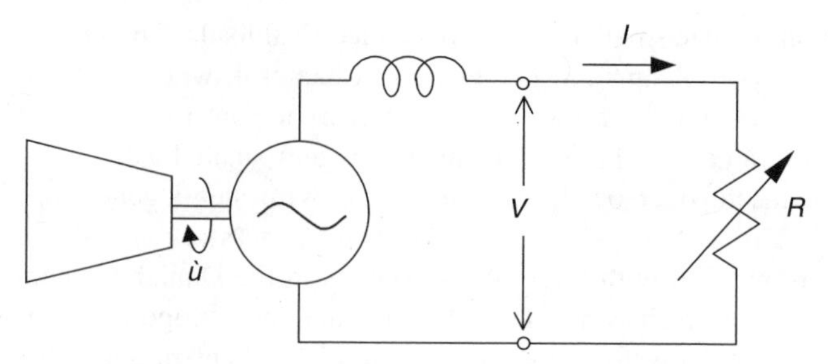

Figure 10.11 Principle of the electronic control by the load.

10.9.5 Control by the Load Versus Control by the Source

Electronic controls may exert their action by the load, so changes in the load may be used to control speed, frequency, and voltage, as illustrated in Figure 10.11. The load represented by R should be seen as regenerative, because the energy may be transferred from the generator to the public network or to a secondary load. This secondary load can be a battery charger, back pumping of water, irrigation, hydrogen production, heating, or fluid tank freezing. Degenerative dissipation can also be achieved by burning out the excess energy in bare resistors to the wind or in a flow of water.

Notice that electronic control by load, compromise is less important with efficiency than with the power generated. That is essentially the case with wind and run-of-river turbines, where the energy available in nature has no means of storage. In these cases, if the energy is not used the instant it is available in nature, it will be wasted.

A more detailed representation of electronic control by load is shown in Figure 10.11, where the primary source may be a turbine, a waterwheel, an ICE, a gas turbine, a diesel set, or another primary source. The electrical load of the rotating generator may be a dc load if the output is rectified, a power converter, or, in per phase terms, represented by *R, RL, RLE*, and *RLI* loads.

The literature discusses many approaches to varying the load through electronic control by load, from the simple step-by-step (discrete modulation) method to a continuous variation of load. For essentially additive loads such as resistors, it is possible to optimize their rated values to obtain a smoother variation of load with a minimum number of resistors by using the 2^n rule, depicted in Figure 10.12*a* for current control by discrete modulation of load. For additive/subtractive loads such as power transformers, optimization may be obtained even with a fewer elements by using the 3^n rule, depicted in Figure 10.12*b* for a series connection of secondary windings. Similar 2^n and 3^n rules can also be used in association with capacitors and inductors. When necessary, these approaches use electronic circuits with commutation under load.

More sophisticated loads, such as computers and electronic devices, cannot cope with large fluctuations of voltage. For these loads, a continuously adjusted duty cycle is used for smooth control of turbine speed and load voltage for secondary

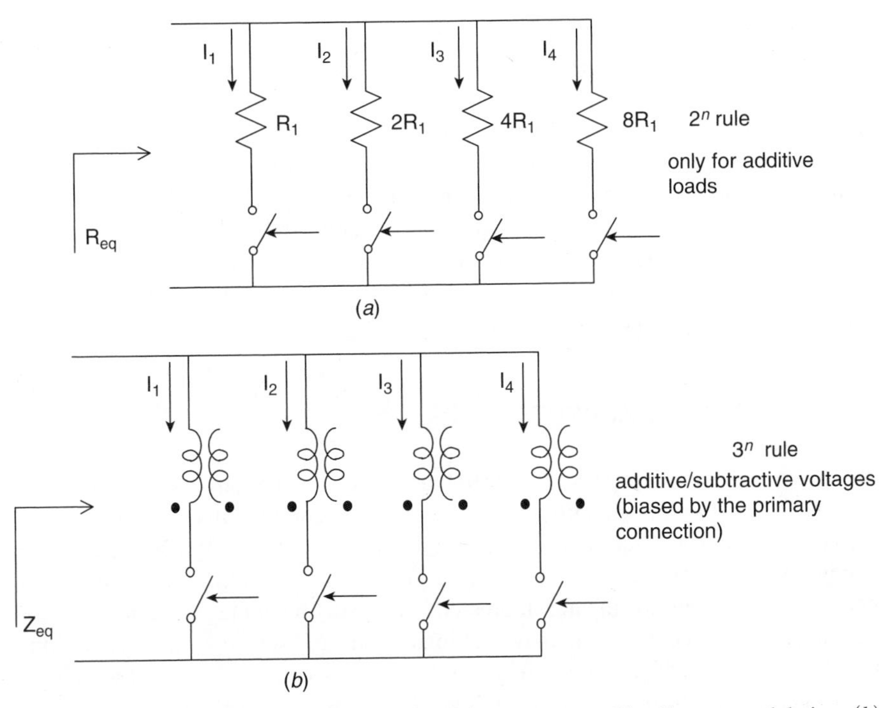

Figure 10.12 Discrete steps of load control: (*a*) current control by discrete modulation; (*b*) voltage control by discrete modulation.

loads. Figure 10.13 illustrates the smooth change of an effectively variable resistor used as a secondary load. It is common to use isolated gate bipolar transistors (IGBTs) for these purposes at switching frequencies of 20 kHz. The advantages of this type of control are the good speed and voltage regulation within certain ranges, use of a single power resistor–IGBT transistor set, ease in modular manufacturing,

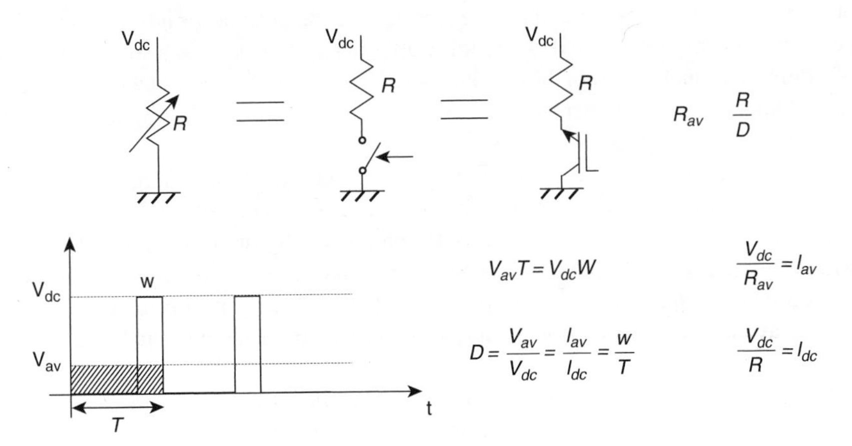

Figure 10.13 Electronically variable resistor.

suitability for ac and dc loads through power electronic ac switches, and an absence of switching surges. Its disadvantages are the electromagnetic interference, power switching complications for power above 50 kW, use of filters, noise, the short life span without maintenance, moderated costs, and needs for qualified servicing.

An example of secondary loads using the additive 2^n rule is the rural irrigation of plantations where water pumps of different sizes, such as 1, 2, 4, and 8 kW, could be used. Therefore, they could be programmed to be activated or deactivated whenever the turbine speed surpasses certain limits of control. Steps of 1-kW power will limit the speed within a reasonable range. An example of the 3^n rule is the voltage control of large loads that are too sensitive to power changes.

10.10 ECONOMIC ASPECTS

To determine economic aspects, several parts of an alternative system must be accessed for a project evaluation. For example, wind energy will consist of a tower, a rotor (with blades, hub, and shaft), a generator, control equipment, and power conditioning and protection equipment. A gearbox is generally used to match the generator speed with the natural impressed velocity by the wind. Some sort of braking is used to protect the system against high winds and severe weather conditions. Combinations with other sources of generation, such as photovoltaic and sometimes diesel generators, form more reliable hybrid systems.

For a modern turbine design, the typical lifetime is assumed to be 20 to 25 years, during which capital costs will be amortized. Currently, a rule of thumb for costs is \$1000 per kilowatt of capacity on average, with the marginal tower cost at \$1600 per meter. Installation costs include foundations, transportation, road construction, utilities, telephones and other communications, substation, transformer, controls, and cabling. Depending on soil conditions and distance to power lines, a margin of safety must be used. Operation and maintenance costs generally increase throughout the life of a wind power project. Costs may range from about \$0.008 to \$0.014 per kilowatthour generated and increase at a rate of about 2.5% per year. The high end (\$0.014 per kilowatthour) would include a major overhaul of equipment such as rotor blades and gearboxes, which are subject to a higher rate of wear and tear. Other operation costs include plant monitoring and periodic semester inspections.

The potential income from wind energy is based on annual energy production, which in turn, is dependent on average annual speed. Wind speed varies globally, regionally, and locally following seasonal patterns. The duration and force of the wind are critical to the lucrative operation of wind turbines, to pay for investment on equipment, maintenance, and operations. Large-scale wind turbines require an annual average wind speed of 5.8 m/s (13 mph) at 10 m height; small-scale wind turbines require 4 m/s (9 mph).

Energy production is also a function of turbine reliability and availability. Even though the induction generator is a small fraction of all the investment required, it is indeed the unit that converts the mechanical energy from shaft to electricity. Therefore, the efficiency of the induction generator pays for even fractions of improvement. Better electrical and magnetic design associated with insulation, thermal, and mechanical design will make the energy converter unit extract more reliable power. Controllers capable of placing the induction generator at the best operating point, of programming the optimum magnetic flux level, or of producing required lagging or leading unit power factor energy contribute to faster amortization and a more productive investment.

Hydropower converts the potential and kinetic energy of water conveyed through a pipeline or canal to the turbine into electrical energy. The energy in the water enters a high pressure, rotates the generator shaft, and leaves at a lower pressure. Power generation can be amplified with an increase of water elevation height, the flow of the river or stream, and the size of the watershed. Small hydropower generators are usually run-of-the-river systems that use a dam or weir for water diversion but not for reservoir storage. The amount of pressure at the turbine is determined by the actual flow of the stream or river and the head, or the height of the water above the turbine.

Controlling the flow of a river can have uses other than power generation, including irrigation, flood control, municipal and industrial water supply, and recreation. These parameters affect the financial return from such projects and sometimes have long-term benefits that may be attractive locally. Hydropower is characterized by extremely high up-front costs, very low operating costs, and very long life cycles. Such capital cost scenarios make projects very sensitive to financial variables, construction timing, interests, and discount rates. Hydro projects usually have long lead times, typically taking 10 years between analysis and deployment. Such projects require resource assessment and environmental and social considerations (such as water use, displacement of homes, and distance to transmission lines).

Small power plants, on the order of maximal (10 to 200 kW) for rural uses of distributed energy, do not have the same requirements as those of large power plants (greater than 1 MW). However, small-scale applications for a target cost of $2000 per kilowatt will require a 35% investment in civil engineering, 40% for plant installations and commission, 7% for overall design and management, 8% for electrical engineering, and contingencies on the order of 10%. For either wind or hydropower energy sources, the induction generator will need a peak power tracking control, a dummy load or any type of storage system, and pump-back schemes are sometimes used in conjunction with water flow needs.

REFERENCES

[1] S. J. Chapman, *Electric Machinery Fundamentals*, 3rd ed., McGraw-Hill, New York, 1999.

[2] A. S. Langsdorf, *Teoría de máquinas de corriente alterna* (Theory of Alternating Current Machines), McGraw-Hill, New York, 1977.

[3] R. L. Lawrence, *Principles of Alternating Current Machinery*, McGraw-Hill, New York, 1953, p. 640.

[4] M. Kostenko and L. Piotrovsky *Electrical Machines*, Vol. II, Mir Publishers, Moscow, 1969.

[5] M. G. Simões and F. A. Farret, *Renewable Energy Systems: Design and Analysis with Induction Generators*, CRC Press, Boca Raton, FL, 2004.

[6] S. S. Murthy, O. P. Malik, and A. K. Tandon, Analysis of self-excited induction generators, *Proceedings of IEE*, Vol. 129, No. 6, pp. 260–265, 1982.

[7] IEEE Std. 1547, *Standard for Interconnecting Distributed Resources with Electrical Power Systems*, IEEE Press, Piscataway, NJ,

[8] IEEE Std. (Draft) P1547.1, *Standard Conformance Test Procedures for Interconnecting Distributed Energy Resources with Electric Power Systems*, IEEE Press, Piscataway, NJ.

[9] IEEE Std. (Draft) P1547.2, *Application Guide for IEEE Standard 1547: Interconnecting Distributed Resources with Electric Power Systems*, IEEE Press, Piscataway, NJ.

[10] IEEE Std. (Draft) P1547.3, *Guide for Monitoring, Information Exchange, and Control of Distributed Resources Interconnected with Electric Power Systems*, IEEE Press, Piscataway, NJ.

[11] IEEE Std. 112-1991, *Standard Test Procedure for Polyphase Induction Motors and Generators*, IEEE Press, Piscataway, NJ.

[12] I. R. Smith and S. Sriharan, Transients in induction motors with terminal capacitors, *Proceedings of IEE*, Vol. 115, pp. 519–527, 1968.

[13] S. S. Murthy, B. Sing, and A. K. Tandon, Dynamic models for the transient analysis of induction machines with asymmetrical winding connections, *Electric Machines and Electromechanics*, Vol. 6, No. 6, pp. 479–492, Nov–December 1981.

[14] D. B. Watson and I. P. Milner, Autonomous and parallel operation of self-excited induction generators, *Electrical Engineering Education*, Vol. 22, pp. 365–374, 1985.

[15] C. Grantham, *Determination of induction motor parameter variations from a frequency stand still test, Electrical Machine Power Systems*, Vol. 10, pp. 239–248, 1985.

[16] S. S. Murthy, H. S. Nagaraj, and A. Kuriyan, Design-based computational procedure for performance prediction and analysis of self-excited induction generators using motor design packages, *Proceedings of IEE*, Vol. 135, No. 1, pp. 8–16, January 1988.

[17] K. E. Hallenius, P. Vas, and J. E. Brown, The analysis of a saturated self-excited induction generator, *IEEE Transactions on Energy Conversion*, Vol. 6, No. 2, June 1991.

[18] N. N. Hancock, *Matrix Analysis of Electric Machinery*, Pergamon, Press, Oxford, 1964, p. 55.

[19] M. Liwschitz-Gärik and C. C. Whipple, *Máquinas de corriente alterna* (Alternating Current Machines), Compannia Editorial Continental, Col Barrio del Nino Jesús, Mexico, 1970.

[20] B. C. Doxey, Theory and application of the capacitor-excited induction generator, *Engineer*, pp. 893–897, November 1963.

[21] C. H. Lee and L. Wang, A novel analysis of parallel operated self-excited induction generators, *IEEE Transactions on Energy Conversion*, Vol. 13, No. 2, June 1998.

[22] C. Grantham, Steady-state and transient analysis of self-excited induction generators, *IEE Proceedings* Vol. 136, Pt. B, No. 2, pp. 61–68, March 1989.

[23] D. Seyoum, C. Grantham, and F. Rahman, The dynamics of an isolated self-excited induction generator driven by a wind turbine, in *Proceedings of the 27th Annual*

Conference of the IEEE Industrial Electronics Society, IECON'01, Denver, CO, December 2001, pp. 1364–1369.

[24] S. P. Singh, B. M. Singh, and P. Jain, Performance characteristics and optimum utilization of a cage machine as capacitor excited induction generator, *IEEE Transactions on Energy Conversion*, Vol. 5, No. 4, pp. 679–684, December 1990.

[25] E. Muljadi, J. Sallan, M. Sanz, and C. P. Butterfield, Investigation of self-excited induction generators for wind turbine applications, *IEEE*, pp. 509–515, 1999.

[26] R. P. Tabosa, G. A. Soares, and R. Shindo, *Motor de alto rendimento* (The High-Efficiency Motor), Technical Handbook of the PROCEL Program, Eletrobrás/Procel/CEPEL, Rio de Janeiro, Brazil, August 1998.

[27] I. Barbi, *Teoria fundamental do motor de indução* Fundamental Theory of the Induction Motor, Universidade Federal de Santa Maria Press–Eletrobrás, Florianópolis, Brazil, 1985.

[28] G. W. Stagg and A. H. El-Abiad, *Computer Methods in Power Systems Analysis*, McGraw-Hill, New York, 1968.

[29] O. I. Elgerd, *Electric Energy Systems Theory: An Introduction*, McGraw-Hill, TMH Edition, New York–New Delhi, 1971.

[30] R. Gasch and J. Twele, *Wind Power Plants: Fundamentals, Design, Construction and Operation*, Solarpraxis, Berlin, 2002, pp. 299–313.

CHAPTER 11

STORAGE SYSTEMS

11.1 INTRODUCTION

Electrical energy storage has long been considered a critical technology. At the beginning of the twentieth century, electrochemical batteries were used to power telephones and telegraphs, and enormous, heavy flywheels were common in rotating generators to smooth out load oscillations. Electric vehicles were more common than gasoline-powered vehicles. Batteries were important in such transportation applications because they could store energy to give reasonable transportation autonomy to vehicles. Even longer ago, water dams were used to store potential energy.

Today, energy storage systems play the important role of unifying, distributing, and augmenting the capabilities of alternative and renewable energy-distributed generating systems. The average central electrical generation, transmission, and distribution system is large enough not to be affected by residential and commercial load changes. Figure 11.1 illustrates a typical household peak day load curve and an individual load. On the other hand, distributed generation (DG) is exposed to the fluctuations of individual loads; it does not experience the averaging effect seen by large generators.

For DG-based industrial applications, base load can be considerably higher, and it is possible that a local generator can operate cost-effectively. Energy storage systems capable of smoothing out load fluctuations, making up for seasonal variations

Integration of Alternative Sources of Energy, by Felix A. Farret and M. Godoy Simões

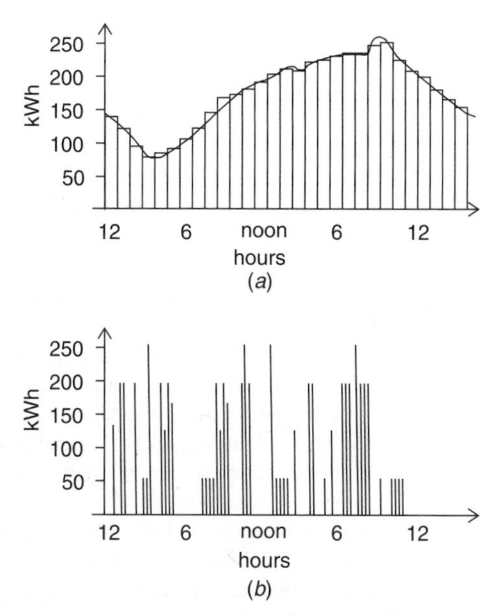

Figure 11.1 Residential load: (*a*) area average; (*b*) one household.

in renewable sources and reacting to fast transient power quality needs may thus contribute to efficient energy management policies and faster economic investments in new projects.

Energy storage technologies are classified according to the energy, time, and transient response required for their operation [1]. It is convenient to define storage capacity in terms of the time that the nominal energy capacity can cover the load at rated power. Storage capacity can then be categorized in terms of energy density requirements (for medium- and long-term needs) or in terms of power density requirements (for short- and very short-term needs). Table 11.1 depicts how storage objectives determine storage features.

Energy storage enhances DG in three ways. First, it stabilizes and permits DG to run at a constant and stable output, despite load fluctuations and required maintenance services. Second, it provides energy to ride through instantaneous lacks of primary energy (such as those of sun, wind, and hydropower sources). Third, it permits DG to operate seamlessly as a dispatchable unit. Energy storage may be designed for rapid damping of peak surges in electricity demand, to counter momentary power disturbances, to provide a few seconds of ride-through while backup generators start in response to a power failure, or to reserve energy for future demand. Those characteristics contribute to different strategies.

For example, long-term storage systems installed on the customer side of a meter may save demand charges on the customer's utility bill. These storage systems charge during off-peak periods and discharge according to a dispatch strategy that minimizes the peak load that is billed monthly. Such a system can also provide energy savings and improve the power factor. Similar to the customer ownership scenario,

TABLE 11.1 Functionality of Storage System in Terms of Time Response

Storage Capacity	Energy Storage Features
Transient (microseconds)	Compensates for voltage sags Rides through disturbances (backup systems) Regenerates electrical motors Improves harmonic distortion and power quality
Very short term (cycles of the grid frequency)	Covers load during startup and synchronization of backup generator Compensates transient response of renewable-based electronic converters Increases system reliability during fault management Keeps computer and telecommunication systems alive for safe electronic data backup
Short term (minutes)	Covers load during short-term load peaks Smoothes renewable energy deficits for online capture of wind or solar power Decreases needs of startup backup generator Improves maintenance needs of fossil fuel–based generators Allows ride-through of critical medical, safety, and financial procedures
Medium term (a few hours)	Stores renewable energy surplus to be used at a later time Compensates for load-leveling policies Allows stored energy to be negotiated on net-metering basis Integrates surplus energy with thermal systems
Long term (several hours to a couple of days)	Stores renewable energy for compensation of weather-based changes Provides reduction in fuel consumption and decreases waste of renewable energy Possible elimination of fossil fuel–based generator backup Requires civil constructions for hydro and air systems Produces hydrogen from renewable sources
Planning (weeks to months)	Includes large power storage systems, such as pumped hydro and compressed air systems Uses fossil fuel storage to offset economic fluctuations Stores hydrogen from biomass or renewable-based systems

peak shaving can be provided by storage technologies owned and operated by the utility. With a transportable storage system, the deferral of transmission and distribution can be provided at multiple sites over the system life. Short-term storage technologies can provide enhanced reliability at a customer site by providing ride-through during momentary utility outages or transition to standby generation.

Storage can also supplement a fuel cell or diesel generator to provide load-following service and enhance generator efficiency (full load versus part load). A stored energy facility can operate in parallel with a generating unit to meet temporary peaks of demand higher than rated generating capacity. A storage system can

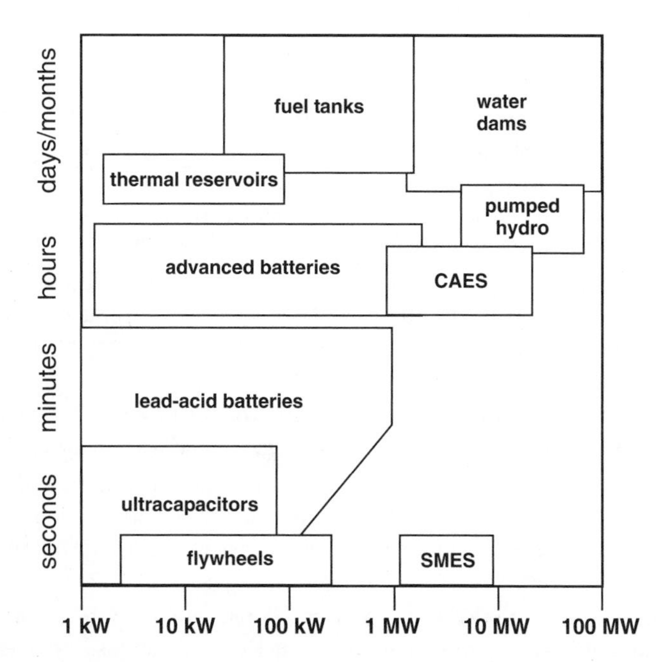

Figure 11.2 Energy management, power quality, and ride-through storage applications.

also support voltage through mechanisms of reactive power control in a microgrid system.

Small-scale storage can provide spinning reserve so that output can be ramped down in a controlled manner without disrupting the grid. On a larger scale, storage systems can provide load leveling or commodity storage capability with controlled dispatchability, in which stored energy is held back for release when prices on the spot market reach a threshold.

Figure 11.2 shows mature and emerging storage technologies organized in categories of ride-through, power quality, and energy management. Applications include lead–acid batteries, advanced batteries, low- and high-energy flywheels, ultracapacitors, superconducting magnetic energy storage (SMES) systems, heating systems, pumped hydro, geothermal underground (see Appendix B), and compressed air energy storage (CAES). Hydrogen storage is not included in this section. Feasibility, economical evaluation studies, and fuel cell technology must be covered fully before discussing this topic.

11.2 ENERGY STORAGE PARAMETERS

Energy storage for DG can be compared by parameters that define performance criteria. These parameters are capacity, specific energy, energy density, specific power, efficiency, recharge rate, self-discharge, lifetime, capital cost, and operating cost.

Capacity: the available energy storage capability. The SI unit of capacity is the joule, but this is a very small unit. Usually, the watthour (i.e., the energy equivalent of working at a power of 1 W for 1 hour, or 3600 J) is used instead.

Specific energy: the electrical energy stored per mass, in units of Wh/kg. Specific energy is ordinarily used when the energy capacity of a battery needed in a certain system is known; it is then divided by the specific energy to give an approximation of battery mass.

Specific power: the amount of power obtained per kilogram of a storage system in W/kg. It is a very sensitive parameter because several storage systems cannot operate at this maximum power for long, so they may affect lifetime or operate very inefficiently.

Energy density: a measure of energy stored per volume in Wh/m^3. It can be used to give an approximation of the energy storage volume for a given application.

Physical efficiency: how much power is stored in a given volume and mass. It is usually considered in batteries for transportation applications, in which industry may accept degradation in electrical efficiency in return for good physical efficiency.

Electrical efficiency: the percentage of power put into a unit that is available to be withdrawn; an important parameter for DG applications. It is measured by the energy capable of being converted into work. A unit with 90% efficiency returns 9 kWh of energy for every 10 kWh put into storage.

Recharge rate: the rate at which power can be pushed for storage. A storage system might take 10 hours to deplete but 14 hours to refill.

Specific gravity: a dimensionless unit defined as the ratio of density of a material to the density of water at a specified temperature. It can be expressed as

$$SG = \frac{\rho}{\rho_{H_2O}}$$

where SG is the specific gravity, ρ is the density of the electrolyte, and ρ_{H_2O} is the density of water at 4°C (because the density of water at this temperature is at its highest).

Self-discharge: indicates how long storage system takes to discharge when unused. This is usually due to current leakage and heat dissipation.

Lifetime: the service life of a unit, which varies with technology and intensity of use, e.g., batteries are noted for having short lifetimes in applications in which they are repeatedly charged and discharged completely.

Capital cost: the initial cost in dollars per kilowatt for design, specification, civil works, and installation.

Operating cost: the costs in dollars per kilowatthour for periodic inspection, fueling, maintenance, parts (bearings and seals) replacement, recalibration, and so on.

Charge/discharge cycle: the number of times the storage system can be loaded and unloaded without altering significantly its storing capabilities.

TABLE 11.2 Typical Lead–Acid Battery Parameters

Specific energy	20–35 Wh/kg
Energy density	50–90 Wh/L
Specific power	About 250 W/kg
Nominal cell voltage	2 V
Electrical efficiency	About 80%, depending on recharge rate and temperature
Recharge rate	About 8 hours (possible to quick recharge 90%)
Self-discharge	1–2% per day
Lifetime	About 800 cycles, depending on the depth of cycle

Table 11.2 shows nominal parameters for lead–acid batteries. As discussed in Section 11.3, they are a common storage solution, but their lifetime is problematic. Therefore, continual over- or undercharging is not recommended, and a charger system should control the operating cycle. Tables 11.3 and 11.4 compare parameters of large-scale energy storage. The parameters in the tables are guidelines only and are useful for broad comparative purposes.

TABLE 11.3 Typical Lead–Acid Battery Parameters

	Specific Energy (Wh/kg)	Energy Density (Wh/L)	Specific Power (W/kg)	Nominal Cell Voltage (V)
Lead–acid	30	75	250	2.0
Nickel–cadmium	50	80	150	1.2
Nickel–metal hydride	65	150	200	1.2
Zebra	100	150	150	High voltage, integrated stack
Lithium ion	90	150	300	3.6
Zinc–air	230	270	105	1.65

TABLE 11.4 Comparison of Large-Scale Energy Storage Systems

		Flywheels				
	Ultracapacitor	Low Speed	High Speed	Pumped Hydro	CAES	SMES
Capital cost/MWh	$25,000,000	$300,000	$25,000,000	$7,000	$2,000	$10,000
Weight/MWh	10,000 kg	7,500 kg	3,000 kg	3,000 kg	2.5 kg	10 kg
Efficiency	95%	90%	95%	80%	85%	95%
Operating cost/MWh	$5	$3	$4	$4	$3	$1
Capacity	0.5 kWh	50 kWh	750 kWh	22,000 MWh	2,400 MWh	0.8 kWh
Lifetime	40 years	20 years	20 years	75 years	30 years	40 years

TABLE 11.5 Cost Projection of Energy Storage Systems

System	Typical Size Range (MW)	\$/kW	\$/kWh
Lead–acid batteries	0.5–100	100–200	150–300
Advanced batteries	0.5–50	200–400	
Ultracapacitors	1–10	300	3,600
Flywheels	1–10	200–500	100–800
SMES	10–1000	300–1000	300–3000
CAES	50–1000	500–1000	10–15
Pumped hydropower	100–1000	600–1000	10–15

Table 11.5 compares cost projections of storage systems. In the following sections, several storage systems are discussed to support the design engineer's understanding of the fundamentals of each technology [2].

11.3 LEAD–ACID BATTERIES

The lead–acid battery is an electrochemical device invented by Planté in 1859. In it, two electrodes react with a sulfuric acid electrolyte. During discharge, both electrodes are converted to lead sulfate, as described by the following charge–discharge reaction [3]:

$$\overset{\text{cathode}}{\mathrm{Pb}} + 2\mathrm{H_2SO_4} + \overset{\text{anode}}{\mathrm{PbO_2}} \underset{\overleftarrow{\text{charge}}}{\xrightarrow{\text{discharge}}} \overset{\text{cathode}}{\mathrm{PbSO_4}} + 2\mathrm{H_2O} + \overset{\text{anode}}{\mathrm{PbSO_4}}$$

When the battery is charged, the anode is restored to lead dioxide and the cathode to metallic lead. However, irreversible changes in the electrodes limit the number of cycles, and failure may occur after a couple of thousand cycles, depending on battery design and depth of discharge.

Many rechargeable batteries are suitable for DG applications, but comprehensive descriptions are outside the scope of this book. The lead–acid battery is still the most common because of its relatively economic power density, and it will remain so for the next few years. However, other technologies may surpass the lead–acid battery on the basis of energy density and lifetime. For example, nickel–cadmium batteries are common in applications that require sealed batteries capable of operating in any position with higher energy density. Other advanced batteries, such as nickel–metal hydride and several lithium technologies, may become cost-effective in the future for residential and commercial applications.

11.3.1 Constructional Features

The basic building block of a lead–acid battery is a 2.12 to 2.15 V single cell. Each battery unit supplies power in case of outages or low production from renewable energy sources. The batteries are wired together in series to produce 12-, 24-, or 48-V strings. These strings are then connected together in parallel to make up a battery bank. The battery bank supplies dc power to an inverter, which produces ac power that can be used to run appliances. Battery bank voltage and current rating are determined by inverter input, the type of battery, and the energy storage required.

The flooded lead–acid battery is used in automobiles, forklifts, and uninterruptible power supply systems. Flooded-cell batteries have two sets of lead plates that are coated with chemicals and immersed in a liquid electrolyte. As the battery is used, the water in the electrolyte evaporates and needs to be replenished with distilled water.

During the mid-1970s, maintenance-free lead–acid batteries were developed by transforming liquid electrolyte into moistened separators and sealing the enclosure. Safety valves were added to allow gas venting during charge and discharge. Two lead–acid systems emerged: the small sealed lead–acid (SLA) battery, also known under the brand name Gel Cell, and the large valve-regulated lead–acid (VRLA) battery.

VRLA batteries do not require water to keep the electrolyte functioning properly or to mix the electrolyte to prevent stratification. The oxygen recombination and the valves of VRLAs prevent the venting of hydrogen and oxygen gases and the ingress of air. However, the battery may need to be replaced more frequently than a flooded lead–acid battery, which can increase the leveled cost of the system. The major advantages of VRLA batteries over flooded lead–acid cells are (1) the dramatic reduction in maintenance, and (2) the reduced footprint and weight because of the sealed construction and immobilized electrolyte, which allow battery cells to be packaged more tightly.

VRLA batteries are less robust than flooded lead–acid batteries, more expensive, and have a shorter lifetime. VRLA batteries are perceived as maintenance-free and safe, and they have become popular as standby power supplies in telecommunications applications and as uninterruptible power supplies in situations in which special rooms cannot be set aside for batteries.

Unlike flooded lead–acid batteries, both SLA and VRLA batteries are designed with a low overvoltage potential to prevent them from reaching their gas-generating potential during charge. Excess charging would cause gassing and water depletion. Consequently, these batteries can never be charged to their full potential. The advantages and limitations of lead–acid batteries are listed in Table 11.6.

11.3.2 Battery Charge–Discharge Cycles

The batteries used to start automobiles are known as *shallow-cycle batteries* because they are designed to supply a large amount of current for a short time and to withstand mild overcharge without losing electrolyte. Unfortunately, they cannot tolerate being deeply discharged. If they are repeatedly discharged more than 20%, their lifetime will be affected. Lead–acid batteries for automotive use are not designed for deep discharge and should always be kept at maximum

TABLE 11.6 Characteristics of Lead–Acid Batteries

Advantages	Limitations
Inexpensive and simple to manufacture	Low energy density; poor weight-to-energy ratio limits use to stationary and wheeled applications
Mature, reliable, and well-understood technology	Cannot be stored in discharged condition; cell voltage should never drop below 2.10 V
When used correctly, durable and provides dependable service	Allows only a limited number of full discharge cycles; well suited for standby applications that require only occasional deep discharges
Self-discharge is among the lowest of rechargeable battery systems	Lead content and acid electrolyte make them environmentally unfriendly
Low maintenance requirements, no memory; no electrolyte to fill on sealed version	Transportation restrictions on flooded lead–acid; environmental concerns regarding spillage; thermal runaway can occur with improper charging
Capable of high discharge rates	

charge using constant voltage at 13.8 V (for six-element car batteries). These batteries are not a good choice for DG systems.

Deep-cycle batteries are designed to be discharged repeatedly by as much as 80% of their capacity, so they are a good choice for DG systems. Although they are designed to withstand deep cycling, these batteries have a longer life if cycles are shallower.

Flooded-cell batteries must be replenished with distilled water. In Gel Cell batteries, the electrolyte is suspended in a gelatinlike material so it will not spill even if the battery is operated on its side. Gel cells are often called *recombinant batteries* because oxygen gas given off at the positive plate is recombined with hydrogen given off at the negative plate to keep the electrolyte moist. This means that distilled water does not need to be added. Absorbed glass mat batteries have a highly porous microfiber spongelike glass mat between the plates to absorb the electrolyte and have no free liquid. Similar to Gel Cells, they are sealed and use recombinant gas effects so they do not lose liquid during use.

A lead–acid battery consists of a lead cathode and a lead oxide (PbO_2) anode immersed in a sulfuric acid solution. The discharging reaction at the anode consists of the exchange of oxygen ions from the anode with sulfate ions of the electrolyte. At the cathode, the discharge involves sulfate ions from the electrolyte combining with lead ions to form lead sulfate. Two electrons must enter the anode terminal and two electrons must leave the cathode terminal via the external circuit for every two sulfate ions that leave the electrolyte; this corresponds to the current supplied by the battery to the external circuit. Removal of sulfate ions from the solution reduces the acidity of the electrolyte. When an external voltage greater than the voltage produced by the reactions at the anode and cathode is applied across the battery terminals, the current will flow into the anode rather than out, thus charging the battery. The chemical processes are then reversed, and sulfate ions are liberated to the solution, which increases the concentration of sulfuric acid in the electrolyte.

Finding the ideal charge voltage limit is critical. A high voltage limit (more than 2.4 V per cell) produces good battery performance but shortens service life because of grid corrosion on the positive plate. The corrosion is permanent. A low voltage (less than 2.4 V per cell) is safe if charged at a higher temperature but is subject to sulfation on the negative plate. If excessive lead sulfate builds up on the electrodes, their effective surface areas are reduced, which affects performance. It is thus important to avoid completely discharging a lead–acid battery. Conversely, if excessive charging is imposed and there is no more sulfate at the cathode to maintain continuity of the charging current, hydrogen is liberated. This is a potential safety hazard. However, an occasional charging to the gassing stage with a slight degree of bubbling is sometimes conducted to clean up the electrodes and provide a mixing action on the electrolyte. The electrolyte freezing point depends on the state of charge. Therefore, fully charged batteries may operate at low temperatures, while batteries in lower state of discharge will need to operate in warmer environments.

A charge controller is essential to maintain batteries under optimized levels. The charge controller shuts down the load when a prescribed level of charge is reached and shuts down the connection to the main source (e.g., a photovoltaic array) when the battery is fully charged. When a charge controller manages a bank of batteries, it performs equalization of charge electronically. Equalization is a charge about 10% more than the normal float or trickle charge. This ensures that the cells are equally charged, and in flooded batteries, it ensures that the electrolyte is fully mixed by the gas bubbles. Gelled and sealed batteries, in general, should be equalized at a much lower rate than flooded batteries. Usually, the final charge cycle on a three-stage charger is sufficient to equalize the cells.

Depending on the depth of discharge and operating temperature, a sealed lead–acid battery provides 200 to 300 discharge–charge cycles. The primary reasons for its relatively short cycle life is grid corrosion of the positive electrode, depletion of the active material, and expansion of the positive plates. These changes are most prevalent at higher operating temperatures. Cycling (repetitive charge/discharge) does not prevent or reverse the trend.

11.3.3 Operating Limits and Parameters

The optimum operating temperature for a lead–acid battery is 25°C (77°F). As a guideline, every 8°C (15°F) rise in temperature will cut the battery life in half. A VRLA battery that would last for 10 years at 25°C (77°F) will only be good for five years if operated at 33°C (95°F). Theoretically the same battery would endure a little more than one year at a desert temperature of 42°C (107°F). Among modern rechargeable batteries, the lead–acid battery family has the lowest energy density, making it unsuitable for handheld devices that demand compact size. In addition, performance at low temperatures is poor.

The charge capacity of a battery is referred to as C. Thus, if a load is connected to a battery such that it will discharge in n hours, the discharge rate is C/n. Figure 11.3

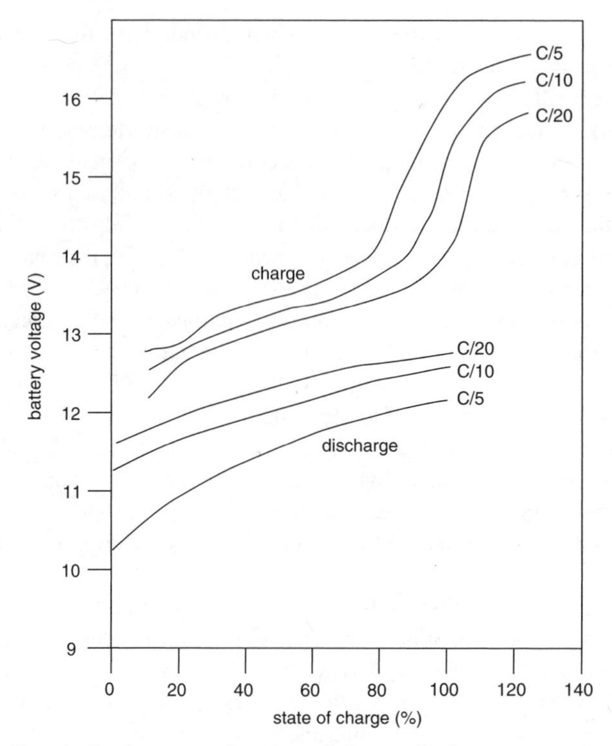

Figure 11.3 Terminal voltage as a function of charge–discharge rate and state of charge.

shows how the terminal voltage depends on the charge or discharge rate and the state of charge for a typical lead–acid battery.

It is extremely difficult to measure accurately the state of charge of a lead–acid battery and predict the remaining capacity. In a battery, the rate at which input or output current is drawn affects the overall energy available from the battery. For example, a 100-Ah battery at a 20-hour rate means that over 20 hours, 100 Ah is available (i.e., the user can expect to draw up to 5 A per hour for up to 20 hours). If a quick discharge is imposed, the effective ampere-hours will be fewer.

There are some commercial state-of-charge meters, called *E-meters*. They sample the rate of discharge every few minutes, recalculate the time remaining before the battery is discharged, and update and display a fuel gauge bar graph. The E-meter also measures kilowatthours and historical battery information such as number of cycles, deepest discharge, and average depth of discharge. If an E-meter is not available, a voltage measurement of the open-circuit voltage is a fair approximation of battery discharge.

The sealed lead–acid battery is rated at a five-hour discharge, or *C*/5. Some batteries are rated at a slow 20-hour discharge. Longer discharge times produce higher-capacity readings. The lead–acid battery performs well on high load currents. Note that if the battery is charged at a *C*/5 rate, full charge will be reached at a terminal voltage of 16 V. However, if the battery is charged at *C*/20, the battery will reach full charge at a terminal voltage of 14.1 V. When the charging current is zeroed, the terminal voltage will drop below 13 V.

As an example, a six-cell battery has the following characteristic voltage set points:

- Quiescent (open-circuit) voltage: 12.6 V
- Discharging end voltage: 11.8 V
- Charge: 13.2 to 14.4 V
- Gassing voltage: 14.4 V
- Recommended floating voltage for charge preservation: 13.2 V

After full charge, the terminal voltage drops quickly to 13.2 V and then slowly to 12.6 V.

11.3.4 Maintenance of Lead–Acid Batteries

A routine checkup of battery banks should be done every month to inspect the water level and corrosion on the terminals. At least two or three times a year, the state of charge should also be evaluated. As the battery discharges, more water is produced. A lower specific gravity is found in discharged batteries. State of charge, or the depth of discharge, can be determined by measuring the voltage or specific gravity of the acid with a hydrometer. These parameters will not support the battery condition; only a sustained load test can do that.

Voltage on a fully charged battery is typically 2.12 to 2.15 V per cell, or 12.7 V for a 12-V battery. At 50%, the reading will be 2.03 V_{pC} (volts per cell), and at 0% it will be 1.75 V_{pC} or less. Specific gravity will be about 1.265 for a fully charged cell and 1.13 or less for a totally discharged cell.

This can vary. The user should measure it with new batteries by fully charging them and leaving them to settle for a while and then taking a reference measurement. Of course, hydrometer readings are not possible for sealed batteries, and only voltage readings are used for evaluation of the depth of discharge. Maintenance personnel should use only distilled water to avoid adding minerals that can reduce the batteries' effectiveness, and the batteries should be brought to a full charge before adding water because the electrolyte expands as the state of charge increases.

11.3.5 Sizing Lead–Acid Batteries for DG Applications

To size lead–acid batteries for DG applications, the designer must decide if the battery bank will be used for transient compensation of source response or autonomous operation for source outage. For the former, battery storage will be used in conjunction with diesel engines, microturbines, or fuel cells and typically connected to the grid. For autonomous operation, energy is stored from renewable sources and used for a long period of time when the main source is not available. Therefore, for the latter, sizing batteries for a DG system will depend on the prescribed autonomy. If the system is grid-connected, battery size can be reduced for users that do not typically sell power to the grid. It must be increased if users sell power through net metering.

The following procedure for sizing batteries for DG applications was adapted from Ref. [4]. The dc link where the battery charger is connected defines the string connection of battery cells. The dc-link voltage usually fluctuates with the intermittence of the main source of energy (solar, wind, hydro, or other). Wind and hydro resources can be connected to the dc link with a rectifier or directly to the ac grid. If only a photovoltaic array is connected to the dc link, it will define the maximum dc-link voltage.

Batteries are specified with a nominal cell voltage equal to the open-circuit voltage (V_{cn}) for a 100% state of charge, and that is the recommended float voltage of the cell. Let the cell voltage at the completion of discharge be denoted by final volts per cell (FV_{pC}), which should be selected as high as possible, typically within 80 to 90% of V_{cn}:

$$\mathrm{FV}_{pC} = 0.8\, V_{cn} \tag{11.1}$$

Let V_{bat} represent the terminal voltage of the battery (consisting of a series string of N_c cells) after discharge. V_{bat} is equal to the minimum dc-link bus voltage, which is related to the minimum voltage at which an inverter connected to the dc link will operate. Then the number of cells can be selected by the following ratio, taking the next-higher integer value:

$$N_c = \frac{V_{\mathrm{bat}}}{\mathrm{FV}_{pC}} \tag{11.2}$$

The decision at to whether the battery should be sized for transient compensation or autonomous operation must be made at this point. When selecting a battery to compensate the main source (e.g., fuel cell or microturbine) for changes in load demand, the typical transient response of the fuel cell or microturbine source must be known empirically and defined by the mathematical function $g(t)$, which starts at zero and ends at the unity value at a time $T_{r,\max}$.

This function represents the source response to a step change in power, and either a ramp or an exponential function usually approximates it. Then the maximum power drawn from the battery to compensate for this transient is

$$P_{b,\max}(t) = P_{\mathrm{fl}}\, g(t) \tag{11.3}$$

where P_{fl} is the full-load power (kilowatts). The maximum energy storage required in the battery can be determined by

$$E_{b,\max} = \int_0^{T_{r,\max}} P_{b,\max}(t)\, dt \tag{11.4}$$

The integral can be approximated by

$$E_{b,\max} = K_1 P_{\mathrm{fl}} T_{r,\max} \tag{11.5}$$

where $K_1 = \int_0^{T_{r,\max}} g(t)\,dt$ is evaluated to be 0.5 and 0.2 for ramp function and exponential function, respectively. The full-load power P_{fl} can be obtained by

$$P_{\mathrm{fl}} = \frac{(\mathrm{kVA_{fl}})\mathrm{PF}}{\eta} \times 10^5 \tag{11.6}$$

where $\mathrm{kVA_{fl}}$ is the full-load kilovolt-amperes of the inverter, PF the power factor of the load connected to the inverter, and η the inverter efficiency.

Combining equations (11.5) and (11.6), the battery energy is given by

$$E_{b,\max} = \frac{K_1(\mathrm{kVA_{fl}})\mathrm{PF}T_{r,\max}}{\eta} \times 10^5 \tag{11.7}$$

The ampere-hour per cell, which is the C rate of the battery, is given by

$$\mathrm{Ah} = \frac{E_{b,\max}}{N_c \mathrm{FV}_{pC}(3600)} = \frac{K_1(\mathrm{kVA_{fl}})\mathrm{PF}T_{r,\max}}{\eta N_c \mathrm{FV}_{pC}(3600)} \times 10^5 \tag{11.8}$$

Considering d to be a multiplier that determines the discharge rate of the battery, we then obtain

$$d = \frac{3600}{T_{r,\max}} \tag{11.9}$$

and the discharge rate of the battery is expressed as dC. The average current drawn from the battery during discharge is $d \cdot \mathrm{Ah}$ for a duration of $T_{r,\max}$.

Batteries should not be discharged excessively. A deep-cycle lead–acid battery (the main battery option) will last longest if it is discharged only 50%. By multiplying the total ampere-hours from equation (11.8) by 2, the optimal battery capacity can be determined.

After this procedure it becomes straightforward to size batteries for autonomous operation. The designer needs to find the "daily energy use" as $\mathrm{kVA_{fl}}$ and to decide the number of days to have power in storage when sun is not shining sufficiently for PV-based systems or when any other sources are not capable of supplying power. Three to five days is recommended. If three days is used, $T_{r,\max} = 72$ hours and K_1 is assumed to be unity. For photovoltaic-based systems installed in a latitude higher than 40°, five days of storage is recommended. The flowchart in Figure 11.4 demonstrates this procedure.

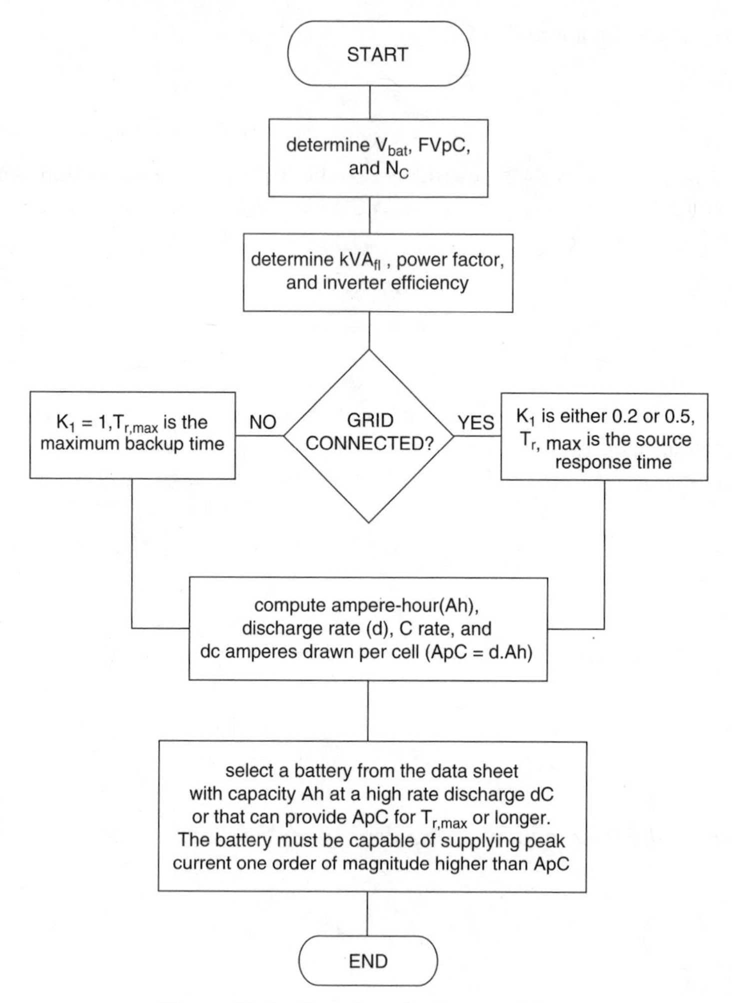

Figure 11.4 Flowchart for battery sizing.

11.4 ULTRACAPACITORS

Ordinary capacitors store energy in the dielectric material at a value of $\frac{1}{2}CV^2$, where C is its capacitance (farads) and V (volts) is the voltage across its terminals. The maximum voltage of a regular capacitor is dependent on the breakdown characteristics of the dielectric material. The charge Q (coulombs) stored in the capacitor is given by $Q = CV$. The capacitance of the dielectric capacitor depends on the dielectric constant (ε) and the thickness (d) of the dielectric material plus its geometric area:

$$C = \varepsilon \frac{A}{d} \tag{11.10}$$

During the past few years, electric double-layer capacitors with very large capacitance values have been developed. Those capacitors are frequently called *supercapacitors*, *ultracapacitors*, or *electrochemical capacitors* [5–7]. The term *ultracapacitor* is used in this book because the power industry seems to use it more frequently. An ultracapacitor is an electrical energy storage device that is constructed much like a battery because it has two electrodes immersed in an electrolyte with a separator between them. The electrodes are fabricated from porous high-surface-area material that has pores of diameter in the nanometer range. Charge is stored in the micropores at or near the interface between the solid electrode material and the electrolyte. The charge and energy stored are given by the same expressions as those for an ordinary capacitor, but the capacitance depends on complex phenomena that occurs in the micropores of the electrode.

11.4.1 Double-Layer Ultracapacitors

Figure 11.5 shows the construction details of a double-layer ultracapacitor. The capacitor contains two particulate–carbon electrodes formed on conductive-polymer films. An ion-conductive membrane separates the two electrodes, and a potassium hydroxide electrolyte permeates the capacitor. The micropores in the carbon particles result in an enormous surface area and yield extremely high capacitance values, which conventional capacitors cannot attain. Energy is stored in the double-layer capacitor as charge separation in the double layer formed at the interface between the solid electrode material surface and the liquid electrolyte in the micropores of the electrodes. The ions displaced in forming the double layers in the pores are transferred between the electrodes by diffusion through the electrolyte. The separator prevents electrical contact between the two electrodes. It is very thin, with high electrical resistance, but ion-permeable, which allows ionic charge transfer. Polymer or paper separators can be used with organic electrolytes, and ceramic or glass fiber separators are often used with aqueous electrolytes.

When stored in batteries as potentially available chemical energy, electrical energy requires faradic oxidation and reduction of the electroactive reagents to release charges. These charges can perform electrical work when they flow between two electrodes of different potentials. A dielectric material sandwiched between the two electrodes stores electrostatic energy by accumulation of charges, but it has a

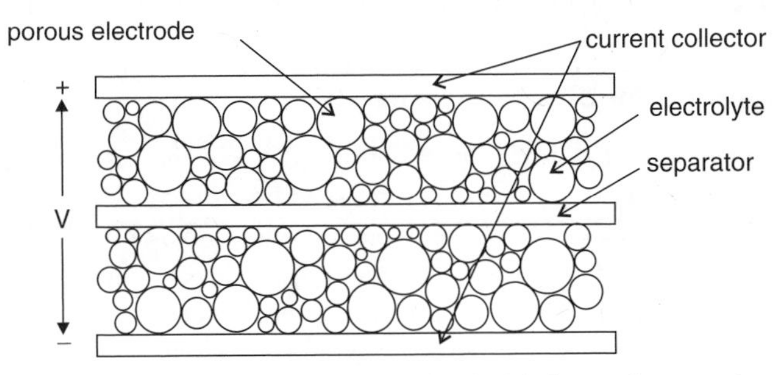

Figure 11.5 Construction details of a double-layer ultracapacitor.

very limited storage capability compared with that of batteries. The energy and charge stored in the electrochemical capacitor are given by the same equations as for ordinary capacitors. However, the capacitance is dependent primarily on the characteristics of the electrode material (i.e., surface area and pore size distribution). The specific capacitance of an electrode material depends on the effective dielectric constant of the electrolyte and the thickness of the double layer formed at the interface. These are very complex phenomena that are not fully understood. The thickness of the double layer is very small (a fraction of a nanometer in liquid electrolytes), which results in a high value for the specific capacitance.

The performance of electrochemical capacitors depends on the specific capacitance (F/g or F/cm^3) of the electrode material and the ionic conductivity of the electrolyte used in the device. The specific capacitance of a particular electrode material depends on whether the material is used in the positive or negative electrode of the device and whether the electrolyte is aqueous or organic. Most carbon materials exhibit higher specific capacitance: in the range 75 to 175 F/g for aqueous electrolytes and 40 to 100 F/g for organic electrolytes, which allow much more ions. Although a lower specific capacitance is achieved with organic electrolytes, they have the advantage of higher operating voltage. For aqueous electrolytes, cell voltage is about 1 V; for organic electrolytes, cell voltage is 3 to 3.5 V.

The following example illustrates the principles of high-capacitance devices. Suppose that two carbon rods are separated from each other and immersed in a thin sulfuric acid solution, which is subjected to a voltage slowly increased from zero. Under such conditions, almost nothing happens up to 1 V. Then, at about 1.2 V, gas bubbles appear on the surface of both electrodes because of the electrical decomposition of water. Just before decomposition occurs, where there is no current flow yet, an *electric double layer* occurs at the boundary of electrode and electrolyte (i.e., electrons are charged across the double layer and form a capacitor). Gassing comes up at voltages above 1 V. This indicates that the capacitor is breaking down by overvoltage, which causes decomposition of the electrolyte. The electrical double layer works as an insulator only below the decomposing voltage because water-soluble or aqueous electrolytes constrain the application at a withstanding operating voltage of about 1 V. Organic electrolytes are used to increase the operating voltage (currently at 2.5 V) and provide overall stability against electrolyte decomposition. In organic electrolytes, there is a salt in the solvent that determines the peak capacity; aqueous electrolytes do not have this limitation.

11.4.2 High-Energy Ultracapacitors

The high energy content of ultracapacitors originates in the activated carbon electrode material, which has an extremely high specific surface area and an extremely short length (on the order of a few nanometers) between the opposite charges of the capacitor. Because the dielectric is extremely thin, consisting only of the phase boundary between electrode and electrolyte, capacitances on the order of a few thousand farads are possible in a very small volume. Researchers have been exploring the possibility of using carbon nanotubes for ultracapacitor electrodes because they have nanopores (about 0.8 nm in diameter), which could, in theory, store much more charge if the nanotubes could be properly assembled into macro-scale units.

When such a device is called an electrochemical capacitor it is because redox (i.e., reduction–oxidation) chemical reactions take place. In a true capacitor, no chemical reactions are involved. An ultracapacitor uses the effect of an electric double layer (EDL) on both positive and negative electrodes. Therefore, it is often called *symmetric EDLC.*

To increase energy density, a combination of a battery and a capacitor has been designed. In it, one electrode is for a battery, and the other electrode is for a capacitor. Such a hybrid configuration is called *asymmetric EDLC*. Asymmetric capacitors tend to be capable of higher capacitance because the battery electrode, physically smaller than the carbon electrode, allows more space. Most EDLCs use aqueous electrolytes, which are limited by the gassing voltage of the water. On the other hand, most symmetrical capacitors use organic electrolytes to achieve voltages rated at 2.5 to 2.7 V.

The leading manufacturers of ultracapacitors are Maxwell Technologies in the United States, NESS Capacitor Co. in South Korea, SAFT in France, Panasonic and Okamura in Japan, and EPCOS in Germany. These companies manufacture carbon–carbon or symmetric ultracapacitors. Some companies in Russia are developing aqueous technology. ESMA produces asymmetrical devices, and ECOND and ELIT produce symmetrical units.

11.4.3 Applications of Ultracapacitors

The energy and power density of ultracapacitors fall between those of batteries and conventional capacitors, as depicted in Figure 11.6. They have more energy than a capacitor but less than a battery and more power than a battery but less than a capacitor. Unlike in batteries, the ultracapacitor voltage varies linearly with the state of charge, as depicted in Figure 11.7. The voltage range between full charge and end of charge is higher for ultracapacitors than for batteries. Because of such drastic

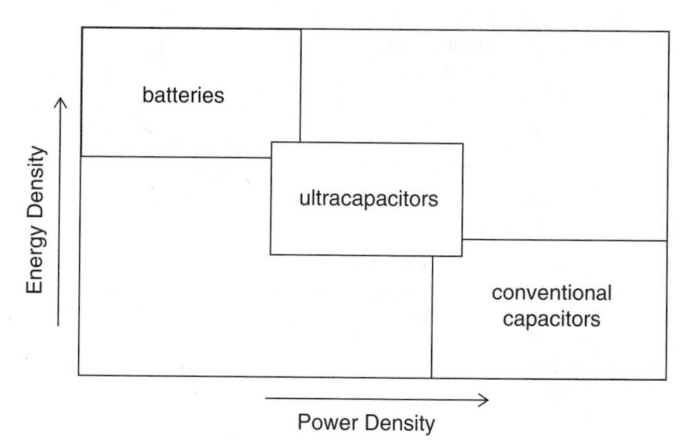

Figure 11.6 Energy and power density of ultracapacitors.

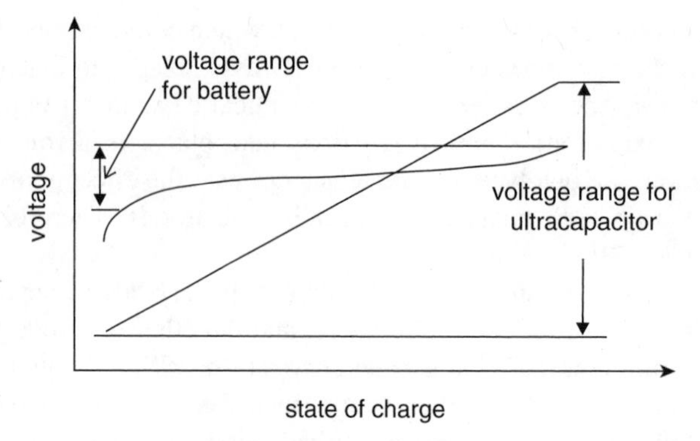

Figure 11.7 Comparison of voltage profile for ultracapacitor and battery.

variation with state of charge, series connection is often required for high-voltage applications, and power electronic circuits must be integrated with the stack to control charge, discharge, and voltage equalization, as portrayed in Figure 11.8.

A current pump charger is a switching voltage regulator that maintains a constant load current under varying terminal voltages. A current pump is inherently suited to the charging requirements of nickel–cadmium and nickel–metal hydride batteries or constant-current requirements of Peltier coolers and in this case is used for an ultracapacitor system.

The following are typical applications for ultracapacitors:

- Pulse power, in which a load receives short, high-current pulses
- Hold-up or bridge power to a device or piece of equipment for seconds, minutes, or days when the main power or battery fails or when the battery is swapped out

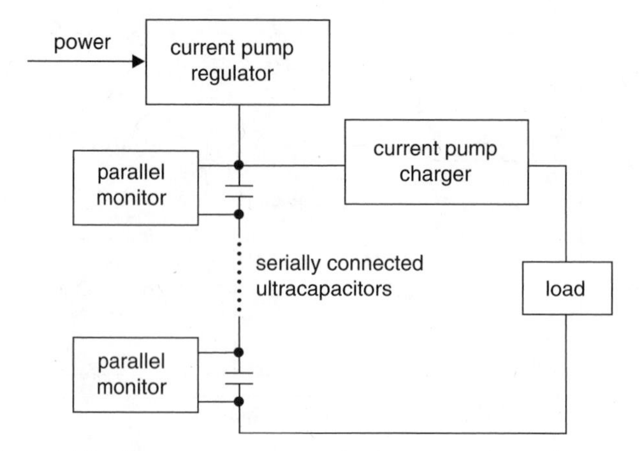

Figure 11.8 Ultracapacitors controlled by power electronic systems.

Ultracapacitors can be used:

- As efficient battery packs, achieved by paralleling a low-impedance, high-power aerogel capacitor and a high-impedance, high-energy battery to create a low-impedance, high-power and high-energy hybrid battery–capacitor
- In wireless communications, in which they are used for pulse power during transmission in GSM cell phones, 1.5- and 2-way pagers, and other data communication devices
- In mobile computing, in which they are used for hold-up and pulse power in portable data terminals, personal digital assistants, and all other portable devices using microprocessors
- For industrial applications such as solenoid and valve actuation, electronic door locks, and uninterruptible power supplies

Ultracapacitors have several advantages over batteries. They tend to have a longer cycle life because the absence of chemical reactions yields a stable electrode matrix and no wear-out. Ultracapacitors typically work more than 100,000 cycles with an energy efficiency greater than 90%. They are currently able to bridge power for seconds in the hundreds of kilowatts power range, and their use for residential peak shaving is under consideration (see the hybrid photovoltaic system depicted in Figure 11.9). The use of ultracapacitors for multimegawatt and transmission and distribution applications that require several hours of energy storage for peak shaving and load leveling is not yet feasible.

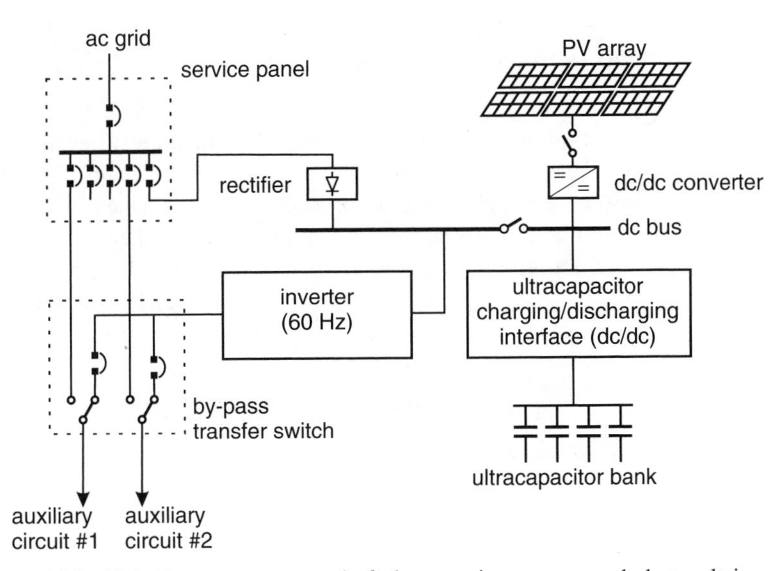

Figure 11.9 Hybrid system composed of ultracapacitor storage and photovoltaic array.

Ultracapacitors have been strongly considered for use in association with uninterruptible power supplies and fuel cells because they respond faster than batteries and have a higher energy density. Ultracapacitors could provide a longer period of autonomy for an uninterruptible power supply and cope with heavier transients than fuel cells are able to handle because of their slow response to current surges.

11.5 FLYWHEELS

Prior to the development of cost-effective power-conversion electronics, adding inertia to a motor–generator set was the primary method of limiting power interruptions to critical loads. Flywheels were first used in steam engines. In the 1970s, when the technology was used for centrifugal enrichment of uranium, it was rediscovered as a potential new energy storage system [8,9].

Flywheels have very simple physics. They store energy in a spinning mass. Through a coupled electric machine, the system can store power by accelerating the shaft, and retrieve power by slowing it. The stored energy depends on the moment of inertia of the rotor, I, which depends on the distribution of mass density around the rotating axis, $\rho(x)$, and geometrical radius of the rotor, r. The energy will depend on the moment of inertia and the square of the angular speed of the flywheel shaft:

$$E = \tfrac{1}{2} I \omega^2 \tag{11.11}$$

$$I = \int \rho(x) r^2 \, dx \tag{11.12}$$

11.5.1 Advanced Performance of Flywheels

Although it appears that only high angular velocity and large masses around the axis of rotation would allow high energy storage, it is important to evaluate how to obtain the highest energy density given a prescribed material strength. Therefore, the tensile strength (i.e., the highest stress not leading to a permanent deformation of the material) should be used to define the optimal flywheel shape. The components in x, y, and z directions of the force per unit volume are related to the stress tensor τ_{ij} by equation (11.13), and the tensile strength σ_i, in a direction specified by i, is given by equation (11.14):

$$f_i = \sum_j \frac{\partial \tau_{ij}}{\partial x_j} \tag{11.13}$$

$$\sigma_i = \max\left(\sum_j \tau_{ij} n_j\right) \tag{11.14}$$

where n_j is a unit vector normal to the cut in the material and the maximum sustainable stress in the direction i is to be sought by varying the angle of slicing.

TABLE 11.7 Flywheel Shape Factor, K

Flywheel Type	Shape	K
Constant selection		1.000
Constant-stress disk	—	0.931
Approx. constant section		0.834
Conic disk		0.806
Flat unpierced disk		0.606
Thin rim		0.500
Rod or circular brush	—	0.333
Flat pierced disk		0.305

The value of the *shape factor* or *form factor* is a measure of the efficiency with which the flywheel geometry uses the material strength, and a general expression for any flywheel made from material of uniform mass density is

$$E_{m,\max} = K\frac{\sigma_{\max}}{\rho} \tag{11.15}$$

where $E_{m,\max}$ is the maximum kinetic energy per unit mass. If the material is isotropic, the maximum tensile strength is independent of directions and may be denoted by $\sigma_{\max}$. By calculating the internal forces in the flywheel and considering a given flywheel shape symmetrical around the axis of rotation, a resulting shape factor K may be calculated as indicated in Table 11.7.

Expression (11.15) can be interpreted as being a given maximum acceptable stress $\sigma_{\max}$ for a given maximum energy storage density (i.e., it is not dependent on rotational speed, and it is maximized for light materials and high stress-based design). For a given geometry, the limiting energy density (energy per unit mass) of a flywheel is proportional to the ratio of material strength to weight density (specific strength).

11.5.2 Applications of Flywheels

Because of their fast actuation compared with electrochemical energy storage, flywheel systems have recently regained consideration as a means of supporting a critical load during grid power interruption. Advances in power electronics and digital controller fields initially provided an alternative to flywheels and have now actually enabled better flywheel designs to deliver a cost-effective alternative to the power quality market. In typical systems, a motor inputs electrical energy to the flywheel, and a generator is coupled on the same shaft and outputs the energy through an electronic converter. It is also possible to design a bidirectional power electronic system with one machine that is capable of motoring and regenerating operations.

Figure 11.10 shows that flywheels can be used in applications that intersect the areas of capacitors and batteries for power greater than 80 kW in a time span of 1 to

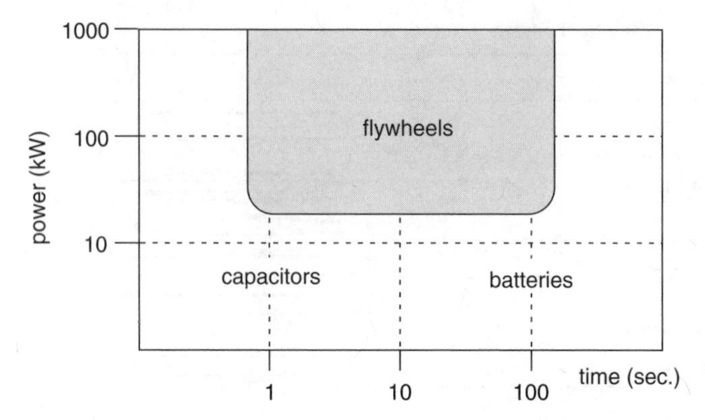

Figure 11.10 Power and time boundaries for flywheels, capacitors, and batteries.

100 seconds. So flywheels cannot store energy for a long time, but the stored energy can be released in a relatively short time. This makes them suitable for uninterruptible power supply and electric vehicle applications.

11.5.3 Design Strategies

The energy storage capability of flywheels can be improved by using the right material and increasing the moment of inertia, or spinning it at higher rotational velocities. Some designs use hollow cylinders for the rotor, which allows the mass to be concentrated at the outer radius of the flywheel and improves storage capability with a smaller weight increase in the storage system. Flywheel energy storage systems have capacities from 0.5 to 500 kWh. The flywheel itself loses less than 0.1% of the stored energy per hour if magnetic bearings are used.

Two strategies have been used in the development of flywheels for power applications. One option is to increase the inertia by using a steel mass with a large radius and rotational velocities up to a few thousand revolutions per minute. A fairly standard motor and power electronic drive can be used as the power conversion interface for this type of flywheel. This design results in relatively large and heavy flywheel systems. The effective increase in run time for such systems rarely exceeds 1 second at rated load, which corresponds to a delivery of less than 5% of the additional stored energy in the wheel. Delivery of more energy would result in reduced rotational speed and hence reduced electrical frequency, which is usually unacceptable. Rotational energy losses also limit the long-term storage ability of this type of flywheel. To decouple the decrease in rotational speed from the stable electrical frequency needed by the grid, a system with variable speed controlled by a power electronic drive with rectification and inversion electronics is usually implemented, as indicated in Figure 11.11. For such standard systems, the rotational speed is limited to approximately 10,000 rpm. Induction machines operating in field-weakening mode can be employed.

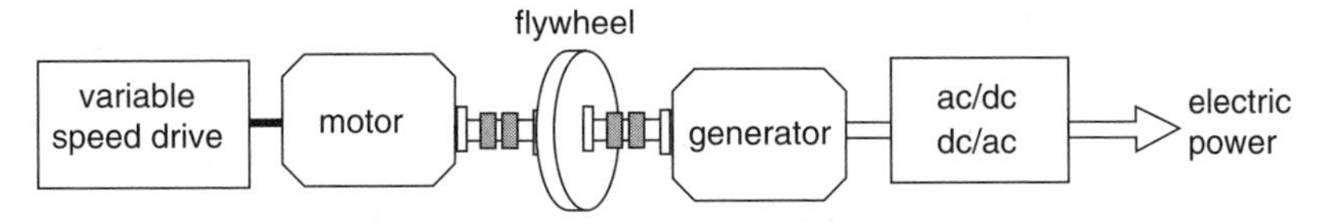

Figure 11.11 Standard flywheel system with motoring and generating sets.

The second design strategy is to produce flywheels with a lightweight rotor that turns at high rotational velocities of up to 100,000 rpm. This approach results in compact, lightweight energy storage devices. Modular designs are possible, with a large number of small flywheels possible as an alternative to a few large flywheels. However, rotational losses because of air drag and bearing losses result in significant self-discharge, which affects long-term energy storage. Recent advances in composite materials technology may allow nearly an order of magnitude advantage in the specific strength of composites when compared with even the best engineering metals. The result of this continuous research in composites has supported the design of flywheels that operate at rotational speeds in excess of 100,000 rpm and with tip speeds in excess of 1000 m/s.

The ultrahigh rotational speeds that are required to store significant kinetic energy in flywheel systems virtually rule out the use of conventional mechanical bearings. High-velocity flywheels are operated in vacuum vessels to eliminate air resistance, and the magnetic bearings must withstand the difference in pressure of the vessel and the outside environment. There is some R&D on superconducting magnetic bearings for extremely high-speed flywheels and magnetic forces to levitate a rotor and eliminate the frictional losses inherent in rolling element and fluid film bearings. Induction machines are not used for this application because the vacuum chamber will preclude the machine rotor to dissipate power (there is no air for heating exchange). Rare-earth permanent-magnet machines are often used. Because the specific strength of these magnets is typically just fractions of that of the composite flywheel, they must spin at lower tip speeds; in other words, they must be placed very near the hub of the flywheel, which compromises the power density of the generator.

Figure 11.12 shows the use of a high-speed flywheel for a dynamic voltage restorer in which an energy control system keeps the balance in the shaft while a sag detector injects the correctional voltage in series with the power system.

Advanced flywheel storage systems promise compact size, no emissions, light weight, and low maintenance. In addition, they offer long life cycles, do not suffer from multiple discharges, and have minimal sensitivity to operating temperature. However, a mass rotating at very high speeds has inherent safety complexities that might be considered in application. When the tensile strength of a flywheel is exceeded, the flywheel will shatter and release all its stored energy at once. It is usual for flywheel systems to be contained in vessels filled with red-hot sand as a safety precaution. Fortunately, composite materials tend to disintegrate quickly once broken (rather than producing large chunks of high-velocity fragments of the vessel).

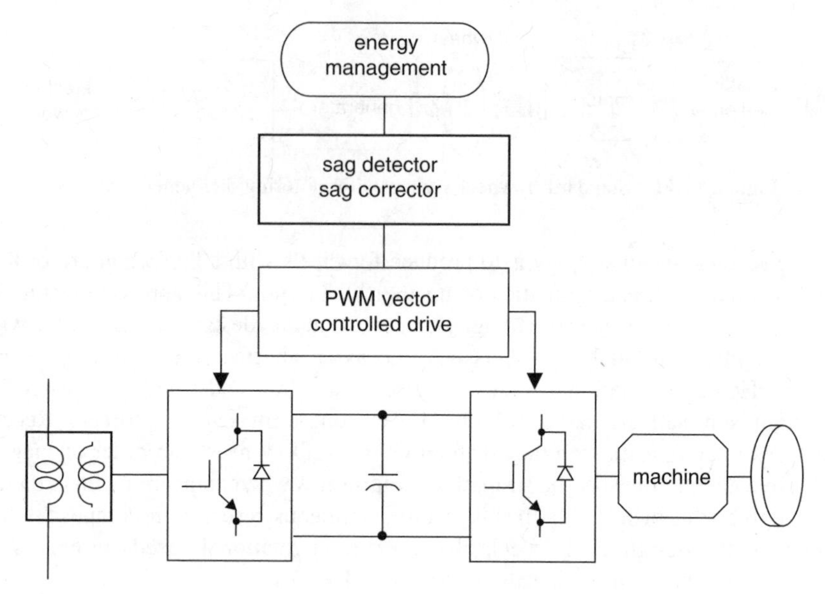

Figure 11.12 Flywheel system for voltage restoration.

11.6 SUPERCONDUCTING MAGNETIC STORAGE SYSTEM

The energy content in an electromagnetic field is determined by the current through the N turns of the magnet coil. The product NI is called *magnetomotive force*. By assuming the electromotive force in the coil as in equation (11.16), the energy is given by equation (11.17). The stored energy is given by integrating the magnetic field strength H over the entire volume in which the induction B is significant. By assuming a linear relationship between H and B, the volume energy density can be obtained from those equations as shown in equation (11.18). The energy density achieved in a magnetic field can be estimated by equations (11.17) and (11.18).

$$e = -N\frac{d\phi}{dt} \tag{11.16}$$

$$E = \int_0^{\phi} Ni(t)d\phi = \int_0^B lHA\ dB = \int_{\text{volume}} \int_0^B H\ dB \tag{11.17}$$

$$E_{\text{volume}} = \frac{B^2}{2\mu} \tag{11.18}$$

As an example, typical ferromagnetic materials at $B = 2T$ are approximately 2×10^6 J/m^3, which is an order of magnitude greater than the electrostatic field but still small compared with electrochemical batteries. An inductor stores energy proportional to the inductance value and to the square of the current, as indicated by equation (11.19). Therefore, it seems plausible that high values of energy density can only be possible by using such media as air or vacuum with very high current values, but electrical resistance of the coil is a limiting factor.

$$E(t) = \tfrac{1}{2}Li(t)^2 \tag{11.19}$$

A *superconductor* is a material whose resistance is zero when it is cooled to a very low temperature, known as a *cryogenic temperature*. In a superconducting magnetic energy storage (SMES) system, an inductor wound with superconducting wire, such as niobium–titanium, is used to create a dc magnetic field [10–12]. The current-carrying capacity of the wire is dependent on temperature and the local magnetic field. A cryogenic system keeps the operating temperature such that the wire becomes a superconductor.

The critical temperature is the point at which the electrical resistance drops dramatically. For all superconductors, this used to be nearly 4 K using liquid helium. After the 1980s, new superconductors made of copper oxide ceramic became available. They only have to be chilled to around 100 K, using liquid nitrogen or special refrigerators. These newer materials are classified as high-temperature superconductors. Low-temperature superconductors must be cooled to around −269°C.

Some research has been conducted in the use of the cable-in-conduit conductor concept. In this design a superconducting cable is placed inside a conduit (or jacket) filled with helium. The conductor is not only the main electrical path but also the helium containment element. This combination of functions allows for more flexibility, with the potential for simplification and lower cost.

11.6.1 SMES System Capabilities

SMES systems are typically able to store up to about 10 MW. Higher capacities are possible for shorter time spans. Because a coil of around 150 to 500 m radius would be able to support a load of 5000 MWh, at 1000 MW it is expected that SMES can potentially store up to 2000 MW.

One problem associated with SMES systems is the requirement of compensation for stray field. The field falls at the inverse cube of the distance from the center of the coil, and 1 km of radius would still keep a few milliteslas. Because Earth's field is about one-twentieth of a millitesla, a guard coil with a magnetic moment opposing the SMES systems must be added around an outer circle to avoid interference on aircraft, power transmission lines, communications, people, and bird navigation. A special requirement for SMES systems is a protection system capable of detecting and avoiding quenches (i.e., loss of superconductivity because of thermodynamic critical coupling between superconductors and cooling tubes), which leads to ohmic heat release that would cause irreversible damage to the superconducting coil.

Figure 11.13 shows the basic idea of controlling the charge and discharge of a superconducting inductor for single- and three-phase systems. The inductor is kept with current flowing in the same direction. As energy is stored, a positive voltage is applied across the superconducting coil, and, as energy is released, the voltage is reversed. For average zero voltage, the system is put in persistent-current mode, so that the only loss is the energy required to run the cryogenic refrigeration and

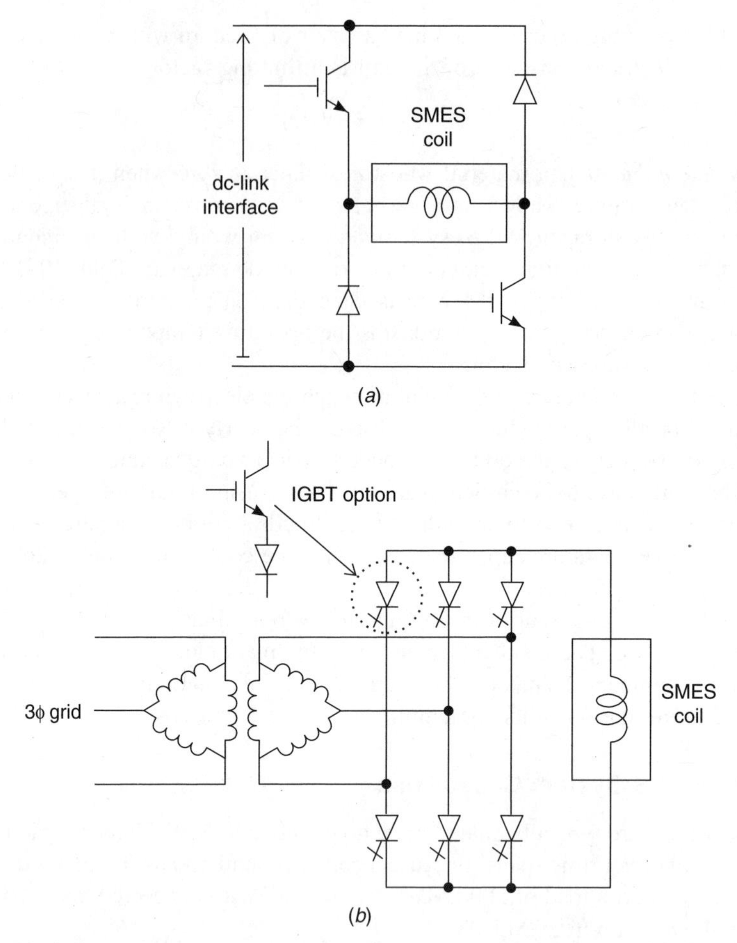

Figure 11.13 Power electronics topology for superconducting inductor: (*a*) single-phase; (*b*) three-phase.

activate the power electronic circuitry. The topology presented in Figure 11.13 requires that the power flow be bidirectional with the grid.

11.6.2 Developments in SMES Systems

The ACCEL Instruments GmbH team in Germany has designed a 2-MJ SMES system to ensure the power quality of a laboratory plant of Dortmunder Elektrizitäts und Wasserwerke. Figure 11.14 shows a SMES system connected by a dc-link converter to the electrical grid of the plant. The system is designed for a carryover time of 8 s at an average power of 200 kW. The SMES system is wound of NbTi mixed matrix superconductor cooled with helium, with a two-stage cryo-cooler. The SMES system has a sophisticated quench protection system, and the interface to the dc link makes the system more flexible than a direct grid connection.

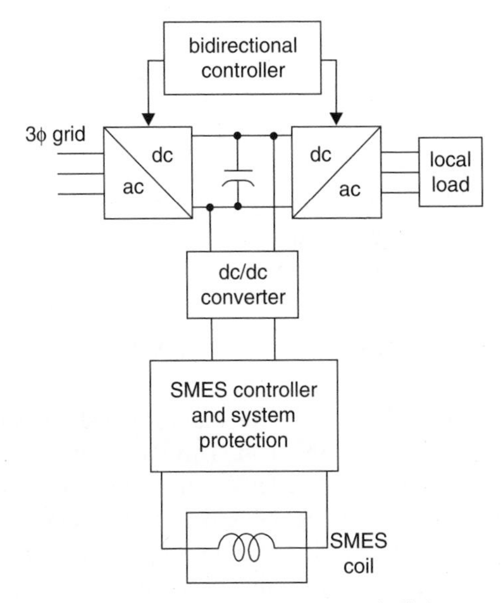

Figure 11.14 SMES integrated to a dc-link system.

Technical improvements and a better knowledge of how to control cryogenic systems have allowed SMES to penetrate the power system market in the past few years. American Superconductor has a commercial system called D-SMES, which is a shunt-connected flexible AC transmission device designed to increase grid stability, improve power transfer, and increase reliability, as illustrated in Figure 11.15. The

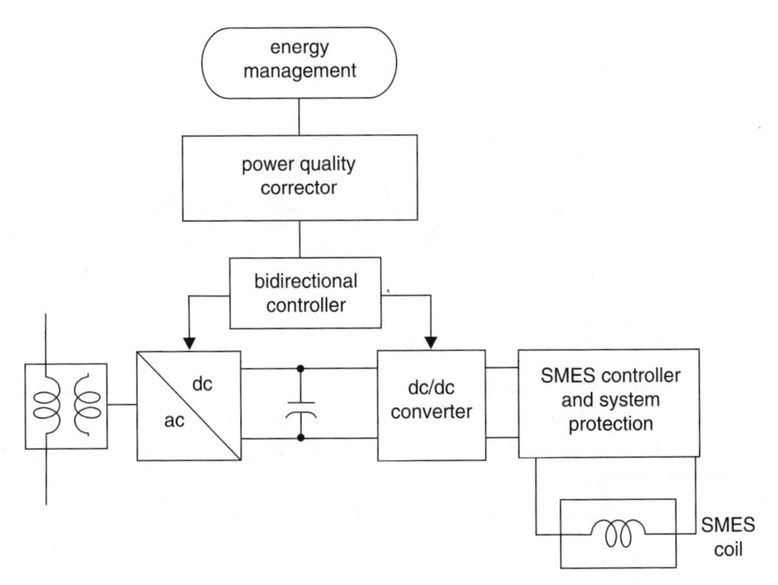

Figure 11.15 SMES integrated to a capacitor bank for voltage restoration.

D-SMES injects real power as well as dynamic reactive power quickly to compensate for disturbances on the utility grid. This system injects in the plant load a correcting voltage in series with the utility grid for power quality improvement.

When selecting a SMES system as a storage solution, several aspects must be considered. Because of its fast response to power demand and high-efficiency features, it has the capability of providing frequency support (spinning reserve) during loss of generation, transient and dynamic stability by damping transmission-line oscillations, and dynamic voltage support.

11.7 PUMPED HYDROELECTRIC ENERGY STORAGE

Water plays an important role in global climate. It is naturally lifted to the atmosphere and released to flow down the land. This can be used to generate power.

Pumped hydroelectric energy storage is the oldest kind of large-scale energy storage. It has been in use since the beginning of the twentieth century. In fact, until 1970, it was the only commercially available option for large energy storage. Pumped hydroelectric stations are in active operation, and new ones are still being built.

As indicated in Figure 11.16, pumped storage projects differ from conventional hydroelectric projects in that they pump water from a lower reservoir to an upper reservoir when demand for electricity is low. The initial potential energy associated with the head is transformed into kinetic energy. One part of this energy is associated with the mass m' moving with velocity. The other is the pressure part, with the enthalpy given by the pressure P over the density of water multiplied by the remaining mass $(m - m')$:

$$E_{\text{initial}}^{\text{potential}} = mgh = m' \cdot \frac{1}{2}u^2 + (m - m')\frac{P}{\rho} = E^{\text{kinetic}} + \text{enthalpy} \tag{11.20}$$

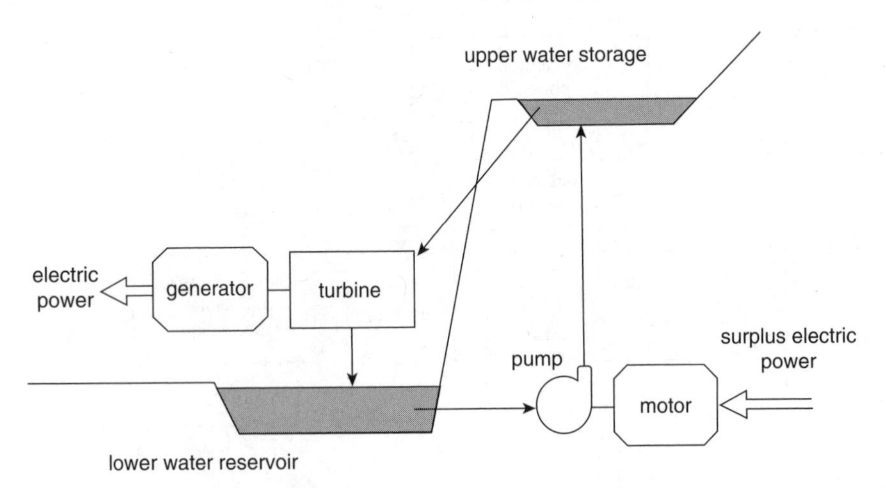

Figure 11.16 Pumped hydroelectric storage system.

There are frictional losses, turbulence, and viscous drag, and the turbine has an intrinsic efficiency. For the final conversion of hydro power to electricity, the generator efficiency must also be considered. Therefore the overall efficiency of pumped hydro systems must consider the ratio of the energy supplied to the consumer and the energy consumed while pumping. The energy used for pumping a volume V of water up to a height h with pumping efficiency η_ρ is given by equation (11.21), and the energy supplied to the grid while generating with efficiency η_g is given by equation (11.22).

$$E_{\text{pumping}} = \frac{\rho g h V}{\eta_p} \tag{11.21}$$

$$E_{\text{generator}} = \rho g h V \eta_g \tag{11.22}$$

During peak demand, water is released from the upper reservoir. It drops downward through high-pressure pipelines. It passes through turbines and ultimately pools in the lower reservoir. The turbines drive power generators that create electricity. Therefore, when releasing energy during peak demand, a pumped hydroelectric storage system works similarly to traditional hydroelectric systems. When production exceeds demand, water is pumped up and stored in the upper reservoir, usually with an early morning surplus.

11.7.1 Storage Capabilities of Pumped Systems

The amount of electricity that can be stored depends on two factors: are the net vertical distance through which the water falls, called the head, h; and the flow rate, Q (or the volume of water per second from the reservoir, and vice versa). A common false assumption is that head and flow are substitutable, but in fact simply increasing the volume per second cannot compensate for the lack of head. This is because high-head plants can quickly be adjusted to meet the electrical demand surge. For a lower head, the diameter of the pipe would have to be enormous to produce the same power except for the run-of-river plants. This would not be feasible economically, which is why most plants pumped hydro systems are of the high-head variety.

In the early days, pumped hydroelectric plants used a separate motor and dynamo, mainly because of the low efficiency of dual generators. This increased cost because separate pipes had to be built. A majority of modern plants use a generator that can be run backward as an electric motor. The efficiency of the generator has increased, and it is now possible to retrieve more than 80% of the input electrical energy. This leads to significant cost savings. Also, modern generators are of suspended vertical construction. This allows better access to the thrust bearing above the rotor for inspection and repair.

Of all large-scale energy storage methods, pumped hydroelectric storage is the most effective. It can store the largest capacity of electricity (more than 2000 MW), and the period of storage is among the longest. A typical plant might store its energy for more than half a year. Because of their rapid response speed, pumped hydroelectric storage systems are particularly useful as backup. An example is the large human-made cavern under the hills of North Wales. Its six huge pump turbines can each deliver 317 MW, producing together up to 1800 MW from the working volume of 6 million cubic meters of water and a head of 600 m. These pump–turbines are second in size to the 337-MW devices installed at Tianhuanping in China, but the North Whales system benefits from more than 50 m extra head compared with the Chinese plant.

The value of the hydroelectric storage system is enhanced by the speed of response of the generators. Any of the turbines can be brought to full power in just 10 seconds if they are spinning initially in air. Even starting from a complete standstill takes only 1 minute. Because of the rapid response speed, pumped hydroelectric storage systems are particularly useful as backup in case of sudden changes in demand.

Partly because of their large scale and relative simplicity of design, pumped hydroelectric storage systems are among the cheapest by operating costs per unit of energy. The cost of storing energy can be an order of magnitude lower in pumped hydroelectric systems than in, for example, superconducting magnetic storage systems.

Unlike hydroelectric dams, pumped hydroelectric systems have little effect on the landscape. They produce no pollution or waste. However, pumped hydroelectric storage systems also have drawbacks. Probably the most fundamental is their dependence on unique geological formations. There must be two large-volume reservoirs and sufficient head for building to be feasible. This is uncommon and often forces location in remote places, such as in the mountains, where construction is difficult and the power grid is not present or is too distant.

11.8 COMPRESSED AIR ENERGY STORAGE

Compressed air energy storage (CAES) is a mature storage technology for high-power, long-term load-leveling applications. Compressed air is the medium that allows elastic energy to be embedded in the gas. The ideal gas law relates the pressure P, the volume V, and temperature T of a gas as indicated by

$$PV = \eta RT \tag{11.23}$$

where η is the number of moles and R is the universal gas constant.

It is very difficult to calculate the energy density for any unusual shape volume. Therefore, if a piston without friction is assumed under an isobaric process a simplified volumetric energy density can be calculate by

$$E_{\text{volume}} = \frac{1}{V_0}\eta RT \int_{V_0}^{V} \frac{dV}{V} = \frac{1}{V_0} P_0 V_0 \ln \frac{V_0}{V} = P_0 \ln \frac{V_0}{V} \tag{11.24}$$

where P_0 is the pressure and V_0 and V are the initial and final volumes.

By assuming a starting volume of 1 m^3 and a pressure of 2×10^5 Pa, if the gas is compressed to 0.4 m^3 at constant temperature, the amount of stored energy is 1.8×10^5 J, which is a much higher energy density than that of magnetic or electric fields.

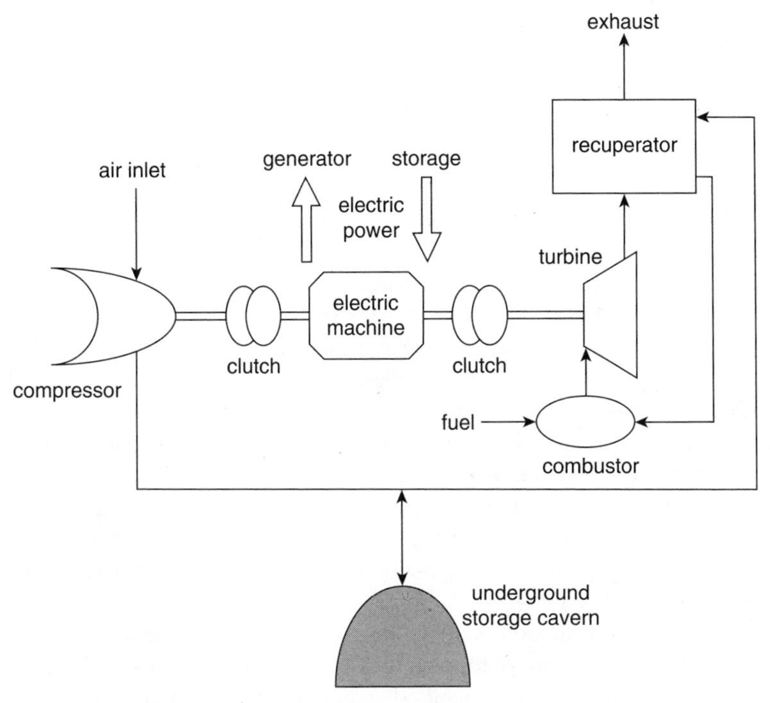

Figure 11.17 Compressed air energy storage system.

In the past few years, research has been conducted to improve the efficiency of turbines and heat transfer mechanisms used to pump and retrieve compressed air. In compressed air energy storage, air is compressed and stored under pressure. Release of the pressurized air is subsequently used to generate electricity, most efficiently in conjunction with a gas turbine. Figure 11.17 shows a typical CAES system, in which air is used to drive the compressor of a gas turbine. This makes up 50 to 60% of the total energy consumed by the gas turbine system.

The most important part of a CAES plant is the storage facility for compressed air. This is usually a human-made rock cavern, salt cavern, or porous rock created by water-bearing aquifers or as a result of oil and gas extraction. Aquifers, in particular, can be attractive as a storage medium because the compressed air will displace water, setting up a constant-pressure storage system. The pressure in alternative systems varies when adding or releasing air.

The largest CAES plant was built at Huntorf, Germany. This plant rated 290 MW and operated for 10 years with 90% availability and 99% reliability. Despite the decommissioning of this plant, CAES technology was highly promoted, and the Alabama Electric Co-operative built a 110-MW commercial project in McIntosh, Alabama. The plant started to operate in May 1991 and has since supplied power during peak demand periods. It is the first in the world to use a fuel-efficient recuperator, which reduces fuel consumption by 25%. Off-peak electricity is used to compress air. The electricity consumed during compression is 0.82 kWh of peak load generation. The plant provides enough electricity to supply the demand of 11,000 homes for 26 hours. The storage is a salt cavern, 1500 ft underground, capable of holding 10 million cubic feet of air. At full charge, air pressure is 1100 pounds per square inch (psi). At full discharge, cavern air pressure is 650 psi. The air flows through the CAES plant generator at a rate of 340 pounds per second, which is as fast as the rate of a wide-body jet engine. Fuel consumption during generation is 4600 Btu higher heating value per kilowatthour of electricity. Italy tested a 25-MW installation, and a 1050-MW project has apparently been proposed in the Donbass region on the Russia–Ukraine border.

The annual investment costs for a CAES system are estimated as between $90 and $120 per kilowatt (uniform cash flow), depending on air storage. With a 9% discount rate and a 10-year life cycle, these correspond to necessary initial investments between $580 and $770 per kilowatt. The development of large-scale CAES is limited by the availability of suitable sites. Therefore, current research is focused on the development of systems with human-made storage tanks.

Although typically it is the high air pressure that drives the turbines, quite often the air is mixed with natural gas. They are burned together in the same way as in a conventional turbine plant. This method is actually more efficient because the compressed air loses less energy.

Many geological formations can be used in this scheme. In general, rock caverns are about 60% more expensive to mine for CAES purposes than salt caverns. Aquifer storage is the least expensive method and is therefore used in most current locations. The other approach to compressed air storage is called *compressed air storage in vessels* (CASV). In a CASV system, air is stored in fabricated high-pressure tanks. However, the technology is not advanced enough to allow the manufacture of high-pressure tanks at a feasible cost.

The components of a basic CAES installation include the motor–generator, which employs clutches to provide for alternative engagement to the compressor or turbine trains; the air compressor, which may require two or more stages; intercoolers and aftercoolers to achieve economy of compression and reduce moisture content; the recuperator; the turbine train; high- and low-pressure turbines and equipment control for operating the combustion turbine; the compressor; and auxiliaries to regulate and control changeover from generation mode to storage mode.

Recently, research has been conducted to make commercially available small-scale CASV systems. In such cases, because air pressure varies widely, a mechanism for peak power tracking and identifying the best operating point is usually employed.

11.9 STORAGE HEAT

The amount of energy stored by heating a material from temperature T_0 to T_1 without changing its phase or chemical composition at constant pressure is

$$E = m \int_{T_0}^{T_1} c_P \, dT \tag{11.25}$$

where m is the mass heated and c_P is the specific heat capacity at constant pressure. Materials suitable for heat storage have a large heat capacity, are chemically stable, and are used to add heat to and to subtract heat from. Water, soil, solid metals, and rock beds are often used for heat storage. Other configurations are also common, such as a solar pond, where a human-made salty lake collects and stores solar energy. Being heavier than fresh water, salt water does not rise or mix by natural convection, creating a large temperature gradient within the pond. Fresh water forms a thin insulating surface layer at the top, and underneath it is the salt water, which becomes hotter with depth, close to 90°C at the bottom. Then heated brine is draw from the bottom and piped into a heat exchanger, where the heat converts a liquid refrigerant into a pressurized vapor that can spin a turbine, eventually generating electricity. An Israeli project covering 62 acres near the Dead Sea produces up to 5 MW of electricity during peak demand periods, and a 6000-m^2 solar pond in Bhuj (India) is capable of delivering 80,000 L of hot water daily at 70°C for a dairy plant.

The fundamentals of heat transfer calculations of thermally isolated reservoirs are discussed in Section 5.3 and a solar tower power station, in Section 5.5. Central receivers or power tower systems use thousands of individual tracking mirrors called *heliostats* to reflect solar energy onto a receiver located at the top of a tall tower. This energy heats molten salt flowing through the receiver, and the salt's heat energy is then used to generate electricity in a steam generator. The molten salt retains heat, so it can be stored for hours or even days before being used to generate electricity. The liquid salt at 290°C is pumped from a cold storage tank through the receiver, where it is heated to 570°C and then to a hot tank for storage. When power is needed, hot salt is pumped to a steam generating system that produces superheated steam to power a turbine and generator. From the steam generator the salt is returned to the cold tank for reuse.

Fuel cells are also conveniently hybridized with heat storage. Lower temperatures can fulfill space heating, and by capturing unused low-temperature heat energy rejected from the electric production process, fuel energy is used more efficiently. Combining heat and power production reduces the net fuel demands for energy generation by supplying otherwise unused heat to residential, commercial, and industrial consumers that have thermal needs. Sections 7.7 and 7.8 cover the integration of fuel cell heat into electric power systems.

11.10 ENERGY STORAGE AS AN ECONOMIC RESOURCE

Any renewable energy project must be assessed on key socioeconomic variables. After deciding on the alternative energy source and corresponding topology, whether grid-connected or stand-alone, the storage system is the next most important technical decision that might affect the overall impact. Issues such as employment, value added, and imports must be considered. For example, in the cases of pumped hydro, CAES, and SMES systems, the neighbor community will be strongly affected, and qualified operating personnel will have to be hired. Biomass

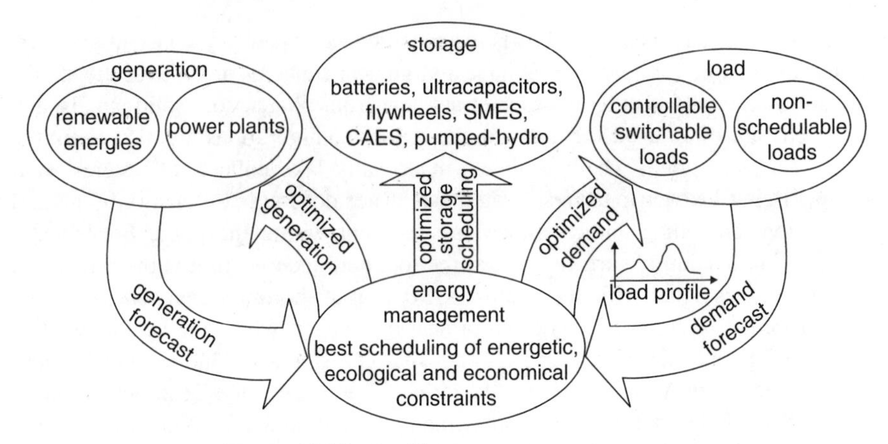

Figure 11.18 Intelligent energy management.

will create important jobs in developing countries, decrease pollution in nearby rivers and the atmosphere, and retain a portion of society in rural areas. Effects on the imports of a country are also relevant because normally, building renewable energy plants requires imports from third countries.

Energy commodities are used to provide energy services such as lighting, space heating, water heating, motive power, and electronic activity. Energy economics studies human use of energy resources and energy commodities and the consequences of that use. Energy resources such as crude oil, natural gas, coal, biomass, uranium, wind, sunlight, and geothermal deposits can be harvested to produce energy commodities. Figure 11.18 is a block diagram of an intelligent energy management system that functions to optimize generation in accordance with forecasting of the renewable energy sources and optimizing demand of controllable and switchable loads in accordance with demand forecasts. The best scheduling of energy, ecological, and economic constraints will define the optimized storage scheduling.

The goal of a module is to optimize energy flows among generation, supply, and points of use to minimize the operating cost or to maximize profit on systems with DG. The output of the module is a set of total energy flows for a given period, issued as a set of vectors that represent recommended energy flows from source to destination during each hour. The optimization module minimizes the cost function for the system. The fixed costs are not considered because they are not influenced by the power dispatch [13,14]. The goal function is made of:

1. The cost of generation by nonrenewable sources and storages
2. The cost associated with the grid
3. The penalty cost

The function of optimization is to minimize the operating cost with DG while maximizing the energy output. As such, it seeks to apply electricity supplied from the generation system or the grid to the load or the grid in the most cost-effective manner. The system can use available storage to offset expensive energy purchases with cheaper energy from an earlier period or to store energy for an anticipated price spike. To capture the dynamics of real-time pricing properly, the system must optimize over a long-enough span of time to ensure reasonable coverage of the real-time energy costs.

The costs of generation include all costs that depend on the energy generated by the renewable sources and storage. The cost of buying or selling energy to a utility is considered while defining the goal function. Violating boundary conditions such as underutilizing a unit or failing to maintain contracts with consumers or a utility causes penalty costs. The penalty cost term in the cost function greatly helps in convergence. Allowing penalty costs for violation of "soft" boundary conditions helps the optimization algorithms find global minimum more rapidly.

Cost of generation:

$$C_G = \sum_{k=1}^{N} \left[\sum_{i=1}^{M} G_{i,k}(\mathrm{CG\$}_{i,k}) \right] \tag{11.26}$$

Cost associated with grid:

$$C_{\mathrm{grid}} = \sum_{k=1}^{N} \left[\mathrm{GP}_k(\mathrm{CGP\$}_k) - \mathrm{GS}_k(\mathrm{CGS\$}_k) \right] \tag{11.27}$$

Penalty costs:

$$C_p = \sum_{k=1}^{N} \left(\sum_{i=1}^{M} \left| G_{i,k}^{\max} - G_{i,k} \right| \mathrm{CP\$}G_{i,k} \right) \tag{11.28}$$

Total costs:

$$C = C_G + C_{\mathrm{grid}} + C_P \tag{11.29}$$

where
C_G = total cost of generation by renewable sources
$G_{i,k}$ = amount of energy generated by source i during period k
$\mathrm{CG\$}_{i,k}$ = unit cost of energy (dollars) generated by source i during period k
C_{grid} = total cost associated with grid
GP_k = amount of energy purchased from grid during period k
$\mathrm{CGP\$}_k$ = unit cost of energy (dollars) purchased from grid during period k
GS_k = amount of energy sold to grid during period k
$\mathrm{CGS\$}_k$ = unit cost of energy (dollars) sold to grid during period k
C_P = total penalty costs
$G_{i,k}^{\max}$ = maximum amount of energy

$CP\$G_{i,k}$ = unit cost of the unused capacity of source i during period k
M = total number of renewable sources
N = total number of periods for which optimization is performed

This minimization process is constrained by the functional and physical limits of the problem. First, enough energy must be supplied from generation or the grid to meet the home load. The storage capacity is constrained by a maximum storage size. The supply of energy from generation is limited to what is available at the time, especially in the case of renewable sources. The transfer of energy to and from the grid is limited by the rating of the electrical service to the household. These constraints can be expressed by the following equations:

$$S_{k+1} = S_k + \sum_{i=1}^{M} G_{i,k} + \mathrm{GP}_k - \mathrm{GS}_k - L_k \tag{11.30}$$

$$S_N = S_0 \tag{11.31}$$

$$0 \leq L_k \leq L^{\max} \tag{11.32}$$

$$0 \leq S_k \leq S^{\max} \tag{11.33}$$

$$0 \leq G_{i,k} \leq G_{i,k}^{\max} \tag{11.34}$$

where S_k and S_{k+1} are storage levels in periods k and $k+1$, L_k the load connected to the microgrid during period k, $L^{\max}$ the maximum possible load, and $S^{\max}$ the maximum amount of storage.

The structure of the constraint equations is such that all energy is routed through storage on its way to a destination. The aim is to include the energy balance, but there are many different ways that could be expressed. Given the complexity of the problem, however, this expression is more intuitively obvious than other methods. Either way, the solution provided by equations (11.25) through (11.34) would be adequate for the situation. It should also be noted that equation (11.31) sets a constraint that storage at the end of N periods should be equal to the starting storage state. This constraint is necessary to provide a bound to the entire sequence of storage states. Where the other constraints provide limits on magnitude of energy in any given period, the constraint in equation (11.30) provides a limit to the entire problem over N periods. The constraint applies only to storage because the energy balance equation is expressed in terms of the state of storage. The cost (objective) function and constraint equations provide the basis for finding an optimal solution. The constraint equations confine the operating region of the system to a N-dimensional hyperspace. Within the hyperspace, every point has an associated cost in accordance with equation (11.29). The goal of the optimization program is to find the single point of operation within that hyperspace that provides the lowest operating cost.

The goal function, C, that is to be minimized emerges from the addition of these three components, as shown in equation (11.29), and it is highly nonlinear, discontinuous, and large in dimensions. The inclusion of storage devices makes the goal function time dependent. The on–off solution, as provided by the mixed integer linear programming, is a popular choice in such an optimization process in a grid-connected network, where it is required to operate the power plants either at the maximum power or switch them off. This cannot be applied to the microgrid, where sometimes the sources, and usually the storages, are to be operated in partial load.

Energy supply is an intricate task that depends on import of goods, seasonal and random differences in energy supply and use, and daily fluctuations in consumption. A very sophisticated management of energy resources and conversion, or energy distribution and resource intermittence, is required. As a result, storage plays a critical role, and energy storage will balance the daily fluctuations and seasonal differences of energy resource availability resulting from physical, economical, or geopolitical constraints.

REFERENCES

[1] J. Jensen and B. Sorensen, Fundamentals of Energy Storage, Wiley, New York, 1984.

[2] S. M. Schoenung and W. V. Hassenzahl, *Long vs. Short-Term Energy Storage Technologies Analysis: A Life-Cycle Cost Study for the DOE Energy Storage Systems Program*, Report SAND2003-2783, Sandia National Laboratories, Albuquerque, NM, and Advanced Energy Analysis, Piedmont, CA, 2003.

[3] Z. M. Salameh, M. A. Casacca, and W. A. Lynch, A mathematical model for lead–acid batteries, *IEEE Transactions on Energy Conversion*, Vol. 7, pp. 93–98, March 1992.

[4] M. S. Illindala and G. Venkataramanan, Battery energy storage for stand-alone microsource distributed generation systems, presented at the 6th IASTED International Conference on Power and Energy Systems, Marina Del Rey, CA, May 13–15, 2002.

[5] M. Okamura and H. Nakamura: Energy capacitor system, parts 1 and 2: Capacitors and their control, presented at the 11th International Seminar on Double Layer Capacitors and Similar Energy Storage Devices, December 3–5, 2001.

[6] A. Rufer and P. Barrade, Key developments for supercapacitive energy storage: power electronic converters, systems and control, in *Proceedings of the International Conference on Power Electronics, Intelligent Motion and Power Quality PCIM 2001*, Nuernberg, Germany, June 19–21, 2001, pp. 17–24.

[7] P. Barrade, S. Pitte, and A. Rufer, Energy storage system using a series connection of supercapacitors with an active device for equalising the voltages, presented at the International Power Electronics Conference, IPEC 2000, Tokyo, April 3–7, 2000.

[8] S. B. Bekiarov, A. Emadi, and M. L. Lazarewicz, Flywheel energy storage systems: competitive alternative of electrochemical batteries, accepted for *IEEE Aerospace and Electronic Systems.*

[9] B. B. Plater and J. A. Andrews, Advances in flywheel energy-storage systems, http://www.powerpulse.net/powerpulse/archive/aa_031901c1.stm.

[10] P. D. Baumaan, Energy conservation and environmental benefits that may be realized from superconducting magnetic energy storage, *IEEE Transactions on Energy Conversion*, Vol. 7, pp. 253–259, June 1992.

[11] V. Karasik, K. Dixon, C. Weber, B. Batchelder, G. Campbell, and P. F. Ribeiro, SMES for power utility applications: a review of technical and cost considerations, *IEEE Transactions on Applied Superconductivity*, Vol. 9, pp. 541–546, June 1999.

[12] A. B. Arsoy, W. Z. Wang, L. Y. Liu, and P. F. Ribeiro, Transient modeling and simulation of a SMES coil and the power electronics interface, *IEEE Transactions on Applied Superconductivity*, Vol. 9, No. 4, pp. 4715–4724, December 1999.

[13] G. Venkataramanan, M. S. Illindala, C. Houle, and R. H. Lasseter, *Hardware Development of a Laboratory-Scale Microgrid Phase 1 Single Inverter in Island Mode Operation*, NREL Report SR-560-32527, National Renewable Energy Laboratory, Golden CO, November 2002.

[14] G. Winkler, C. Meisenback, M. Hable, and P. Meier, Intelligent energy management of electrical power systems with distributed feeding on the basis of forecasts of demand and generation, presented at CIRED 2001, June 18–21, 2001.

CHAPTER 12

INTEGRATION OF ALTERNATIVE SOURCES OF ENERGY

12.1 INTRODUCTION

Integration of alternative sources of energy into a network for distributed generation (DG) requires small-scale power generation technologies located close to the loads served. The move toward on-site distributed power generation has been accelerated because of deregulation and restructuring of the utility industry and the feasibility of alternative energy sources. DG technologies can improve power quality, boost system reliability, reduce energy costs, and defray utility capital investment. Four major issues related to DG are covered in this and next three chapters: hardware and control, effect on the grid, interconnection standards, and economic evaluation.

Alternative sources of energy are either fossil fuel–based (systems that use diesel, gas, or hydrogen from reforming hydrocarbons) or renewable-based (systems that use solar, wind, hydro, tidal, geothermal energy or hydrogen produced from renewable sources). For an energy source to be considered renewable, the recovery cycle of the source must be considered. For example, a sugarcane plantation used for the production of alcohol has a recovery cycle of one year. In comparison, fossil fuels can take millions of years to recover.

Integration of Alternative Sources of Energy, by Felix A. Farret and M. Godoy Simões

The integration of renewable sources of energy poses a challenge because their output is intermittent and variable and must be stored for use when there is demand. If only one renewable energy source is considered, the electric power system is simple. The source can be connected to a storage system to deliver electricity for stand-alone use or interconnected with the grid, as discussed in Chapter 11. In the grid-connected application, the grid acts as energy storage. However, if multiple renewable energy sources are used, the electric power system can be rather complex, and a microgrid will be formed.

The concept of a *microgrid* was developed recently. It is defined as a cluster of loads and DG units that operate so as to improve the reliability and quality of the power system in a controlled manner. For consumers, microgrids can provide power and heat in a reliable way. For the distribution network, microgrids are dispatchable cells that can respond autonomously or from signals from the power system operator. Information technology achievements, along with new DG systems with intelligent controllers, will allow system operators and microgrid operators to interact in an optimal manner [1,2].

In this chapter we first present the concept of integrating one renewable source of energy into the grid and provide a fundamental understanding of interactive inverters. The integration of multiple sources into a microgrid is then covered, and three types of electrical buses for integration (dc, ac, and high-frequency ac) are discussed.

Power electronics is the enabling technology that allows the conversion of energy and injection of power from renewable energy sources to the grid. Therefore, it is suggested that the reader review some background information on power electronics from the technical literature to complement this material.

12.2 PRINCIPLES OF POWER INJECTION

When applying electrical energy conversion technology to renewable energy systems, two classes of electrical systems must be considered: stationary and rotatory. The stationary type usually provides direct current. Photovoltaic arrays and fuel cells are the main renewable energy sources in this group. The rotatory type usually provides alternating current. Induction, synchronous, and permanent-magnet generators are the main drivers for hydropower, wind, and gas turbine energy sources. Dc machines are the rotatory type but are not usually employed because of their high cost, bulky size, and maintenance needs.

12.2.1 Converting Technologies

Figure 12.1 illustrates the main electrical conversion technologies required for injection of renewable energy power. If only photovoltaic and fuel cell systems are used, a dc-link bus might be used to aggregate them. Ac power can be integrated through dc–ac conversion (inverter) systems. If only hydro or wind power is used, a variable-frequency ac voltage control must be aggregated into an ac link through an

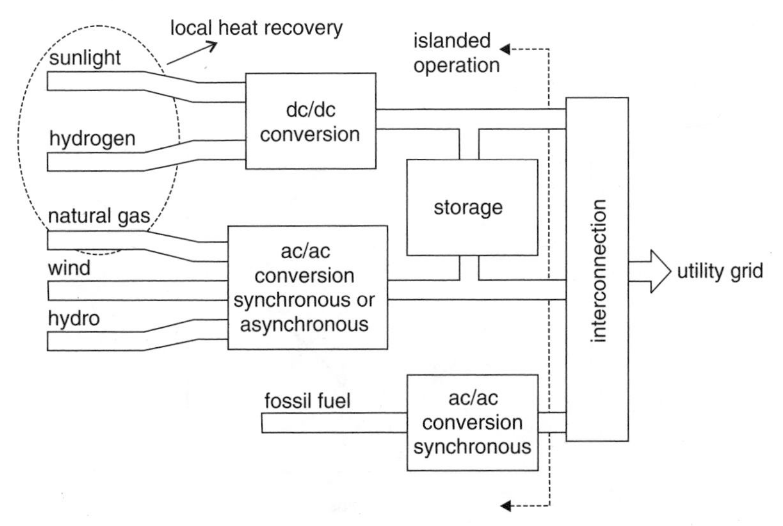

Figure 12.1 Alternative energy conversion technologies for injection of alternative energy power to the grid.

ac–ac conversion system. Ac–ac conversion systems can be put together by several approaches discussed later in the chapter.

Of course, alternative energy sources such as diesel and gas can also be integrated with renewables. They have a consistent and constant fuel supply, and the decision to schedule their operation is based more on straightforward economic reasons than any others. Diesel generators, which are commercially available with synchronous generators that supply 60 Hz, and direct interconnection with the grid, are typically easier to implement. Gas microturbines are most frequently implemented with a power electronic inverter–based technology. When integrating and mixing all those sources, a microgrid has to be based on a dc or ac link. However, it seems promising to use a high-frequency ac link (HFAC), for reasons discussed later in the chapter.

Figure 12.1 illustrates the use of some alternative energy conversion technologies for injection of alternative energy power to the grid, but it does not include how energy storage is integrated. As discussed in Chapter 11, the ultimate design of a microgrid must incorporate energy storage with seamless control and integration of source, storage, and demand and allow for system expansion or interconnection.

The stationary group of renewable energy sources does not absorb power; the energy flows uniquely outward to the load. On the other hand, the rotatory group of renewable energy sources require bidirectional power flow, either to rotate in motoring mode (for pumping, braking, or startup) or to absorb reactive power (for induction generators and transformers). The storage systems can also be stationary (e.g., batteries, supercapacitors, and magnetic coils) or rotatory (generators, flywheels, and pumping hydro). Therefore, a generalized concept of power

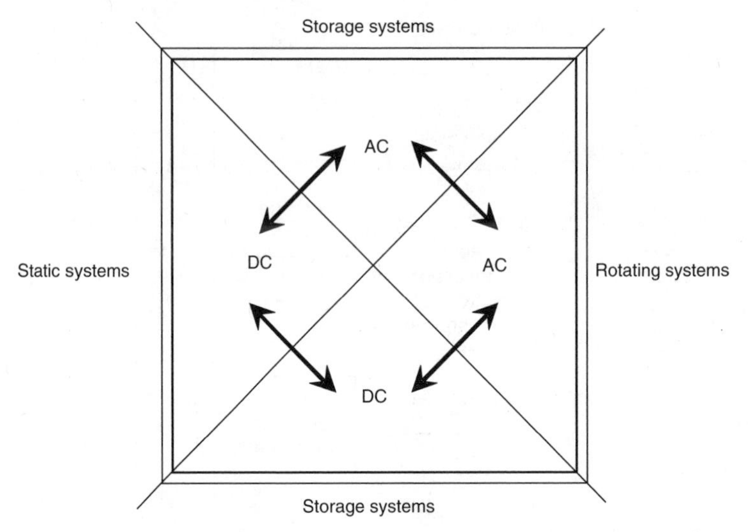

Figure 12.2 Ac and dc conversion needs for renewable energy systems.

electronics is broadly needed for renewable energy systems to embrace such categories, as portrayed in Figure 12.2.

Static systems appear as a dc input converted to a dc or ac output and unidirectional power flow (no moving parts, low time constant, silent, no vibrations, minimally polluting, and low power up to now). The rotating systems appear as an ac input converted to a dc or ac output and bidirectional power flow (angular speed, noise, mechanical inertia, weight and volume, and inrush). Storage systems are necessarily bidirectional systems with requirements of ac and dc conversion (depending on the application). The voltage generated by variable-speed wind power generators, PV generators, and fuel cells cannot be directly connected with the grid. Therefore, power electronics technology plays a vital role in matching the characteristics of the dispersed generation units and the requirements of grid connection, including frequency, voltage, active and reactive power controls, and harmonic minimization.

12.2.2 Power Converters for Power Injection into the Grid

A power electronic converter uses a matrix of power semiconductor switches to enable it to convert sources of stationary electric power into rotating electrical power, and vice versa, at high efficiency. The standard definition of a converter assumes one input source connected to an output source or a load, as indicated in Figure 12.3.

The inherent regenerative capabilities of renewable energy systems require fewer restrictions for input–output or supply–load when connecting two sources. The links between a power converter and the outside world can be perceived as a possibility of power reversal, where the sink can be considered as a negative source,

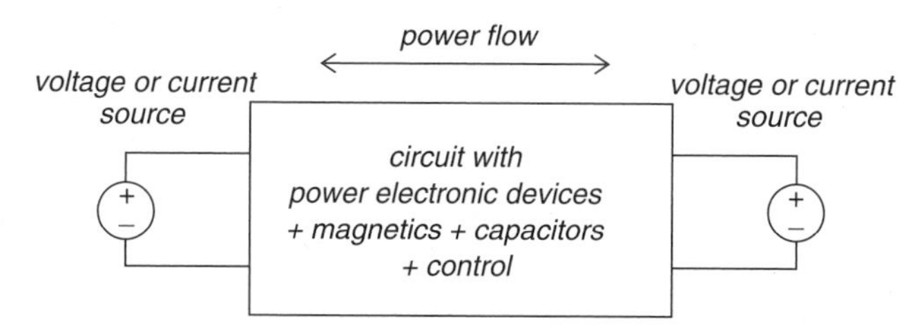

Figure 12.3 Connection of two dc sources.

in accordance with Figure 12.3. Therefore, it is a better fit to define voltage source (VS) and current source (CS) when describing the immediate connection of converters to entry and exit ports.

The *voltage source* is driven to maintain a prescribed voltage across its terminals, irrespective of the magnitude or polarity of the current flowing through the source. The prescribed voltage may be constant dc, sinusoidal ac, or a pulse train. The *current source* maintains a prescribed current flowing through its terminals, irrespective of the magnitude or polarity of the voltage applied across these terminals. Again, the prescribed current may be constant dc, sinusoidal ac, or a pulse train.

Current source inverters (CSIs) have several advantages over variable-voltage inverters because they can operate under a wide input voltage range and a boost converter stage may not be required. They provide protection against short circuits in the output stage and can easily handle temporary overrating situations such as wind gusts or faulty turbines. In addition, CSIs have relatively simple control circuits and good efficiency. As disadvantages, CSIs produce torque pulsations at low speed, cannot handle undersized motors, and are large and heavy. The phase-controlled bridge rectifier CSI is less noisy than its chopper-controlled counterpart, losses are smaller, and it does not need high-speed switching devices, but it cannot operate efficiently from direct dc voltage. The chopper-controlled CSI can operate from batteries and produces more noise as a result of its need for high-speed switching devices. With the advent of fast, high-power controlled devices, VS inverters have become popular, but much R&D is needed before they are applied to very high power systems.

Figure 12.3 shows that a power electronic circuit with magnetic devices and capacitors can control the power flow between two voltage or current sources. Usually, those power-converting circuits are based on the two conventional topologies: VS and CS. A VS converter is typically a dc–dc converter—such as buck, boost, and sepic [4]—or a three-phase inverter [5]. A CS converter is typically used for very high power GTO-based applications with an inductor imposing a constant current in a dc link [6]. For most small- and medium-power applications, VS inverters are used, with some inherent constraints. For example, a VS inverter might be a buck converter that cannot produce an output voltage greater than the source voltage. This limitation often requires an additional dc–dc boost converter for stationary applications such as batteries, photovoltaics, and fuel cells,

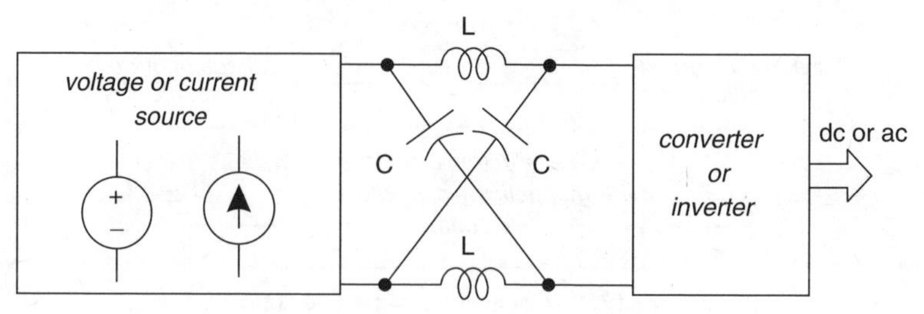

Figure 12.4 Z-source converter system.

which makes the power conversion have lower efficiency, larger size, and higher cost.

A new type of power conversion circuit, known as an *impedance-source* (*Z-source*) *power converter*, is designed to overcome the characteristics associated with the conventional V-source and I-source converters [7]. A Z-source inverter can produce any desired voltage regardless of its input voltage, which greatly reduces system complexity, cost, size, and power loss.

Figure 12.4 shows the concept of a Z-source converter. In it, there is a unique LC network in the dc link, making it possible for buck and boost operation. Even though such topology has gained attention in the past few years (with some interesting features such as application to almost all the conversion circuits: dc-to-ac inversion, ac-to-ac conversion, and dc-to-dc conversion), in this chapter we concentrate on the analysis of dc-link voltage source conversion because of a vast and solid application to the field of renewable energy-based conversion systems. In addition, a Z-source converter has more passive components in the link, and an economical breakthrough analysis must be carried out for high-power applications.

Figure 12.5 shows a renewable energy conversion that consists of a standard VS inverter. L_s represents the inductance between the inverter ac voltage V_{RI} and the ac system voltage V_s, and I_s is the current injected into the grid. The output waveform of a VS inverter can be decomposed into a fundamental component plus harmonics, but for the analysis of power transfer, the resistance and harmonics can be neglected.

12.2.3 Power Flow

Decoupling active and reactive power flow is a common practice in load flow studies. The variation of active power is strongly coupled with power angle and frequency variation, whereas reactive power is strongly coupled with voltage amplitude. The fundamentals of power flow control can be evaluated by the single-phase equivalent circuit of Figure 12.6. Considering for simplicity that $V_{s1} = V_{r1}$, one gets the phasor diagram of Figure 12.7*a* and *b*, and when the output voltage V_{c1} leads by 90° with respect to V_{s1}, current I_1 is in phase with V_{s1}, which causes active power flow from the source to the load. When V_{c1} is in phase with V_{s1}, current I_1 will lag by 90° with respect to V_{s1}, which results in reactive power

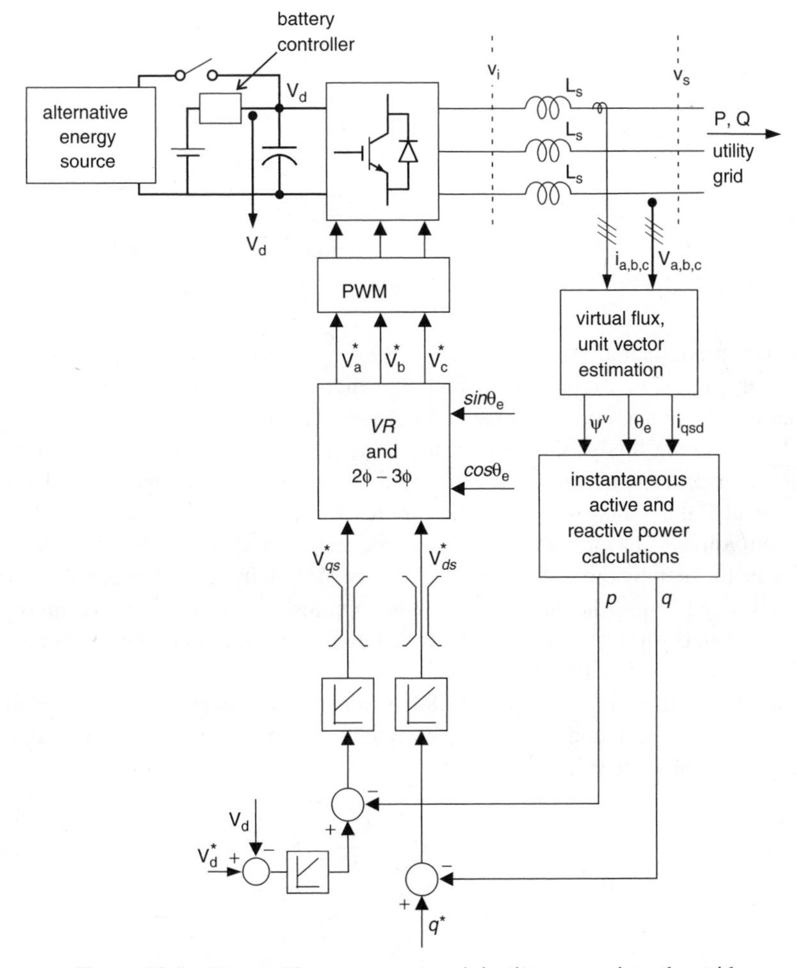

Figure 12.5 Renewable energy system injecting power into the grid.

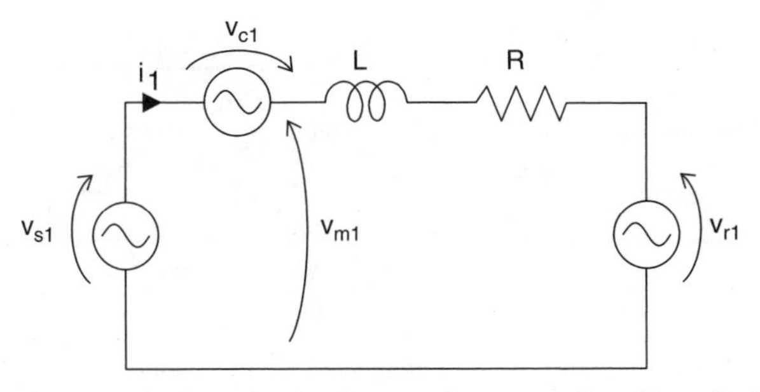

Figure 12.6 Equivalent circuit of power flow controller. (From Ref. [18].)

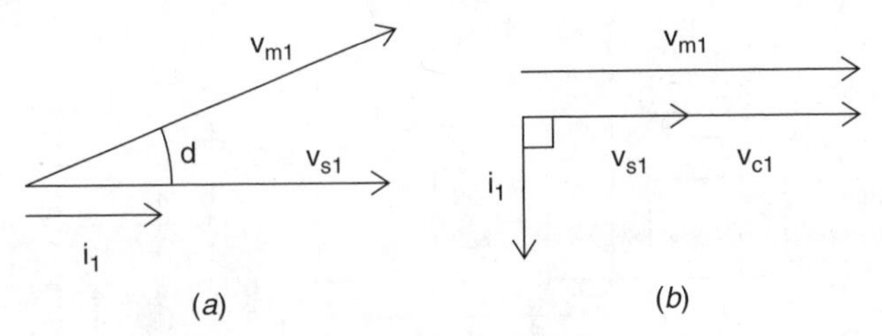

Figure 12.7 Simplified phasor diagrams for power flow: (*a*) active power control; (*b*) reactive power control.

flow. The fundamental voltage and current components can be represented by the phasor diagrams in Figure 12.8*a* and *b*, which correspond, respectively, to the fundamental voltage control mode and current control mode.

The following analysis considers the voltage source inverter (VSI) connected with the grid, as indicated by Figure 12.5. v_s is the grid voltage; v_s and i_s are, respectively, the voltage and current impressed by the inverter; δ is the angle between voltage phasors; and ϕ is the angle between v_s and i_s (i.e., $\cos\phi$ is the power factor seen by the utility). For the sake of simplicity, all voltages and currents are considered sinusoidal, and the discussion applies only to the fundamental of any of the distorted variables. If a more detailed analysis is required, the harmonic distortion must be included in the analysis.

The phasor diagram of Figure 12.8*a* is similar to that of typical synchronous generators. The real and active power injected into the grid by a VSI can be expressed in per-unit values as

$$P_s = P_i = \frac{V_s V_i}{X_s} \sin\delta \tag{12.1}$$

$$Q_s = \frac{V_s V_I}{X_s} \cos\delta - \frac{V_s^2}{X_s} \tag{12.2}$$

where $X_s = 2\pi f_s L_s$ and f_s is the grid frequency.

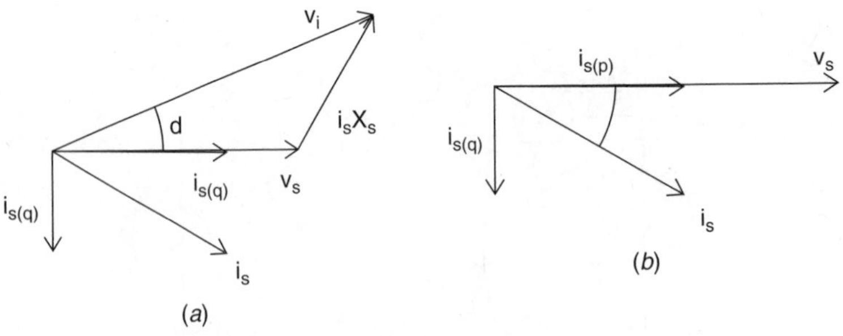

Figure 12.8 Phasor diagrams of grid-connected voltage source inverter: (*a*) voltage-controlled VSI; (*b*) current-controlled VSI.

Equation (12.1) shows that for a given grid voltage V_s and inductance L_s, the real power is P_s. Equation (12.2) shows that reactive power Q_s can be regulated by the magnitude of V_i and power angle δ. In accordance with equation (12.1), the reference for real power P^* and reactive power Q^* can be obtained by setting the following reference power angle δ^* and inverter ac terminal voltage V_i^*:

$$\delta^* = \tan^{-1} \frac{P^*}{Q^* + V_s^2/X_s} \tag{12.3}$$

$$V_i^* = \frac{P^* X_s}{V_s \sin \delta^*} \tag{12.4}$$

A VSI can also operate in closed-loop current-controlled mode [11]. The power control of a current-controlled VSI can be explained with the phasor diagram in Figure 12.8*b*. The power injected into the grid can be expressed in per-unit values as

$$P_s = V_s I_s \cos \phi \tag{12.5}$$

$$Q_s = V_s I_s \sin \phi \tag{12.6}$$

and the reference real and active power can be controlled by regulating the current magnitude and the angle:

$$\phi^* = \tan^{-1} \frac{P^*}{Q^*} \tag{12.7}$$

$$I_s^* = \frac{P^*}{V_s \cos \phi^*} \tag{12.8}$$

12.3 INSTANTANEOUS ACTIVE AND REACTIVE POWER CONTROL APPROACH

If the exact amount of power generated by the renewable energy source is transferred to the grid, as illustrated in Figure 12.9, the power absorbed by the load and the power injected into the grid are generated by the same source. However, if the ac load is absorbing active power P_L, and the reactive power Q_L is not supplied by the VS inverter, the power factor seen by the grid may fall out of the prescribed limits allowed by the utility and possibly force overcharges and penalties to the consumer. To solve this problem, the VSI should supply active power and at the same time, compensate reactive power to restrain the power factor within the limitations. To have a continuous controlled active and reactive power as demanded by a variable load and source conditions, the theory developed by Akagi [13] and applied to renewable energy power injection by Watanabe [14,15] helps the design of a power electronic control system.

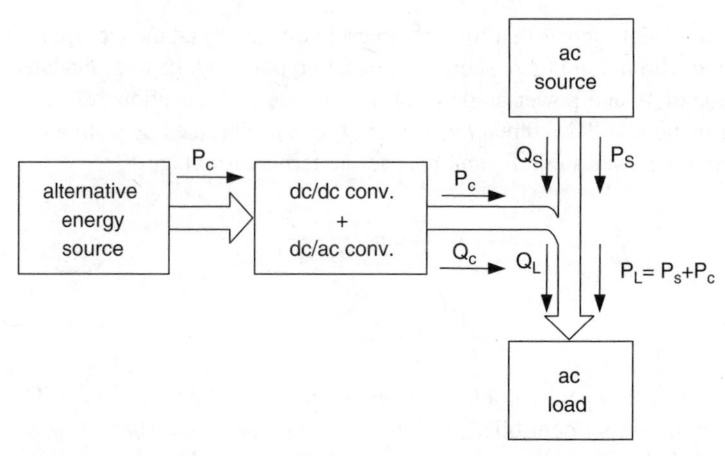

Figure 12.9 Active and reactive power flows from the energy system into the grid.

For a three-phase balanced system, equation (12.9) states that p and q [i.e., the instantaneous real power (W) and instantaneous imaginary power (VA)] are given by the average (i.e., overbar) and oscillatory (i.e., overcaret) components. The oscillatory components are present because of harmonics in the system. The average $\bar{p}$ and $\bar{q}$ are scalar values with same magnitudes as of the phasor quantities P and Q. The voltages and currents in equation (12.9) are obtained from the power invariance of the Clarke transformation, whose terms are indicated by equations (12.10) and (12.11).

$$\begin{bmatrix} p \\ q \end{bmatrix} = \begin{bmatrix} \bar{p} \\ \bar{q} \end{bmatrix} + \begin{bmatrix} \tilde{p} \\ \tilde{q} \end{bmatrix} = \begin{bmatrix} v_\alpha & v_\beta \\ -v_\beta & v_\alpha \end{bmatrix} \begin{bmatrix} i_\alpha \\ i_\beta \end{bmatrix} \tag{12.9}$$

$$\begin{bmatrix} v_0 \\ v_\alpha \\ v_\beta \end{bmatrix} = \sqrt{\frac{2}{3}} \begin{bmatrix} \frac{1}{2} & 1 & 0 \\ \frac{1}{2} & -\frac{1}{2} & \frac{\sqrt{3}}{2} \\ \frac{1}{2} & -\frac{1}{2} & -\frac{\sqrt{3}}{2} \end{bmatrix} \begin{bmatrix} v_a \\ v_b \\ v_c \end{bmatrix} \tag{12.10}$$

$$\begin{bmatrix} v_a \\ v_b \\ v_c \end{bmatrix} = \sqrt{\frac{2}{3}} \begin{bmatrix} \frac{1}{2} & \frac{1}{2} & \frac{1}{2} \\ 1 & -\frac{1}{2} & -\frac{1}{2} \\ 0 & \frac{\sqrt{3}}{2} & -\frac{\sqrt{3}}{2} \end{bmatrix} \begin{bmatrix} v_0 \\ v_\alpha \\ v_\beta \end{bmatrix} \tag{12.11}$$

Equation (12.9) gives the definition of instantaneous real and imaginary power assuming measured voltages and currents. Therefore, to control the VSI connected to the renewable energy source, the current references in the α–β frame must be imposed as defined by equation (12.12) for given set-point values of p_c and q_c. Inversion of the voltage matrix in equation (12.9) yields

$$\begin{bmatrix} i_\alpha^* \\ i_\beta^* \end{bmatrix} = \frac{1}{(v_m)^2} \begin{bmatrix} v_\alpha & -v_\beta \\ v_\beta & v_\alpha \end{bmatrix} \begin{bmatrix} p_c \\ q_c \end{bmatrix} \tag{12.12}$$

where $(v_m)^2 = (v_\alpha)^2 + (v_\beta)^2$.

The pulse-width modulation scheme of the VSI can produce i_α^* and i_β^* and then generate the switching pulses for the transistor by using a hysteresis current controller [14,15], an extra synchronous reference frame *d–q* transformation to impress the voltage command with sinusoidal pulse width modulation through proportional-integral (PI) controllers or space vector modulation [9]. Details of such implementations are outside the scope of this book. The reader can learn more from the available technical literature on power electronics and drives.

The renewable energy source should be injected into the ac system by the pulse-width modulation VSI through appropriate control of the reference currents. Neglecting the harmonics $p_c = p^*$, the real power set point p^* must be impressed by feedforward path (very hard for variable renewable energy source input) or a controller tied to the dc-link capacitor, which is responsible for maintaining the power balance. Of course, a hybrid system with storage would be able to compensate for any power mismatch. This scheme considers the ac grid to be the storage under power import/export net-metering capabilities. The energy stored E_c in the dc-link capacitor C is given by

$$E_c = \int_{-\infty}^{t} (v_d i_s - p_c) d\lambda = \frac{1}{2} C v_d^2 \qquad (12.13)$$

where v_d is the instantaneous dc capacitor voltage (V), i_s is the source current supplied by the renewable energy source, p_c is the instantaneous power at the pulse-width-modulation VSI terminals (W), and λ is a dummy variable for time integration.

Instantaneous active power control can be generated by a feedback loop, in which the capacitor voltage is used to indicate power mismatch. The small signal-transfer function dv/dp indicated in equation (12.14) considers, for a constant input current, how the dc-link voltage deviation responds to small variations of the real power flowing through the dc bus.

$$\frac{\Delta \tilde{v}_d(s)}{\Delta \tilde{p}_c(s)} = \frac{1}{sCV_d} \qquad (12.14)$$

V_d is the quiescent voltage at the dc-link capacitor. The stationary renewable energy source is typically associated with a dc–dc converter. For example, photovoltaic arrays are connected through a maximum peak power-tracking converter, and fuel cells are connected through a boost converter with optimized balance of plant control. Therefore, it is possible to assume that the i_s input source current can be modeled by derivation of the control law as a current source. Therefore, dc-link voltage raising from a reference value indicates a surplus of generated input power, and decreasing voltage indicates a lack of input power. The PI controller depicted in Figure 12.10 shows that the measured dc-link voltage v_d is compared with the set point v_d^*, and the error is fed back through a PI controller that automatically produces the tracking real power $p* = p_c$. By properly fine-tuning the PI gains, the

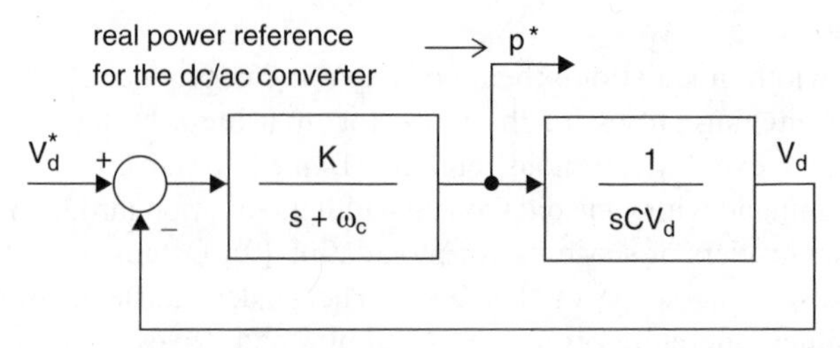

Figure 12.10 Real power tracking by dc-link voltage control.

intrinsic low-pass filter behavior will eliminate the switching harmonics, and the dc-link capacitor voltage will have a time response compatible with the power balance.

The VSI must also program the amount of reactive power by impressing a q^* set point in accordance with

$$q^* = q_{\text{load}} - (p_{\text{load}} - p^*) \tan \phi^* \tag{12.15}$$

in which $\cos \phi^*$ is a prescribed power factor as seen by the utility grid. The pulse-width modulation VSI must be rated in accordance with the apparent power required by the total power flow. Equation (12.16) can be used to specify the capacity of such converters:

$$S_{\text{converter}} = \sqrt{[q_{\text{load}} - (p_{\text{load}} - p^*) \tan \phi^*]^2 + (p^*)^2} \tag{12.16}$$

12.4 INTEGRATION OF MULTIPLE RENEWABLE ENERGY SOURCES

The integration of multiple renewable energy sources can be considered a facet of DG. As introduced in Chapter 1, DG systems consist of small generators, typically 1 kW to 10 MW, scattered throughout the system to provide electrical and, sometimes, heat energy close to consumers. When interconnected with distribution systems, these small, modular generation technologies can form a new type of power system, the microgrid. Microgrids are broadly illustrated in Figure 12.11.

The microgrid concept assumes a cluster of loads and microsources operating as a single controllable system that can provide power and heat to the local area. The electrical connection of sources and loads can be done through a dc link, an ac link, or an HFAC link. Converters are usually connected in parallel, although series arrangements of sources and loads are possible to allow better use of high voltages and currents. An example of series connection is a group of small run-of-river hydroelectric systems that drive induction generators connected in. All configurations require a controlled voltage at the load link bus that is capable of supplying

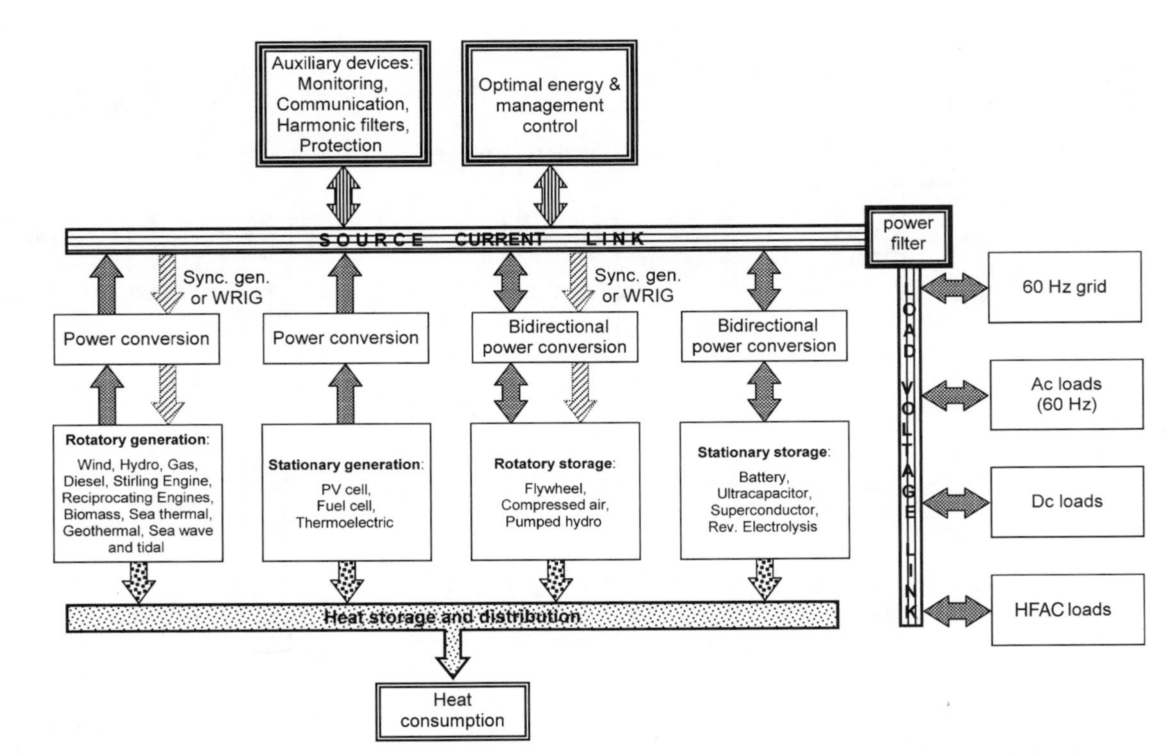

Figure 12.11 General schemes for integration of electrical sources and loads and thermal dissipation.

power to all the loads with reasonable power quality. On the other hand, a controlled current source is more convenient to adjust the voltage of every electrical source to the common voltage of the source bus, according to the strategy of power control.

In the configurations depicted in Figure 12.12, thyristors or GTOs are still recommended for their robustness, guaranteed reliability, handling of higher power levels, low cost, easy control, and high efficiency. The major problem with these

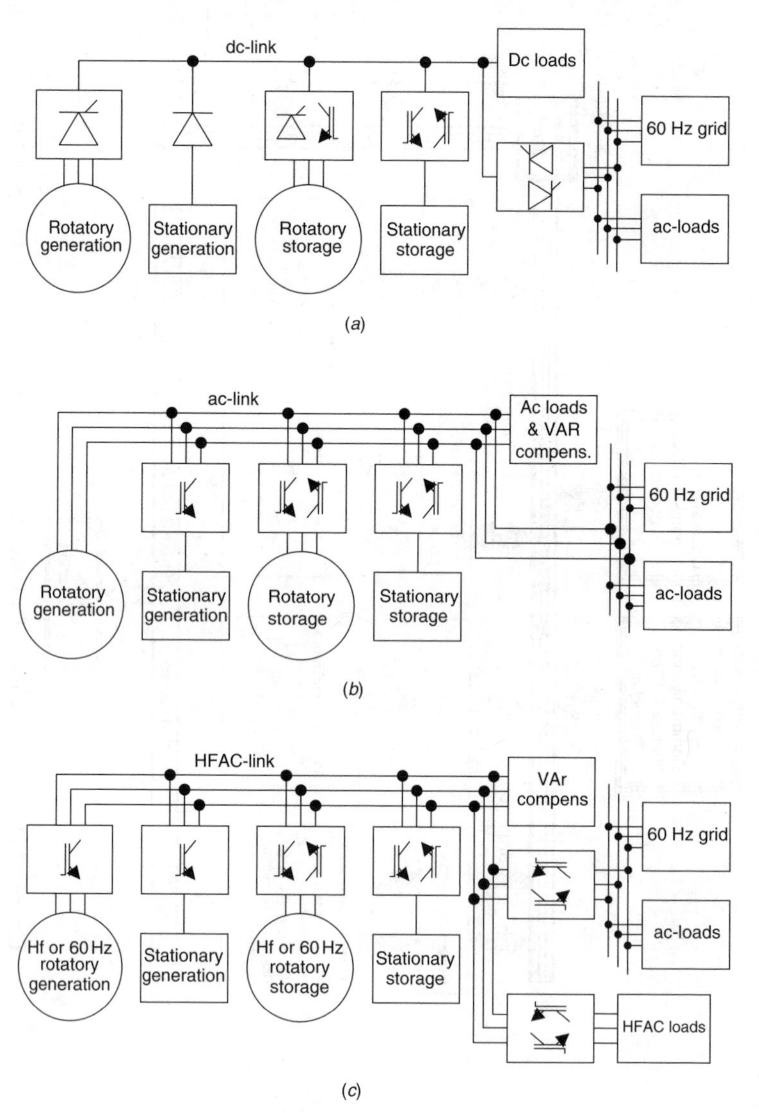

Figure 12.12 Varieties of energy integration: (*a*) dc-link integration; (*b*) ac-link integration; (*c*) HFAC-link integration.

components is that they cannot operate easily with frequencies much higher than 1 kHz. Furthermore, GTOs demand a high control current to interrupt their load currents. Even so, for very high-power applications, they are absolute in the industry, and they will be preferred in the discussions below.

12.4.1 DC-Link Integration

The simplest and oldest type of electrical energy integration is through a dc link (see Figure 12-12*a*). An example is the straightforward connection of a DC source to a battery and load scheme. Most of the first cars used this type of integration. With the advent of power diode rectifiers and controlled rectifiers, dc-link integration has widened its capabilities with ac–dc connections, controlled links, voltage-matching levels, and dc transmission and distribution.

The advantages of dc-link integration include:

1. Synchronism is not required.
2. There are lower distribution and transmission losses than with ac-link.
3. It has high reliability because of parallel sources.
4. Although the terminal needs of a dc link are more complex, the dc transmission infrastructure per kilometer is simpler and cheaper than in ac links.
5. Long-distance transmission (for high-voltage links) is possible, which enables integration of offshore wind turbines and other energies with inland networks.
6. The converters required are easily available.
7. Single-wired connections allow balanced terminal ac systems.

The disadvantages are:

1. The need for careful compatibility of voltage levels to avoid current recirculation between the input sources
2. The need for robust forced commutation capabilities in circuits at high power levels
3. Corrosion concerns with the electrodes
4. A large number of components and controls
5. More complex galvanic isolation
6. Higher costs of terminal equipment
7. Difficulties with multiterminal or multi-voltage-level operation for transmission and distribution

In the rotatory generation with synchronous or induction generators of Figure 12.12*a*, two separate converters would be necessary to feed the dc link. For stationary generation, it is used only in a diode scheme to prevent backfeeding to the dc source. In some cases, such as for fuel cell stacks, a dc–dc converter with filtering features is required for voltage matching and smoothing out undesirable ripples. The dc–ac converter connecting the dc link to the 60-Hz grid is necessary if the grid is used to store energy; otherwise, it could be an ordinary thyristor inverter. Control is exerted individually on each source because current contribution to the dc link can occur just by raising the output voltage above other sources.

The same solution can be used to store energy in a battery, making sure that the battery can supply energy when the bus voltage is lower or receive energy when the bus voltage is higher. A typical case of a current control source is that of a small wind power plants for a boat or remote load. Other alternatives have to be adapted according the needs of the generating system.

12.4.2 AC-Link Integration

Another possibility for integrating renewable energy sources is the ac-link bus operating at either 50 or 60 Hz, as indicated in Figure 12.12*b*. This bus can be the public grid or a local grid for islanded operation. In this case, the interconnection of a local bus to the public bus must follow the standards and requirements for interconnection (Institute of Electrical and Electronics Engineers standards and others; see Chapter 14).

Positive features of this configuration include:

1. Utility regulation and maintenance of the operational voltage, which makes it easier to inject power into the grid
2. The possibility, in some cases, of eliminating the electronic converters (e.g., by using synchronous generators or induction generators that establish their own operating point)
3. Easy multivoltage and multiterminal matchings
4. Easy galvanic isolation
5. Well-established scale economy for consumers and existing utilities

Negative points include:

1. The need for rigorous synchronism, voltage-level matching, and correct phase sequence between sources during interconnection as well as during operation
2. Leakage inductances and capacitances in addition to the skin and proximity effects that cause losses in long distributions
3. Electromagnetic compatibility concerns
4. The possibility of current recirculation between sources
5. The need for power factor and harmonic distortion correction
6. Reduced limits for transmission and distribution

The rotatory generation in Figure 12.12*b* is feeding the ac link to the grid directly. The stationary generation will need a dc–ac inverter or, in some cases, a dc–dc booster and then a dc–ac converter to allow minimum levels of efficient conversion. No dc loads are assumed in this case. Volt-amperes reactive compensation and harmonic compensation are easily implemented in this scheme using the storage facilities and the *p–q* theory concepts, as explained in Section 12.3.3.

12.4.3 HFAC Link Integration

An HFAC microgrid system is a power electronics solution that is promising as an interface for the utility grid and stand-alone operation. It provides embedded fault protection, small dimensions, and a configuration to serve various power quality functions. Frequencies in industrial use are 400 Hz, as is the case in spaceships, boats, buses, planes, submarines, and other loads of the kind. Researchers are investigating increased levels of these frequencies for high-power electronic circuits and devices.

Power electronic-based systems connected with the grid can be controlled and monitored remotely to allow real-time optimization of power generation and power flow, which allows aggregation of distributed power generation resources into a "virtual utility." Several HFAC power distribution systems were implemented for aerospace applications, and NASA evaluated this scheme for space station applications [10,11]. It has been proposed as a workable power distribution system for hybrid electric vehicles as well [12] using traditional ac–ac conversion through semiconductor switches. This is performed in two stages, from ac–dc and then from dc–ac (dc-link converters), or directly by cycloconverters. A microgrid is illustrated in Figure 12.12*c*, and details are provided in Figure 12.13.

Historically, cycloconverters have been used in high-power applications driving induction and synchronous motors, using phase-controlled thyristors, because of their easy zero-current commutation features [13]. The cycloconverter output waveforms have complex harmonics. Higher-order harmonics are usually filtered by machine leakage inductance. In a cycloconverter, there are no storage devices; therefore, instantaneous output power equals input power minus losses. To maintain this balance on the input side with sinusoidal voltages, the input current is expected to have complex harmonic patterns.

The matrix converter is another technology under development. First proposed in the early 1980s, it consists of a matrix of ac switches that connects the three input phases to the three output phases directly [14]. This circuit allows any of the input phases to be connected to any output phase at any instant, depending on the control strategy. Power semiconductor manufacturers are expending efforts to come up with modern, fast-power semiconductor devices (ac switches) capable of symmetric-voltage blocking, gate controlled or commanded by either zero-current or zero-voltage conditions. It is expected that these devices will accelerate the widespread use of matrix converters to convert HFAC into utility voltage and frequency.

Series resonant converters, using zero-voltage or zero-current switching, can be used with each of the sources to generate the HFAC link [10]. In this way,

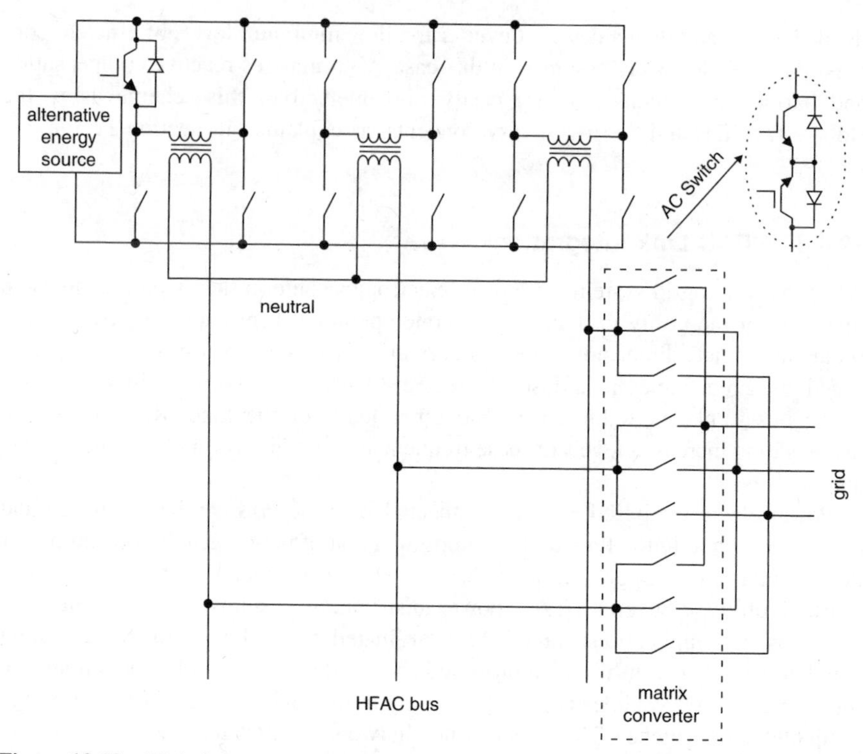

Figure 12.13 High-frequency ac microgrid for integration of alternative sources of energy.

the overall losses in the converters can be reduced. An HFAC microgrid system has the following inherent advantages:

1. The harmonics are of higher orders and are easily filtered out.
2. Fluorescent lighting will experience improvement because, with higher frequency, the luminous efficiency is improved, flicker is reduced, and dimming is accomplished directly. The ballast inductance is reduced proportionally to the frequency, with a corresponding reduction in size and weight.
3. High-frequency induction motors can be used for compressors, high-pressure pumps, high-speed applications, and turbines. Ac frequency changers based on matrix converters can be used to soft start high-frequency induction motors. A safe operating area is not a restriction for soft switching, and therefore modern power electronic devices will be advantageous.
4. Harmonic ripple current in electric machines will decrease, improving efficiency.
5. High-frequency power transformers, harmonic filters for batteries, and other passive circuit components become smaller.
6. Auxiliary power supply units are easily available by tapping the ac link. They would be smaller with higher efficiencies.

7. Batteries have been the traditional energy storage source, but in HFAC microgrids, dynamic storage is an alternative.

The disadvantages are:

1. The high cost of transformers
2. The large number of devices (because of the use of bipolar ac switches)
3. Very complex control
4. The dependence on future advances of power electronic components
5. Concerns about electromagnetic compatibility
6. Extremely reduced limits and technological problems for transmission and distribution at high frequencies

When a high-frequency ac microgrid is adopted, all the active and reactive power flow inside the grid must be considered, and a high-level hierarchical control, such as the one depicted in Figure 12.14, must be adopted to control the voltage and reactive power exchange with the interacting grid. For each energy source, HFAC loads and the 60-Hz grid or other HFAC microgrids will have a power electronic-controlled unit. At the coupling point, the active and reactive power injection or demand can be calculated by Akagi's *p–q* theory or by measuring the voltage angles

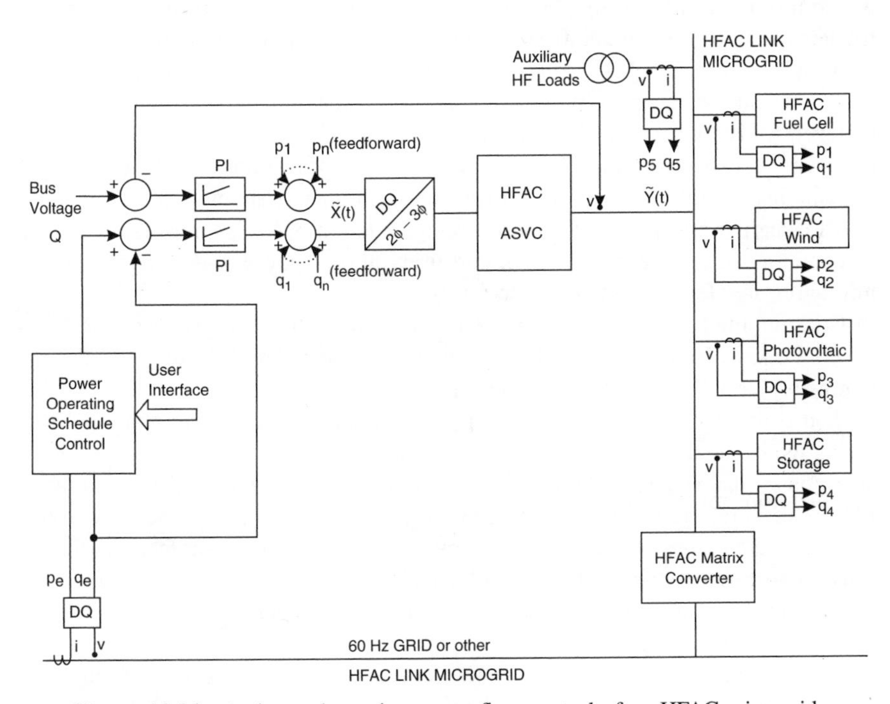

Figure 12.14 Active and reactive power flow control of an HFAC microgrid.

and computing the equations for incoming bus i:

$$P_i = \sum_{\substack{j=1 \\ j \neq i}}^{n} V_i V_j B_{ij} \sin(\delta_i - \delta_j) \tag{12.17}$$

$$Q_i = \sum_{j=1}^{n} V_i V_j B_{ij} \cos(\delta_i - \delta_j) \tag{12.18}$$

where V_i and V_j are the voltage magnitudes at buses i and j, B_{ij} is the line susceptance (inverse of reactance) between buses i and j, and δ_i and δ_j are the voltage angles at buses i and j.

12.5 ISLANDING AND INTERCONNECTION CONTROL

An important issue related to microgrids is *islanding*. This condition occurs when the microgrid continues to energize a section of the main grid after that section has been isolated from the main utility. Unintentional islanding is a concern for the utility because sources connected to the system but not controlled by the utility pose a possibility of harm to utility personnel and damage from uncontrolled voltage and frequency excursions. In addition, utility equipment such as surge arresters may be damaged by overvoltages during a shift-neutral reference or resonance. Furthermore, special synchronization and voltage-level matching measures will have to be taken for reconnection of the utility power. Islanding can lead to asynchronous reclosure, which can damage equipment. It is therefore important that microgrid systems incorporate methods to prevent unintentional islanding. However, intentional islanding may occur when a secure aggregation of loads and sources capable of operating in parallel with, or to, end users allows power flow only within the microgrid or safely exports it to the utility grid.

Consider the configuration shown in Figure 12.15. The microgrid is connected to a feeder line with a local load, which is, in turn, connected to the utility grid through a transformer and some sort of switch (a recloser, breaker, or fuse). If the switch is opened under certain conditions, it is possible for the microgrid to continue energizing the isolated section of the grid and to supply power to the local load. This is an unintentional island, and the isolated section of the utility being powered by the PV system is referred to as an *island of supply* or, simply, an *island*. Because of this, grid-connected alternative energy systems are required to have protective relays to sense conditions of over and under voltage and frequency and to signal the unit to disconnect the microgrid from the utility in the event that the magnitude or frequency goes beyond limits.

Now consider Figure 12.15. When the separation device is closed, real and reactive power P_{mic} and Q_{mic} flow from the microgrid to node a. Power P_{load} and Q_{load}

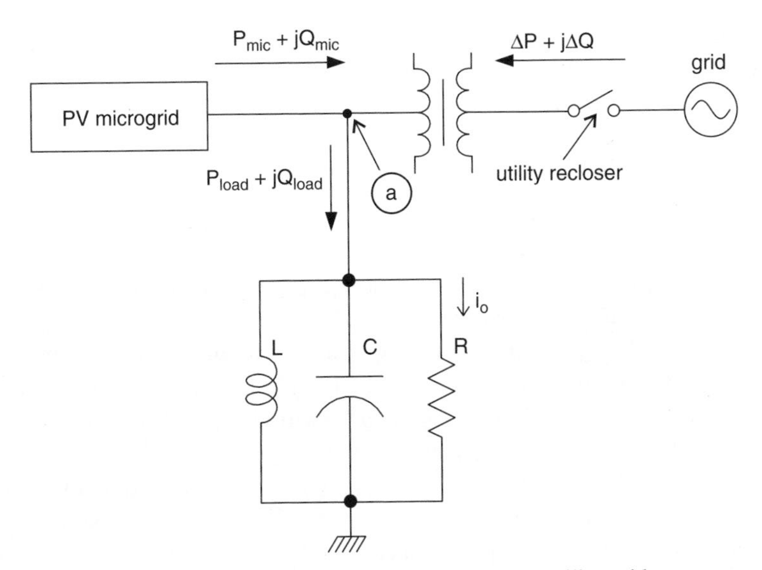

Figure 12.15 Microgrid power flow to the utility grid.

flow from a to the load, and the utility provides

$$\Delta P = P_{\text{load}} - P_{\text{mic}} \tag{12.19}$$

$$\Delta Q = Q_{\text{load}} - Q_{\text{mic}} \tag{12.20}$$

The microgrid and any local load are considered a lumped load by the utility. The frequency dependence on the load can be approximated by equation (12.21), where D is typically within 4% of the base frequency. The utility generators will have a frequency variation in accordance with equation (12.22). In it, the constant R is defined by utility operational policies, and Δf is the frequency regulation constant or frequency droop, normally given in MW/Hz.

$$\Delta P_{\text{load}} = D\Delta f \tag{12.21}$$

$$\Delta P_{\text{grid}} = -R\Delta f \tag{12.22}$$

Assuming that the load may have a leading or lagging power factor, an RLC circuit, as shown in Figure 12.15, can represent such an operation. The following equations relate the active and reactive power at the load with the voltage at node a:

$$P_{\text{load}} = \frac{V_a^2}{R_{\text{load}}} \tag{12.23}$$

$$Q_{\text{load}} = \frac{V_a^2}{\omega L} - \frac{V_a^2}{1/\omega C} \tag{12.24}$$

When the separation device opens, the grid active power change ΔP and reactive power change ΔQ will go to zero. The behavior of the isolated system will depend on ΔP and ΔQ at the instant before the separation device opens to form the island, denoted by ΔP_{-} and ΔQ_{-}.

There are four cases in which the overvoltage relay (OVR), undervoltage relay (UVR), overfrequency relay (OFR), or underfrequency relay (UFR) can prevent unintentional islanding:

1. $\Delta P_{-} > 0$, in which the microgrid system produces less real power than is required by the local load ($P_{\text{load}} > P_{\text{mic}}$). Therefore, when the switch opens and ΔP becomes zero, P_{load} will decrease, and voltage V_a at node a will decrease because R_{load} can be assumed to be constant over this short time span. This decrease can be detected by the UVR, and islanding is prevented.
2. $\Delta P_{-} < 0$, in which power flows into the utility ($P_{\text{load}} < P_{\text{mic}}$). When ΔP becomes zero, P_{load} will decrease, and V_a will increase. This condition can be detected by the OVR, and again islanding is prevented.
3. $\Delta Q_{-} > 0$, which corresponds to a lagging power factor load or a load whose reactive component is inductive. After the separation device opens, $\Delta Q = 0$. For a microgrid unity power factor operation at the point of common coupling, $Q_{\text{mic}} = 0$ and $Q_{\text{load}} = 0$. In accordance with equation (12.24), the right-hand term has to become zero, meaning that the inductive term must drop and the capacitive term must increase. That occurs for the frequency ω increasing, which can be detected by the OFR.
4. $\Delta Q_{-} < 0$, which corresponds to a leading power factor load or one that is primarily capacitive. When ΔQ becomes zero, the inductive and capacitive terms of equation (12.24) must balance, requiring the frequency ω to decrease, which can be detected by the UFR.

There are other possible cases when ΔP_{-} or ΔQ_{-} is zero. One is a case in which the microgrid power production is matched to the load power requirement and the load displacement factor is unity. In this case, when the switch is opened, no change occurs in the isolated system, and the OVR/UVR and OFR/UFR will not detect any voltage or frequency deviation. The magnitude and frequency of the utility voltage can be expected to deviate slightly from nominal values, and therefore the thresholds for the four relays cannot be set arbitrarily small, or the microgrid system will be subject to nuisance trips. This limitation leads to the formation of a nondetection zone. The probability of ΔP_{-} or ΔQ_{-} falling into the nondetection zone of the OVR/UVR and OFR/UFR is significant. It is therefore important that microgrid systems incorporate methods to prevent unintentional islanding in the case in which the microgrid is injecting a lagging or leading power factor or when $\Delta P_{-} \approx 0$ or $\Delta Q_{-} \approx 0$.

When the separation device opens to isolate a utility fault, the DG units indicated in Figure 12.16 must immediately share the increased power demand in a predetermined manner among themselves to continue supplying power adequately to all critical loads within the microgrid. This sharing of power can be achieved with no physical communication links between the DG systems by introducing artificial *real power versus supply frequency* and *reactive power versus voltage* droop characteristics into the DG controllers.

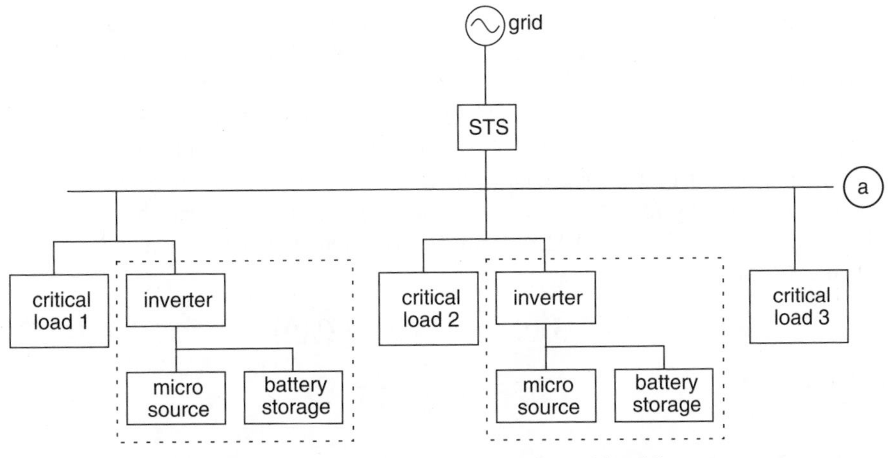

Figure 12.16 Microgrid interaction with grid.

As an illustration, consider the droop characteristics of the two DG systems represented in Figure 12.17. The droop characteristics should be coordinated to make each DG system supply real and reactive power in proportion to its power ratings. It is expressed mathematically as

$$\omega_j(t) = \omega^* - R\lfloor P_j^* - P_j(t)\rfloor \tag{12.25}$$

$$R = \frac{\omega^* - \omega_{\min}}{P_j^* - P_{j,\max}} \tag{12.26}$$

where $P_j(t)$ is the actual real power output of the DG system j (in accordance with Figure 12.16, $j = 1$ or 2), and $P_{j,\max}$ and $\omega_{\min}$ are the maximum real power output of the DG system and the minimum allowable operating frequency. P_j^* and $\omega*$

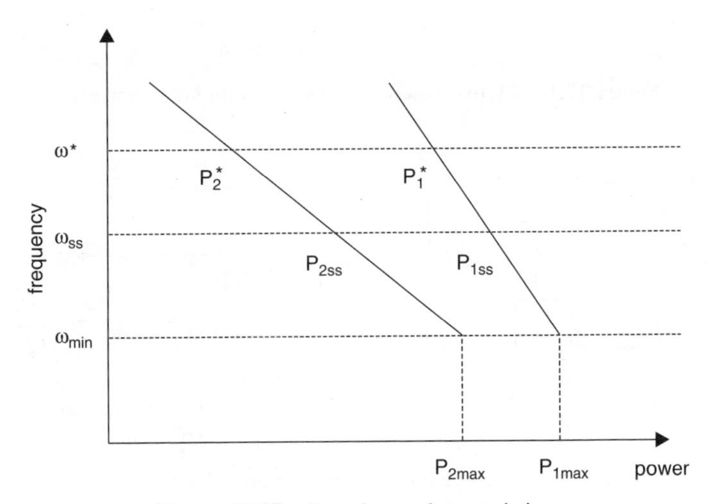

Figure 12.17 *P*–ω droop characteristics.

are, respectively, the dispatched real power and operating frequency of the DG system j when in the grid-connected mode, and $R < 0$ is the slope of the droop characteristic.

A block diagram of equations (12.25) and (12.26) is shown in Figure 12.18. Similarly, the magnitude set point of each DG output voltage can be tuned according to a specified Q–V droop characteristic to control the flow of reactive power within the microgrid. Mathematically, those characteristics can be expressed by equations (12.27) and (12.28), with a block realization function indicated in Figure 12.20.

$$V_j(t) = V^* + \varepsilon \lfloor Q_j^* - Q_j(t) \rfloor \tag{12.27}$$

$$\varepsilon = \frac{V^* - V_{\min}}{Q_{j,\max} - Q_j^*} \tag{12.28}$$

When the utility grid returns to normal operating conditions, the microgrid has to resynchronize with it for reconnection. Synchronization can be achieved by aligning the voltage phasors at the microgrid and utility ends of the separation device. This can be implemented conveniently by adding two separate synchronization compensators to the external real and reactive power control loops, as shown in the dashed frames in Figures 12.18 and 12.19. Inputs to these synchronization

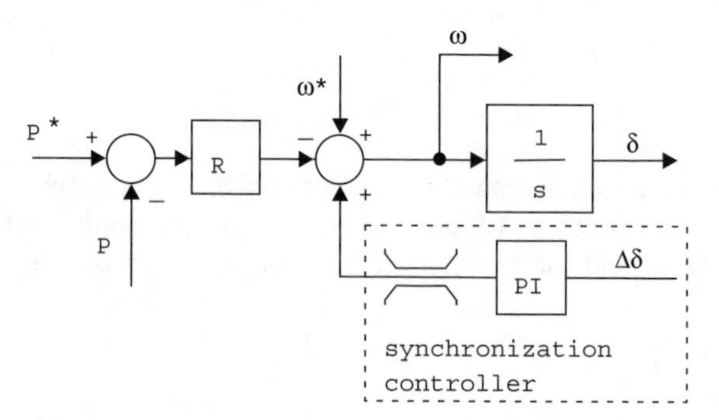

Figure 12.18 Active power compensator with synchronization.

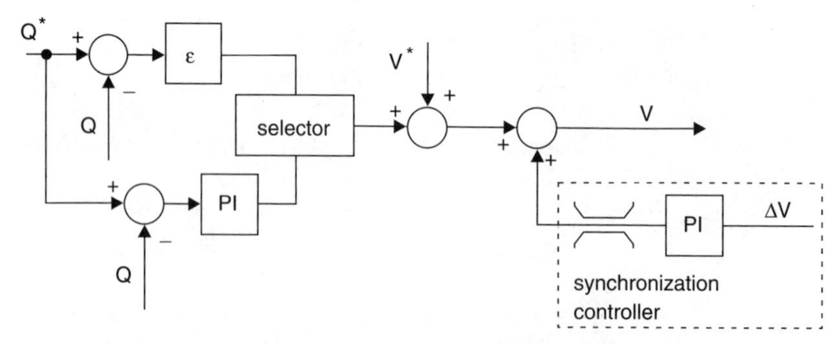

Figure 12.19 Reactive power compensator with synchronization.

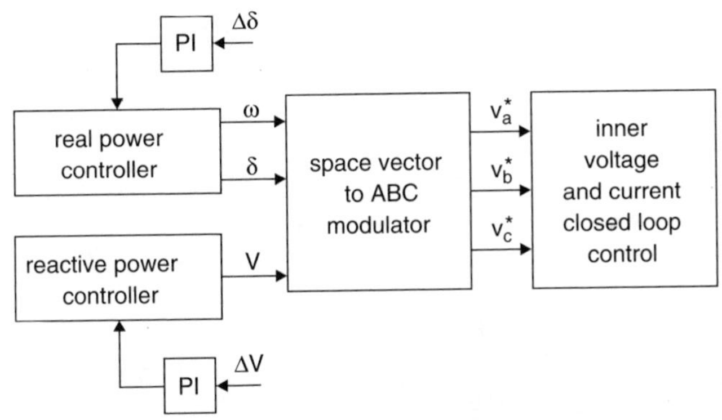

Figure 12.20 Unified power and reactive microgrid power controller with synchronization.

compensators are the magnitude and phase errors of the two voltage phasors at both ends of the separation device. Their outputs are fed to the real and reactive power loops to make the voltage phasor at the microgrid end closely to track the phasor at the utility end (in both magnitude and frequency). Once synchronized and upon closing the separation device, the synchronization compensators must be deactivated by setting their outputs to zero so they will not interfere with proper operation of the real and reactive power control loops in grid-connected mode. The droop control requires that each DG output voltage be different from that of the utility grid to allow for proper reactive power control in the grid-connected mode.

The inverter at the microgrid side can incorporate the active and reactive power controller, as depicted in Figure 12.20. When connected to the utility grid, only the PI compensator is selected for reactive power control, which forces the DG reactive power output to track its desired value to a zero steady-state error. For this PI compensator, low gains are used to give a sufficiently long response time, which, in turn, allows the dynamics of the inner voltage and current loops and external power loops to be decoupled. When a utility fault occurs and the microgrid "islands," the control switches over to select the droop characteristics for reactive power sharing between the DG systems, which ensures a smooth transition from grid-connected to islanding mode.

12.6 DG CONTROL AND POWER INJECTION

A microgrid can be implemented by combining loads with sources, allowing for intentional islanding and using available waste heat. Figure 12.21 shows how the traditional role of central generation, transmission, and distribution is transformed by aggregation of distributed resources, which results in a microgrid architecture. There is a single point of connection to the utility called the *point of common coupling*. In the microgrid, some feeders can have sensitive loads that require local generation. Intentional islanding from the grid is provided by static switches that

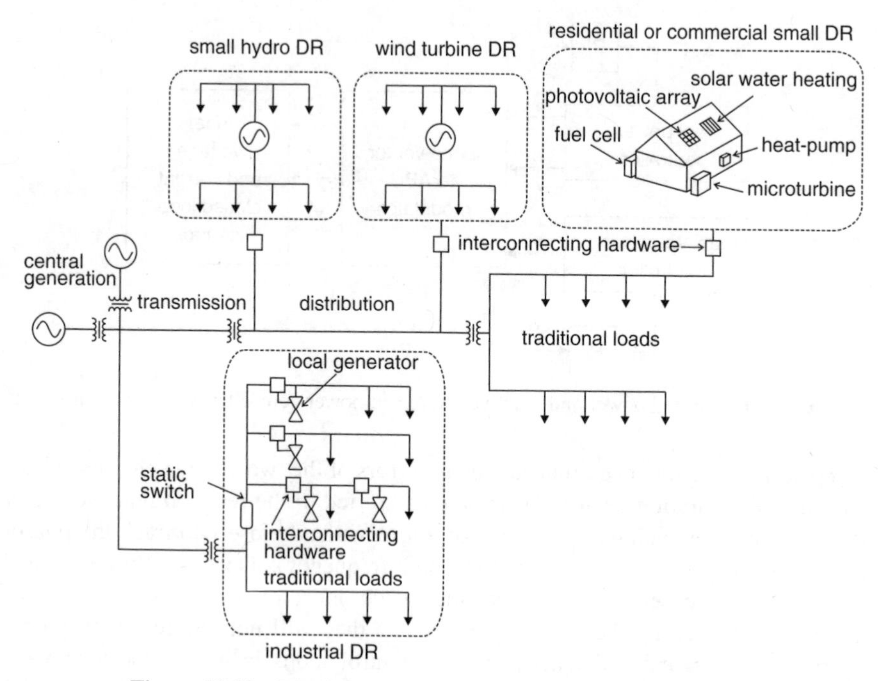

Figure 12.21 Distributed resources shaping the traditional grid.

can separate them in less than a cycle. When the microgrid is connected, power from local generation can be directed to the feeder with noncritical loads or be sold to the utility if agreed or allowed by net metering. In addition to allowing better efficiency, waste recovery, and tailored reliability, a microgrid is designed for the requirements of end users, a stark difference from the central generation paradigm. Key to this characteristic is the reliance on the flexibility of advanced power electronics that control the interface between distributed resources and their surrounding ac system. In Figure 12.21, the term *distributed resource* (DR) is used for units that participate in a microgrid.

In addition to the power electronics control, a microgrid requires operational control to ensure economic commitment and dispatch within environmental and other constraints. In a DG power system, there are needs of coordination of control layers. A hierarchical system consists of several decision-making components and has an overall goal, which is distributed among its individual components. The levels of the hierarchy exchange information (usually vertically) among themselves in an iterative mode, and as the level increases, the time horizon increases (i.e., lower-level components or modules are usually faster than their higher-level counterparts). In a multilayer hierarchical structure, the first layer acts as the regulation or direct control layer. It is followed by optimization, adaptation, and self-organization functions.

A multiechelon structure (Figure 12.22) consists of a number of subsystems situated in levels such that each one can coordinate lower-level units while it is itself coordinated by a higher-level unit. Among the particular tasks of higher-

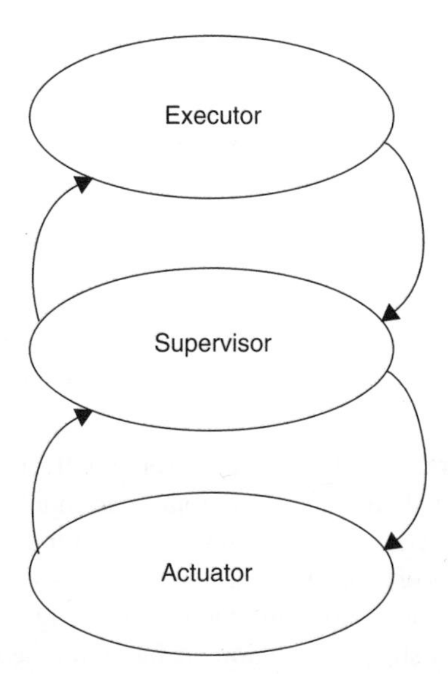

Figure 12.22 Hierarchical control of DG systems.

level echelons is conflict resolution in achieving specified objectives. The most time-demanding level of control, the first layer (actuator), is related to the switching of high-power transistors in the power electronic converters. This task requires high-speed, real-time control with sampling rates on the order of microseconds. Vector control, space vector, pulse-width modulation, current regulation, suppression of harmonics, and power electronic device protection are within this layer.

A second level of control (supervisor) is required to manage the system; to generate power set points to control the power flow among the energy source, energy storage, and load; to control dc or ac bus voltage; and to monitor faulty signals. The particular signals to be controlled depend on the specific DG technology, but sampling rates on the order of milliseconds are typically required.

The third level of control (executor) is related to communications to external equipment and the outside world, which provides a variety of remote monitoring and control. The executor level is responsible for implementation of control schemes to produce as much energy from the system as possible to recover the installation cost. Policies related to the costs of fuel, maintenance, and negotiation with neighbor sites are implemented in this level. Typically, sampling times are on the order of minutes, and hours are required to implement such policies.

A crucial and usually overlooked aspect of DG system control is the operation of these three levels under a real-time structure. The concepts of software design and hardware integration are very complex and require close interaction and cooperation of engineers with very diverse areas of expertise. If done well, a high-performance and highly integrated system will result.

Conventional power systems with multiple generators use a load-sharing technique (droop scheme) in which the generators share the system load by drooping the

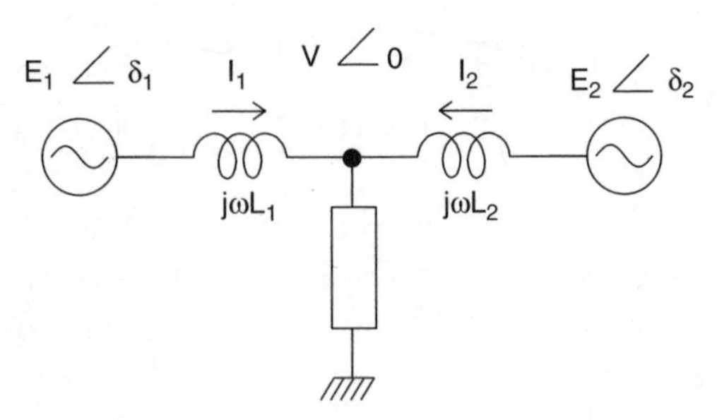

Figure 12.23 Parallel connection of two power converters with a common load.

frequency of each generator with the real power P delivered by the generator. This allows each generator to share changes in total load in a manner determined by its frequency droop characteristic. This simple idea essentially uses the system frequency as a communication link between the generator control systems. Similarly, a droop in the voltage amplitude with reactive power Q is used to ensure reactive power sharing. This load-sharing technique is based on the power flow theory in an ac system. This theory states that flow of active power P and reactive power Q between two sources can be controlled by adjusting the power angle and voltage magnitude of each system. This means that the active power flow P is controlled predominantly by the power angle, while the reactive power Q is controlled predominantly by the voltage magnitude.

For simplified analysis, Figure 12.23 indicates variables for load sharing of two power converters arranged in parallel configuration. In it, two inverters are represented by two voltage sources connected to a load through line impedances represented by pure inductances L_1 and L_2. The complex power at the load because of the ith inverter is given by

$$S_i = P_i + jQ_i = VI_i^* \tag{12.29}$$

where $i = 1, 2$ and I_i^* is the complex conjugate of the inverter current I and is given by

$$I_i^* = \left[\frac{E_i \cos \delta_i + jE_i \sin \delta_i - V}{j\omega L_i}\right]^* \tag{12.30}$$

$$S_i = V\left[\frac{E_i \cos \delta_i + jE_i \sin \delta_i - V}{j\omega L_i}\right]^* \tag{12.31}$$

Such equations allow the calculation of active and reactive power flowing from the ith inverter as

$$P_i = \frac{VE_i}{\omega L_i} \sin \delta_i \tag{12.32}$$

$$Q_i = \frac{VE_i \cos \delta_i - V^2}{\omega L_i} \tag{12.33}$$

Equations (12.29) to (12.33) show that if δ_1 and δ_2 are small enough, the real power flow is influenced primarily by the power angles δ_1 and δ_2, whereas the reactive power flow depends predominantly on the inverter output voltages E_1 and E_2. This means that to a certain extent, real and reactive power flow can be controlled independently. Because controlling the frequencies dynamically controls the power angles, the real power flow control can be achieved equivalently by controlling the frequencies of the voltages generated by the inverters. Therefore, as mentioned above, the power angle and the inverter output voltage magnitude are critical variables that can control the real and reactive power flow directly for proper load sharing of power converters connected in parallel.

Figure 12.24 is a simplified diagram that represents a DG control model with frequency and voltage command. The inverters respond instantaneously to decoupled current references (p and q) to impress current into the grid with a prescribed amplitude and angle. The same approach can be applied to parallel operation of distributed energy systems in a stand-alone ac power supply application.

In general, there is a large distance between inverter output and load bus, so each distributed resource is required to operate independently using only locally measurable voltage and current information. There is also a long distance between DR units, and proper load sharing between each unit must be guaranteed despite impedance mismatches. Voltage and current measurement error mismatches and interconnection tie-line impedance can heavily affect the performance of load sharing.

Conceptually, the isolated microgrid is like a scaled-down version of a large-scale utility grid, and most of the technical requirements are the same. To supply reliable good-quality power, the microgrid must have mechanisms to regulate voltage and frequency in response to changes in consumer loads and disturbances.

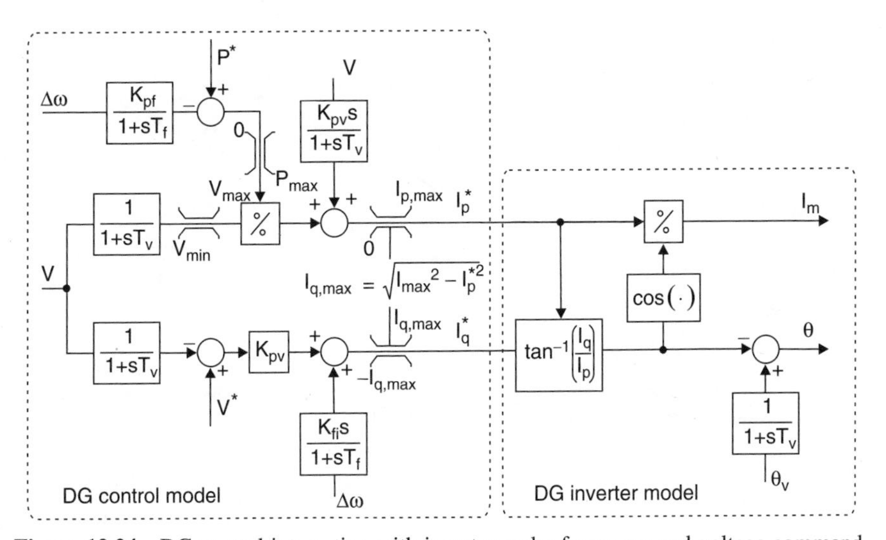

Figure 12.24 DG control interacting with inverter under frequency and voltage command.

The grid-connected microgrid should be designed and operated such that it presents the appearance of a single predictable and orderly load or generator to the grid at the point of interconnection. This arrangement provides several advantages. For example, the DG owners may be able to rate and operate their generation more economically by being able to export (and import) power to the microgrid. Load consumers may be able to have continued service (possibly at a reduced level) when connection to the host utility is lost. The host utility may be able to depend on the microgrid to serve load consumers in such a way that the substation and bulk power infrastructure need not be rated (or expanded) to meet the entire load. The microgrid could be controlled in such a fashion as to be an active asset to bulk system reliability (e.g., by providing spinning reserve or black-start services).

Another form of simple and effective control, perhaps suitable up to 100 kW, is heuristic hill-climbing control. It is based on the idea that all conversion systems for alternative sources of energy, either rotatory or stationary, have a similar power versus current characteristic, as presented in Figure 12.25. Therefore, it is possible to adapt them to operate with hill-climbing control or with fuzzy control for maximum power-point tracking [23]. An example application of this method with combined speed and voltage electronic control is explained in Section 10.6.5. The most obvious applications of hill-climbing control are wind, solar, and hydroelectric power plants. This method is not suitable for fuel cells because their maximum power does not coincide with their best efficiency ranges. In this case, the coincidence of maximum power and efficiency ranges is an important feature for hydrogen consumption and longer life span of cells, as discussed in Chapter 7.

Chapter 13 deals with DG and its basic concepts, applications, and optimization. It assumes that the means of injection of energy into the grid discussed in this

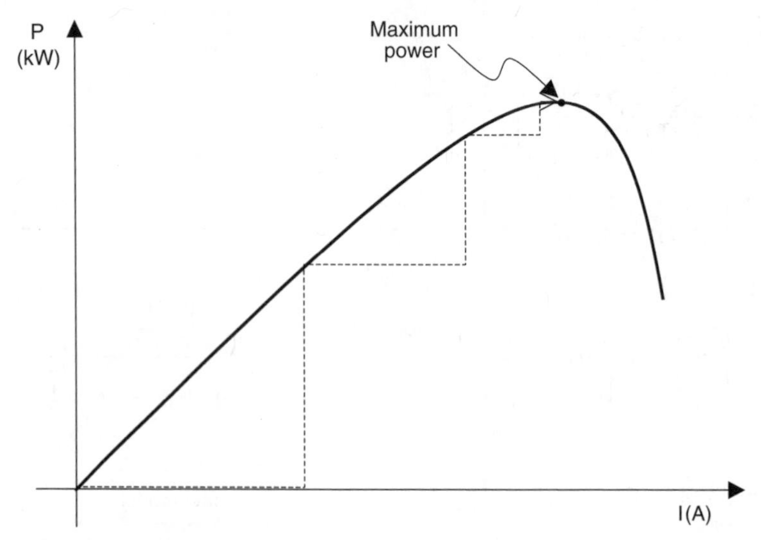

Figure 12.25 Typical shape of the power characteristic of electrical generators.

chapter serve as the connection between the renewable/alternative source of energy and the grid.

REFERENCES

[1] G. Venkataramanan and M. Illindala, Microgrids and sensitive loads, in *Proceedings of the IEEE Power Engineering Society Winter Meeting*, January 27–31, 2002, Vol. 1, pp. 315–322.

[2] R. Dugan and S. Price, Issues for distributed generation in the U.S., in *Proceedings of the IEEE Power Engineering Society Winter Meeting*, January 27–31, 2002, Vol. 1, pp. 121–126.

[3] H. Stemmler and P. Guggenbach, Configurations of high-power voltage source inverter drives, in *Proceedings of the 5th European Conference on Power Electronics and Applications*, Brighton, East Sussex, England, 1993, Vol. 5, pp. 7–14.

[4] N. Mohan, T. M. Undeland, W. P. Robbins, *Power Electronics: Converters, Application and Design*, Wiley, New York, 2003.

[5] B. K. Bose, *Modern Power Electronics and AC Drives*, Prentice Hall, Upper Saddle River, NJ, 2001.

[6] S. Nonaka and Y. Neba, A PWM GTO current source converter–inverter system with sinusoidal inputs and outputs, *IEEE Transactions on Industry Applications*, Vol. 25, No. 1, pp. 76–85, January–February 1989.

[7] F. Z. Peng, X. Yuan, X. Fang and Z. Qian, Z-source inverter, *IEEE Transactions on Industry Applications*, Vol. 38, pp. 504–510, March–April 2003.

[8] P. C. Loh, D. M. Vilathgamuwa, Y. S. Lai, G. T. Chua and Y. Li, Pulse-width modulation of Z-source inverters, in *Conference Record of the IEEE 39th Industry Applications Annual Meeting Conference*, October 3–7, 2004, Vol. 1, pp. 155–162.

[9] V. Vlatkovic, Alternative energy: state of the art and implications on power electronics, in *Proceedings of the IEEE Applied Power Electronics Conference and Exposition, APEC '04*, 2004, Vol. 1, pp. 45–50.

[10] Z. Chen and E. Spooner, Voltage source inverters for high-power variable-voltage dc power sources, *IEE Proceedings: Generation, Transmission and Distribution*, Vol. 148, No. 5, pp. 439–447, September 2001.

[11] M. A. Rahman, R. S. Radwan, A. M. Osheiba and A. E. Lashine, Analysis of current controllers for a voltage source inverter, *IEEE Transactions on Industrial Electronics*, Vol. 44, No. 4, pp. 477–XXX, August 1997.

[12] M. Mohr, B. Bierhoft and F. W. Fuchs, Dimensioning of a current source inverter for the feed-in of electrical energy from fuel cells to the mains, Paper 41, presented at the Nordic Workshop on Power and Industrial Electronics, NORPIE 2004, Trondheim, Norway, June 14–16, 2004.

[13] H. Akagi, Y. Kanazawa and A. Nabae, Instantaneous reactive power compensators comprising switching devices without energy storage components, *IEEE Transactions on Industrial Applications*, Vol. 20, No. 3, pp. 625–630, 1984.

[14] E. H. Watanabe, R. M. Stephan and M. Aredes, New concepts of instantaneous active power in electrical systems with generic loads, *IEEE Transactions on Power Delivery*, Vol. 8, No. 2, pp. 697–703, April 1993.

[15] P. G. Barbosa, L. G. B. Rolim, E. H. Watanabe and R. Hanitsch, Control strategy for grid connected dc–ac converters with load power factor correction, *IEE Proceedings: Generation, Transmission and Distribution*, Vol. 145, No. 5, pp. 487–491, September 1998.

[16] M. Malinowski, M. P. Kazmierkowski, S. Hansen, F. Bllabjerb and G.D. Marques, Virtual flux-based direct power control of three-phase PWM rectifiers, *IEEE Transactions on Industry Applications*, Vol. 37, No.4, pp. 1019–1027, July–August 2001.

[17] J. L. Duarte, A. Van Zwam, C. Wijnands and A. Vandenput, Reference frames fit for controlling PWM rectifiers, *IEEE Transactions on Industrial Electronics*, Vol. 46, No. 3, pp. 628–630, June 1999.

[18] H. Fujita, Y. Watanabe and H. Akagi, Control and analysis of a unified power flow controller, *IEEE Transactions on Power Electronics*, Vol. 14, No. 6, pp. 1021–1027, November 1999.

[19] M. E. Ropp, M. Begovic, A. Rohatgi, Prevention of islanding in grid-connected photovoltaic systems, *Progress in Photovoltaics: Research and Applications*, Vol. 7, pp. 39–50, 1999.

[20] V. John, Y. Zhihong and A. Kolwalkar, Investigation of anti-islanding protection of power converter based distributed generators using frequency domain analysis, *IEEE Transactions on Power Electronics*, Vol. 19, No. 5, pp. 1177–1183, September 2004.

[21] Y. Zhihong, A. Kolwalkar, and Y. Zhang, Evaluation of anti-islanding schemes based on nondetection zone concept, *IEEE Transactions on Power Electronics*, Vol. 19, No. 5, pp. 1171–1176, September 2004.

[22] Y. Li, M. Vilathgamuwa and P. C. Loh, Design, analysis, and real time testing of a controller for a multibus microgrid system, *IEEE Transactions on Power Electronics*, Vol. 19, No. 5, pp. 1195–1204, September 2004.

[23] M. G. Simões and Farret, F. A. *Renewable Energy Systems: Design and Analysis with Induction Generators*, CRC Press, Boca Raton, FL, 2004.

CHAPTER 13

DISTRIBUTED GENERATION

13.1 INTRODUCTION

Economic and industrial growth in the twentieth century allowed electricity to be generated centrally and transported over long distances. Economies of scale in electricity generation led to an increase in power output and massive power systems. A balance of demand and supply was possible by the average combination of large, instantaneously varying loads. Security of supply increased because the failure of one power plant was compensated by other power plants in an interconnected system. For the purposes of this book, the electric power system grid is composed of the generation system, the transmission system, the distribution system, and the loads. Although the electric power market had experienced steady growth for several decades, changes in fuel economy, congestion, and required investments in transmission, distribution, and generation were required to meet ever-increasing demand. This demand eventually reached a critical level that threatened system integrity, reliability, and efficiency.

Deregulation of the power industry made the transmission network accessible in a nondiscriminatory manner. Distributed generation made remote central power plants a feasible and cost-effective strategy for power generation. This type of clean, full-time on-site power generation is based on technologies such as turbines

Integration of Alternative Sources of Energy, by Felix A. Farret and M. Godoy Simões

and engines powered by natural gas, biogas, propane, wind, and small-scale hydropower as well as hydrogen-powered fuel cells and photovoltaic panels.

The overall U.S. electricity system is only about 30% efficient [i.e., there is an average input of four units of energy (coal, nuclear, natural gas, or oil) to deliver one unit of electricity]. DG has the potential of being less costly, more efficient, and more reliable. Most existing power plants, central or distributed, deliver electricity to user sites at an overall fuel-to-electricity efficiency in the range 28 to 32%. This represents a loss of around 70% of the primary energy provided to the generator.

To reduce energy loss, it is necessary to increase the fuel-to-electricity efficiency of the generation plant or to use the waste heat. The use of waste heat in DG close to the user increases further the overall efficiency for space heating or industrial processes. The ability to avoid transmission losses and make effective use of waste heat makes on-site cogeneration or combined heat and power (CHP) systems 70 to 80% efficient.

Industrial, commercial, and residential DG systems can be tailored for efficiency improvement by using, for example, heat exchangers, absorption chillers, or desiccant dehumidification to reach overall fuel-to-useful energy efficiencies of more than 80%. For example, Capstone manufactures a 60-kW microturbine that uses waste heat to heat water. This system has an energy efficiency of fuel to useful energy that approaches 90%. The use of waste heat through cogeneration or combined cooling, heating, and power implies an integrated energy system that delivers both electricity and useful heat from an energy source.

Unlike electricity, heat cannot be transported over long distances easily or economically, so CHP systems typically provide heat for local use. Because electricity is transported more readily than heat, the generation of heat close to the load usually makes more sense than the generation of heat close to the generator site. Figure 13.1 illustrates total energy efficiency as a function of loading ratio. Three

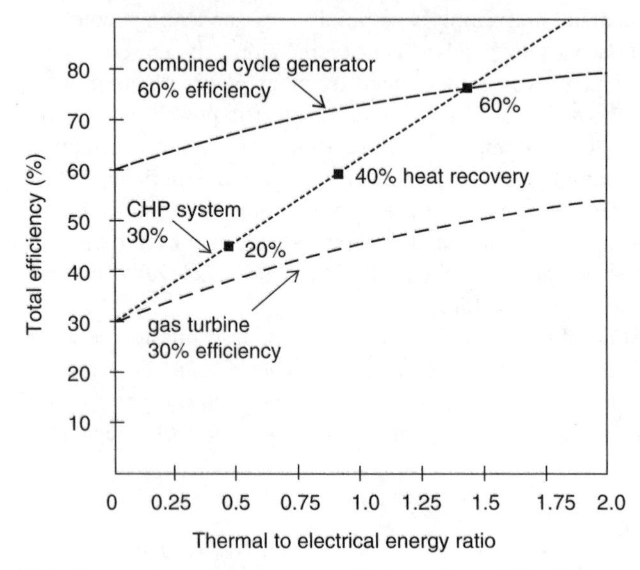

Figure 13.1 Comparison of total energy efficiency for combined power cycles.

thermal recovery efficiencies are shown on the plot. Two systems assume separate generation of electricity and heat; the third is a CHP system. The assumed thermal generation efficiency for the non-CHP examples is 85%; the electrical efficiencies are 60% and 30%, respectively. If a loading ratio of 1 is assumed, the overall efficiencies of the separate systems are 70% and 44%. For the case in which the waste heat is near the heat load, it can be used instead of fuel to provide the required heat.

Typical thermal recovery efficiencies range from 20 to 80%. The maximum ratio of heat to electricity is limited. For example, if the electrical efficiency is 30%, 70% of the fuel will result in waste heat. If this waste heat can be converted to useful heat, assuming a thermal recovery efficiency of 40%, the total energy efficiency is 58%, and the ratio of thermal energy to electrical energy is 1. This is the maximum loading and maximum total efficiency for this system. If the system is not loaded to this level, the total efficiency drops linearly. As observed in [5], CHP systems can greatly improve total energy efficiency through loading levels and thermal recovery efficiencies, as depicted in Figure 13.1.

13.2 THE PURPOSE OF DISTRIBUTED GENERATION

Distributed generation (DG) is the application of small generators from 10 to 10,000 kW scattered throughout a power system and either interacting with the grid or providing power to isolated sites. *Dispersed generation* is sometimes used as an interchangeable term, but it should be used for very small generation units, in the range 1 to 100 kW, sized to serve individual households or small businesses. DG technologies may be renewable (e.g., photovoltaic, thermal, wind, geothermal, and ocean-source systems) or nonrenewable (e.g., internal combustion engines, combined cycle engines, combustion turbines, microturbines, and fuel cells).

Household and rural users are concerned with the deployment of DG because of the overwhelming investments required to connect to a distant grid. For these users, DG is more economical than the central station system plus associated transmission and distribution expansion. Because of cost and reliability, industrial and commercial institutions may decide to install DG as a match with the electric utility system. This can happen when the particular application is of very high reliability and high cost or very low reliability and low cost, as discussed in [10]. The trend of DG to win at either end of the cost-reliability spectrum is depicted in Figure 13.2.

Reliability concerns (e.g., in medical and financial institutions) require uninterruptible power supply systems for conditioning input voltage. A backup generation system based on a rotating machine can be used as an alternative to a UPS system to supply power to sensitive loads during temporary interruptions in grid power. In this case, a full-fledged DG approach can be taken.

The following benefits are associated with DG projects:

Modularity Central-station power relies on economy of scale to increase efficiency and reduce the cost of power generation (i.e., with the approach of building

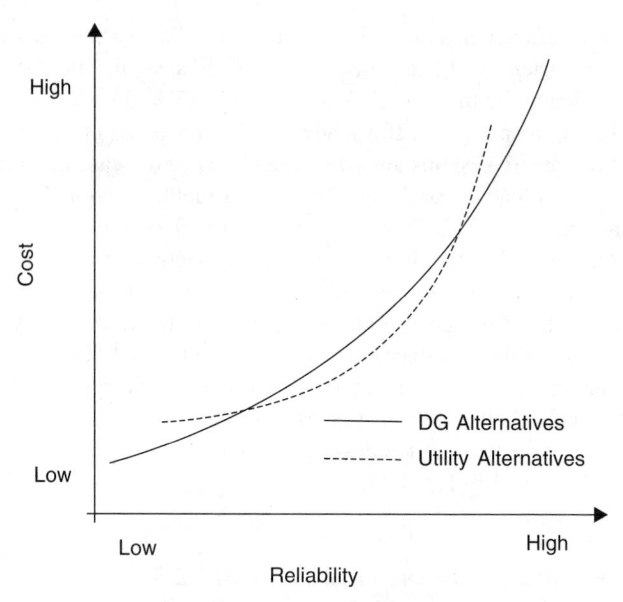

Figure 13.2 Flexibility of cost and reliability for DG applications.

larger generators to amortize costs over a larger consumer base). Distributed power relies on economy of production to reduce the cost of power generation. The capital investment made in a manufacturing plant produces thousands of low-cost units, which can penetrate markets that large-scale generation cannot reach. This model naturally leads to modular unit sizing. Modular system design increases the ability of a system designer to respond quickly and locally to near-term load growth without risking large sums of capital on capacity that may not be used in the future.

Efficiency Central-station generation has developed energy-efficient technologies such as natural gas combined-cycle systems. However, these technologies require continuous, reliable, and inexpensive access to a natural gas fuel source and are unsuited for small-scale deployment. DG technologies generally use renewable resources with high energy efficiencies. Renewable energy technologies use fundamentally local energy supplies, and some nonrenewable DG technologies use easily transported fuels other than natural gas. Other resources—such as energy storage, cogeneration, and demand-control devices—help improve the efficiency of whatever source of generation is being used. Increased energy efficiency decreases both energy costs and greenhouse gas emissions per unit of generation.

Low or No Emissions DG that uses renewable resources is inherently emission-free. However, some advanced distributed power technologies can also reduce emissions of conventional fossil fuels such as oil, natural gas, biogas, and propane. They accomplish this through increased efficiency and alternative energy conversion processes, such as found in a fuel cell, CO sequestration reform, and production of gas.

Security This includes system reliability and power quality issues. Distributed power provides inherent redundancy. When one on-site generator fails, the spare capacity in the remaining system resources can provide instantaneous reserve power (typically known as *spinning reserve*). Even if the primary generator fails, critical loads can be supported from on-site generators or overall system capacity. Distributed resources support power quality by preventing systemwide problems and mitigating line problems before a consumer load detects them.

Load Management This implies modifying the load profile by peak-load clipping, valley filling, load shifting, reducing voltage (brownout), reducing load, and load building. Conservation of energy involves reducing the entire energy load. Energy-efficiency measures fall under this category. Conservation can also be interpreted as reducing waste and reusing waste for production of energy.

Demand-side management (DSM) means modifying energy use to maximize energy efficiency, as discussed in Section 13.4. In contrast to supply-side strategies, which increase or redistribute supplies (by building new power plants or changing system reconfigurations), the goal of DSM is to smooth out the peaks and valleys in electric (or gas) demand. It makes the most efficient use of energy resources and defers the need to develop new power plants. This may entail shifting energy use to off-peak hours, reducing overall energy requirements, or increasing demand for energy during off-peak hours. DSM strategies can be classified as peak-clipping or valley-filling strategies. In peak-clipping strategy, a controller seeks to reduce energy consumption at the time of peak load. In programs to reduce peak load, the utility or consumer generally exerts control over appliances such as air conditioners or water heaters.

DG systems such as photovoltaic and solar–thermal power supplies can play a role in clipping peak demand if it coincides with their output. In valley-filling strategies, the goal is to build up off-peak loads to smooth out the load and improve the economic efficiency of the utility. An example of valley filling is charging electric vehicles or electrolyzing water to produce hydrogen and oxygen at night, when the utility is not required to generate as much power as during the day. Large battery storage can also be operated at night, and the energy stored can be used for peak clipping during the day. Economic assessment, including charging and discharging losses, must be accounted for to validate the feasibility of this option.

Another possible strategy is load shifting, in which thermal energy storage enables a consumer to use electricity to make ice or chilled water at night, when overall electricity consumption is low. The ice or chilled water is then used to cool buildings during the day, when overall electricity consumption is high. Water heating is another application that is widely used during off-peak hours.

Figure 13.3 illustrates a reason that utilities invest in small DG modules, which can be added closely in step with demand. The shaded areas show construction and financing times required by central units. In addition to avoiding overshooting, small units have short lead times and reduce the risk of buying technology that will become obsolete. Distributed resources also exploit agile manufacturing

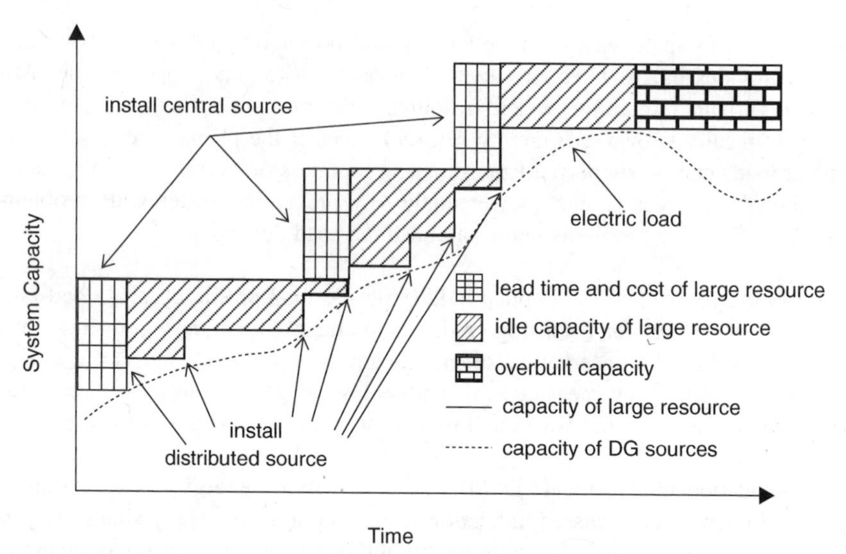

Figure 13.3 Addition of central or distributed sources to match demand.

techniques with a standardized workforce, so they have fewer turnovers, less retraining, and better management than that for a very long project.

13.3 SIZING AND SITING OF DISTRIBUTED GENERATION

In several U.S. states and in other countries, consumers can install small utility grid-connected renewable energy systems, such as solar or wind systems, to reduce their electricity bills using a protocol called *net metering*. Under net metering, electricity produced by renewable energy systems can flow into the utility grid and spin the electricity meter backward when the production of energy is more than local consumption. Other than the renewable energy system, no special equipment is needed.

Even in the absence of net metering, consumers can use the electricity they produce to offset their electricity demand instantaneously. But if the consumer produces excess electricity (beyond what is needed to meet his own needs at the time), the utility purchases that electricity at the wholesale or "avoided cost" price, which is lower than the retail price. Net metering simplifies this arrangement by allowing the consumer to use excess electricity to offset electricity used at other times during the billing period.

There are three reasons to support net metering procedures. First, as more residential consumers install renewable energy systems, they are getting used to standardized protocol for connecting their systems with the electricity grid and ensuring safety and power quality. Second, most residential consumers are not at home to use electricity during the day, and net metering allows them to receive full value for the electricity they produce then without expensive battery storage systems. Third, net metering provides a simple, inexpensive, and easily administered mechanism for encouraging the use of renewable energy systems, which provide important local, national, and global benefits such as economic development and a cleaner environment.

As a result, microgrid architectures can be alternative source–based, in which local generators supply electricity for industrial and commercial needs, or renewable-based, in which residential consumers install dispersed generation under net-metering policies. For utility companies, a major economical drawback of net metering is accounting in tariffs for the costs of installation and maintenance of the transmission and distribution systems used by energy-producing consumers.

The addition of DG to the energy matrix has significant effects on the power quality of the system. The power quality issues related to DG include sustained interruptions, voltage regulation, overall stability, and harmonics. To size DG on preexisting grids, the voltage profile and losses can be analyzed. Voltage regulation plays a vital role in determining how much DG can be accommodated on a distribution feeder while keeping the voltage level of the system within specified limits. By studying the effects of DG on voltage level and system losses, the optimal size and location of DG can be identified. Under this approach, a model of the system with all the operating scenarios during open and closed states of the circuit breakers must be formed. DG injection is assumed in different scenarios, and power flow analysis is conducted for the scenarios. The result of the power flow study gives the voltage level of the buses and losses in the system. Voltage level and voltage regulation at the consumer end are then tabulated in order of their degree of impact to find voltage regulation and system losses for critical cases. Then DG optimal size and location are identified.

13.4 DEMAND-SIDE MANAGEMENT

Conventional operation of grid power systems depended on changing generation to match demand. As suggested in Figure 13.4, DG can be applied for contingency capacity support at only one feeder (higher priority) or both feeders. Various methods can then be used to change load according to the needs of generation (e.g., DSM).

The most obvious example is a low-price electricity tariff to maintain load for nuclear power and large coal-burning stations at night. There are several options for implementing nighttime tariff measures:

- Consumer meters are switched to different energy-flow registers by a local clock or a radio signal (such as in the United Kingdom, often named "Economy 7"), which allows all load to change tariff. Consumers' bills show energy consumed at each tariff, and they are charged accordingly.
- Specific loads (e.g., storage heaters and water-tank heaters) are enabled by time clocks or ripple control (a signal "down the wire"), as is common in Australia for water heaters and in the United Kingdom for space heaters. These loads are wired on a separate circuit with a separate meter register.

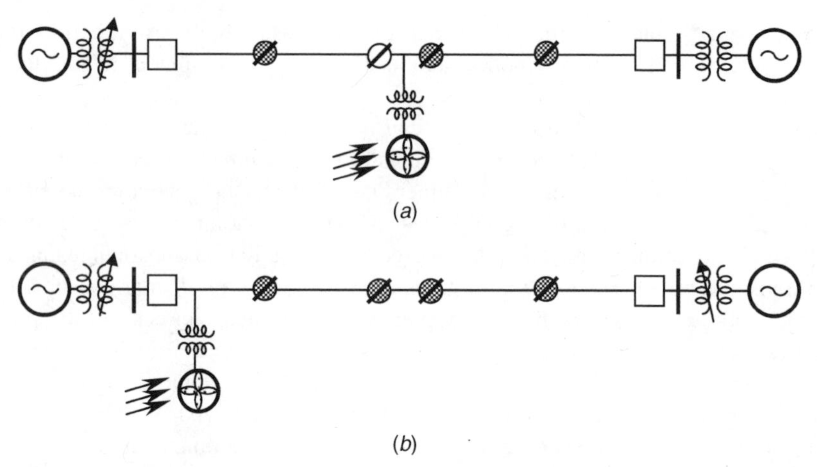

Figure 13.4 DG applied for contingency capacity support: (*a*) support for both feeders; (*b*) support for only one feeder.

The technology for remote switching (ripple control) is widely available (e.g., from Landis + Gyr and Siemens). There are many ways to communicate such control, including through long-wave radios, mobile telephone networks, and signals sent on power lines. Modern digital electronics offer accurate, low-cost communication for tariff information or direct control. Also, modern communication methods allow remote metering to measure and monitor electricity consumption at short intervals (e.g., minutes). Analysis of such information allows recommendations for DSM and reduced consumption. The communication technologies for remote metering are similar to those needed for remote switching.

The range of options for DSM, whereby consumers reduce electricity costs by responding over short periods to options offered by suppliers, is large. Most attention is given to peak shaving, but a range of techniques, tariffs, and other methods (e.g., on-site generation, interruptible loads, real-time pricing, and peak pricing) are lumped together as demand response.

Storage systems for electricity regeneration absorb electrical power from the grid, store it as potential energy, and then release it to regenerate electrical power back into the grid. The operation is usually a commercial enterprise. Examples are large-scale pumped-hydro reservoirs, compressed gas accumulators, and flywheel motor/generators. Such systems are called *dedicated storage*. They are judged by their efficiency and the cost of absorbing electrical energy and regenerating it back into the grid when required.

13.5 OPTIMAL LOCATION OF DISTRIBUTED ENERGY SOURCES

Advances in generation technology and new directions in electricity industry regulation should cause a significant increase in DG use all over the world

TABLE 13.1 Factors Influencing Realized Load Reductions

Aspect	Examples of Effect on Savings
Lighting quality	Lower lumens or light quality from a compact fluorescent bulb may result in the owner turning on more light fixtures than were used previously.
Comfort level	A more efficient device lowers a customer's bill, but the lower bill may encourage the customer to set the thermostat to a more comfortable setting and, thus reconsume some of the expected energy savings.
Output level	An industrial customer's use may vary with plant production levels. Changes because of plant production have to be separated from energy saving because of DSM measures.
Weather	Moderate weather may result in greater-than-expected energy savings. Similarly, extreme weather conditions could reduce expected savings by triggering more energy-consuming alternatives.
System balance	Benefits from DSM measures often depend on decisions made regarding other aspects of a user facility. Power consumption decreases with the square of voltage reduction, but it may be compensated by an increase in bulb wattage to improve illumination levels. Better housing insulation may be compensated by more efficient energy supply.

Source: J. Diamond, California Energy Commission, consultant report prepared by Outside Energy Corporation, USA, Jan/2002.

(Table 13.1). It is predicted that small sources of energy may account for up to 25% of all new generation units going online by 2010. It is critical that distribution system effects be assessed as accurately as possible to avoid degradation in efficiency, power quality, reliability, and control.

The energy-saving politics used in past years have resulted in significant growth reduction of electric loads. To continue this progress, a reformulation of the traditional strategies of centralization and increased generating unit sizes, very typical during the 1970s and 1980s, is needed. Of course, the traditional way demands massive investments in large power plants with relatively long return periods and an increase in the length of networks, which, in some ways, causes negative effects on the environment. The appearance of new and efficient generation technologies increases the cost of electric power transmission and distribution. Restructuring of the energy sector, together with the considerations above, will lead to growth of energy consumption in the near future, which will be satisfied primarily through generation resources (including alternative sources) with relatively low output powers located along distribution systems [11].

Most alternative-source generating units belong to independent producers, who began to have relatively free access to the power system after the energy sector's deregulation. However, this causes some conflicts of interest. The energy-supply companies do not have an interest in reducing their energy revenues. Also, if installed in a random way, distributed energy resources may hinder the operational control of distribution systems. It is therefore necessary to emphasize that DG located at strategic points of distribution systems allows energy distributors to reduce investments in development (e.g., in the reinforcement of distribution lines or additional control equipment) and cuts operational costs (through reductions of power and energy losses and increases in the reliability of supply and quality of marketable energy).

In the following sections we analyze important aspects of DG optimal location for power distribution companies.

13.5.1 DG Influence on Power and Energy Losses

Power losses in distribution systems can be reduced by altering load flows through certain network sections, located between substations and the DG installation point. For a quantitative estimation of loss alterations, a hypothetical feeder with uniformly distributed load is analyzed. Suppose that a linear distribution of load along the distribution line such that $J_0 = I_0/l$ is the feeder load per unit of length of a unitary line resistance R_0. In any line section of the feeder, as depicted in Figure 13.5*a*, with a current J_0 through it, the three-phase power losses of this configuration, during any random period of time and at a distance x from the beginning of the section, are defined by

$$\Delta P = \int_0^l 3(J_0 x)^2 R_0\, dx = 3J_0^2 R_0 \int_0^l x^2 dx = J_0^2 R_0 l^3 \qquad (13.1)$$

where J_0 is the linear density of phase current in A/km and x is the distance from the line end in kilometers.

Assume that the DG power is equivalent to a load on the distribution system and that it is installed a distance y from the network line end. In this case, power losses are calculated by

$$\begin{aligned}\Delta P &= \int_0^y 3(J_0 x)^2 R_0\, dx + \int_y^l 3(J_0 x - I)^2 R_0\, dx \\ &= J_0^2 l^3 R_0 - 3J_0 I l^2 R_0 + 3J_0 I y^2 R_0 + 3I^2 l R_0 - 3I^2 y R_0\end{aligned} \qquad (13.2)$$

where I is the phase current in amperes and l is the length of the line in kilometers. The coordinates of the DG connecting point may then be defined to guarantee a minimum of power losses:

$$\frac{\partial(\Delta P)}{\partial y} = 6J_0 I y R_0 - 3I^2 R_0 = 0$$

By this condition it follows that

$$y = \frac{I^2 R_0}{2J_0 I R_0} = \frac{1}{2J_0} \qquad (13.3)$$

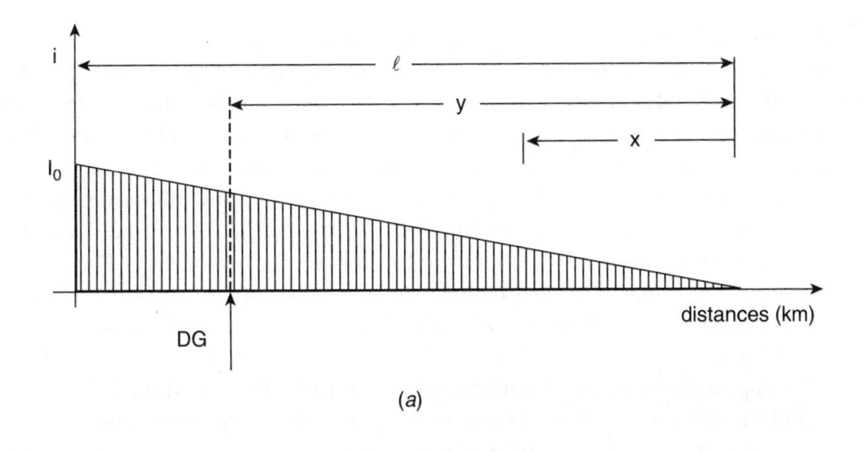

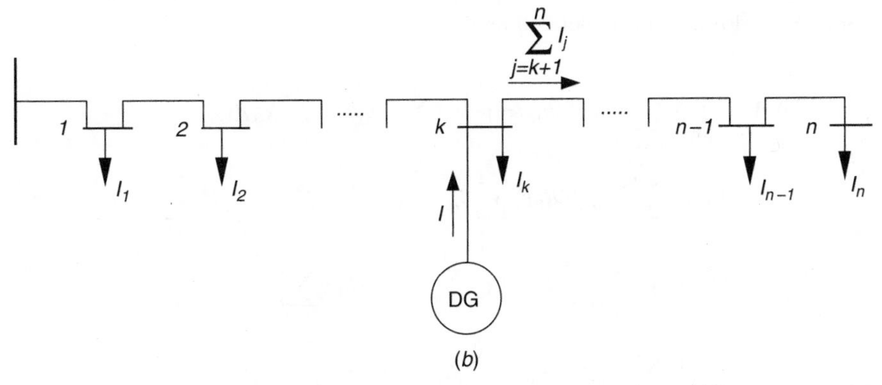

Figure 13.5 Examples of distributed loads and concentrated DG: (*a*) a feeder with uniformly distributed loads and DG integration; (*b*) a feeder with discrete loads.

Equation (13.3) means that in the case of a uniformly distributed load along the line, its middle point would be the best location for the alternative source. That is, when half the current injected by the renewable source fully compensates or minimizes the current going to each half of the feeder, the power losses will be minimum because the load current will be minimum. The example analyzed here does not correspond to a real load distribution in distribution networks with discrete locations of random intensity and uneven dispersal. Nevertheless, the results can be used to construct rules that define the best location of a DG for power loss minimization. For example, point k in Figure 13.5b, which may be a considerable point for connecting DG, is defined for minimum power losses as

$$\left| \frac{I - I_k}{2} \sum_{i=k+1}^{n} I_i \right| \rightarrow \min \tag{13.4}$$

where I is the DG injection current into the network and I_k is the load node under consideration.

The DG location for power loss minimization becomes appropriate when the resources are used for a limited time (e.g., only during the load peak period). Considering the daily alterations of node loads, their heterogeneity, and that DG can generate energy with constant power during the entire day, the condition presented in equation (13.4), in general, does not guarantee minimum energy losses in the same feeder for a long period (e.g., for an entire day). Naturally, considering node load alterations and a DG constant output, the solution for optimal location, based on equation (13.4), may be different for distinct periods of time. The difference between the connecting point defined by the analysis of the maximum demand state and the point that guarantees minimum energy losses depends on the consumer load curve.

Returning to the analysis of a feeder with uniformly distributed load, the coordinates of the DG point of connection can be defined for energy loss minimization, assuming that the generator operates at constant load during the entire day. In this case, energy losses can be defined as

$$\begin{aligned}\Delta W &= \int_0^y 3\sum_{t=1}^{T}(J_{0t}x)^2 R_0 dx + \int_y^l 3\sum_{t=1}^{T}(J_{0t}x - I)^2 R_0\, dx \\ &= R_0 y^3 \sum_{t=1}^{T} J_{0t}^2 + R_0 l^3 \sum_{t=1}^{T} J_{0t}^2 - R_0 y^3 \sum_{t=1}^{T} J_{0t}^2 - 3R_0 l^2 \sum_{t=1}^{T} J_{0t} I \\ &\quad + 3R_0 y^2 \sum_{t=1}^{T} J_{0t} I + 3R_0 l \sum_{t=1}^{T} I^2 - 3R_0 y \sum_{t=1}^{T} I^2 \\ &= R_0 l^3 \sum_{t=1}^{T} J_{0t}^2 - 3R_0 l^2 \sum_{t=1}^{T} J_{0t} I + 3R_0 y^2 \sum_{t=1}^{T} J_{0t} I \\ &\quad + 3R_0 l \sum_{t=1}^{T} I^2 - 3R_0 y \sum_{t=1}^{T} I^2 \end{aligned} \tag{13.5}$$

where T is the time duration of DG injection.

Similar to equation (13.3), the condition of a DG connection is defined to guarantee minimum daily energy losses in a feeder with uniformly distributed loads:

$$\frac{\partial(\Delta W)}{\partial y} = 6R_0 y + I\sum_{t=1}^{T} J_{0t} - 3R_0 T I^2 = 0 \tag{13.6}$$

From equation (13.6) it follows that

$$y = \frac{TI}{2\sum_{t=1}^{T} J_{0t}} \tag{13.7}$$

Assuming now that the voltage across the FD and the power factor across the terminals of the distribution transformers do not change during the time period T (e.g., one day), equation (13.7) can be transformed into the following form:

$$y = \frac{\sqrt{3}V_n \cos\varphi\, TI}{2\sqrt{3}V_n \cos\varphi \sum_{t=1}^{T} J_{0t}} = \frac{W_d}{2\,W_0} \tag{13.8}$$

where W_d is the DG generated energy during period T and W_0 is the energy consumed in the distribution system.

So the optimal point for a DG connection depends on the amount of energy it generates and on the energy consumed by the loads connected across the FD. To define the coordinates of the best connecting point, initially it is necessary to find the energy flow using consumption data (for the entire period T) of the distribution transformer loads in the FD analyzed. In this way, under the point of view of minimization of energy losses, it is reasonable to connect the DG across the terminal of the distribution transformer k for which the following condition is satisfied:

$$\left| \frac{W_d - W_k}{2} - \sum_{i=k+1}^{n} W_i \right| \to \min \tag{13.9}$$

where W_k is the energy consumption of the distribution transformer k during the period T; $\sum_{i=k+1}^{n} W_i$ is the summation of the energy consumption during period T of all distribution transformers located after point k, for which installation of DG is foreseen.

As an example, assume that the FD contains 15 uniformly distributed distribution transformers with identical daily load curves like the ones represented in Figure 13.6. The DG used in this example has the following parameters: $S_n = 600\,\text{kVA}$, $I = 25\,\text{A}$, $\cos\varphi = 1$, and $V_n = 13.8\,\text{kV}$. The calculation presented in Table 13.2 is only for adjacent points 9 to 14. The minimum $P_{\max}$ (kW) [equation (13.4)] and minimum W (kWh) [equation (13.9)] show that for power loss minimization within the load peak period, it is reasonable to locate the source at node 13.

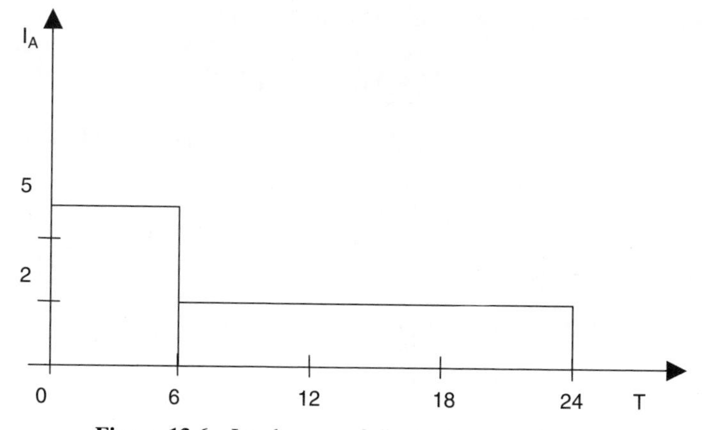

Figure 13.6 Load curve of distribution transformers.

TABLE 13.2 Power and Energy Losses Related to Various DG Locations

	DG Installation Point						
	9	10	11	12	13	14	Equation
P_{max} (kW)	3.56	3.3	3.11	3.0	2.96	2.99	(13.4)
W (kWh)	25.07	23.63	23.18	23.72	25.25	27.77	(13.9)

Considering that daily consumption in the network nodes is 1450 kWh, and supposing that the DG source is working 24 hours a day at constant load (which corresponds to 14,324 kWh), the optimal point of DG location, taking into account the minimization of energy losses, would be node 11. This solution is based on equation (13.7) and verified with the data of Table 13.2.

13.5.2 Estimation of DG Influence on Power Losses of Subtransmission Systems

Load alteration or redistribution in distribution networks causes alterations in the operational state of subtransmission systems. Particularly, the installation of DG on distribution systems not only causes reduction of loads in the same networks and substation system transformers, but also alters the load flow (and consequently, power losses) in the networks of subtransmission systems. For this reason, when selecting the optimal location of DG, it is desirable to estimate its influence on subtransmission system losses. According to Ref. [12], by the classical definition of complex power, power losses in the system are defined by

$$\Delta P + j\Delta Q = 3[V]^t[I]^* \tag{13.10}$$

where * is the conjugated symbol and t indicates transposed matrix.

Considering that $[V] = [Z][I]$ and that the matrix $[Z]$ is symmetrical, the expression above can be transformed into equation (13.10):

$$\Delta P + j\Delta Q = 3[I]^t[Z][I]^* \tag{13.11}$$

Assuming that $[Z] = [R] + j[X]$ and $[I] = \lfloor I_p \rfloor + j\lfloor I_q \rfloor$, equation (13.11) becomes

$$\Delta P + j\Delta Q = 3([I_p] + j[I_q])^t([R] + j[X])([I_p] - j[I_q]) \tag{13.12}$$

where the term of interest is

$$\begin{aligned}\Delta P &= 3([I_p]^t[R][I_p] + [I_p]^t[X][I_q] + [I_q]^t[R][I_q] - [I_q]^t[X][I_p]) \\ &= 3([I_p]^t[R][I_p] + [I_q]^t[R][I_q])\end{aligned} \tag{13.13}$$

where subindexes p and q refer to active and reactive components, respectively.

Again, assuming that I is the vector of the initial loads of each transformer in the substation system (SS) and I' is the load vector considering the introduction of the DG in the distribution network of one of the SS, the reduction in power losses in the subtransmission system lines will be given by

$$\delta(\Delta P) = 3([I_p]^t[R][I_p] + [I_q]^t[R][I_q] - [I'_p]^t[R][I'_p] - [I'_q]^t[R][I'_q]) \tag{13.14}$$

or, after some simplifications,

$$\delta(\Delta P) = 3([I]^t[R][I] - [I']^t[R][I']) \tag{13.15}$$

For instance, suppose that

$$I = \begin{bmatrix} I_1 \\ I_2 \\ I_3 \end{bmatrix} \qquad I' = \begin{bmatrix} I_1 \\ I_2 \\ I_3 - \Delta I \end{bmatrix} \tag{13.16}$$

where ΔI is the load reduction in one of the system transformers related to the DG implantation on the distribution network. In this case,

$$\begin{aligned}
\delta(\Delta P)_3 &= 3\left([I_1 I_2 I_3] \begin{bmatrix} R_{11} & R_{12} & R_{13} \\ R_{21} & R_{22} & R_{23} \\ R_{31} & R_{32} & R_{33} \end{bmatrix} \begin{bmatrix} I_1 \\ I_2 \\ I_3 \end{bmatrix} \right) \\
&\quad - 3\left([I_1 I_2 I_3 - \Delta I] \begin{bmatrix} R_{11} & R_{12} & R_{13} \\ R_{21} & R_{22} & R_{23} \\ R_{31} & R_{32} & R_{33} \end{bmatrix} \begin{bmatrix} I_1 \\ I_2 \\ I_3 - \Delta I \end{bmatrix} \right) \\
&= 3\{\Delta I^2(R_{31} + R_{32} + R_{33}) - \Delta I[I_1(R_{31} + R_{11} \\
&\quad + R_{32} + R_{12} + R_{33} + R_{13}) + I_2(R_{31} + R_{21} + R_{32} \\
&\quad + R_{22} + R_{33} + R_{23}) + I_3(2R_{31} + 2R_{32} + 2R_{33})]\}
\end{aligned} \tag{13.17}$$

where R_{ij} for $i, j = 1, 2, 3$ are the line resistances of phases 1, 2, and 3 when $i = j$ or the resistance between lines i and j when $i \neq j$.

Generalizing the results obtained, it is possible to build an equation to calculate the alterations of power losses in the lines of the subtransmission system regarding the use of a DG source located in the distribution system belonging to the arbitrary SS m as

$$\delta(\Delta P)_m = 3\left\{ \Delta I_m^2 \sum_{l=1}^{n} R_{ml} - \Delta I_m \left[\sum_{\substack{i=1 \\ i \neq m}}^{n} I_i \left(\sum_{j=1}^{n} R_{ij} + \sum_{j=1}^{n} R_{mj} \right) + 2I_m \sum_{l=1}^{n} R_{ml} \right] \right\} \tag{13.18}$$

where n is the total number of SS in the network.

However, in most cases, power and distribution systems are operated separately. This causes difficulties in obtaining the necessary information related to subtransmission systems during the strategic or operational planning of the distribution systems. At the same time, to solve these problems, it is often necessary to perform multiple experimental calculations. For this reason it is reasonable to create a simplified equivalent of the subtransmission system that can be used by the distribution department to assist its own interests. Evidently, this equivalence cannot have a universal character, but it should be suitable for the solution of an exact functional problem. As a rule, it is necessary to build a mathematical model [based on equation (13.11)] to estimate power loss alterations in the subtransmission lines and choose the optimal DG location. To realize this task, a mathematical tool known as *experimental design* can be used.

13.5.3 Equivalent of Subtransmission Systems Using Experimental Design

The choice of experimental design [13] as an instrument for the construction of functional models of subtransmission systems is based on three considerations. First, this approach demands a minimum number of experimental calculations to construct a multifactor model. Second, unlike multiple regression analysis, this method produces a well-formalized statistical analysis that includes the significant estimation of each factor together with the adequacy of the model. Finally, in case the simplest (linear) models are not appropriate, experimental design allows the construction of more complex (nonlinear) models using the results of preliminary stages. Of course, these circumstances simplify the construction of complex system models. Experimental design is based on the study of changes in the response function due to factor alterations.

In general, two factor levels are considered, which correspond to their maximum and minimum values (x_p^+, x_p^-). When considering all possible combinations of factor levels, it is necessary to realize $N = 2^k$ tests (where k is the number of factors) to allow construction of a linear model. In general, the normalized values of the given factors are used as

$$\tilde{x}_p = \frac{x_p - x_p^0}{\Delta x_p} \qquad p = 1, \ldots, k$$
$$x_p^0 = \frac{x_p^- + x_p^+}{2} \qquad \Delta x_p = \frac{x_p^+ - x_p^-}{2} \qquad p = 1, \ldots, k \tag{13.19}$$

Factors x_p^0 are the common factors, and Δx_p are the differential factors. This simplifies model construction and statistical analysis. Such an approach permits one to build a model as follows:

$$y = b_0 + \sum_{p=1}^{k} b_p \tilde{x}_p + \sum_{\substack{p=1 \\ p<q}}^{k} b_{pq} \tilde{x}_p \tilde{x}_q + \sum_{\substack{p=1 \\ p<q<r}}^{k} b_{pqr} \tilde{x}_p \tilde{x}_q \tilde{x}_r + \cdots \tag{13.20}$$

where the coefficients of equation (13.20) are defined as

$$\begin{aligned}
\tilde{b}_0 &= \frac{1}{N}\sum_{n=1}^{N}\tilde{x}_{np}y_n \\
\tilde{p}_{pq} &= \frac{1}{N}\sum_{n=1}^{N}\tilde{x}_{np}\tilde{x}_{nq}y_n \\
\tilde{b}_{pqr} &= \frac{1}{N}\sum_{n=1}^{N}\tilde{x}_{np}\tilde{x}_{nq}\tilde{x}_{nr}y_n \qquad p, q, r = 0, 1, \ldots, s, \quad p \neq q \neq r
\end{aligned} \tag{13.21}$$

Obviously, the increase in factors quickly increases the tests necessary for construction of the model. At the same time, changes inside the model increase the number of members that reflect interactions among the factors. According to early experiences in the process of following statistical analysis, such members become insignificant most of the time. This serves as a base for the use of fractional experimental design. In this case, the interactions among factors that can be insignificant are used for presentation of new factors.

As an example, a simplified subtransmission system is presented in Figure 13.7. Analysis of equation (13.18) allows one to define the following factors that can be included in the model for estimation of the alteration of power losses in relation to the system presented in Figure 13.7. Assuming that one neglects interactions among the significance factors to construct the model, it is possible to use fractional experimental design 2^{7-4}, shown in Table 13.3. To adjust the factor values in

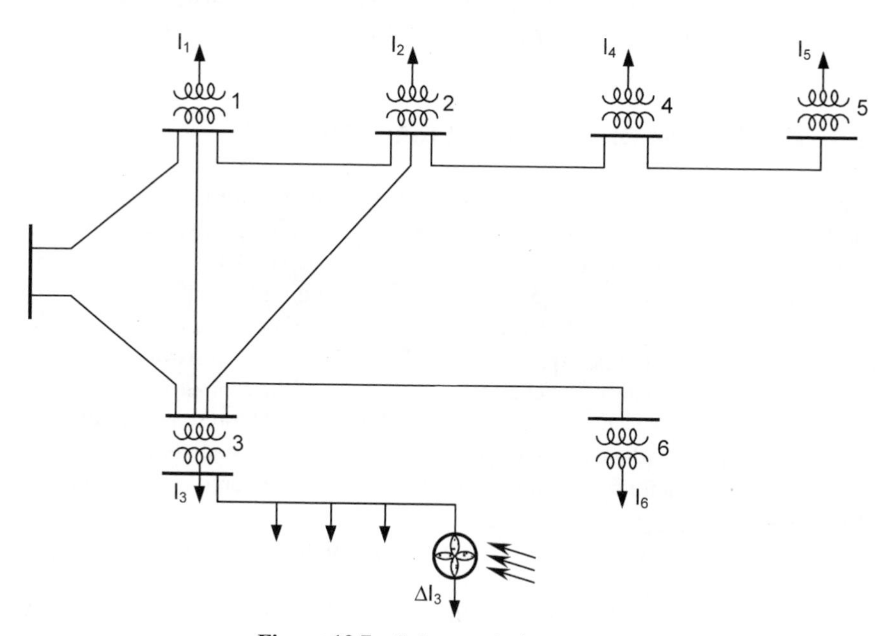

Figure 13.7 Subtransmission system.

TABLE 13.3 Matrix of the Experimental Design 2^{7-4}

N	x_0	x_1	x_2	x_3	$x_6 = x_1x_2$	$x_7 = x_1x_3$	$x_5 = x_2x_3$	$x_4 = x_1x_2x_3$	Y
1	+1	−1	−1	−1	+1	+1	+1	−1	y_1
2	+1	+1	−1	−1	−1	−1	+1	+1	y_2
3	+1	−1	+1	−1	−1	+1	−1	+1	y_3
4	+1	+1	+1	−1	+1	−1	−1	−1	y_4
5	+1	−1	−1	+1	+1	−1	−1	+1	y_5
6	+1	+1	−1	+1	−1	+1	−1	−1	y_6
7	+1	−1	+1	+1	−1	−1	+1	−1	y_7
8	+1	+1	+1	+1	+1	+1	+1	+1	y_8

each test, it is first necessary to define the possible alterations of substation system loads (considering their daily variations). The model has the following form:

$$\delta(\Delta P) = b_0 + b_1\tilde{x}_1 + b_2\tilde{x}_2 + b_3\tilde{x}_3 + b_4\tilde{x}_4 + b_5\tilde{x}_5 + b_6\tilde{x}_6 + b_7\tilde{x}_7 \tag{13.22}$$

It is important to emphasize that computer experiments do not permit the use of formal statistical analysis. In this case, to estimate the significance factors and verify model adequacy, some adaptation, as proposed in Ref. [14], is used.

$$\Delta I_3 = x_1 \qquad \Delta I_3 I_3 = x_2 \qquad \Delta I_3 I_1 = x_3 \qquad \Delta I_3 I_2 = x_4$$
$$\Delta I_3 I_4 = x_5 \qquad \Delta I_3 I_5 = x_6 \qquad \Delta I_3 I_6 = x_7$$

13.6 ALGORITHM OF MULTICRITERIAL ANALYSIS

Reduction of power and energy losses is just one positive factor associated with the appropriate location of DG in the distribution system. In most cases, this indicator cannot serve as the only criterion for decision making. The DG influences several operational characteristics of an electric network, including the voltage and reactive power operating modes, the loading level of the elements, and system reliability. Each of these can be defined quantitatively. Energy quality—in particular, voltage levels—can be estimated by the energy consumed out of the standard voltage deviations. The quantitative characteristics of reliability can be presented through an integral indicator. Evidently, calculations of each of these characteristics and the analysis of their relationships with DG demands special research and development of appropriate methods.

In the same way, building DG projects is very complex. In the process of decision making, it is impossible to neglect cogeneration efficiency, necessary infrastructure, and safety of operation. In the preliminary analysis, there is no possibility of accomplishing a detailed project for each alternative point of DG installation. Expert estimates can be used, in particular, under a linguistic form. So during the process of decision making, several quantitative and qualitative criteria should be taken into account.

This book uses an approach based on an algorithm of Bellman–Zadeh [15]:

$$\tilde{A}_j = \{X, \mu_{A_j}(X)\} \qquad X \in D_x, \quad j = 1, \ldots, n$$

where $\mu_{A_j}(X)$ = membership function of $\tilde{A}_j$. Initially, with this algorithm, all objective functions $F_j(X)$, $X \in D_x$, $j = 1, \ldots, n$ are represented by fuzzy objective functions. The primary problem with the Bellman–Zadeh approach is the formation of membership functions, which should be concave and reflect the proximity level of each objective function with respect to its own optimal solution. Experience with this approach shows the efficiency of the use of membership functions determined according to equations (13.23) and (13.24) [15]:

$$\mu_{A_j}(x) = \frac{F_j(x) - \min_{x \in D_x} F_j(x)}{\max_{x \in D_x} F_j(x) - \min_{x \in D_x} F_j(x)} \tag{13.23}$$

for objective functions that should be maximized and

$$\mu_{A_j}(x) = \frac{\max_{x \in D_x} F_j(x) - F_j(x)}{\max_{x \in D_x} F_j(x) - \min_{x \in D_x} F_j(x)} \tag{13.24}$$

for objective functions that should be minimized.

As presented in Ref. [15], equation (13.25) defines a fuzzy solution $\tilde{D}$ of the initial problem:

$$\tilde{D} = \bigcap_{j=1}^{n} \tilde{A}_j \tag{13.25}$$

In this case, the membership function of the fuzzy solution is calculated by

$$\mu_D(x) = \bigwedge_{j=1}^{n} \mu_{A_j}(x) = \min_{j=1,\ldots,n} \mu_{A_j}(x) \qquad x \in D_x \tag{13.26}$$

According to the algorithm proposed, the optimal solution is the one that presents a maximum value of the membership function:

$$\max_{x \in D_x} \mu_D(x) = \max_{x \in D_x} \min_{j=1,\ldots,n} \mu_{A_j}(x) \tag{13.27}$$

A general multicriterial approach to resolve the problem of DG optimal location in distribution systems shows that the optimal location of sources of DG allows a general increase in the efficiency of the generating units, including capability and energy quality. Special attention has to be given to the reduction of power and energy losses. The model of a subtransmission system presented in this section reflects loss reduction on DG sources located in distribution systems. Other legal and technical restrictions related to the integration and interconnection of DG are dealt with in Chapters 12 and 15, respectively.

REFERENCES

[1] P. G. Barbosa, L. G. B. Rolim, E. H. Watanabe, and R. Hanitsch, Control strategy for grid-connected dc–ac converters with load power factor correction, *IEE Proceedings: Generation, Transmission and Distribution*, Vol. 145, No. 5, pp. 487–491, 1998.

[2] G. Perpermans, J. Driesen, D. Haeseldonckx, W. D'haeseleer, and R. Belmans, *Distributed Generation: Definition, Benefits and Issues*, Working Paper Series 2003–8, K.U. Leuven Energy Institute, August 2003.

[3] M. N. Marwali and A. Keyhani, Control of distributed generation systems, part I: Voltages and currents control, *IEEE Transactions on Power Electronics*, Vol. 19, No. 6, November 2004.

[4] M. N. Marwali, J. W. Jung, and A. Keyhani, Control of distributed generation systems, part II: Load sharing control, *IEEE Transactions on Power Electronics*, Vol. 19, No. 6, November 2004.

[5] R. H. Lasseter and P. Piagi, Micro-grids: a conceptual solution, in *Proceedings of the 35th Annual IEEE Power Electronics Specialists Conference*, Aachen, Germany, June 20–25, 2004, pp. 4285–4290.

[6] H. Zang, M. Chnadorkar, and G. Venkataramanan, Development of static switchgear for utility interconnection in a microgrid, in *Proceedings of the Conference on Power and Energy Systems, PES 2003*, Palm Springs, CA, February 24–26, 2003, pp. 235–240.

[7] S. W. Park, I. Y. Chung, J. H. Choi, S. I. Moon, and J. E. Kim, Control schemes of the inverter interfaced multi-functional dispersed generation, in *Proceedings of the IEEE Power Engineering Society General Meeting*, July 13–17, 2003, Vol. 3, pp. 1924–1929.

[8] F. Blaabjerg, Z. Chen, and S. B. Kjaer, Power electronics as efficient interface in dispersed power generation systems, *IEEE Transactions on Power Electronics*, Vol. 19, No. 5, pp. 1184–1194, September 2004.

[9] Z. Chen and E. Spooner, Voltage source inverters for high-power variable-voltage dc power sources, *IEE Proceedings: Generation, Transmission and Distribution*, Vol. 148, No. 5, pp. 439–447, 2001.

[10] H. L. Willis and W. G. Scott, *Distributed Power Generation: Planning and Evaluation*, Marcel Dekker, New York, 2000.

[11] L. N. Canha, V. A. Popov, A. R. Abaide, F. A. Farret, A. L. König, D. P. Bernardon, and L. Comassetto, Multicriterial analysis for optimal location of distributed energy sources considering the power system reaction, presented at the 9th CIGRÉ Symposium of Specialists in Electric Operational and Expansion Planning, Rio de Janeiro, Brazil, May 2004.

[12] O. E. Elgerd, *Electric Energy Systems: An Introduction*, McGraw-Hill, New Delhi, 1975.

[13] B. Barros-Neto, I. S. Scarmino, and R. E. Bruns, *Planejamento e Otimização de Experimentos*, Editora Unicamp, Campinas, Brazil, 1995.

[14] F. G. Guseinov and S. M. Mamediarov, *Experimental Design in Problems of Electrical Engineering*, Energoatomizdat, Moscow, 1988 (in Russian).

[15] R. Bellman and L. Zadeh, Decision making in a fuzzy environment, *Management Sciences*, Vol. 17, No. 4, pp. 141–164, 1970.

CHAPTER 14

INTERCONNECTION OF ALTERNATIVE ENERGY SOURCES WITH THE GRID

BENJAMIN KROPOSKI, THOMAS BASSO, and RICHARD DEBLASIO
National Renewable Energy Laboratory

N. RICHARD FRIEDMAN
Resource Dynamics Corp.

14.1 INTRODUCTION

Historically, utility electric power systems (EPSs) were not designed to accommodate active generation and storage at the distribution level. In addition, a multitude of utility companies employed numerous EPS architectures based on differing designs and choices of equipment. As a result, major factors must be addressed when interconnecting alternative energy sources (AESs) with the utility grid. The interconnection system consists of the hardware and software that make up the physical link between AESs and the area EPS (usually the local electric grid). The interconnection system is the means by which an AES unit connects electrically with the outside EPS. It can also provide monitoring, control, metering, and dispatch of the AES unit.

Figure 14.1 shows the major functional components required for the interconnection of an AES with the utility grid (area EPS). The *interconnection system* is composed functionally of the components within the dashed lines. These functions are not necessarily independent or discrete objects as shown in the figure. For example, some of the functions may be co-located in the AES. Similarly, some of the interconnection functions shown as discrete objects may be combined in

Integration of Alternative Sources of Energy, by Felix A. Farret and M. Godoy Simões

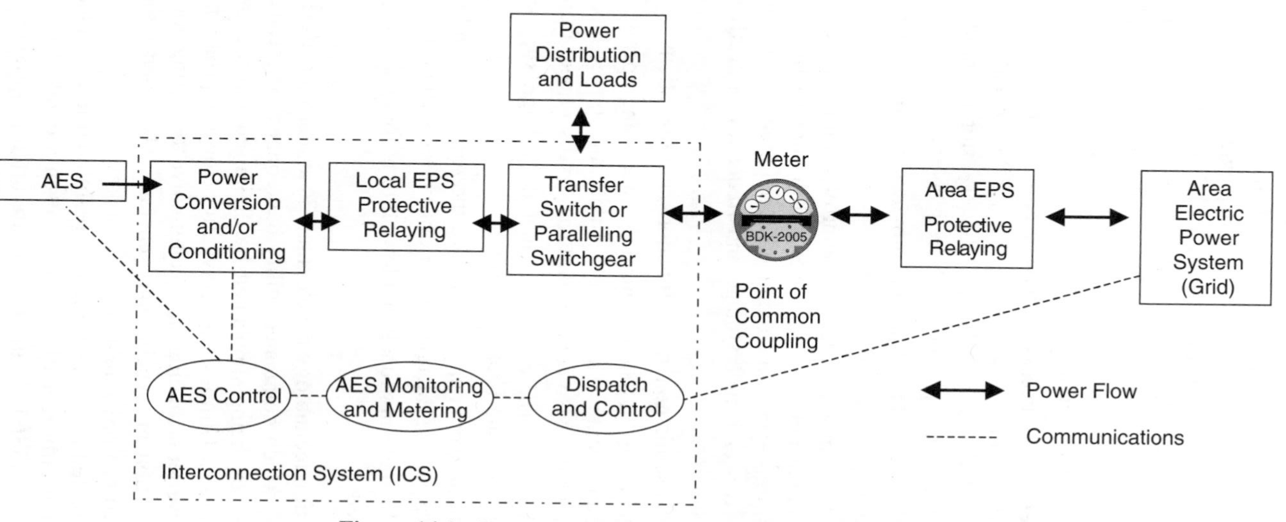

Figure 14.1 Interconnection system functional schematic.

equipment. In Figure 14.1 the boxes shown within the interconnection system are generally associated with power equipment functions, and the ellipses are associated with monitoring, information exchange, communications, and command and control functions. Therefore, there may not be independent, discrete demarcations among these power and communication functions in the interconnection system.

Functions that may be included in whole or in part in an interconnection system include:

1. Power conversion and conditioning
   a. Power conversion. If necessary, power conversion functions change one type of electricity into another to make it EPS-compatible. For example, photovoltaics (PVs), fuel cells, and battery storage produce dc power, whereas microturbines produce high-frequency ac.
   b. Power conditioning. This function provides the basic power quality to supply clean ac power to the load.
2. Protection functions monitor the EPS point of common coupling (PCC) and the input and output power of the AES and disconnect from the EPS when operating conditions exceed interconnection requirements. Examples of these functions are over- and undervoltage and frequency protective settings and anti-islanding schemes.
3. Autonomous and semiautonomous functions and operations
   a. AES and load controls. These control the status and operation of the AES and local loads. Status can include on/off and power-level commands. These functions can also control hardware to disconnect from the EPS.
   b. Ancillary services. These services include voltage support, regulation, operating reserve, and backup supply.
   c. Communications. Communications allow the AES and local loads to interact and operate as part of a larger network of power systems or microgrids.
   d. Metering. The metering function allows billing for AES energy production and local loads.

AES interconnection technology development is at a crossroads. Electromechanical discrete relays, which dominated utility interconnection, protection, and coordination for years, are being supplanted by digital equipment, frequently with multifunction capability. Utilities are gravitating toward programmable digital relays. The rise of inverter technology as an alternative to rotating power conversion technology (i.e., induction and synchronous generators) has opened the door to integrated, inverter-based protective relaying.

This trend has created a major hurdle to streamlined interconnection because utility engineers have only recently begun to reach a comfort level with digital circuitry. In addition , digital circuit designs are often proprietary, which makes the approval process challenging for utilities and the limited group of third-party certification organizations.

14.2 INTERCONNECTION TECHNOLOGIES

The *prime mover* or *engine* most often differentiates one AES from another. It is typically the most expensive subsystem of an AES technology. Prime movers may be powered by traditional fuel or renewable energy, and they can change energy from one form to another. For example, prime movers may use direct physical energy to rotate a shaft (e.g., in a hydro-turbine or wind energy device), or they may use chemical energy conversion to change thermodynamic energy to physical energy (e.g., by burning diesel fuel, natural gas, or propane to move a piston or turbine to rotate a shaft).

The electric generator converts prime mover energy to electrical energy. Historically, most electrical generation has been accomplished via rotating machines. The prime mover drives an electric generator that is synchronous or induction (asynchronous). The generator has a stator that consists of a set of ac windings and a rotor with windings that are either ac or dc. The stator is stationary, and the rotor is rotated by the prime mover. The means of power conversion is through interaction of the magnetic fields of the stator and rotor circuits.

Synchronous generators are used when power production from the prime mover is relatively constant (e.g., for internal combustion engines). Most large utility generators are synchronous. Induction generators, on the other hand, are well suited for rotating systems in which prime mover power is not constant (e.g., for wind turbines and small hydro-turbines). However, they can also be used with engines and combustion turbines. Synchronous and induction generator electrical output then undergoes another power conversion to achieve the sinusoidal power-quality level compatible with utility grid interconnection.

Inverters are electronic systems that convert power statically. Inverters are based on power semiconductor devices, microprocessor or digital signal processor technologies, and control and communications algorithms (see Chapter 12). Generally, inverters are used with prime movers that provide dc electricity (such as PV or fuel cell units) or microturbines, which are small, high-speed, rotating combustion turbines coupled directly to a synchronous-type electric generator. In the case of the microturbine, the generator output voltage waveform is very high frequency, well beyond the utility grid's 50 or 60 Hz. The high-frequency waveform is rectified to dc electricity, and the dc electricity is then synthesized to a sinusoidal waveform suitable for connection with the grid.

14.2.1 Synchronous Interconnection

Most generators in service today are synchronous. A synchronous generator is an ac machine in which the rotational speed of normal operation is constant and in synchronism with the frequency of the area EPS with which it is connected. In synchronous generators, field excitation is supplied by a separate motor–generator set, a directly coupled self-excited dc generator, or a brushless exciter that does not require an outside electrical source. Therefore, this type of generator can run in stand-alone mode or interconnected with the area EPS. When interconnected, the generator output is exactly in step with area EPS voltage and frequency. Note that separately excited synchronous generators can supply sustained fault current under nearly all operating conditions.

Synchronous generators require more complex control than grid-connected induction generators to synchronize with the area EPS and control field excitation. They also require special protective equipment to isolate them from the area EPS under fault conditions. An advantage is that this type of machine can provide power during area EPS outages. In addition, it permits AES owners to control the power factor at their facilities by adjusting the dc field current.

14.2.2 Induction Interconnection

Induction generators are asynchronous machines that require an external source to provide the magnetizing (reactive) current necessary to establish the magnetic field across the airgap between the generator rotor and stator. Without such a source, induction generators cannot supply electric power and must always operate in parallel with an area EPS, a synchronous machine, or a capacitor that can supply the reactive requirements.

In certain instances, an induction generator may continue to generate electric power after the area EPS source is removed. This phenomenon, known as *self-excitation*, can occur when there is sufficient capacitance in parallel with the induction generator to provide excitation and the load connected has certain resistive characteristics (see Chapter 10). This external capacitance may be part of the AES system or may consist of power factor correction capacitors located on the area EPS circuit with which the AES is directly connected.

Induction generators operate at a rotational speed determined by the prime mover and slightly higher than required for exact synchronism. Below synchronous speed, these machines operate as induction motors and thus become a load on the area EPS. Some advantages of induction generators are:

- Grid-connected induction generators need only basic control systems because their operation is relatively simple.
- They do not require special procedures to synchronize with the area EPS because this occurs essentially automatically.
- They will normally cease to operate when an area EPS outage occurs.

A disadvantage of induction generators is their response when connected with the area EPS at speeds significantly below synchronous speed. Depending on the machine class, potentially damaging inrush currents and associated torques can result. Regardless of load, induction generators draw reactive power from the area EPS and may adversely affect the voltage regulation on the circuit with which it is connected. Induction generators then consume vars from the system. It is important to consider the addition of capacitors to improve the power factor and reduce reactive power draw.

14.2.3 Inverter Interconnection

Some AESs produce electric power with voltages not in synchronism with those of the area EPS with which they are to be connected. An electric power converter provides an interface between a nonsynchronous AES and an area EPS so that the two may be properly interconnected.

The two categories of nonsynchronous AES output voltage are:

1. Direct-current (dc) voltages generated by dc generators, fuel cells, PV devices, storage batteries, or ac generators through rectifiers.
2. Alternating-current (ac) voltages generated by synchronous generators running at nonsynchronous speed or by asynchronous generators.

The two categories of electric power converters that can connect AESs with area EPSs are:

1. Dc-to-ac power converters (inverters). In this case, the input voltage to the device is generally a nonregulated dc voltage. The output of the device is at the frequency and voltage magnitude specified by the local utility or grid operator. This is the dominant means of small and renewable energy interconnection.
2. Ac-to-dc electric power converters. In this case, the input frequency, voltage magnitude to the device, or both do not meet area EPS requirements. The output of the converter device is at the appropriate frequency and voltage magnitude as specified by the area EPS in cases in which dc power can be used. This approach is not widely used.

The profusion of data centers and other customers using dc power supplies (such as the power supplied by electronic ballasts) has opened the door to either a direct dc or dc-to-ac converter designed to deliver the dc output of small AES units directly to the application.

Static power converters are built with diodes, transistors, and thyristors and have ratings compatible with AES applications. These solid-state devices are configured into rectifiers (to convert ac voltage into dc voltage), inverters (to convert dc voltage into ac voltage), and cycloconverters (to convert ac voltage at one frequency into ac voltage at another frequency). Some types require the area EPS to operate; others may continue to function normally after area EPS failure. The major advantages of solid-state converters are their higher efficiency and potentially higher reliability than rotating machine converters. In addition, this technology offers increased flexibility with the incorporation of protective relaying, coordination, and communications options.

14.3 STANDARDS AND CODES FOR INTERCONNECTION

Recent advancements in AESs have allowed their increased use and integration into the EPS. When interconnected with the EPS at or near load centers, these technologies can provide increased efficiency, availability, and reliability; better power quality; and a variety of economic and power system benefits.

The interconnection of AESs with the distribution system is regulated by codes and standards to address performance, safety, and power quality issues. Three organizations are major players in the interconnection codes and standards arena: (1) The Institute of Electrical and Electronics Engineers (IEEE), (2) The National Fire Protection Association (NFPA), and (3) Underwriters' Laboratories (UL).

14.3.1 IEEE 1547

IEEE develops voluntary consensus standards for electrical and electronic equipment. These standards are developed with the involvement of equipment manufacturers, users, utilities, and general interest groups. Most state public utility commission guidelines and utility interconnection requirements reference IEEE standards. In addition, most UL standards that relate to interconnection ensure that the equipment is manufactured to comply with IEEE standards.

IEEE approved IEEE 1547, Standard for Interconnecting Distributed Resources With Electric Power Systems, which established technical requirements for using electricity from dispersed sources, including alternative energy sources. The 1547 standard focuses on the technical specifications for, and testing of, the interconnection itself. It provides requirements relevant to the performance, operation, testing, safety, and maintenance of the interconnection. It covers general requirements, response to abnormal conditions, power quality, islanding, and test specifications and requirements for design, production, installation evaluation, commissioning, and periodic tests. These requirements are needed universally for the interconnection of AESs (including synchronous generators, induction generators, and power inverters or converters) and are sufficient for most installations. The requirements are applicable to all AES technologies with an aggregate capacity of 10 MVA or less at the PCC that are interconnected with EPSs at typical primary or secondary distribution voltages. The installation of AESs on radial primary and secondary distribution systems is the emphasis of the 1547 standard, but installation on primary and secondary network distribution systems is also considered.

IEEE Standards Coordinating Committee (SCC) 21 is developing a series of 1547 standards. IEEE P1547.1 will provide detailed test procedures for proving or validating that interconnection specifications and equipment conform to the functional and test requirements of IEEE 1547. IEEE P1547.2 will provide technical background and application details to make IEEE 1547 easier to use. It will characterize distributed resource technologies and associated interconnection issues. IEEE P1547.3 will aid interoperability by offering guidelines for monitoring, information exchange, and control among fuel cells, PV, wind turbines, and other distributed generators interconnected with an EPS. IEEE P1547.4 will address engineering aspects of how local facilities can function as "electrical islands" to provide power when utility power is not available.

14.3.2 National Electrical Code (NEC)

NFPA has been a worldwide leader in the provision of fire, electrical, and life safety to the public since 1896. The mission of this nonprofit organization is to reduce the worldwide burden of fire and other hazards on quality of life by providing and advocating scientifically based consensus codes and standards, research, training, and education. The NFPA publishes the National Electrical Code (NFPA-70), which covers electrical equipment wiring and safety on the customer side of the PCC. The NFPA also publishes other standards relating to AES interconnection.

NFPA 70: National Electrical Code The National Electrical Code covers electric conductors and equipment installed within or on public and private buildings or other structures, including mobile homes and recreational vehicles, floating buildings, and premises such as yards, carnivals, parking and other lots, and industrial substations; conductors that connect the installations to a supply of electricity and other outside conductors and equipment on the premises; optical fiber cable; and buildings used by the electric utility, such as office buildings, warehouses, garages, machine shops, and recreational buildings that are not an integral part of a generating plant, substation, or control center. Some National Electrical Code articles related to interconnection are described below.

Article 230: Services This article includes provisions and requirements for electric service—including emergency, backup, and parallel power—to a building.

Article 690: Solar Photovoltaic Systems Article 690 mentions interconnection to the grid but focuses on descriptions of components and proper system wiring.

Article 692: Fuel Cells This article covers stationary fuel cells for power production.

Article 700: Emergency Systems Article 700 includes provisions that apply to emergency power systems and information about interconnection (such as references to transfer switches).

Article 701: Legally Required Standby Systems Article 701 includes provisions that apply to standby power systems. It also has information about interconnection (such as references to transfer switches, UPSs, and generators).

Article 702: Optional Standby Systems This article includes provisions that apply to standby systems that are not required legally. It has some information about interconnection (such as references to transfer switches, grounding, and circuit wiring).

Article 705: Interconnected Electrical Power Production Systems This article broadly covers the interconnection of AESs (other than PV systems and fuel cells).

NFPA 853: Standard for the Installation of Stationary Fuel Cell Power Plants Article 853 applies to the design and installation of the following stationary fuel cell power plant applications: (1) a singular prepackaged, self-contained power plant unit; (2) a combination of prepackaged, self-contained units; and (3) power plant units composed of two or more factory-matched modular components intended to be assembled in the field.

14.3.3 UL Standards

UL is an independent, not-for-profit product safety testing and certification organization. UL has tested products for public safety for more than a century and is the leader in U.S. electrical product safety and certification. UL has a number of certifications that apply to AES interconnection equipment.

UL 1741: Inverters, Converters, and Controllers for Use in Independent Power Systems These requirements cover inverters, converters, charge controllers, and output controllers intended for use in stand-alone (not grid-connected) or utility-interactive (grid-connected) power systems. Utility-interactive inverters and converters are intended to be installed in parallel with an area EPS. An electric utility supplies common loads. This standard is harmonized with IEEE 1547 interconnection requirements and IEEE P1547.1 test procedures.

UL 1008: Transfer Switch Equipment These requirements cover automatic, nonautomatic (manual), and bypass/isolation transfer switches intended to provide for lighting and power in ordinary locations. They include:

1. Automatic transfer switches and bypass/isolation switches for use in emergency systems in accordance with Articles 517, Health Care Facilities; 700, Emergency Systems; 701, Legally Required Standby Systems; and 702, Optional Standby Systems of the National Electrical Code and the NFPA Standard for Health Care Facilities, ANSI/NFPA 99.
2. Transfer switches for use in optional standby systems in accordance with Article 702 of the National Electrical Code.
3. Transfer switches for use in legally required standby systems in accordance with Article 701 of the National Electrical Code.
4. Automatic transfer switches and bypass/isolation switches for use in accordance with the NFPA Standard for Centrifugal Fire Pumps, ANSI/NFPA 20.
5. Nonautomatic transfer switches for use in accordance with Articles 517, Health Care Facilities; and 702, Optional Standby Systems of the National Electrical Code and the NFPA Standard for Health Care Facilities.

An automatic transfer switch for use in a legally required standby system is identical to that used for an emergency system. These requirements cover transfer switch equipment rated at 6000 A or less and 600 V or less. These requirements also cover transfer switches with their associated control devices, including voltage-sensing relays, frequency-sensing relays, and time-delay relays.

An automatic transfer switch, as covered by these requirements, is a device that automatically transfers a common load from a normal supply to an alternative supply in the event of failure of the normal supply. It also automatically returns the load to the normal supply when it is restored. An automatic transfer switch is allowed to be provided with a logic control circuit that inhibits automatic operation of the device from either a normal to an alternative supply or from an alternative to a normal supply when the switch reverts to automatic operation upon loss of power to the load. Automatic transfer switches may be open or closed transition transfer.

As covered by these requirements, a nonautomatic transfer switch is a device operated manually by a physical action or electrically by a remote control for transferring a common load between normal and alternative supplies. A transfer switch may incorporate overcurrent protection for the main power circuits. UL 1008 requirements cover completely enclosed transfer switches and open types intended for mounting in other equipment, such as switchboards.

Transfer switches are rated in amperes and are generally considered to be acceptable for total system transfer, which includes control of motors, electric-discharge lamps, electric-heating loads, and tungsten-filament lamp loads. A transfer switch intended for total system transfer is considered to be acceptable for the control of tungsten-filament lamp loads not exceeding 30% of the switch ampere rating unless the switch has been investigated for a higher percentage of lamp load and marked accordingly. A transfer switch may be limited to use with one or more specific types of load if investigated accordingly and marked appropriately.

These requirements also cover bypass/isolation switches that can be used to manually select an available power source to feed load circuits and provide for total isolation of an automatic transfer switch. These switches may be completely enclosed, enclosed with a transfer switch, or be of the open type, which is intended for mounting in other equipment.

A product that contains features, characteristics, components, materials, or systems new or different from those covered by the requirements in this standard and that involve a risk of fire or electric shock or injury to persons are to be evaluated using appropriate additional components and end-product requirements to maintain the level of safety as anticipated originally by the intent of the standard. A product whose features, characteristics, components, materials, or systems conflict with specific requirements or provisions of this standard does not comply with this standard. Revision of requirements can be proposed and adopted in conformance with the methods employed for development, revision, and implementation of the standard.

UL 2200: Standard for Safety for Stationary Engine Generator Assemblies These requirements cover stationary engine generator assemblies rated 600 V or less that are intended for installation and use in ordinary locations in accordance with the National Electrical Code; the Standard for the Installation and Use of Stationary Combustion Engines and Gas Turbines, NFPA 37; the Standard for Health Care Facilities, NFPA 99; and the Standard for Emergency and Standby Power Systems, NFPA 110. These requirements do not cover generators for use in hazardous (classified) locations. That equipment is covered by the Standard for Electric Motors and Generators for Hazardous (Classified) Locations, UL 674. In turn, these requirements do not cover UPS equipment. That equipment is covered by the Standard for Uninterruptible Power Supply Equipment, UL 1778. These requirements do not cover generators for marine use, which are covered by the Standard for Marine Electric Motors and Generators, UL 1112.

14.4 INTERCONNECTION CONSIDERATIONS

When interconnecting an AES with an area EPS, several issues should be examined to ensure that the area EPS and AES integrate in a safe and reliable manner. These issues are discussed in the following sections.

14.4.1 Voltage Regulation

Voltage regulation describes the process and equipment used by an area EPS operator to maintain approximately constant voltage to users despite normal variations in voltage caused by changing loads. Voltage regulation and voltage stability are important factors that affect the operation of a power distribution system. If a system is not well regulated or stable, machines that receive power from this system will not operate efficiently.

AES interconnection equipment should not degrade the voltage provided to the customers of an area EPS to service voltages outside the limits of ANSI C84.1, Range A. Apart from the effect on the voltage of the area EPS because of the real power generation of the AES, the AES should not attempt to oppose or regulate changes in the prevailing voltage level of the area EPS at the PCC. An exception is that AES generators can use automatic voltage regulation when it is accomplished without detriment to either the area EPS or local EPS.

Voltage regulation is based on radial power flows from the substation to the loads. AESs introduce "meshed" power flows that may interfere with the effectiveness of standard voltage regulation practice. The effect of AESs on area EPS voltage regulation can cause changes in power system voltage by (1) the generator offsetting the load current or (2) the AESs attempting to regulate voltage. Most types of AES generators and utility-interactive inverters should strive to maintain an approximately constant power factor at any voltage within their rating; accordingly, the primary effect of AESs on voltage regulation is the result of an AES offsetting the load current.

14.4.2 Integration with Area EPS Grounding

A grounding system consists of all interconnected grounding connections in a specific power system and is defined by its isolation or lack of isolation from adjacent grounding systems. The isolation is provided by transformer primary and secondary windings that are coupled only by magnetic means.

The interconnection of an AES with an area EPS needs to be coordinated with the neutral grounding method in use on the area EPS. Use of an AES that does not appear as an effectively grounded source connected to such systems may lead to overvoltages during line-to-ground faults on the area EPS. This condition is especially dangerous if a generation island develops and continues to serve a group of customers on a faulted distribution system. Customers on the unfaulted phases could, in the worst case, see their voltage increase to 173% of the prefault voltage level for an indefinite period. At this high level, utility and customer equipment would almost certainly be damaged. Saturation of distribution transformers will help slightly limit this voltage rise. Nonetheless, the voltage can still become quite high (150% or higher).

14.4.3 Synchronization

To synchronize an AES with an area EPS, the output of the AES and the input of the area EPS must have the same voltage magnitude, frequency, phase rotation, and phase angle. Synchronization is the act of checking that these four variables are within an acceptable range (or acceptable ranges). For synchronism to occur, the output variables of the AES must match the input variables of the area EPS. With polyphase machines, the direction of phase rotation must also be the same. This is typically checked at the time of installation. The phases should be connected to the switches such that the phase rotation will always be correct. Phase rotation is not usually checked again unless wiring changes have been made on the generator or inverter or the area EPS.

14.4.4 Isolation

When required by area EPS operating practices, a readily accessible, lockable, visible-break isolation should be located between the area EPS and the AES. Strategically located disconnect switches are an integral part of any EPS. These switches provide visible isolation points to allow for safe work practices. The National Electrical Code dictates the requirements for disconnect devices, which allow for safe operation and maintenance of the EPSs within public or private buildings and structures. This requirement deals specifically with disconnect switches required to ensure safe work practices on the area EPS and not addressed by the National Electrical Code.

All electric utilities have established practices and procedures similar to the National Electrical Code to ensure safe operation of the EPS under normal and abnormal conditions. Several of these procedures identify methods to ensure that the electrical system has been configured properly to provide safe working conditions for area EPS line and service personnel. Although these procedures may vary somewhat between utilities, the underlying intent of the procedures is to establish *safe work area clearances* to allow area EPS line and service personnel to operate safely in proximity to the EPS. To achieve this, electric utilities have developed procedures that require visible isolation, protective grounding, and jurisdictional tagging of the portion of the EPS where clearance is to be gained. These procedures, in unison with other safety procedures and sound judgment based on knowledge and experience, have resulted in an essentially hazard-free work environment for area EPS personnel.

In an AES installation, some equipment and fuses or breakers may be energized from two or more directions. Thus, disconnect switches should be strategically installed to permit disconnection from all sources. Typically, the load-side contacts (switch blades) of a disconnect switch are deenergized when the switch is open. However, this is not necessarily the case when an AES is connected with the area EPS, so a safety label should be placed on the switch to warn that the load-side contacts may still be energized when the switch is in the open position. Also, a means should be provided for fuse replacement (in fused switches) without exposing workers to energized parts.

14.4.5 Response to Voltage Disturbance

The protection functions of the interconnection system should measure the effective or fundamental frequency value of each phase-to-neutral or, alternatively, each phase-to-phase voltage. When any of the measured voltages is in any voltage range noted in Table 14.1, the AES should cease to energize the area EPS within the clearing time indicated. Clearing time is the time between the start of the abnormal condition and the AES ceasing to energize the area EPS. For AESs less than or equal to 30 kW in peak capacity, the voltage set points and clearing times should be either fixed or field-adjustable. For AESs more than 30 kW, the voltage set points should be field-adjustable.

TABLE 14.1 Interconnection System Response to Abnormal Voltages per IEEE 1547

Voltage Range (% of base voltage*)	Clearing Time** (s)
$V < 50$	0.16
$50 \leq V < 88$	2.0
$110\ V < 120$	1.0
$V \leq 120$	0.16

*Base voltages are the nominal system voltages stated in ANSI C84.1, Table 1.

**AES ≤ 30 kW, maximum clearing times; AES > 30 kW, default clearing times.

The voltages should be measured at the point of AES connection when any of the following conditions exist:

1. The aggregate capacity of AES systems connected to a single PCC is less than or equal to 30 kW.
2. The interconnection equipment is certified to pass a nonislanding test.
3. The aggregate AES capacity is less than 50% of the total local EPS minimum electrical demand, and export of real or reactive power to the area EPS is not permitted.

The purpose of the time delay is to ride through short-term disturbances to avoid excessive nuisance tripping. For systems less than 30 kW in peak capacity, the set points are to be protected against unauthorized adjustment. Adjustment by a qualified person (or automatic adjustment for prevailing conditions) is desirable to allow compensation for voltage differences between the inverter and the PCC.

14.4.6 Response to Frequency Disturbance

Under- and overfrequency protective functions are among the most important means of preventing an AES island. It is desirable for these protections to operate promptly, but nuisance trips need to be avoided. At the point of generation, the frequency in a typical area EPS is very stable. However, voltage phase-angle swings can occur in transmission and distribution lines because of sudden changes in feeder loading and load current. Over a short enough time, these voltage swings can cause nuisance trips of under- or overfrequency protective functions.

Frequency excursions typically occur on the area EPS during distribution system operations. Maintaining stable area EPS operations depends on the AES clearing off line when area EPS voltage or frequency is out of agreed-upon operating ranges. Table 14.2 gives the interconnection system response times per IEEE 1547. AES units of less than 30 kW potentially have less effect on system operations and typically can disconnect from the area EPS well within 10 cycles. AES units larger than 30 kW can have an effect on distribution system security. The IEEE 1547 requirement takes this into account by allowing the area EPS operator to specify the frequency setting and time delay for underfrequency trips down to 57 Hz.

TABLE 14.2 Interconnection System Response to Abnormal Frequencies (60-Hz Base) per IEEE 1547

AES Size	Frequency Range (Hz)	Clearing Time* (s)
≤ 30 kW	> 60.5	0.16
	< 59.3	0.16
> 30 kW	> 60.5	0.16
	< {59.8–57.0} (adjustable set point)	Adjustable 0.16–300
	< 57.0	0.16

*AES ≤ 30 kW, maximum clearing times; AES > 30 kW, default clearing times.

Area EPS security depends on the system's ability to withstand the outage of certain lines or equipment without being forced into a system emergency. Security also depends on the proper matching of system load and generation. When generation is matched inadequately with system load, the area EPS frequency will decline. When this happens, the system operator seeks to match load quickly with available generation. Underfrequency relays are installed on the distribution system to shed load automatically to stabilize operations. This is the purpose of allowing the area EPS to determine the setting of the AES underfrequency trip relay.

Some underfrequency relays are sensitive to the rate of area EPS frequency decay and provide information to the system operator to assist in the timing of load shedding. Similar problems on the area EPS can occur when generation exceeds available load. An example is when a large block load is suddenly lost or when the tie lines exporting power relay are quickly closed. Overfrequency is much less of a problem to system operations than underfrequency.

In large power systems, frequency changes are rare. However, with installed AESs, some frequency change is unavoidable when blocks of load are switched. If a modern synchronous governor or static transfer switch is used on a distribution system feeder, these disturbances should be under 5% frequency change and less than 5 seconds in duration, even for full-load switching.

Both the frequency and voltage trip pickup settings for induction generators and static power converters may be relaxed at the discretion of the area EPS if it appears that the AES will experience too many nuisance trips. Synchronous generator trip settings can also be relaxed—but not too much because of the increased threat of islanding.

The frequency trip points should be adjustable in increments with a setting resolution of 0.5 Hz or better. Internal microprocessor protection functions in static power converter units may be substituted for external relays if they provide suitable accuracy. External test ports for periodic utility testing of the trip pickup settings should be included in the interconnection package.

14.4.7 Disconnection for Faults

Short-circuit currents on distribution circuits in the United States are from more than 20,000 A to values less than 1 A for high-impedance single-phase-to-ground faults. The maximum fault can be controlled by system design. Area EPSs are designed not to exceed the rating of distribution line equipment. Maximum faults are limited by restricting substation transformer size, impedance, or both by installing bus or circuit reactors or inserting reactance or resistance in the transformer neutral. Minimum fault magnitude is largely dependent on fault resistance that cannot be controlled. These faults are the most dangerous and difficult to detect.

Clearing times for short circuits on distribution circuits vary widely and depend on magnitude and the type of protective equipment installed. In general, large current faults will clear in 0.1 second or faster. Low-current faults frequently clear in 5 to 10 seconds or longer, and some very low-level but potentially dangerous ground faults may not clear at all except by manual disconnection of the circuit.

The AES system should be designed with adequate protection and control equipment and include an interrupting device that will disconnect the generator if the area EPS that connects to the AES system or the AES system itself experiences a fault. The AES system should have, as a minimum, an interrupting device that is:

1. Of sufficient capacity to interrupt maximum available fault current at its location
2. Sized to meet all applicable ANSI and IEEE standards
3. Installed to meet all local, state, and federal codes

A failure of the AES system protection and control equipment, including loss of control power, should open the disconnecting device automatically and thus disconnect the AES system from the area EPS.

14.4.8 Loss of Synchronism

A synchronous generator typically employs three-phase stator winding that when connected to the area EPS three-phase source creates a rotating magnetic field inside the stator and cuts through the rotor. The rotor is excited with a dc current that creates a fixed field. If spun around at the speed of the stator field, the rotor will "lock" its fixed field into synchronism with the rotating stator field. Force (torque) applied to the rotor in this synchronous state will cause power to be generated as long as the force is not so great that the rotor pulls out of step with the stator field.

An island is formed when a relay-initiated trip causes a section of the area EPS that contains AESs to separate from the main section of the area EPS. The main section of the area EPS and the island then operate out of synchronism. If an isolation is reclosed between the main section of the area EPS and the island, a voltage and current transient will occur while the island is brought into synchronism with the remainder of the area EPS. The severity of this transient will depend on the voltage phase-angle separation magnitude across the isolation when the reclosing event occurs.

14.4.9 Feeder Reclosing Coordination

Experience has shown that 70 to 95% of line faults are temporary if the faulted circuit is quickly disconnected from the system. Most line faults are caused by lightning or contact with tree limbs. If the resulting arc at the fault does not continue long enough to damage conductors or insulators, the line can be returned to service quickly.

Modern distribution feeders reclose (reenergize the feeder) automatically after a trip resulting from a feeder fault. This trip–reclose sequence may be initiated by reclosing relays that control the corresponding feeder breaker at the substation or by pole-mounted reclosers or sectionalizers located on the feeder away from the substation. Pole-mounted reclosers or sectionalizers are strategically placed to limit the customers affected per given feeder fault. Automatic reclosing allows immediate testing of a previously faulted portion of the feeder and makes it possible to restore service if the fault is no longer present. Depending on the fault magnitude, the first reclosing try can occur very fast, sometimes within 0.2 second. This short interval assumes settings of an instantaneous trip followed by an instantaneous reclosing.

It is common practice for utilities to attempt to reclose their circuit breakers automatically following a relay-initiated trip. The time delay between tripping and the initial reclose attempt can range from 0.2 second (12 cycles) to 15 seconds (or more). For radial feeders, this initial attempt is usually followed by two more time-delayed attempts, normally with 30- to 90-second intervals. If none of the reclose attempts is successful, the feeder will lock out. The reclose attempts are normally performed without any synchronism-check supervision because the feeders are radial in design, with the area EPS being the only source of power.

In the case of an area EPS protection function initiating a trip of an area EPS protective device in reaction to a fault, the AES must be designed to coordinate with the area EPS reclosing practices of the protective device. The response of the AES must be coordinated with the reclosing strategy of the isolations within the area EPS. Coordination is required to prevent possible damage to area EPS equipment and equipment connected to the area EPS. The AES and the area EPS reclosing strategy will be coordinated if one or more of the following conditions are met for all reclosing events:

1. The AES is designed to cease energizing the area EPS before the reclosing event.
2. The reclosing device is designed to delay the reclosing event until after the AES has ceased to energize the area EPS.
3. The AES is controlled to ensure that the voltage phase-angle separation magnitude across the isolation is less than one-fourth of a cycle when the reclosing event occurs.
4. The reclosing device is controlled to ensure that the voltage phase-angle separation magnitude across the isolation is less than one-fourth of a cycle when the reclosing event occurs.
5. The AES capacity is less than 33% of the minimum load on the feeder.

14.4.10 DC Injection

Dc injection produces a dc offset in the basic power system waveform. This offset increases the peak voltage of one half of the power system waveform (and decreases the peak voltage in the other half of the waveform). The increased half-cycle voltage has the potential to increase saturation of magnetic components, such as cores of distribution transformers. This saturation, in turn, causes increased power system distortion.

Dc injection is an issue because of the economics of magnetic component design. These economics dictate using the smallest amount of magnetic core material possible to accomplish the needed task. This results in the magnetic circuit of the component operating near that part of the *B–H* curve where the curve begins to become very nonlinear.

There is a concern that transformerless inverters may inject sufficient current into distribution circuits to cause distribution transformer saturation. Distribution transformers range from 25 kVA to more than 100 kVA. A 25-kVA transformer would typically supply power for four to six houses. A 100-kVA unit typically supplies power for 14 to 18 houses. These numbers vary depending on the amount of electric heating used but average about 5 kVA per residence.

14.4.11 Voltage Flicker

Determining the risk of flicker problems because of basic generator starting conditions or output fluctuations is fairly straightforward using the flicker curve approach. This is particularly true if the rate of these fluctuations is well defined, the fluctuations are step changes, and there are no complex dynamic interactions of equipment. The dynamic behavior of machines and their interactions with upstream voltage regulators and generators can complicate matters considerably. For example, it is possible for output fluctuations of an AES (even smoother ones from solar or wind systems) to cause hunting of an upstream regulator, and, although the AES fluctuations alone may not create visible flicker, the hunting regulator may create visible flicker. Thus, flicker can involve factors beyond starting and stopping generation machines or their basic fluctuations. Dealing with these interactions requires an analysis that is far beyond the ordinary voltage drop calculation performed for generator starting. Identifying and solving these types of flicker problems when they arise can be difficult, and the engineer must have a keen understanding of the interactions between the AES and the area EPS. In short, the AES should not create objectionable flicker for other customers on the area EPS.

14.4.12 Harmonics

When an AES is serving balanced linear loads, harmonic current injection into the area EPS at the PCC should not exceed the limits stated in Table 14.3. The harmonic current injections should be exclusive of any harmonic currents because of harmonic voltage distortion present in the area EPS without the AES connected.

TABLE 14.3 Maximum Harmonic Current Distortion in Percent of Current (*I*) per IEEE 1547*

Individual Harmonic Order (Odd Harmonics)**	<11	$11 \leq h < 17$	$17 \leq h < 23$	$23 \leq h < 35$	$35 \leq h$	TDD
Percent (%)	4.0	2.0	1.5	0.6	0.3	5.0

**I* is the greater of the local EPS maximum load current integrated demand (15 or 30 min) without the AES unit or the AES unit rated current capacity (transformed to the PCC when a transformer exists between the AES unit and the PCC).

**Even harmonics are limited to 25% of the odd harmonic limits noted.

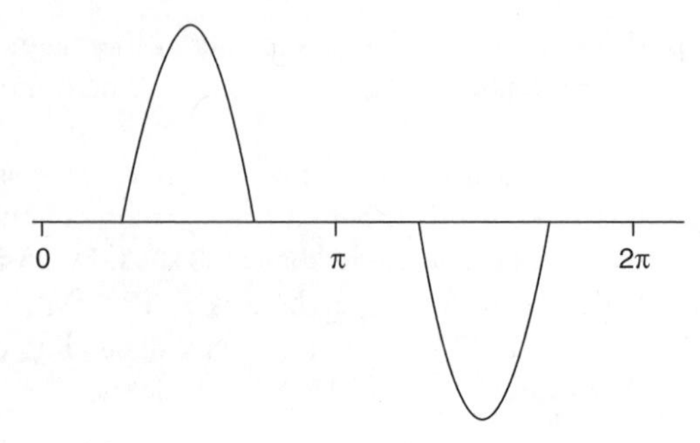

Figure 14.2 Current wave of switched-mode power supply.

Harmonic distortion is a form of electrical noise. Harmonics are electrical signals at multiple frequencies of the power line frequency. Many electronic devices, including personal computers, adjustable speed drives, and other types of equipment that use just part of the sine wave by drawing current in short pulses, cause harmonics.

Equipment with this operating characteristic is dominated by switched power supplies. These power supplies are an economical way to provide voltage to the equipment being served and are not affected by minor voltage changes in the power system. Switched power supplies feed a capacitor that supplies the voltage to the electronic circuitry. Because the load is a capacitor as seen from the power system, the current to the power supply is discontinuous. That is, current flows for only part of the half-cycle. Figure 14.2 shows the current waveform of such a power supply.

Linear loads, those that draw current in direct proportion to the voltage applied, do not generate large levels of harmonics. The nonlinear load of a switched power supply superimposes signals at multiples of the fundamental power frequency in the power sine wave and creates harmonics. The uses of nonlinear loads connected to area EPSs include static power converters, arc discharge devices, saturated magnetic devices, and to a lesser degree, rotating machines. Static power converters of electric power are the largest nonlinear loads. Harmonic currents cause transformers to overheat, which, in turn, overheats neutral conductors. This overheating may cause erroneous tripping of circuit breakers and other equipment malfunctions. The voltage distortion created by nonlinear loads may create voltage distortion beyond the premise's wiring system, through the area EPS system, to another user.

The type and severity of harmonic contributions from an AES unit depend on the power converter technology, its filtering, and its interconnection configuration. There has been particular concern about the possible harmonic current contributions inverters may make to the area EPS. Fortunately, these concerns are in part because of older, SCR-type power inverters that are line-commutated and produce high levels of harmonic current. Most new inverter designs are based on solid-state technology that uses pulse-width modulation to generate the injected alternating current and are capable of generating very clean outputs. When powered by small synchronous generators with high impedance, nonlinear loads can result in voltage distortion.

In general, harmonic contributions from AES units are less of an issue than problems associated with other equipment on the distribution system. In some cases, equipment at the AES site may need to be derated because of added heating caused by harmonics elsewhere on the system. Filters and other mitigation approaches are sometimes required.

14.4.13 Unintentional Islanding Protection

Islanding occurs when an AES (or a group of AESs) continues to energize a portion of the area EPS that has been separated from the rest of the area EPS. This separation could be because of the operation of an upstream breaker, fuse, or automatic sectionalizing switch. Manual switching or "open" upstream conductors could also lead to islanding. Islanding can occur only if the AES continues to serve a load in the islanded section.

In most cases, it is not desirable for an AES to island any part of the area EPS unplanned. This can lead to safety and power quality problems that affect the area EPS and local loads. During utility repair operations, AES islanding can expose utility workers to circuits that otherwise would be deenergized (and that the workers believe to be deenergized). This situation can pose a threat to the public as well. Service restoration can also be delayed as line crews seek to ensure that AES islanding is not a problem. IEEE 1547 requires that the AES cease to energize the area EPS within 2 seconds of the formation of an island. Several approaches can prevent unintentional islands from occurring.

The installed AES capacity can be limited to less than one-third of the minimum customer load. By limiting the installed AES capacity, the system protective functions for over- and undervoltage and frequency will be able to respond correctly to an area EPS outage. Reverse power protection can also be employed for unintentional islanding. This is done by placing a reverse power–current relay at the PCC. If the area EPS experiences an outage, any power that the AES tries to feed back onto the area EPS will cause the relay to trip.

AESs can be certified as nonislanding by UL. This means that the interconnection system can pass the nonislanding test specified in UL 1741 and IEEE 1547.1. This test sets up worst-case balanced island conditions and confirms that the AES will disconnect in the required 2 seconds. Inverter-based systems typically have some type of active anti-islanding method with which they try to force the voltage or frequency outside normal trip limits. When the area EPS is connected, the inverter cannot move these parameters. When the area EPS experiences an outage, however, the inverter's anti-islanding algorithm forces the voltage or frequency outside the trip limits and disconnects from the area EPS. Direct transfer trip is often used for large AES systems. This involves a dedicated communications line that will trip the AES system when the area EPS has experienced an outage.

14.5 INTERCONNECTION EXAMPLES FOR ALTERNATIVE ENERGY SOURCES

This section gives examples of the interconnection systems used for two types of AESs. Figure 14.3 shows a typical radial EPS. In a radial system, power flows from

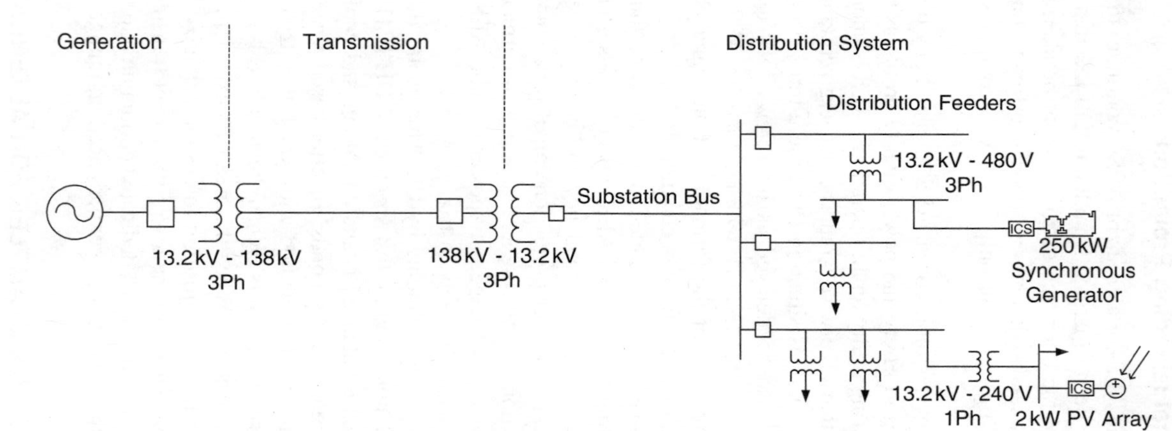

Figure 14.3 Typical EPS with AES and interconnection systems installed at the distribution system level.

large central-station generators through transmission lines to the distribution system, where customer loads are located. New AES systems are typically installed at the customer load site. This figure shows a synchronous generator and PV system installed at the distribution system level.

14.5.1 Synchronous Generator for Peak Demand Reduction

A typical application for a synchronous generator AES is peak shaving. In this application, the generator is installed at the customer site. Electric utilities typically sell power to commercial customers with a usage and demand charge. The usage charge is based on the energy (in kilowatthours) used, and the demand charge is based on the peak load (in kilowatts), typically over a 15-minute period. Figure 14.4 shows how running a 250-kW generator during times of peak load can reduce demand. Figure 14.5 shows the interconnection system functions that allow the customer or the area EPS to run the unit on demand. Currently, most utility dispatchable systems use a local utility remote terminal unit to communicate with the AES. The AES can be operated to assume all customer load or to assume customer load and export excess power back to the utility. This type of interconnection system contains the paralleling switch as well as the generator and local EPS protective relays.

14.5.2 Small Grid-Connected PV System

Another common application is a small PV array that uses inverter-based interconnection systems. These systems are typically less than 10 kW. A functional diagram is given in Figure 14.6. Some state public utility commissions require utilities to

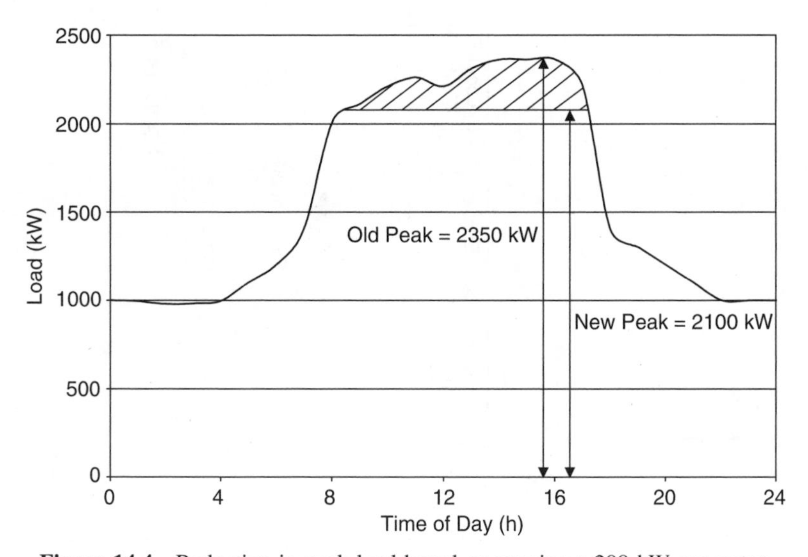

Figure 14.4 Reduction in peak load based on running a 200-kW generator.

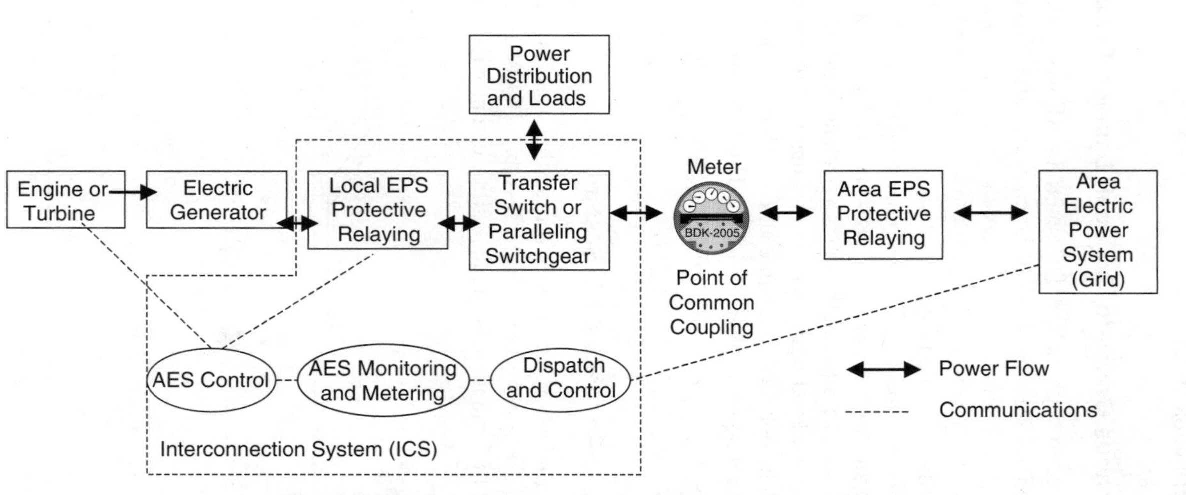

Figure 14.5 Synchronous generator interconnection system.

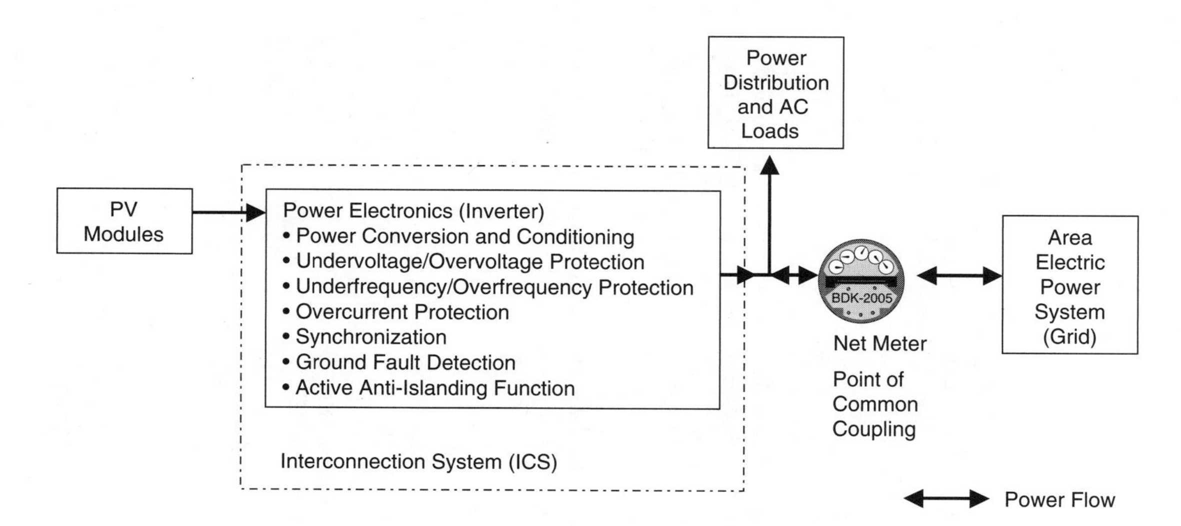

Figure 14.6 Small PV system with net metering.

allow net-metering systems. In a net-metered system, the customer is allowed to put power onto the grid when his system is producing energy. The meter keeps track of the net import of energy. The customer is typically allowed to have a monthly or yearly net energy balance of zero. Any energy in excess of the total customer load is fed back to the utility for free. In the PV system, the interconnection system consists of an inverter to convert dc power to ac power and all the necessary protective functions for compliance with interconnection standards.

REFERENCES

[1] B. Kroposki, T. Basso, and R. DeBlasio, Interconnection testing of distributed resources, presented at the 2004 PES General Meeting, Denver, CO, June 2004.

[2] T. Basso and R. DeBlasio, IEEE 1547 series of standards: interconnection issues, *IEEE Transactions on Power Electronics*, pp. 1159–1162, September 2004.

[3] IEEE Std. 1547, *Standard for Interconnecting Distributed Resources with Electric Power Systems*, IEEE Press, Piscataway, NJ, July 2003.

[4] IEEE P1547.1, *Draft Standard for Conformance Test Procedures for Interconnecting Distributed Resources with Electric Power Systems*, IEEE Press, Piscataway, NJ, February 2005.

[5] IEEE P1547.2, *Draft Application Guide for IEEE 1547 Standard for Interconnecting Distributed Resources with Electric Power Systems*, IEEE Press, Piscataway, NJ, July 2004.

[6] IEEE P1547.3, *Draft Guide for Monitoring, Information Exchange, and Control of Distributed Resources Interconnected with Electric Power Systems*, IEEE Press, Piscataway, NJ.

[7] IEEE P1547.4, *Draft Guide for Design, Operation, and Integration of Distributed Resource Island Systems with Electric Power Systems*, IEEE Press, Piscataway, NJ.

[8] Electrical Generating Systems Association, *On-Site Power Generation: A Reference Book, 4th ed., EGSA*, 2002.

[9] N. R. Friedman, *Distributed Energy Resources Interconnection Systems: Technology Review and Research Needs*, NREL/SR-560-32459, National Renewable Energy Laboratory, Golden, CO, Sept. 2002.

CHAPTER 15

MICROPOWER SYSTEM MODELING WITH HOMER

TOM LAMBERT
Mistaya Engineering Inc.

PAUL GILMAN and PETER LILIENTHAL
National Renewable Energy Laboratory

15.1 INTRODUCTION

The HOMER Micropower Optimization Model is a computer model developed by the U.S. National Renewable Energy Laboratory (NREL) to assist in the design of micropower systems and to facilitate the comparison of power generation technologies across a wide range of applications. HOMER models a power system's physical behavior and its life-cycle cost, which is the total cost of installing and operating the system over its life span. HOMER allows the modeler to compare many different design options based on their technical and economic merits. It also assists in understanding and quantifying the effects of uncertainty or changes in the inputs.

A *micropower system* is a system that generates electricity, and possibly heat, to serve a nearby load. Such a system may employ any combination of electrical generation and storage technologies and may be grid-connected or autonomous, meaning separate from any transmission grid. Some examples of micropower systems are a solar–battery system serving a remote load, a wind–diesel system serving an isolated village, and a grid-connected natural gas microturbine providing electricity and heat to a factory. Power plants that supply electricity to a high-voltage transmission system do not qualify as micropower systems because they are not

Integration of Alternative Sources of Energy, by Felix A. Farret and M. Godoy Simões

dedicated to a particular load. HOMER can model grid-connected and off-grid micropower systems serving electric and thermal loads, and comprising any combination of photovoltaic (PV) modules, wind turbines, small hydro, biomass power, reciprocating engine generators, microturbines, fuel cells, batteries, and hydrogen storage.

The analysis and design of micropower systems can be challenging, due to the large number of design options and the uncertainty in key parameters, such as load size and future fuel price. Renewable power sources add further complexity because their power output may be intermittent, seasonal, and nondispatchable, and the availability of renewable resources may be uncertain. HOMER was designed to overcome these challenges.

HOMER performs three principal tasks: simulation, optimization, and sensitivity analysis. In the simulation process, HOMER models the performance of a particular micropower system configuration each hour of the year to determine its technical feasibility and life-cycle cost. In the optimization process, HOMER simulates many different system configurations in search of the one that satisfies the technical constraints at the lowest life-cycle cost. In the sensitivity analysis process, HOMER performs multiple optimizations under a range of input assumptions to gauge the effects of uncertainty or changes in the model inputs. Optimization determines the optimal value of the variables over which the system designer has control such as the mix of components that make up the system and the size or quantity of each. Sensitivity analysis helps assess the effects of uncertainty or changes in the variables over which the designer has no control, such as the average wind speed or the future fuel price.

Figure 15.1 illustrates the relationship between simulation, optimization, and sensitivity analysis. The optimization oval encloses the simulation oval to represent the fact that a single optimization consists of multiple simulations. Similarly, the sensitivity analysis oval encompasses the optimization oval because a single sensitivity analysis consists of multiple optimizations.

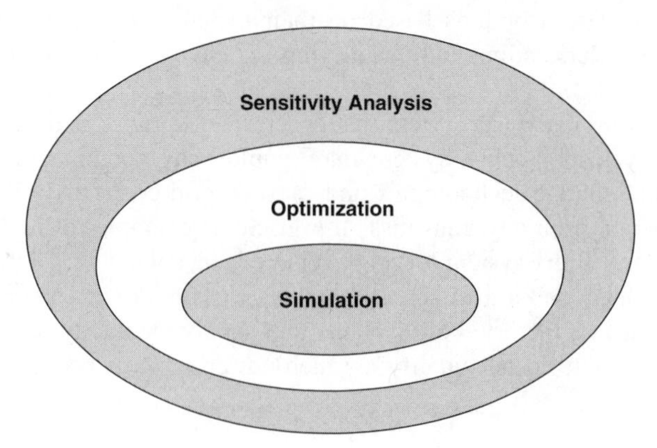

Figure 15.1 Conceptual relationship between simulation, optimization, and sensitivity analysis.

To limit input complexity, and to permit fast enough computation to make optimization and sensitivity analysis practical, HOMER's simulation logic is less detailed than that of several other time-series simulation models for micropower systems, such as Hybrid2 [1], PV-DesignPro [2], and PV*SOL [3]. On the other hand, HOMER is more detailed than statistical models such as RETScreen [4], which do not perform time-series simulations. Of all these models, HOMER is the most flexible in terms of the diversity of systems it can simulate.

In this chapter we summarize the capabilities of HOMER and discuss the benefits it can provide to the micropower system modeler. In Sections 15.2 through 15.4 we describe the structure, purpose, and capabilities of HOMER and introduce the model. In Sections 15.5 and 15.6 we discuss in greater detail the technical and economic aspects of the simulation process. A glossary defines many of the terms used in the chapter.

15.2 SIMULATION

HOMER's fundamental capability is simulating the long-term operation of a micropower system. Its higher-level capabilities, optimization and sensitivity analysis, rely on this simulation capability. The simulation process determines how a particular *system configuration*, a combination of system components of specific sizes, and an operating strategy that defines how those components work together, would behave in a given setting over a long period of time.

HOMER can simulate a wide variety of micropower system configurations, comprising any combination of a PV array, one or more wind turbines, a run-of-river hydro-turbine, and up to three generators, a battery bank, an ac–dc converter, an electrolyzer, and a hydrogen storage tank. The system can be grid-connected or autonomous and can serve ac and dc electric loads and a thermal load. Figure 15.2 shows schematic diagrams of some examples of the types of micropower systems that HOMER can simulate.

Systems that contain a battery bank and one or more generators require a dispatch strategy, which is a set of rules governing how the system charges the battery bank. HOMER can model two different dispatch strategies: load-following and cycle-charging. Under the load-following strategy, renewable power sources charge the battery but the generators do not. Under the cycle-charging strategy, whenever the generators operate, they produce more power than required to serve the load with surplus electricity going to charge the battery bank.

The simulation process serves two purposes. First, it determines whether the system is feasible. HOMER considers the system to be feasible if it can adequately serve the electric and thermal loads and satisfy any other constraints imposed by the user. Second, it estimates the life-cycle cost of the system, which is the total cost of installing and operating the system over its lifetime. The life-cycle cost is a convenient metric for comparing the economics of various system configurations. Such comparisons are the basis of HOMER's optimization process, described in Section 15.3.

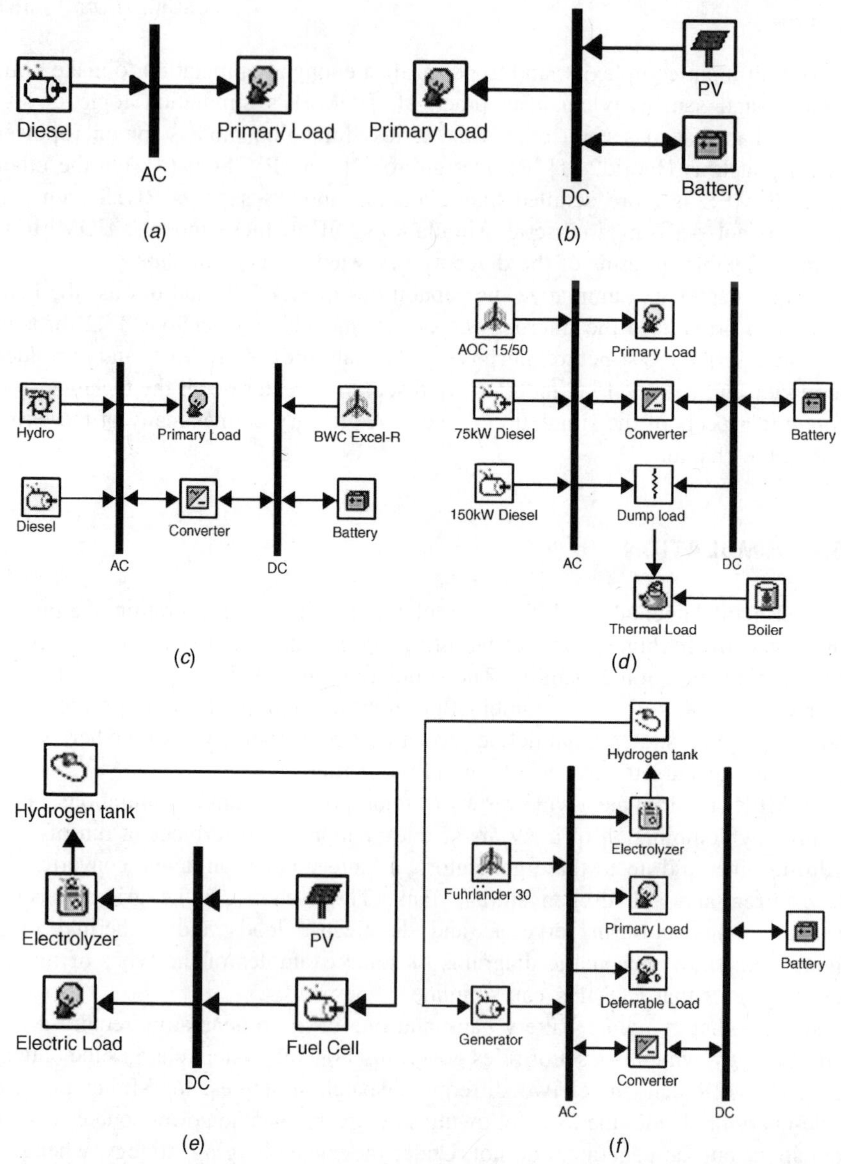

Figure 15.2 Schematic diagrams of some micropower system types that HOMER models: (*a*) a diesel system serving an ac electric load; (*b*) a PV–battery system serving a dc electric load; (*c*) a hybrid hydro–wind–diesel system with battery backup and an ac–dc converter; (*d*) a wind–diesel system serving electric and thermal loads with two generators, a battery bank, a boiler, and a dump load that helps supply the thermal load by passing excess wind turbine power through a resistive heater; (*e*) a PV–hydrogen system in which an electrolyzer converts excess PV power into hydrogen, which a hydrogen tank stores for use in a fuel cell during times of insufficient PV power; (*f*) a wind-powered system using both batteries and hydrogen for backup, where the hydrogen fuels an internal combustion engine generator; (*g*) a grid-connected PV system; (*h*) a grid-connected combined heat and power (CHP) system in which a microturbine produces both electricity and heat; (*i*) a grid-connected CHP system in which a fuel cell provides electricity and heat.

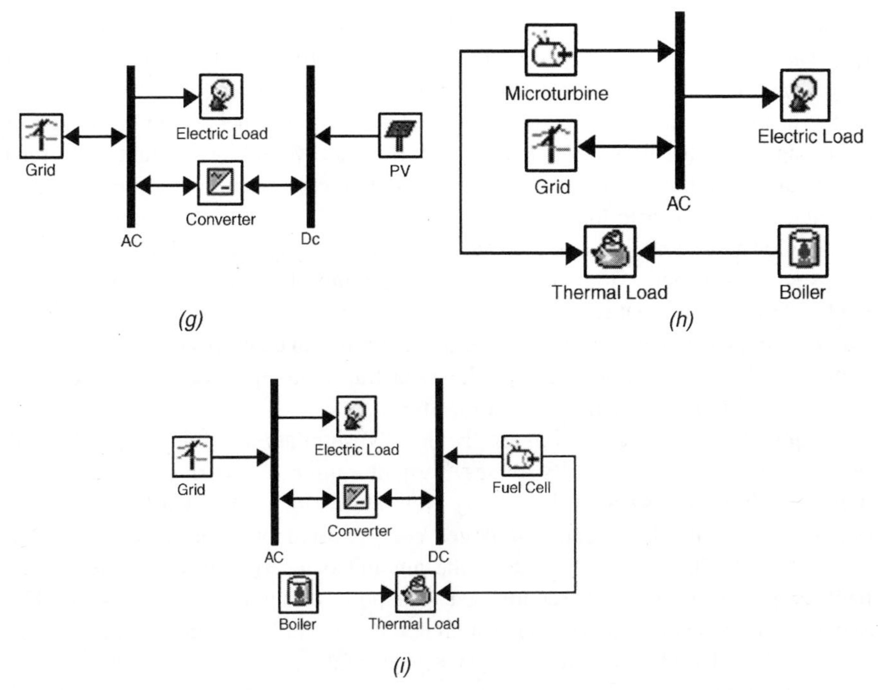

Figure 15.2 (*Continued*)

HOMER models a particular system configuration by performing an hourly time series simulation of its operation over one year. HOMER steps through the year one hour at a time, calculating the available renewable power, comparing it to the electric load, and deciding what to do with surplus renewable power in times of excess, or how best to generate (or purchase from the grid) additional power in times of deficit. When it has completed one year's worth of calculations, HOMER determines whether the system satisfies the constraints imposed by the user on such quantities as the fraction of the total electrical demand served, the proportion of power generated by renewable sources, or the emissions of certain pollutants. HOMER also computes the quantities required to calculate the system's life-cycle cost, such as the annual fuel consumption, annual generator operating hours, expected battery life, or the quantity of power purchased annually from the grid.

The quantity HOMER uses to represent the life-cycle cost of the system is the *total net present cost* (NPC). This single value includes all costs and revenues that occur within the project lifetime, with future cash flows discounted to the present. The total net present cost includes the initial capital cost of the system components, the cost of any component replacements that occur within the project lifetime, the cost of maintenance and fuel, and the cost of purchasing power from the grid. Any revenue from the sale of power to the grid reduces the total NPC. In Section 15.6 we describe in greater detail how HOMER calculates the total NPC.

For many types of micropower systems, particularly those involving intermittent renewable power sources, a one-hour time step is necessary to model the behavior of the system with acceptable accuracy. In a wind–diesel–battery system, for example, it is not enough to know the monthly average (or even daily average) wind power output, since the timing and the variability of that power output are as important as its average quantity. To predict accurately the diesel fuel consumption, diesel operating hours, the flow of energy through the battery, and the amount of surplus electrical production, it is necessary to know how closely the wind power output correlates to the electric load, and whether the wind power tends to come in long gusts followed by long lulls, or tends to fluctuate more rapidly. HOMER's one-hour time step is sufficiently small to capture the most important statistical aspects of the load and the intermittent renewable resources, but not so small as to slow computation to the extent that optimization and sensitivity analysis become impractical. Note that HOMER does not model electrical transients or other dynamic effects, which would require much smaller time steps.

Figure 15.3 shows a portion of the hourly simulation results that HOMER produced when modeling a PV–battery system similar to the one shown in Figure 15.2*b*. In such a system, the battery bank absorbs energy when the PV power output exceeds the load, and discharges energy when the load exceeds the PV power output. The graph shows how the amount of energy stored in the battery bank drops during three consecutive days of poor sunshine, October 24–26. The depletion of the battery meant that system could not supply the entire load on October 26 and 27. HOMER records such energy shortfalls and at the end of the simulation determines whether the system supplied enough of the total load to be considered feasible according to user-specified constraints. HOMER also uses the simulation results to calculate the battery throughput (the amount of energy that cycled through the battery over the year), which it uses to calculate the lifetime of the battery. The lifetime of the battery affects the total net present cost of the system.

HOMER simulates how the system operates over one year and assumes that the key simulation results for that year (such as fuel consumption, battery throughput, and surplus power production) are representative of every other year in the project

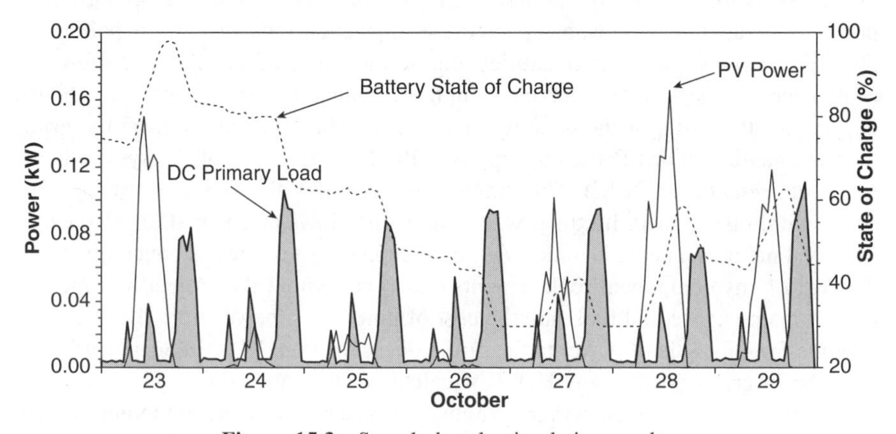

Figure 15.3 Sample hourly simulation results.

lifetime. It does not consider changes over time, such as load growth or the deterioration of battery performance with aging. The modeler can, however, analyze many of these effects using sensitivity analysis, described in Section 15.4. In Sections 15.5 and 15.6 we discuss in greater detail the technical and economic aspects of HOMER's simulation process.

15.3 OPTIMIZATION

Whereas the simulation process models a particular system configuration, the optimization process determines the best possible system configuration. In HOMER, the best possible, or *optimal*, system configuration is the one that satisfies the user-specified constraints at the lowest total net present cost. Finding the optimal system configuration may involve deciding on the mix of components that the system should contain, the size or quantity of each component, and the dispatch strategy the system should use. In the optimization process, HOMER simulates many different system configurations, discards the infeasible ones (those that do not satisfy the user-specified constraints), ranks the feasible ones according to total net present cost, and presents the feasible one with the lowest total net present cost as the optimal system configuration.

The goal of the optimization process is to determine the optimal value of each decision variable that interests the modeler. A decision variable is a variable over which the system designer has control and for which HOMER can consider multiple possible values in its optimization process. Possible decision variables in HOMER include:

- The size of the PV array
- The number of wind turbines
- The presence of the hydro system (HOMER can consider only one size of hydro system; the decision is therefore whether or not the power system should include the hydro system)
- The size of each generator
- The number of batteries
- The size of the ac–dc converter
- The size of the electrolyzer
- The size of the hydrogen storage tank
- The dispatch strategy (the set of rules governing how the system operates)

Optimization can help the modeler find the optimal system configuration out of many possibilities. Consider, for example, the task of retrofitting an existing diesel power system with wind turbines and batteries. In analyzing the options for redesigning the system, the modeler may want to consider the arrangement of components shown in Figure 15.4, but would not know in advance what number of wind turbines, what number of batteries, and what size of converter minimize

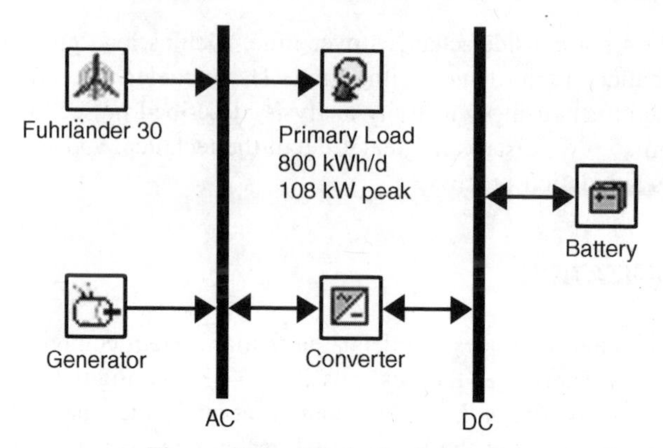

Figure 15.4 Wind–diesel system.

the life-cycle cost. These three variables would therefore be decision variables in this analysis. The dispatch strategy could also be a decision variable, but for simplicity this discussion will exclude the dispatch strategy. In Section 15.5.4 we discuss dispatch strategy in greater detail.

HOMER allows the modeler to enter multiple values for each decision variable. Using a table like the one shown in Figure 15.5, the user enters any number of values for each decision variable. The spacing between values does not have to be regular. In this example, the modeler chose to simulate five quantities of wind turbines, ranging from zero to four; the one existing generator size; seven quantities of batteries, ranging from zero to 128; and four converter sizes, ranging from zero to 120 kW. This table shows the *search space*, which is the set of all possible system configurations over which HOMER can search for the optimal system configuration. This search space includes 140 distinct system configurations because the possible values of the decision variables comprise 140 different combinations: five quantities of wind turbines multiplied by seven quantities of batteries, multiplied by four sizes of converter.

In the optimization process, HOMER simulates every system configuration in the search space and displays the feasible ones in a table, sorted by total net present

	FL30 (Quantity)	Gen (kW)	Batteries (Quantity)	Converter (kW)
1	0	135.00	0	0.00
2	1		16	30.00
3	2		32	60.00
4	3		48	120.00
5	4		64	
6			96	
7			128	
8				

Figure 15.5 Search space comprising 140 system configurations ($5 \times 1 \times 7 \times 4 = 140$).

cost. Figure 15.5 shows the results of the sample wind–diesel retrofit analysis. Each row in the table represents a feasible system configuration. The first four columns contain icons indicating the presence of the different components, the next four columns indicate the number or size of each component, and the next five columns contain a few of the key simulation results: namely, the total capital cost of the system, the total net present cost, the levelized cost of energy (cost per kilowatthour), the annual fuel consumption, and the number of hours the generator operates per year. The modeler can access the complete simulation results, including hourly data, for any particular system configuration; this table is a summary of the simulation results for many different configurations.

The first row in Figure 15.6 is the optimal system configuration, meaning the one with the lowest net present cost. In this case, the optimal configuration contains one wind turbine, the 135-kW generator, 64 batteries, and a 30-kW converter. The second-ranked system is the same as the first except that it contains two wind turbines instead of one. The third-ranked system is the same as the first except that it contains fewer batteries. The eighth- and tenth-ranked systems contain no wind turbines.

HOMER can also show a subset of these overall optimization results by displaying only the least-cost configuration within each system category or type. In the overall list shown in Figure 15.6, the top-ranked system is the least-cost configuration within the wind–diesel–battery system category. Similarly, the eighth-ranked system is the least-cost configuration within the diesel–battery system category.

FL30	Gen (kW)	Batt.	Conv. (kW)	Initial Capital	Total NPC	COE ($/kWh)	Diesel (L)	Gen (hrs)
1	135	64	30	$ 216,500	$ 849,905	0.273	75,107	4,528
2	135	64	30	$ 346,500	$ 854,660	0.274	54,434	3,350
1	135	48	30	$ 200,500	$ 855,733	0.275	78,061	4,910
2	135	48	30	$ 330,500	$ 856,335	0.275	57,654	3,685
2	135	32	30	$ 314,500	$ 873,322	0.280	62,394	4,139
2	135	96	60	$ 401,000	$ 878,370	0.282	48,139	2,603
2	135	64	60	$ 369,000	$ 880,421	0.282	52,999	3,195
	135	64	30	$ 86,500	$ 885,175	0.284	101,290	5,528
1	135	96	30	$ 248,500	$ 887,379	0.285	74,193	4,346
	135	48	30	$ 70,500	$ 888,528	0.285	104,009	6,067
1	135	32	30	$ 184,500	$ 889,688	0.285	85,310	5,615
2	135	96	30	$ 378,500	$ 890,504	0.286	52,442	3,136
2	135	48	60	$ 353,000	$ 891,896	0.286	57,316	3,615
2	135	32	60	$ 337,000	$ 905,959	0.291	62,312	4,080
2	135	128	60	$ 433,000	$ 907,508	0.291	45,596	2,226
1	135	64	60	$ 239,000	$ 911,667	0.292	77,753	4,613
	135	96	30	$ 118,500	$ 912,410	0.293	101,003	5,330

Figure 15.6 Overall optimization results table showing system configurations sorted by total net present cost.

	FL30	Gen (kW)	Batt.	Conv. (kW)	Initial Capital	Total NPC	COE ($/kWh)	Diesel (L)	Gen (hrs)
	1	135	64	30	$ 216,500	$ 849,905	0.273	75,107	4,528
		135	64	30	$ 86,500	$ 885,175	0.284	101,290	5,528
		135			$ 0	$ 996,273	0.320	132,357	8,760
	1	135			$ 130,000	$ 1,130,637	0.363	127,679	8,740

Figure 15.7 Categorized optimization results table.

The categorized optimization results list, shown in Figure 15.7, makes it easier to see which is the least-cost configuration for each category by eliminating the need to scroll through the longer list of systems displayed in the overall list.

The results in Figure 15.7 show that under the assumptions of this analysis, adding wind turbines and a battery bank would indeed reduce the life-cycle cost of the system. The initial investment of $216,500 for the optimal system leads to a savings in net present cost of $146,000 compared to the existing diesel system, shown in the third row. (Note that the diesel-only system has a capital cost of zero because the existing diesel generator requires no capital investment.) The optimization results tables allow this kind of comparison because they display more than just the optimal system configuration. The overall optimization results table in particular tends to show many system configurations whose total net present cost is only slightly higher than that of the optimal configuration. The modeler may decide that one of these suboptimal configurations is preferable in some way to the configuration that HOMER presents as optimal. For example, the simulation results may show that a system configuration with a slightly higher total net present cost than the optimal configuration does a better job of avoiding deep and extended discharges of the battery bank, which can dramatically shorten battery life in real systems. This is a technical detail that is beyond the scope of the model, but one that the modeler can take into consideration in making a final design decision.

15.4 SENSITIVITY ANALYSIS

In Section 15.3 we described the optimization process, in which HOMER finds the system configuration that is optimal under a particular set of input assumptions. In this section we describe the sensitivity analysis process, in which HOMER performs multiple optimizations, each using a different set of input assumptions. A sensitivity analysis reveals how sensitive the outputs are to changes in the inputs.

In a sensitivity analysis, the HOMER user enters a range of values for a single input variable. A variable for which the user has entered multiple values is called a *sensitivity variable*. Almost every numerical input variable in HOMER that is not a decision variable can be a sensitivity variable. Examples include the grid power price, the fuel price, the interest rate, or the lifetime of the PV array. As described in Section 15.4.2, the magnitude of an hourly data set, such as load and renewable resource data, can also be a sensitivity variable.

The HOMER user can perform a sensitivity analysis with any number of sensitivity variables. Each combination of sensitivity variable values defines a distinct sensitivity case. For example, if the user specifies six values for the grid power price and four values for the interest rate, that defines 24 distinct sensitivity cases. HOMER performs a separate optimization process for each sensitivity case and presents the results in various tabular and graphic formats.

One of the primary uses of sensitivity analysis is in dealing with uncertainty. If a system designer is unsure of the value of a particular variable, he or she can enter several values covering the likely range and see how the results vary across that range. But sensitivity analysis has applications beyond coping with uncertainty. A system designer can use sensitivity analysis to evaluate trade-offs and answer such questions as: How much additional capital investment is required to achieve 50% or 100% renewable energy production? An energy planner can determine which technologies, or combinations of technologies, are optimal under different conditions. A market analyst can determine at what price, or under what conditions, a product (e.g., a fuel cell or a wind turbine) competes with the alternatives. A policy analyst can determine what level of incentive is needed to stimulate the market for a particular technology, or what level of emissions penalty would tilt the economics toward cleaner technologies.

15.4.1 Dealing with Uncertainty

A challenge that often confronts the micropower system designer is uncertainty in key variables. Sensitivity analysis can help the designer understand the effects of uncertainty and make good design decisions despite uncertainty. For example, consider the wind–diesel system analysis in Section 15.3. In performing this analysis, the modeler assumed that the price of diesel fuel would be $0.60 per liter over the 25-year project lifetime. There is obviously substantial uncertainty in this value, but many other inputs may be uncertain as well, such as the lifetime of the wind turbine, the maintenance cost of the diesel engine, the long-term average wind speed at the site, and even the average electric load. Sensitivity analysis can help the modeler to determine the effect that variations in these inputs have on the behavior, feasibility, and economics of a particular system configuration; the robustness of a particular system configuration (in other words, whether it is nearly optimal in all scenarios, or far from optimal in certain scenarios); and how the optimal system configuration changes across the range of uncertainty.

The spider graph in Figure 15.8 shows the results of a sensitivity analysis on three variables. In this analysis, the modeler fixed the system configuration to the wind–diesel system that appears in the first row of the optimization table in Figure 15.6, but entered multiple values for three input variables: the diesel fuel price, the wind turbine lifetime, and the generator O&M (operating and maintenance) cost. For each variable, the modeler entered values ranging from 30% below to 30% above a best estimate. Figure 15.8 shows how sensitive the total net present cost is to each of these three variables. The relative steepness of the three curves shows that the total NPC is more sensitive to the fuel price than to the other two

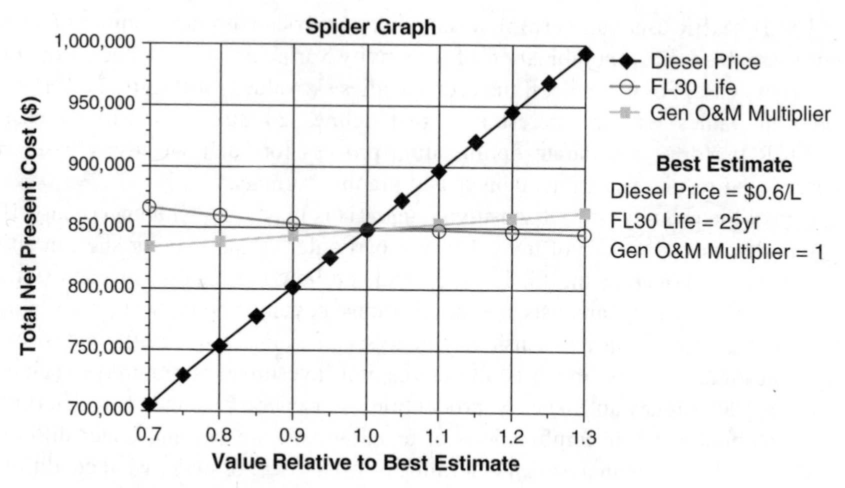

Figure 15.8 Spider graph showing the effect of changes in three sensitivity variables.

variables. Such information can help a system designer to establish the bounds of a confidence interval or to prioritize efforts to reduce uncertainty.

A sensitivity analysis can also incorporate optimization. Figure 15.9 shows the results of a second sensitivity analysis on the wind–diesel system from Section 15.3. The modeler performed this analysis to see whether the fuel price, which Figure 15.8 identified as the most important of the uncertain variables, affects the optimal system configuration. This time the modeler used the search space shown in Figure 15.5 and entered a range of fuel prices above and below the best-estimate

Diesel ($/L)	FL30	Gen (kW)	Batt.	Conv. (kW)	Total NPC
0.420		135	48	30	$ 688,679
0.450		135	48	30	$ 721,987
0.480	1	135	64	30	$ 753,695
0.510	1	135	64	30	$ 777,748
0.540	1	135	64	30	$ 801,800
0.570	1	135	64	30	$ 825,852
→ 0.600	1	135	64	30	$ 849,905
0.630	2	135	64	30	$ 872,093
0.660	2	135	64	30	$ 889,525
0.690	2	135	64	30	$ 906,957
0.720	2	135	64	30	$ 924,389
0.750	2	135	64	30	$ 941,821
0.780	2	135	64	30	$ 959,253

Figure 15.9 Tabular sensitivity results showing optimal system configuration changing with fuel price.

FL30	Gen (kW)	Batt.	Conv. (kW)	Total NPC
	135	48	30	$ 688,679
	135	64	30	$ 690,550
1	135	64	30	$ 705,590
1	135	48	30	$ 705,704
	135	32	30	$ 708,090
	135	96	30	$ 718,337

Figure 15.10 Optimization results for the $0.42 per liter fuel price sensitivity case.

value of $0.60 per liter. The tabular results in Figure 15.9 show that as the fuel price increases, the optimal system configuration changes from a diesel–battery system to a wind–diesel–battery system comprising one wind turbine, and as the fuel price increases further, to a wind–diesel–battery system comprising two wind turbines.

The arrow in Figure 15.9 highlights the best estimate scenario with a fuel price of $0.60 per liter. The optimal system configuration for this scenario (one wind turbine, a 135-kW generator, 64 batteries, and a 30-kW converter) is optimal for five of the 13 sensitivity cases. An investigation of the overall optimization results tables for the other sensitivity cases shows that this system configuration is nearly optimal for those sensitivity cases as well. For example, Figure 15.10 shows the optimization results table for the lowest fuel price scenario ($0.42 per liter). The system configuration that was optimal in the $0.60 per liter scenario ranks third in this scenario, with a total NPC of $705,590, which is only 2.4% higher than the total NPC of the optimal configuration. For the most expensive fuel scenario, it ranks fifth, with a total NPC only 3.6% higher than the optimal configuration. It is therefore a fairly robust solution in that it performs well across the given range of fuel prices. By contrast, the diesel–battery system that is optimal in the lowest fuel price scenario ranks thirty-sixth in the highest fuel price scenario, with a total NPC 13.5% higher than the optimal configuration. The analysis shows that the diesel–battery system would have a higher risk than the wind–diesel–battery system because it is far from optimal under some fuel price scenarios.

With this kind of information, a modeler can make informed decisions despite uncertainty in important variables. As in the example above, a sensitivity analysis can reveal how different system configurations perform over a wide range of possible scenarios, helping the designer assess the risks associated with each.

15.4.2 Sensitivity Analyses on Hourly Data Sets

One of HOMER's most powerful features is its ability to do sensitivity analyses on hourly data sets such as the primary electric load or the solar, wind, hydro, or biomass resource. HOMER's use of scaling variables enables such sensitivity analyses. Each hourly data set comprises 8760 values that have a certain average value. But each hourly data set also has a corresponding scaling variable that the modeler can use to scale the entire data set up or down. For example, a user may specify hourly primary load data with an annual average of 120 kWh/day, then specify 100, 150, and 200 kWh/day for the primary load scaling variable. In the course

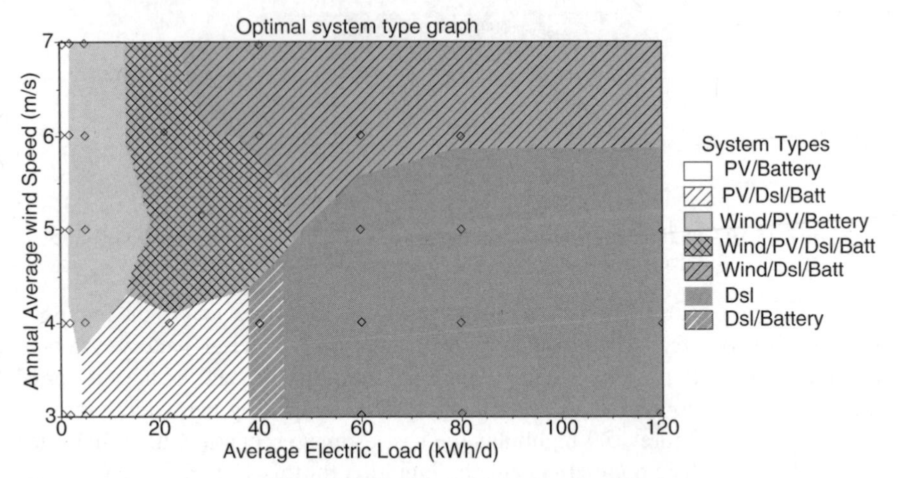

Figure 15.11 Optimal system graph.

of the sensitivity analysis, HOMER will scale the load data so that it averages first 100 kWh/day, then 150 kWh/day, and finally, 200 kWh/day. This scaling process changes the magnitude of the load data set without affecting the daily load shape, the seasonal pattern, or any other statistical properties. HOMER scales renewable resource data in the same manner.

Figure 15.11 shows the results of a sensitivity analysis over a range of load sizes and annual average wind speeds. The modeler specified eight values for the average size of the electric load and five values for the annual average wind speed. The axes of the graph correspond to these two sensitivity variables. At each of the 40 sensitivity cases, HOMER performed an optimization over a search space comprising more than 5000 system configurations. The diamonds in the graph indicate these sensitivity cases, and the color of each diamond indicates the optimal system type for that sensitivity case. At an average load of 22 kWh/day and an average wind speed of 4 m/s, for example, the optimal system type was PV–diesel–battery. At an average load of 40 kWh/day and the same wind speed, the optimal system type was diesel–battery. HOMER uses two-dimensional linear interpolation to determine the optimal system type at all points between the diamonds.

The graph in Figure 15.11 shows that for the assumptions used in this analysis, PV–battery systems are optimal for very small systems, regardless of the wind speed. At low wind speeds, as the load size increases the optimal system type changes to PV–diesel–battery, diesel–battery, then pure diesel. At high wind speeds, as the load size increases, the optimal system type changes to wind–PV–battery, wind–PV–diesel–battery, and finally, wind–diesel–battery. To an energy planner intending to provide electricity to many unelectrified communities in a developing country, such an analysis could help decide what type of micropower system to use for different communities based on the load size and average wind speed of each community.

15.5 PHYSICAL MODELING

In Section 15.2 we discussed the role of simulation and briefly described the process HOMER uses to simulate micropower systems. In this section we provide greater detail on how HOMER models the physical operation of a system. In HOMER, a micropower system must comprise at least one source of electrical or thermal energy (such as a wind turbine, a diesel generator, a boiler, or the grid), and at least one destination for that energy (an electrical or thermal load, or the ability to sell electricity to the grid). It may also comprise conversion devices such as an ac–dc converter or an electrolyzer, and energy storage devices such as a battery bank or a hydrogen storage tank.

In the following subsections we describe how HOMER models the loads that the system must serve, the components of the system and their associated resources, and how that collection of components operates together to serve the loads.

15.5.1 Loads

In HOMER, the term *load* refers to a demand for electric or thermal energy. Serving loads is the reason for the existence of micropower systems, so the modeling of a micropower system begins with the modeling of the load or loads that the system must serve. HOMER models three types of loads. Primary load is electric demand that must be served according to a particular schedule. Deferrable load is electric demand that can be served at any time within a certain time span. Thermal load is demand for heat.

Primary Load Primary load is electrical demand that the power system must meet at a specific time. Electrical demand associated with lights, radio, TV, household appliances, computers, and industrial processes is typically modeled as primary load. When a consumer switches on a light, the power system must supply electricity to that light immediately—the load cannot be deferred until later. If electrical demand exceeds supply, there is a shortfall that HOMER records as *unmet load.*

The HOMER user specifies an amount of primary load in kilowatts for each hour of the year, either by importing a file containing hourly data or by allowing HOMER to synthesize hourly data from average daily load profiles. When synthesizing load data, HOMER creates hourly load values based on user-specified daily load profiles. The modeler can specify a single 24-hour profile that applies throughout the year, or can specify different profiles for different months and different profiles for weekdays and weekends. HOMER adds a user-specified amount of randomness to synthesized load data so that every day's load pattern is unique. HOMER can model two separate primary loads, each of which can be ac or dc.

Among the three types of loads modeled in HOMER, primary load receives special treatment in that it requires a user-specified amount of operating reserve. Operating reserve is surplus electrical generating capacity that is operating and can respond instantly to a sudden increase in the electric load or a sudden decrease in the renewable power output. Although it has the same meaning as the more common term *spinning reserve*, we call it *operating reserve* because batteries, fuel cells, and the grid can provide it, but they do not spin. When simulating the operation of the system, HOMER attempts to ensure that the system's operating capacity is always sufficient to supply the primary load and the required operating reserve. Section 15.5.4 covers operating reserve in greater detail.

Deferrable Load Deferrable load is electrical demand that can be met anytime within a defined time interval. Water pumps, ice makers, and battery-charging stations are examples of deferrable loads because the storage inherent to each of those loads allows some flexibility as to when the system can serve them. The ability to defer serving a load is often advantageous for systems comprising intermittent renewable power sources, because it reduces the need for precise control of the timing of power production. If the renewable power supply ever exceeds the primary load, the surplus can serve the deferrable load rather than going to waste.

Figure 15.12 shows a schematic representation of how HOMER models the deferrable load. The power system puts energy into a "tank" of finite capacity, and energy drains out of that tank to serve the deferrable load. For each month, the user specifies the average deferrable load, which is the rate at which energy drains out of the tank. The user also specifies the storage capacity in kilowatthours (the size of the tank), and the maximum and minimum rate at which the power system can put energy into the tank. Note that the energy tank model is simply an analogy; the actual deferrable load may or may not make use of a storage tank.

When simulating a system serving a deferrable load, HOMER tracks the level in the deferrable load tank. It will put any excess renewable power into the tank, but as long as the tank level remains above zero, HOMER will not use a dispatchable power source (a generator, the battery bank, or the grid) to put energy into the tank. If the level in the tank drops to zero, HOMER temporarily treats the deferrable load as a primary load, meaning that it will immediately use any available power source to put energy into the tank and avoid having the deferrable load go unmet.

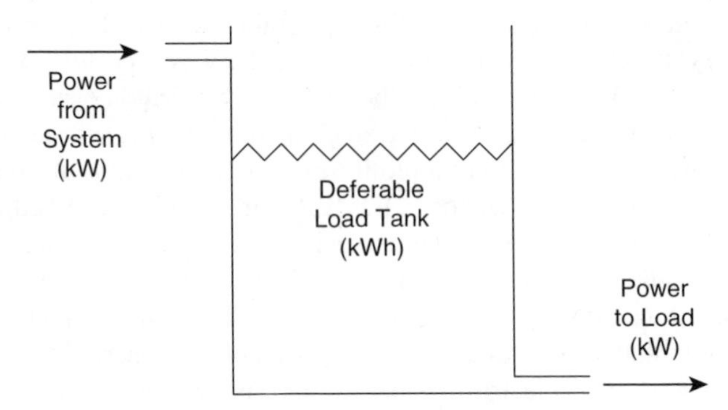

Figure 15.12 Deferrable load tank analogy.

Thermal Load HOMER models thermal load in the same way that it models primary electric load, except that the concept of operating reserve does not apply to the thermal load. The user specifies the amount of thermal load for each hour of the year, either by importing a file containing hourly data or by allowing HOMER to synthesize hourly data from 24-hour load profiles. The system supplies the thermal load with either the boiler, waste heat recovered from a generator, or resistive heating using excess electricity.

15.5.2 Resources

The term *resource* applies to anything coming from outside the system that is used by the system to generate electric or thermal power. That includes the four renewable resources (solar, wind, hydro, and biomass) as well as any fuel used by the components of the system. Renewable resources vary enormously by location. The solar resource depends strongly on latitude and climate, the wind resource on large-scale atmospheric circulation patterns and geographic influences, the hydro resource on local rainfall patterns and topography, and the biomass resource on local biological productivity. Moreover, at any one location a renewable resource may exhibit strong seasonal and hour-to-hour variability. The nature of the available renewable resources affects the behavior and economics of renewable power systems, since the resource determines the quantity and the timing of renewable power production. The careful modeling of the renewable resources is therefore an essential element of system modeling. In this section we describe how HOMER models the four renewable resources and the fuel.

Solar Resource To model a system containing a PV array, the HOMER user must provide solar resource data for the location of interest. Solar resource data indicate the amount of global solar radiation (beam radiation coming directly from the sun, plus diffuse radiation coming from all parts of the sky) that strikes Earth's surface in a typical year. The data can be in one of three forms: hourly average global solar radiation on the horizontal surface (kW/m^2), monthly average global solar radiation on the horizontal surface (kWh/m^2·day), or monthly average clearness index. The clearness index is the ratio of the solar radiation striking Earth's surface to the solar radiation striking the top of the atmosphere. A number between zero and 1, the clearness index is a measure of the clearness of the atmosphere.

If the user chooses to provide monthly solar resource data, HOMER generates synthetic hourly global solar radiation data using an algorithm developed by Graham and Hollands [7]. The inputs to this algorithm are the monthly average solar radiation values and the latitude. The output is an 8760-hour data set with statistical characteristics similar to those of real measured data sets. One of those statistical properties is autocorrelation, which is the tendency for one day to be similar to the preceding day, and for one hour to be similar to the preceding hour.

Wind Resource To model a system comprising one or more wind turbines, the HOMER user must provide wind resource data indicating the wind speeds the turbines would experience in a typical year. The user can provide measured hourly wind speed data if available. Otherwise, HOMER can generate synthetic hourly data from 12 monthly average wind speeds and four additional statistical parameters: the Weibull shape factor, the autocorrelation factor, the diurnal pattern strength, and the hour of peak wind speed. The Weibull shape factor is a measure of the distribution of wind speeds over the year. The autocorrelation factor is a measure of how strongly the wind speed in one hour tends to depend on the wind speed in the preceding hour. The diurnal pattern strength and the hour of peak wind speed indicate the magnitude and the phase, respectively, of the average daily pattern in the wind speed. HOMER provides default values for each of these parameters.

The user indicates the anemometer height, meaning the height above ground at which the wind speed data were measured or for which they were estimated. If the wind turbine hub height is different from the anemometer height, HOMER calculates the wind speed at the turbine hub height using either the logarithmic law, which assumes that the wind speed is proportional to the logarithm of the height above ground, or the power law, which assumes that the wind speed varies exponentially with height. To use the logarithmic law, the user enters the surface roughness length, which is a parameter characterizing the roughness of the surrounding terrain. To use the power law, the user enters the power law exponent.

The user also indicates the elevation of the site above sea level, which HOMER uses to calculate the air density according to the U.S. Standard Atmosphere, described in Section 2.3 of White [6]. HOMER makes use of the air density when calculating the output of the wind turbine, as described in Section 15.5.3.

Hydro Resource To model a system comprising a run-of-river hydro turbine, the HOMER user must provide stream flow data indicating the amount of water available to the turbine in a typical year. The user can provide measured hourly stream flow data if available. Otherwise, HOMER can use monthly averages under the assumption that the flow rate remains constant within each month. The user also specifies the residual flow, which is the minimum stream flow that must bypass the hydro turbine for ecological purposes. HOMER subtracts the residual flow from the stream flow data to determine the stream flow available to the turbine.

Biomass Resource The biomass resource takes various forms (e.g., wood waste, agricultural residue, animal waste, energy crops) and may be used to produce heat or electricity. HOMER models biomass power systems that convert biomass into electricity. Two aspects of the biomass resource make it unique among the four renewable resources that HOMER models. First, the availability of the resource depends in part on human effort for harvesting, transportation, and storage. It is consequently not intermittent, although it may be seasonal. It is also often not free. Second, the biomass feedstock may be converted to a gaseous or liquid fuel, to be consumed in an otherwise conventional generator. The modeling of the biomass resource is therefore similar in many ways to the modeling of any other fuel.

The HOMER user can model the biomass resource in two ways. The simplest way is to define a fuel with properties corresponding to the biomass feedstock and then specify the fuel consumption of the generator to show electricity produced versus biomass feedstock consumed. This approach implicitly, rather than explicitly, models the process of converting the feedstock into a fuel suitable for the generator, if such a process occurs. The second alternative is to use HOMER's biomass resource inputs, which allow the modeler to specify the availability of the feedstock throughout the year, and to model explicitly the feedstock conversion process. In the remainder of this section we focus on this second alternative. In the next section we address the fuel inputs.

As with the other renewable resource data sets, the HOMER user can indicate the availability of biomass feedstock by importing an hourly data file or using monthly averages. If the user specifies monthly averages, HOMER assumes that the availability remains constant within each month.

The user must specify four additional parameters to define the biomass resource: price, carbon content, gasification ratio, and the energy content of the biomass fuel. For greenhouse gas analyses, the carbon content value should reflect the net amount of carbon released to the atmosphere by the harvesting, processing, and consumption of the biomass feedstock, considering the fact that the carbon in the feedstock was originally in the atmosphere. The gasification ratio, despite its name, applies equally well to liquid and gaseous fuels. It is the fuel conversion ratio, indicating the ratio of the mass of generator-ready fuel emerging from the fuel conversion process to the mass of biomass feedstock entering the fuel conversion process. HOMER uses the energy content of the biomass fuel to calculate the thermodynamic efficiency of the generator that consumes the fuel.

Fuel HOMER provides a library of several predefined fuels, and users can add to the library if necessary. The physical properties of a fuel include its density, lower heating value, carbon content, and sulfur content. The user can also choose the most appropriate measurement units, either L, m^3, or kg. The two remaining properties of the fuel are the price and the annual consumption limit, if any.

15.5.3 Components

In HOMER, a *component* is any part of a micropower system that generates, delivers, converts, or stores energy. HOMER models 10 types of components. Three generate electricity from intermittent renewable sources: photovoltaic modules, wind turbines, and hydro turbines. PV modules convert solar radiation into dc electricity. Wind turbines convert wind energy into ac or dc electricity. Hydro turbines convert the energy of flowing water into ac or dc electricity. HOMER can only model run-of-river hydro installations, meaning those that do not comprise a storage reservoir.

Another three types of components, generators, the grid, and boilers, are dispatchable energy sources, meaning that the system can control them as needed. Generators consume fuel to produce ac or dc electricity. A generator may also produce thermal power via waste heat recovery. The grid delivers ac electricity to a grid-connected system and may also accept surplus electricity from the system. Boilers consume fuel to produce thermal power.

Two types of components, converters and electrolyzers, convert electrical energy into another form. Converters convert electricity from ac to dc or from dc to ac. Electrolyzers convert surplus ac or dc electricity into hydrogen via the electrolysis of water. The system can store the hydrogen and use it as fuel for one or more generators. Finally, two types of components store energy: batteries and hydrogen storage tanks. Batteries store dc electricity. Hydrogen tanks store hydrogen from the electrolyzer to fuel one or more generators.

In this section we explain how HOMER models each of these components and discuss the physical and economic properties that the user can use to describe each.

PV Array HOMER models the PV array as a device that produces dc electricity in direct proportion to the global solar radiation incident upon it, independent of its temperature and the voltage to which it is exposed. HOMER calculates the power output of the PV array using the equation

$$P_{\mathrm{PV}} = f_{\mathrm{PV}} Y_{\mathrm{PV}} \frac{I_T}{I_S} \tag{15.1}$$

Where, f_{PV} is the PV derating factor, Y_{PV} the rated capacity of the PV array (kW), I_{T} the global solar radiation (beam plus diffuse) incident on the surface of the PV array (kW/m^2), and I_{S} is 1 kW/m^2, which is the standard amount of radiation used to rate the capacity of the PV array. In the following paragraphs we describe these variables in more detail.

The rated capacity (sometimes called the peak capacity) of a PV array is the amount of power it would produce under standard test conditions of 1 kW/m^2 irradiance and a panel temperature of 25°C. In HOMER, the size of a PV array is always specified in terms of rated capacity. The rated capacity accounts for both the area and the efficiency of the PV module, so neither of those parameters appears explicitly in HOMER. A 40-W module made of amorphous silicon (which has a relatively low efficiency) will be larger than a 40-W module made of polycrystalline silicon (which has a relatively high efficiency), but that size difference is of no consequence to HOMER.

Each hour of the year, HOMER calculates the global solar radiation incident on the PV array using the HDKR model, explained in Section 2.16 of Duffie and Beckmann [5]. This model takes into account the current value of the solar resource (the global solar radiation incident on a horizontal surface), the orientation of the PV array, the location on Earth's surface, the time of year, and the time of day. The orientation of the array may be fixed or may vary according to one of several tracking schemes.

The derating factor is a scaling factor meant to account for effects of dust on the panel, wire losses, elevated temperature, or anything else that would cause the output of the PV array to deviate from that expected under ideal conditions. HOMER does not account for the fact that the power output of a PV array decreases with increasing panel temperature, but the HOMER user can reduce the derating factor to (crudely) correct for this effect when modeling systems for hot climates.

In reality, the output of a PV array does depend strongly and nonlinearly on the voltage to which it is exposed. The maximum power point (the voltage at which the power output is maximized) depends on the solar radiation and the temperature. If the PV array is connected directly to a dc load or a battery bank, it will often be exposed to a voltage different from the maximum power point, and performance will suffer. A maximum power point tracker (MPPT) is a solid-state device placed between the PV array and the rest of the dc components of the system that decouples the array voltage from that of the rest of the system, and ensures that the array voltage is always equal to the maximum power point. By ignoring the effect of the voltage to which the PV array is exposed, HOMER effectively assumes that a maximum power point tracker is present in the system.

To describe the cost of the PV array, the user specifies its initial capital cost in dollars, replacement cost in dollars, and operating and maintenance (O&M) cost in dollars per year. The replacement cost is the cost of replacing the PV array at the end of its useful lifetime, which the user specifies in years. By default, the replacement cost is equal to the capital cost, but the two can differ for several reasons. For example, a donor organization may cover some or all of the initial capital cost but none of the replacement cost.

Wind Turbine HOMER models a wind turbine as a device that converts the kinetic energy of the wind into ac or dc electricity according to a particular power curve, which is a graph of power output versus wind speed at hub height. Figure 15.13 is an example power curve. HOMER assumes that the power curve

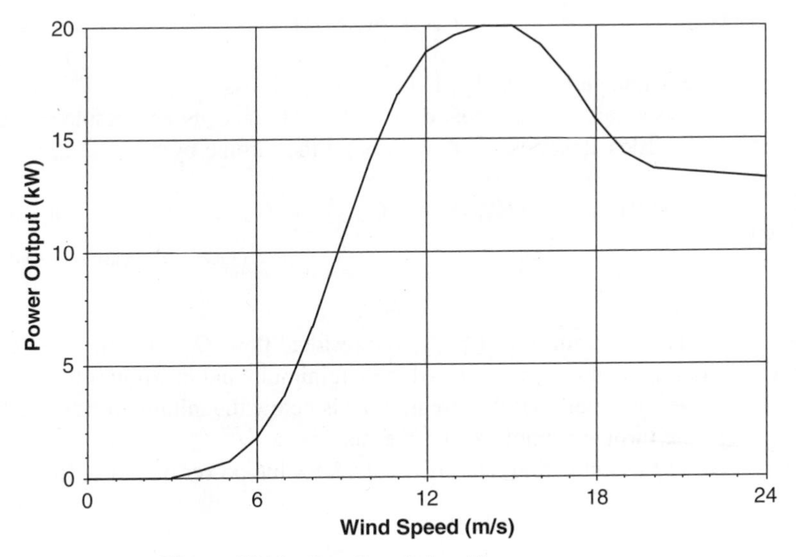

Figure 15.13 Sample wind turbine power curve.

applies at a standard air density of 1.225 kg/m^3, which corresponds to standard temperature and pressure conditions.

Each hour, HOMER calculates the power output of the wind turbine in a four-step process. First, it determines the average wind speed for the hour at the anemometer height by referring to the wind resource data. Second, it calculates the corresponding wind speed at the turbine's hub height using either the logarithmic law or the power law. Third, it refers to the turbine's power curve to calculate its power output at that wind speed assuming standard air density. Fourth, it multiplies that power output value by the air density ratio, which is the ratio of the actual air density to the standard air density. As mentioned in Section 15.5.2, HOMER calculates the air density ratio at the site elevation using the U.S. Standard Atmosphere [6]. HOMER assumes that the air density ratio is constant throughout the year.

In addition to the turbine's power curve and hub height, the user specifies the expected lifetime of the turbine in years, its initial capital cost in dollars, its replacement cost in dollars, and its annual O&M cost in dollars per year.

Hydro Turbine HOMER models the hydro turbine as a device that converts the power of falling water into ac or dc electricity at a constant efficiency, with no ability to store water or modulate the power output. The power in falling water is proportional to the product of the stream flow and the head, which is the vertical distance through which the water falls. Information on the stream flow available to the hydro turbine each hour comes from the hydro resource data. The user also enters the available head and the head loss that occurs in the intake pipe due to friction. HOMER calculates the net head, or effective head, using the equation

$$h_{\text{net}} = h(1 - f_h) \tag{15.2}$$

where h is the available head and f_h is the pipe head loss.

The user also enters the turbine's design flow rate and its acceptable range of flow rates. HOMER calculates the flow through the turbine by

$$\dot{Q}_{\text{turbine}} = \begin{cases} \min\left(\dot{Q}_{\text{stream}} - \dot{Q}_{\text{residual}}, w_{\max}\dot{Q}_{\text{nom}}\right) & \dot{Q}_{\text{stream}} - \dot{Q}_{\text{residual}} \geq w_{\min}\dot{Q}_{\text{nom}} \\ 0 & \dot{Q}_{\text{stream}} - \dot{Q}_{\text{residual}} < w_{\min}\dot{Q}_{\text{nom}} \end{cases} \tag{15.3}$$

where $\dot{Q}_{\text{stream}}$ is the stream flow, $\dot{Q}_{\text{residual}}$ the residual flow, $\dot{Q}_{\text{nom}}$ the turbine design flow rate, and $w_{\min}$ and $w_{\max}$ are the turbine's minimum and maximum flow ratios. The turbine does not operate if the stream flow is below the minimum, and the flow rate through the turbine cannot exceed the maximum.

Each hour of the simulation, HOMER calculates the power output of the hydroturbine as

$$P_{\text{hyd}} = \eta_{\text{hyd}}\rho_{\text{water}}\, g\, h_{\text{net}}\dot{Q}_{\text{turbine}} \tag{15.4}$$

where η_{hyd} is the turbine efficiency, ρ_{water} the density of water, g the gravitational acceleration, h_{net} the net head, and $\dot{Q}_{turbine}$ the flow rate through the turbine. The user specifies the expected lifetime of the hydro-turbine in years, as well as its initial capital cost in dollars, replacement cost in dollars, and annual O&M cost in dollars per year.

Generators A generator consumes fuel to produce electricity, and possibly heat as a by-product. HOMER's generator module is flexible enough to model a wide variety of generators, including internal combustion engine generators, microturbines, fuel cells, Stirling engines, thermophotovoltaic generators, and thermoelectric generators. HOMER can model a power system comprising as many as three generators, each of which can be ac or dc, and each of which can consume a different fuel.

The principal physical properties of the generator are its maximum and minimum electrical power output, its expected lifetime in operating hours, the type of fuel it consumes, and its fuel curve, which relates the quantity of fuel consumed to the electrical power produced. In HOMER, a generator can consume any of the fuels listed in the fuel library (to which users can add their own fuels) or one of two special fuels: electrolyzed hydrogen from the hydrogen storage tank, or biomass derived from the biomass resource. It is also possible to cofire a generator with a mixture of biomass and another fuel.

HOMER assumes the fuel curve is a straight line with a y-intercept and uses the following equation for the generator's fuel consumption:

$$F = F_0 Y_{gen} + F_1 P_{gen} \tag{15.5}$$

where F_0 is the fuel curve intercept coefficient, F_1 is the fuel curve slope, Y_{gen} the rated capacity of the generator (kW), and P_{gen} the electrical output of the generator (kW). The units of F depend on the measurement units of the fuel. If the fuel is denominated in liters, the units of F are L/h. If the fuel is denominated in m^3 or kg, the units of F are m^3/h or kg/h, respectively. In the same way, the units of F_0 and F_1 depend on the measurement units of the fuel. For fuels denominated in liters, the units of F_0 and F_1 are L/h·kW.

For a generator that provides heat as well as electricity, the user also specifies the heat recovery ratio. HOMER assumes that the generator converts all the fuel energy into either electricity or waste heat. The heat recovery ratio is the fraction of that waste heat that can be captured to serve the thermal load. In addition to these properties, the modeler can specify the generator emissions coefficients, which specify the generator's emissions of six different pollutants in grams of pollutant emitted per quantity of fuel consumed.

The user can schedule the operation of the generator to force it on or off at certain times. During times that the generator is neither forced on or off, HOMER decides whether it should operate based on the needs of the system and the relative costs of the other power sources. During times that the generator is forced on, HOMER decides at what power output level it operates, which may be anywhere between its minimum and maximum power output.

The user specifies the generator's initial capital cost in dollars, replacement cost in dollars, and annual O&M cost in dollars per operating hour. The generator O&M cost should account for oil changes and other maintenance costs, but not fuel cost because HOMER calculates fuel cost separately. As it does for all dispatchable power sources, HOMER calculates the generator's fixed and marginal cost of energy and uses that information when simulating the operation of the system. The fixed cost of energy is the cost per hour of simply running the generator, without producing any electricity. The marginal cost of energy is the additional cost per kilowatthour of producing electricity from that generator.

HOMER uses the following equation to calculate the generator's fixed cost of energy:

$$c_{\text{gen,fixed}} = c_{\text{om,gen}} + \frac{C_{\text{rep,gen}}}{R_{\text{gen}}} + F_0 Y_{\text{gen}} c_{\text{fuel,eff}} \tag{15.6}$$

where $c_{\text{om,gen}}$ is the O&M cost in dollars per hour, $C_{\text{rep,gen}}$ the replacement cost in dollars, R_{gen} the generator lifetime in hours, F_0 the fuel curve intercept coefficient in quantity of fuel per hour per kilowatt, Y_{gen} the capacity of the generator (kW), and $c_{\text{fuel,eff}}$ the effective price of fuel in dollars per quantity of fuel. The effective price of fuel includes the cost penalties, if any, associated with the emissions of pollutants from the generator.

HOMER calculates the marginal cost of energy of the generator using the following equation:

$$c_{\text{gen,mar}} = F_1 c_{\text{fuel,eff}} \tag{15.7}$$

where F_1 is the fuel curve slope in quantity of fuel per hour per kilowatthour and $c_{\text{fuel,eff}}$ is the effective price of fuel (including the cost of any penalties on emissions) in dollars per quantity of fuel.

Battery Bank The battery bank is a collection of one or more individual batteries. HOMER models a single battery as a device capable of storing a certain amount of dc electricity at a fixed round-trip energy efficiency, with limits as to how quickly it can be charged or discharged, how deeply it can be discharged without causing damage, and how much energy can cycle through it before it needs replacement. HOMER assumes that the properties of the batteries remain constant throughout its lifetime and are not affected by external factors such as temperature.

In HOMER, the key physical properties of the battery are its nominal voltage, capacity curve, lifetime curve, minimum state of charge, and round-trip efficiency. The capacity curve shows the discharge capacity of the battery in ampere-hours versus the discharge current in amperes. Manufacturers determine each point on this curve by measuring the ampere-hours that can be discharged at a constant current out of a fully charged battery. Capacity typically decreases with increasing discharge current. The lifetime curve shows the number of discharge–charge cycles the battery can withstand versus the cycle depth. The number of cycles to failure typically decreases with increasing cycle depth. The minimum state of charge is the

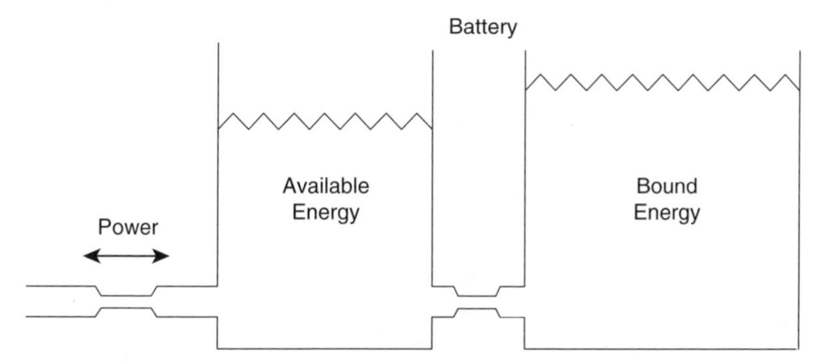

Figure 15.14 Kinetic battery model concept.

state of charge below which the battery must not be discharged to avoid permanent damage. In the system simulation, HOMER does not allow the battery to be discharged any deeper than this limit. The round-trip efficiency indicates the percentage of the energy going into the battery that can be drawn back out.

To calculate the battery's maximum allowable rate of charge or discharge, HOMER uses the kinetic battery model [8], which treats the battery as a two-tank system as illustrated in Figure 15.14. According to the kinetic battery model, part of the battery's energy storage capacity is immediately available for charging or discharging, but the rest is chemically bound. The rate of conversion between available energy and bound energy depends on the difference in "height" between the two tanks. Three parameters describe the battery. The maximum capacity of the battery is the combined size of the available and bound tanks. The capacity ratio is the ratio of the size of the available tank to the combined size of the two tanks. The rate constant is analogous to the size of the pipe between the tanks.

The kinetic battery model explains the shape of the typical battery capacity curve, such as the example shown in Figure 15.15. At high discharge rates, the available tank empties quickly, and very little of the bound energy can be converted to available energy before the available tank is empty, at which time the battery can no longer withstand the high discharge rate and appears fully discharged. At slower discharge rates, more bound energy can be converted to available energy before the available tank empties, so the apparent capacity increases. HOMER performs a curve fit on the battery's discharge curve to calculate the three parameters of the kinetic battery model. The line in Figure 15.15 corresponds to this curve fit.

Modeling the battery as a two-tank system rather than a single-tank system has two effects. First, it means the battery cannot be fully charged or discharged all at once; a complete charge requires an infinite amount of time at a charge current that asymptotically approaches zero. Second, it means that the battery's ability to charge and discharge depends not only on its current state of charge, but also on its recent charge and discharge history. A battery rapidly charged to 80% state of charge will be capable of a higher discharge rate than the same battery rapidly discharged to 80%, since it will have a higher level in its available tank. HOMER tracks the levels in the two tanks each hour, and models both these effects.

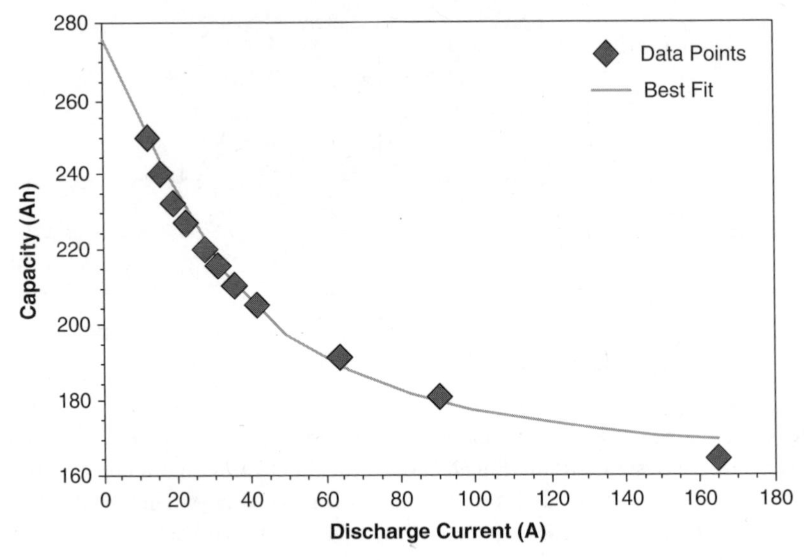

Figure 15.15 Capacity curve for deep-cycle battery model US-250 from U.S. Battery Manufacturing Company (www.usbattery.com).

Figure 15.16 shows a lifetime curve typical of a deep-cycle lead–acid battery. The number of cycles to failure (shown in the graph as the lighter-colored points) drops sharply with increasing depth of discharge. For each point on this curve, one can calculate the lifetime throughput (the amount of energy that cycled through the battery before failure) by finding the product of the number of cycles, the depth of discharge, the nominal voltage of the battery, and the aforementioned maximum capacity of the battery. The lifetime throughput curve, shown in Figure 15.16 as black dots, typically shows a much weaker dependence on the cycle depth. HOMER makes the simplifying assumption that the lifetime throughput is independent of the depth of discharge. The value that HOMER suggests for this lifetime throughput is the average of the points from the lifetime curve above the minimum state of charge, but the user can modify this value to be more or less conservative.

The assumption that lifetime throughput is independent of cycle depth means that HOMER can estimate the life of the battery bank simply by monitoring the amount of energy cycling through it, without having to consider the depth of the various charge–discharge cycles. HOMER calculates the life of the battery bank in years as

$$R_{\text{batt}} = \min\left(\frac{N_{\text{batt}} Q_{\text{lifetime}}}{Q_{\text{thrpt}}}, R_{\text{batt},f}\right) \tag{15.8}$$

where N_{batt} is the number of batteries in the battery bank, Q_{lifetime} the lifetime throughput of a single battery, Q_{thrpt} the annual throughput (the total amount of

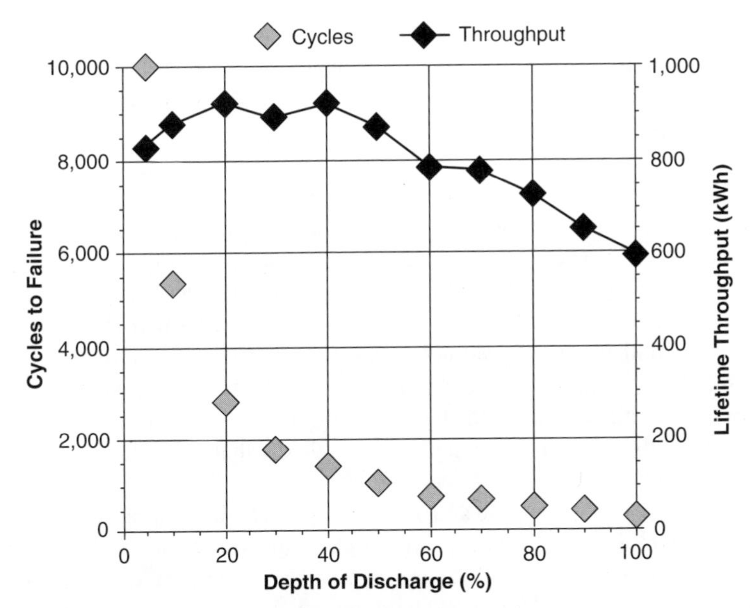

Figure 15.16 Lifetime curve for deep-cycle battery model US-250 from U.S. Battery Manufacturing Company (www.usbattery.com).

energy that cycles through the battery bank in one year), and $R_{\text{batt},f}$ the float life of the battery (the maximum life regardless of throughput).

The user specifies the battery bank's capital and replacement costs in dollars, and the O&M cost in dollars per year. Since the battery bank is a dispatchable power source, HOMER calculates its fixed and marginal cost of energy for comparison with other dispatchable sources. Unlike the generator, there is no cost associated with "operating" the battery bank so that it is ready to produce energy; hence its fixed cost of energy is zero. For its marginal cost of energy, HOMER uses the sum of the battery wear cost (the cost per kilowatthour of cycling energy through the battery bank) and the battery energy cost (the average cost of the energy stored in the battery bank). HOMER calculates the battery wear cost as

$$c_{\text{bw}} = \frac{C_{\text{rep,batt}}}{N_{\text{batt}} Q_{\text{lifetime}} \sqrt{\eta_{\text{rt}}}} \tag{15.9}$$

where $C_{\text{rep,batt}}$ is the replacement cost of the battery bank (dollars), N_{batt} is the number of batteries in the battery bank, Q_{lifetime} is the lifetime throughput of a single battery (kWh), and η_{rt} is the round-trip efficiency.

HOMER calculates the battery energy cost each hour of the simulation by dividing the total year-to-date cost of charging the battery bank by the total year-to-date amount of energy put into the battery bank. Under the load-following dispatch strategy, the battery bank is only ever charged by surplus electricity, so the cost associated with charging the battery bank is always zero. Under the cycle-charging strategy, however, a generator will produce extra electricity (and hence consume additional fuel) for the express purpose of charging the battery bank, so the cost associated with charging the battery bank is not zero. In Section 15.5.4 we discuss dispatch strategies in greater detail.

Grid HOMER models the grid as a component from which the micropower system can purchase ac electricity and to which the system can sell ac electricity. The cost of purchasing power from the grid can comprise an energy charge based on the amount of energy purchased in a billing period and a demand charge based on the peak demand within the billing period. HOMER uses the term *grid power price* for the price (in dollars per kilowatthour) that the electric utility charges for energy purchased from the grid, and the *demand rate* for the price (in dollars per kilowatt per month) the utility charges for the peak grid demand. A third term, the *sellback rate*, refers to the price (in dollars per kilowatthour) that the utility pays for power sold to the grid.

The HOMER user can define and schedule up to 16 different rates, each of which can have different values of grid power price, demand rate, and sellback rate. The schedule of the rates can vary according to month, time of day, and weekday/weekend. For example, HOMER could model a situation where an expensive rate applies during weekday afternoons in July and August, an intermediate rate applies during weekday afternoons in June and September and weekend afternoons from June to September, and an inexpensive rate applies at all other times.

HOMER can also model *net metering*, a billing arrangement whereby the utility charges the customer based on the net grid purchases (purchases minus sales) over the billing period. Under net metering, if purchases exceed sales over the billing period, the consumer pays the utility an amount equal to the net grid purchases times the grid power cost. If sales exceed purchases over the billing period, the utility pays the consumer an amount equal to the net grid sales (sales minus purchases) times the sellback rate, which is typically less than the grid power price, and often zero. The billing period may be one month or one year. In the unusual situation where net metering applies to multiple rates, HOMER tracks the net grid purchases separately for each rate.

Two variables describe the grid's capacity to deliver and accept power. The maximum power sale is the maximum rate at which the power system can sell power to the grid. The user should set this value to zero if the utility does not allow sellback. The maximum grid demand is the maximum amount of power that can be drawn from the grid. It is a decision variable because of the effect of demand charges. HOMER does not explicitly consider the demand rate in its hour-by-hour decisions as to how to control the power system; it simply calculates the demand charge at the end of each simulation. As a result, when modeling a grid-connected generator, HOMER will not turn on the generator simply to save demand charges. But it will turn on a generator whenever the load exceeds the maximum grid demand. The maximum grid demand therefore acts as a control parameter that affects the operation and economics of the system. Because it is a decision variable, the user can enter multiple values and HOMER can find the optimal one.

The user also enters the grid emissions coefficients, which HOMER uses to calculate the emissions of six pollutants associated with buying power from the grid, as well as the avoided emissions resulting from the sale of power to the grid. Each emissions coefficient has units of grams of pollutant emitted per kilowatthour consumed.

Because it is a dispatchable power source, HOMER calculates the grid's fixed and marginal cost of energy. The fixed cost is zero, and the marginal cost is equal to the current grid power price plus any cost resulting from emissions penalties. Since the grid power price can change from hour to hour as the applicable rate changes, the grid's marginal cost of energy can also change from hour to hour. This can have important effects on HOMER's simulation of the system's behavior. For example, HOMER may choose to run a generator only during times of high grid power price, when the cost of grid power exceeds the cost of generator power.

Boiler HOMER models the boiler as an idealized component able to provide an unlimited amount of thermal energy on demand. When dispatching generators to serve the electric load, HOMER considers the value of any waste heat that can be recovered from a generator to serve the thermal load, but it will not dispatch a generator simply to serve the thermal load. It assumes that the system can always rely on the boiler to serve any thermal load that the generators do not. To avoid situations that violate this assumption, HOMER ensures that a boiler exists in any system serving a thermal load, it does not allow any consumption limit on the boiler fuel, and it does not allow the boiler to consume biomass or stored hydrogen (since either of those fuels could be unavailable at times).

The idealized nature of HOMER's boiler model means that the user must specify only a few physical properties of the boiler. The user selects the type of fuel the boiler consumes and enters the efficiency with which it converts that fuel into heat. The only other properties of the boiler are its emissions coefficients, which are in units of grams of pollutant emitted per quantity of fuel consumed.

As it does for all dispatchable energy sources, HOMER calculates the fixed and marginal cost of energy from the boiler. The fixed cost is zero. HOMER calculates the marginal cost using the equation

$$c_{\text{boiler,mar}} = \frac{3.6 c_{\text{fuel,eff}}}{\eta_{\text{boiler}} \text{LHV}_{\text{fuel}}} \tag{15.10}$$

where $c_{\text{fuel,eff}}$ is the effective price of the fuel (including the cost of any penalties on emissions) in dollars per kilogram, η_{boiler} is the boiler efficiency, and LHV_{fuel} is the lower heating value of the fuel in MJ/kg.

Converter A converter is a device that converts electric power from dc to ac in a process called *inversion*, and/or from ac to dc in a process called *rectification*. HOMER can model the two common types of converters: solid-state and rotary. The converter size, which is a decision variable, refers to the inverter capacity, meaning the maximum amount of ac power that the device can produce by inverting dc power. The user specifies the rectifier capacity, which is the maximum amount of dc power that the device can produce by rectifying ac power, as a percentage of the inverter capacity. The rectifier capacity is therefore not a separate decision variable. HOMER assumes that the inverter and rectifier capacities are not surge capacities that the device can withstand for only short periods of time, but rather, continuous capacities that the device can withstand for as long as necessary.

The HOMER user indicates whether the inverter can operate in parallel with another ac power source such as a generator or the grid. Doing so requires the inverter to synchronize to the ac frequency, an ability that some inverters do not have. The final physical properties of the converter are its inversion and rectification efficiencies, which HOMER assumes to be constant. The economic properties of the converter are its capital and replacement cost in dollars, its annual O&M cost in dollars per year, and its expected lifetime in years.

Electrolyzer An electrolyzer consumes electricity to generate hydrogen via the electrolysis of water. In HOMER, the user specifies the size of the electrolyzer, which is a decision variable, in terms of its maximum electrical input. The user also indicates whether the electrolyzer consumes ac or dc power, and the efficiency with which it converts that power to hydrogen. HOMER defines the electrolyzer efficiency as the energy content (based on higher heating value) of the hydrogen produced divided by the amount of electricity consumed. The final physical property of the electrolyzer is its minimum load ratio, which is the minimum power input at which it can operate, expressed as a percentage of its maximum power input. The economic properties of the electrolyzer are its capital and replacement cost in dollars, its annual O&M cost in dollars per year, and its expected lifetime in years.

Hydrogen Tank In HOMER, the hydrogen tank stores hydrogen produced by the electrolyzer for later use in a hydrogen-fueled generator. The user specifies the size of the hydrogen tank, which is a decision variable, in terms of the mass of hydrogen it can contain. HOMER assumes that the process of adding hydrogen to the tank requires no electricity, and that the tank experiences no leakage.

The user can specify the initial amount of hydrogen in the tank either as a percentage of the tank size or as an absolute amount in kilograms. It is also possible to require that the year-end tank level must equal or exceed the initial tank level. If the user chooses to apply this constraint, HOMER will consider infeasible any system configuration whose hydrogen tank contains less hydrogen at the end of the simulation than it did at the beginning of the simulation. This ensures that the system is self-sufficient in terms of hydrogen. The economic properties of the hydrogen tank are its capital and replacement cost in dollars, its annual O&M cost in dollars per year, and its expected lifetime in years.

15.5.4 System Dispatch

In addition to modeling the behavior of each individual component, HOMER must simulate how those components work together as a system. That requires hour-by-hour decisions as to which generators should operate and at what power level, whether to charge or discharge the batteries, and whether to buy from or sell to the grid. In this section we describe briefly the logic HOMER uses to make such decisions. A discussion of operating reserve comes first because the concept of operating reserve significantly affects HOMER's dispatch decisions.

Operating Reserve Operating reserve provides a safety margin that helps ensure reliable electricity supply despite variability in the electric load and the renewable power supply. Virtually every real micropower system must always provide some amount of operating reserve, because otherwise the electric load would sometimes fluctuate above the operating capacity of the system, and an outage would result.

At any given moment, the amount of operating reserve that a power system provides is equal to the operating capacity minus the electrical load. Consider, for example, a simple diesel system in which an 80-kW diesel generator supplies an electric load. In that system, if the load is 55 kW, the diesel will produce 55 kW of electricity and provide 25 kW of operating reserve. In other words, the system could supply the load even if the load suddenly increased by 25 kW. In HOMER, the modeler specifies the required amount of operating reserve, and HOMER simulates the system so as to provide at least that much operating reserve.

Each hour, HOMER calculates the required amount of operating reserve as a fraction of the primary load that hour, plus a fraction of the annual peak primary load, plus a fraction of the PV power output that hour, plus a fraction of the wind power output that hour. The modeler specifies these fractions by considering how much the load or the renewable power output is likely to fluctuate in a short period, and how conservatively he or she plans to operate the system. The more variable the load and renewable power output, and the more conservatively the system must operate, the higher the fractions the modeler should specify. HOMER does not attempt to ascertain the amount of operating reserve required to achieve different levels of reliability; it simply uses the modeler's specifications to calculate the amount of operating reserve the system is obligated to provide each hour.

Once it calculates the required amount of operating reserve, HOMER attempts to operate the system so as to provide at least that much operating reserve. Doing so may require operating the system differently (at a higher cost) than would be necessary without consideration of operating reserve. Consider, for example, a wind–diesel system for which the user defines the required operating reserve as 10% of the hourly load plus 50% of the wind power output. HOMER will attempt to operate that system so that at any time, it can supply the load with the operating generators even if the load suddenly increased by 10% and the wind power output suddenly decreased by 50%. In an hour where the load is 140 kW and the wind power output is 80 kW, the required operating reserve would be 14 kW + 40 kW = 54 kW. The diesel generators must therefore provide 60 kW of electricity plus 54 kW of operating reserve, meaning that the capacity of the operating generators must be at least 114 kW. Without consideration of operating reserve, HOMER would assume that a 60-kW diesel would be sufficient.

HOMER assumes that both dispatchable and nondispatchable power sources provide operating capacity. A dispatchable power source provides operating capacity in an amount equal to the maximum amount of power it could produce at a moment's notice. For a generator, that is equal to its rated capacity if it is operating, or zero if it is not operating. For the grid, that is equal to the maximum grid demand. For the battery, that is equal to its current maximum discharge power, which depends on state of charge and recent charge–discharge history, as described in Section 15.5.3. In contrast to the dispatchable power sources, the operating capacity of a nondispatchable power source (a PV array, wind turbine, or hydro turbine) is equal to the amount of power the source is currently producing, as opposed to the maximum amount of power it could produce.

For most grid-connected systems, the concept of operating reserve has virtually no effect on the operation of the system because the grid capacity is typically more than enough to cover the required operating reserve. Unlike a generator, which must be turned on and incurring fixed costs to provide operating capacity, the grid is always "operating" so that its capacity (which is usually very large compared to the load) is always available to the system. Similarly, operating reserve typically has little or no effect on autonomous systems with large battery banks, since the battery capacity is also always available to the system, at no fixed cost. Nevertheless, HOMER still calculates and tracks operating reserve for such systems.

If a system is ever unable to supply the required amount of load plus operating reserve, HOMER records the shortfall as *capacity shortage*. HOMER calculates the total amount of such shortages over the year and divides the total annual capacity shortage by the total annual electric load to find the capacity shortage fraction. The modeler specifies the maximum allowable capacity shortage fraction. HOMER discards as infeasible any system whose capacity shortage fraction exceeds this constraint.

Control of Dispatchable System Components Each hour of the year, HOMER determines whether the (nondispatchable) renewable power sources by themselves are capable of supplying the electric load, the required operating reserve, and the thermal load. If not, it determines how best to dispatch the dispatchable system components (the generators, battery bank, grid, and boiler) to serve the loads and operating reserve. This determination of how to dispatch the system components each hour is the most complex part of HOMER's simulation logic. The nondispatchable renewable power sources, although they necessitate complex system modeling, are themselves simple to model because they require no control logic—they simply produce power in direct response to the renewable resource available. The dispatchable sources are more difficult to model because they must be controlled to match supply and demand properly, and to compensate for the intermittency of the renewable power sources.

The fundamental principle that HOMER follows when dispatching the system is the minimization of cost. HOMER represents the economics of each dispatchable energy source by two values: a fixed cost in dollars per hour, and a marginal cost of energy in dollars per kilowatthour. These values represent all costs associated with

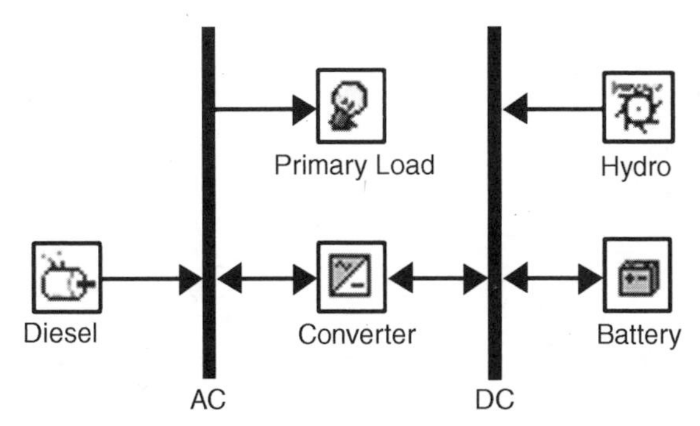

Figure 15.17 Hydro–diesel–battery system.

producing energy with that power source that hour. The sections above on the generator, battery bank, grid, and boiler detail how HOMER calculates the fixed and marginal costs for each of these components. Using these cost values, HOMER searches for the combination of dispatchable sources that can serve the electrical load, thermal load, and the required operating reserve at the lowest cost. Satisfying the loads and operating reserve is paramount, meaning that HOMER will accept any cost to avoid capacity shortage. But among the combinations of dispatchable sources that can serve the loads equally well, HOMER chooses the one that does so at the lowest cost.

For example, consider the hydro–diesel–battery system shown in Figure 15.17. This system comprises two dispatchable power sources, the battery bank and the diesel generator. Whenever the net load is negative (meaning the power output of the hydro turbine is sufficient to serve the load), the excess power charges the battery bank. But whenever the net load is positive, the system must either operate the diesel or discharge the battery, or both, to serve the load. In choosing among these three alternatives, HOMER considers the ability of each source to supply the ac net load and the required operating reserve, and the cost of doing so. If the diesel generator is scheduled off or has run out of fuel, it has no ability to supply power to the ac load. Otherwise, it can supply any amount of ac power up to its rated capacity. The battery's ability to supply power and operating reserve to the ac load is constrained by its current discharge capacity (which depends on its state of charge and recent charge–discharge history, as described in Section 15.5.3) and the capacity and efficiency of the ac–dc converter. If both the battery bank and the diesel generator are capable of supplying the net load and the operating reserve, HOMER decides which to use based on their fixed and marginal costs of energy.

Figure 15.18 shows one possible cost scenario, where the diesel capacity is 80 kW and the battery can supply up to 40 kW of power to the ac bus, after conversion losses. This scenario is typical in that the battery's marginal cost of energy exceeds that of the diesel. But because of the diesel's fixed cost, the battery can supply small amounts of ac power more cheaply than the diesel. In this case the crossover point is around 20 kW. Therefore, if the net load is less than 20 kW, HOMER will serve the load by discharging the battery. If the net load is greater

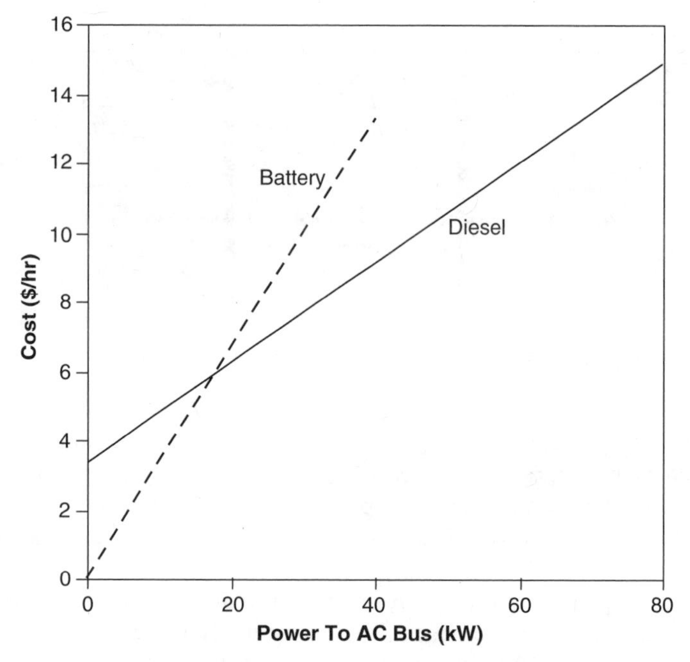

Figure 15.18 Cost of energy comparison.

than 20 kW, HOMER will serve the load with the generator instead of the battery, even if the battery is capable of supplying the load.

HOMER uses the same cost-based dispatch logic regardless of the system configuration. When simulating a system comprising multiple generators, HOMER will choose the combination of generators that can most cheaply supply the load and the required operating reserve. When simulating a grid-connected microturbine supplying both heat and electricity, HOMER will operate the microturbine whenever doing so would save money compared to the alternative, which is to buy electricity from the grid and produce heat with a boiler.

HOMER's simulation is idealized in the sense that it assumes the system controller will operate the system so as to minimize total life-cycle cost, when in fact a real system controller may not. But HOMER's "economically optimal" scenario serves as a useful baseline with which to compare different system configurations.

Dispatch Strategy The economic dispatch logic described in the preceding section governs the production of energy to serve loads and hence applies to all systems that HOMER models. But for systems comprising both a battery bank and a generator, an additional aspect of system operation arises, which is whether (and how) the generator should charge the battery bank. One cannot base this battery-charging logic on simple economic principles, because there is no deterministic way to calculate the value of charging the battery bank. The value of charging the battery in one hour depends on what happens in future hours. In a wind–diesel–battery system, for example, charging the battery bank with diesel power in one hour would be of some value if doing so allowed the system to avoid operating the diesel in some subsequent hour. But it would be of no value whatsoever if the system experienced more than enough excess wind power in subsequent hours to charge the battery bank fully. In that case, any diesel power put into the battery bank would be wasted because the wind power would have fully charged the battery bank anyway.

Rather than using complicated probabilistic logic to determine the optimal battery-charging strategy, HOMER provides two simple strategies and lets the user model them both to see which is better in any particular situation. These dispatch strategies are called *load-following* and *cycle-charging*. Under the load-following strategy, a generator produces only enough power to serve the load, and does not charge the battery bank. Under the cycle-charging strategy, whenever a generator operates, it runs at its maximum rated capacity (or as close as possible without incurring excess electricity) and charges the battery bank with the excess. Barley and Winn [9] found that over a wide range of conditions, the better of these two simple strategies is virtually as cost-effective as the ideal predictive strategy. Because HOMER treats the dispatch strategy as a decision variable, the modeler can easily simulate both strategies to determine which is optimal in a given situation.

The dispatch strategy does not affect the decisions described in the preceding section as to which dispatchable power sources operate each hour. Only after these decisions are made does the dispatch strategy come into play. If the load-following strategy applies, whichever generators HOMER selects to operate in a given hour will produce only enough power to serve the load. If the cycle-charging strategy applies, those same generators will run at their rated output, or as close as possible without causing excess energy.

An optional control parameter called the set-point state of charge can apply to the cycle-charging strategy. If the modeler chooses to apply this parameter, once the generator starts charging the battery bank, it must continue to do so until the battery bank reaches the set-point state of charge. Otherwise, HOMER may choose to discharge the battery as soon as it can supply the load. The set-point state of charge helps avoid situations where the battery experiences shallow charge–discharge cycles near its minimum state of charge. In real systems, such situations are harmful to battery life.

Load Priority HOMER makes a separate set of decisions regarding how to allocate the electricity produced by the system. The presence of both an ac and a dc bus complicates these decisions somewhat. HOMER assumes that electricity produced on one bus will go first to serve primary load on the same bus, then primary load on the opposite bus, then deferrable load on the same bus, then deferrable load on the opposite bus, then to charge the battery bank, then to grid sales, then to serve the electrolyzer, and then to the dump load, which optionally serves the thermal load.

15.6 ECONOMIC MODELING

Economics play an integral role both in HOMER's simulation process, wherein it operates the system so as to minimize total net present cost, and in its optimization process, wherein it searches for the system configuration with the lowest total net present cost. This section describes why life-cycle cost is the appropriate metric with which to compare the economics of different system configurations, why HOMER uses the total net present cost as the economic figure of merit, and how HOMER calculates total net present cost.

Renewable and nonrenewable energy sources typically have dramatically different cost characteristics. Renewable sources tend to have high initial capital costs and low operating costs, whereas conventional nonrenewable sources tend to have low capital and high operating costs. In its optimization process, HOMER must often compare the economics of a wide range of system configurations comprising varying amounts of renewable and nonrenewable energy sources. To be equitable, such comparisons must account for both capital and operating costs. Life-cycle cost analysis does so by including all costs that occur within the life span of the system.

HOMER uses the total net present cost (NPC) to represent the life-cycle cost of a system. The total NPC condenses all the costs and revenues that occur within the project lifetime into one lump sum in today's dollars, with future cash flows discounted back to the present using the discount rate. The modeler specifies the discount rate and the project lifetime. The NPC includes the costs of initial construction, component replacements, maintenance, fuel, plus the cost of buying power from the grid and miscellaneous costs such as penalties resulting from pollutant emissions. Revenues include income from selling power to the grid, plus any salvage value that occurs at the end of the project lifetime. With the NPC, costs are positive and revenues are negative. This is the opposite of the net present value. As a result, the net present cost is different from net present value only in sign.

HOMER assumes that all prices escalate at the same rate over the project lifetime. With that assumption, inflation can be factored out of the analysis simply by using the real (inflation-adjusted) interest rate rather than the nominal interest rate when discounting future cash flows to the present. The HOMER user therefore enters the real interest rate, which is roughly equal to the nominal interest rate minus the inflation rate. All costs in HOMER are real costs, meaning that they are defined in terms of constant dollars.

For each component of the system, the modeler specifies the initial capital cost, which occurs in year zero, the replacement cost, which occurs each time the component needs replacement at the end of its lifetime, and the O&M cost, which occurs each year of the project lifetime. The user specifies the lifetime of most components in years, but HOMER calculates the lifetime of the battery and generators as described in Section 15.5.3. A component's replacement cost may differ from its initial capital cost for several reasons. For example, a modeler might assume that a wind turbine nacelle will need replacement after 15 years, but the tower and foundation will last for the life of the project. In that case, the replacement cost would be considerably less than the initial capital cost. Donor agencies or buy-down programs might cover some or all of the initial capital cost of a PV array but none of the replacement cost. In that case, the replacement cost may be greater than the initial capital cost. When analyzing a retrofit of an existing diesel system, the initial capital cost of the diesel engine would be zero, but the replacement cost would not.

To calculate the salvage value of each component at the end of the project lifetime, HOMER uses the equation

$$S = C_{\text{rep}} \frac{R_{\text{rem}}}{R_{\text{comp}}} \tag{15.11}$$

where S is the salvage value, C_{rep} the replacement cost of the component, R_{rem} the remaining life of the component, and R_{comp} the lifetime of the component. For example, if the project lifetime is 20 years and the PV array lifetime is also 20 years, the salvage value of the PV array at the end of the project lifetime will be zero because it has no remaining life. On the other hand, if the PV array lifetime is 30 years, at the end of the 20-year project lifetime its salvage value will be one-third of its replacement cost.

For each component, HOMER combines the capital, replacement, maintenance, and fuel costs, along with the salvage value and any other costs or revenues, to find the component's annualized cost. This is the hypothetical annual cost that if it occurred each year of the project lifetime would yield a net present cost equivalent to that of all the individual costs and revenues associated with that component over the project lifetime. HOMER sums the annualized costs of each component, along with any miscellaneous costs, such as penalties for pollutant emissions, to find the total annualized cost of the system. This value is an important one because HOMER uses it to calculate the two principal economic figures of merit for the system: the total net present cost and the levelized cost of energy.

HOMER uses the following equation to calculate the total net present cost:

$$C_{\text{NPC}} = \frac{C_{\text{ann,tot}}}{\text{CRF}\left(i, R_{\text{proj}}\right)} \tag{15.12}$$

where $C_{\text{ann,tot}}$ is the total annualized cost, i the annual real interest rate (the discount rate), R_{proj} the project lifetime, and $\text{CRF}(\cdot)$ is the capital recovery factor, given by the equation

$$\text{CRF}(i, N) = \frac{i(1+i)^N}{(1+i)^N - 1} \tag{15.13}$$

where i is the annual real interest rate and N is the number of years.

HOMER uses the following equation to calculate the levelized cost of energy:

$$\text{COE} = \frac{C_{\text{ann,tot}}}{E_{\text{prim}} + E_{\text{def}} + E_{\text{grid,sales}}} \tag{15.14}$$

where $C_{\text{ann,tot}}$ is the total annualized cost, E_{prim} and E_{def} are the total amounts of primary and deferrable load, respectively, that the system serves per year, and

$E_{grid,sales}$ is the amount of energy sold to the grid per year. The denominator in equation (15.14) is an expression of the total amount of useful energy that the system produces per year. The levelized cost of energy is therefore the average cost per kilowatthour of useful electrical energy produced by the system.

Although the levelized cost of energy is often a convenient metric with which to compare the costs of different systems, HOMER uses the total NPC instead as its primary economic figure of merit. In its optimization process, for example, HOMER ranks the system configurations according to NPC rather than levelized cost of energy. This is because the definition of the levelized cost of energy is disputable in a way that the definition of the total NPC is not. In developing the formula that HOMER uses for the levelized cost of energy, we decided to divide by the amount of electrical load that the system actually serves rather than the total electrical demand, which may be different if the user allows some unmet load. We also decided to neglect thermal energy but to include grid sales as useful energy production. Each of these decisions is somewhat arbitrary, making the definition of the levelized cost of energy also somewhat arbitrary. Because the total NPC suffers from no such definitional ambiguity, it is preferable as the primary economic figure of merit.

REFERENCES

[1] J. F. Manwell and J. G. McGowan, A combined probabilistic/time series model for wind diesel systems simulation, *Solar Energy*, Vol. 53, pp. 481–490, 1994.

[2] Maui Solar Energy Software Corporation, PV-DesignPro, http://www.mauisolarsoftware.com, accessed February 2, 2005.

[3] PV*SOL, http://www.valentin.de, accessed February 2, 2005.

[4] RETScreen International http://www.retscreen.net, accessed February 2, 2005.

[5] J. A. Duffie and W. A. Beckman, *Solar Engineering of Thermal Processes* 2nd ed., Wiley, New York, 1991.

[6] F. M. White, *Fluid Mechanics*, 2nd ed., McGraw-Hill, New York, 1986.

[7] V. A. Graham and K. G. T. Hollands, A method to generate synthetic hourly solar radiation globally, *Solar Energy*, Vol. 44, No. 6, pp. 333–341, 1990.

[8] J. F. Manwell and J. G. McGowan, Lead acid battery storage model for hybrid energy systems, *Solar Energy*, Vol. 50, pp. 399–405, 1993.

[9] C. D. Barley and C. B. Winn, Optimal dispatch strategy in remote hybrid power systems, *Solar Energy*, Vol. 58, pp. 165–179, 1996.

GLOSSARY

Annualized cost: the hypothetical annual cost value that if it occurred each year of the project lifetime would yield a net present cost equivalent to the actual net present cost. HOMER calculates the annualized cost of each system component and of the entire system.

Autonomous: not connected to a larger power transmission grid. Autonomous power systems are often called *off-grid systems*. An autonomous system must be controlled carefully to match electrical supply and demand.

Battery energy cost: the average cost per kilowatthour of the energy stored in the battery bank.

Battery wear cost: the cost per kilowatthour of cycling energy through the battery bank. HOMER can calculate this value because it assumes that a certain amount of energy can cycle through the battery bank before it needs replacement. Every kilowatthour of energy that cycles through the battery bank reduces its life span by a known amount.

Beam radiation: solar radiation that travels from the sun to Earth's surface without any scattering by the atmosphere. Beam radiation is also called *direct radiation*.

Capacity shortage: a shortfall that occurs between the required amount of operating capacity (load plus required operating reserve) and the actual operating capacity the system can provide.

Clearness index: the fraction of the solar radiation striking the top of the atmosphere that makes it though the atmosphere to strike the surface of the Earth.

Constraint: a condition that system configurations must satisfy. HOMER discards systems that do not satisfy the applicable constraints. For example, the HOMER user may impose a constraint that all system configurations must serve at least 99% of the electrical demand over the year.

Decision variable: a variable over which the system designer has control, and for which HOMER can consider multiple values in its optimization process. Examples include the number of batteries in the system or the size of the ac–dc converter.

Demand rate: the fee the electric utility applies each month to the peak hourly grid demand. If the demand rate is $5/kW per month and the peak hourly demand in January is 75 kW, the demand charge for January will be $375.

Diffuse radiation: solar radiation that has been scattered by the atmosphere.

Dispatch strategy: a set of rules that controls how a system charges the battery bank.

Extraterrestrial radiation: the solar radiation striking the top of EARTH's atmosphere. One can calculate the quantity of extraterrestrial radiation precisely for any location on earth at any time.

Feasibility: the state of being feasible or infeasible. A feasible system configuration is one that satisfies the constraints imposed by the user. An infeasible system is one that violates one or more constraints.

Grid power price: the price the utility charges for power purchased from the grid. In HOMER, this price can vary according to month, day of the week, and time of day.

Levelized cost of energy: the average cost per kilowatthour of electricity produced by the system.

Life-cycle cost: the total cost of installing and operating a component or system over a specified time span, typically many years.

Micropower system: a system that produces electrical and possibly thermal power to serve a nearby load.

Net load: the load minus the renewable power available to serve the load.

Net metering: a billing scheme by which the utility charges a micropower system operator for the net grid purchases (purchases minus sales) over the billing period.

Operating capacity: the total amount of electrical generation capacity that is operating (and ready to produce electricity) at any one time. A generator that is not operating provides no operating capacity. A 200-kW generator that is operating provides 200 kW of operating capacity.

Operating reserve: surplus electrical generation capacity (above that required to meet the current electric load) that is operating and able to respond instantly to a sudden increase in the electric load or a sudden decrease in the renewable power output.

Replacement cost: the cost of replacing a component at the end of its useful lifetime. The replacement cost may differ from the initial capital cost for several reasons: only part of the component may need replacement, a donor organization may cover the initial capital cost but not the replacement cost, fixed costs may be shared among many components initially but not at the time of replacement, and so on.

Residual flow: the minimum stream flow that must bypass the hydro turbine for ecological purposes.

Search space: the set of all system configurations over which HOMER searches for the optimum during the optimization process.

Sellback rate: the price the utility pays for power sold to the grid. In HOMER, this value can vary according to month, day of the week, and time of day.

Sensitivity variable: an input variable for which the modeler enters multiple values rather than just one.

System configuration: a combination of particular numbers and sizes of components, plus an operating strategy that defines how those components work together.

Unmet load: electrical load that the power system is unable to serve. Unmet load occurs when the electrical demand exceeds the supply.

APPENDIX A

DIESEL POWER PLANTS

A.1 INTRODUCTION

By the end of the nineteenth century there mountains of useless coal dust had piled up in the Ruhr valley in Germany, and Rudolf Diesel started to work to develop an engine that would burn coal dust. The attempts to design such an engine failed, but in 1892, Diesel was issued a patent for a proposed system in which air would be so greatly compressed that the temperature would far exceed the ignition temperature of an oil fuel. Ever since, internal combustion engines have been providing shaft power.

There are basically two main types of diesel power plants combustion engines categorized by the type of fuel used: gasoline or diesel. The vast majority of those engines power automobiles, but they have been also used for ships, boats, agricultural processing machinery, and many other industrial applications. During the last quarter of the twentieth century abundant fossil fuel production and distribution made possible commercial application of diesel-powered electricity generation for several applications. In addition, hybrid schemes were deployed to integrate and complement intermittent distributed generation systems [1,2].

Diesel-based low power generation consists basically of a diesel engine coupled to an electric power generator and a field-exciting generator. The arrangement is

Integration of Alternative Sources of Energy, by Felix A. Farret and M. Godoy Simões

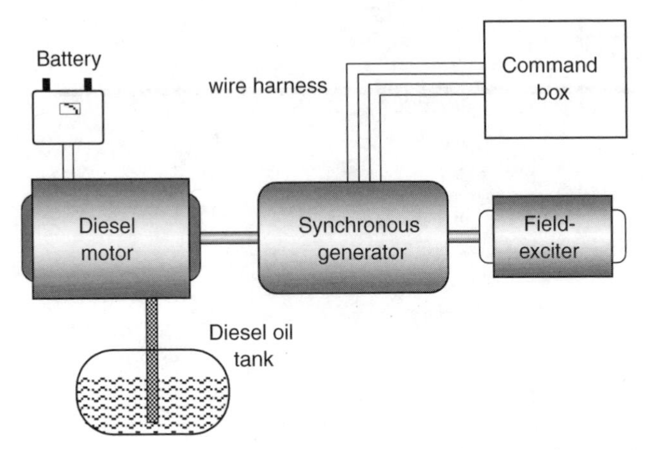

Figure A.1 Block diagram of a diesel oil power plant.

very compact, it goes online in a very short lead time, requires only routine maintenance, and is easily available through practicing professionals in mechanical workshops and garages. Diesel engines have some disadvantages: They are noisy, polluting, driven economically by fuel costs (and thus captive to world politics), require fuel storage close to the power plant, and require logistics and infrastructure for transportation.

Figure A.1 depicts a typical arrangement of a small diesel engine–powered plant. Since a conventional synchronous generator is used, special attention to frequency and synchronization is required (see Chapter 10).

A.2 DIESEL ENGINE

A diesel engine converts the chemical energy of fuel into thermal energy that charges a cylinder in consequence of self-ignition and combustion of the fuel after compression of air. A slider crank mechanism converts the thermal energy into mechanical work using a crankshaft. The combustion process has some significant features. In a gasoline engine the fuel and air mixture is drawn into the cylinder, compressed from 4 : 1 to 10 : 1, and ignited by a spark; in a diesel engine air alone is draw into the cylinder, compressed to a higher ratio (14 : 1 to 25 : 1), causing a rise in the air temperature in the range of 700 to 900°C, then diesel fuel is injected by a nozzle and ignites spontaneously. A diesel engine has some advantages over spark ignition motors (gasoline or alcohol), such as better fuel economy and longer engine life (i.e., over the life of the engine, less money is spent with diesel, but the initial high cost of the engine must be taken into account). Therefore, only long-time operation enables with a favorable fuel economy will overcome the increased price of the engine.

Because of the weight to compression ratio, diesel engines have lower maximum rpm ranges than gasoline engines, making them suitable for high torque rather than high acceleration. This is a good feature for generators that have to rather operate at constant high speed. Diesel engines are also considered to have high efficiency compared to spark ignition engines because of a higher compression rate, the calorific energy conversion into mechanical work has lower losses at higher compression rate.

A.3 PRINCIPAL COMPONENTS OF A DIESEL ENGINE

The basic diesel engine parts are described below.

A.3.1 Fixed Parts

Bedplate/base: a foundation mounted on vibration absorbers providing support for the main bearings and engine crankcase. It is called a bedplate if an oil pan is bolted on it and a base if the oil pan is an integral part of the assembly.

Main bearing caps: cover the crankshaft from the top as an upper part of the engine. They form the cylinder lid and must withstand the peak pressure of the piston in the combustion chamber below the piston.

Cylinder: requires die casting because of its cylindrical format. It is manufactured from melted iron, and the base is made of manganese or nickel–chrome–molybdenum to provide hardness and corrosion resistance.

Crankcase: serves as a housing for the crankshaft and is located between the bedplate and the cylinder block. It usually incorporates the main bearing saddles and a reservoir for the lubricating oil. In some engines the crankcase consists of one piece of cast iron, in others it is constructed of welded steel.

A.3.2 Moving Parts

Piston: receives the direct impact of the combustion for transmission to the connecting rod. To protect the piston against seizure, the diameter must be reduced with an optimal gap. There are troughs in the head of the piston with camped segment rings. The top rings act as a pressure seal, the middle rings remove oil film, and the bottom rings ensure even deposition of oil on the cylinder walls.

Connecting rod: loaded alternately in compression and tension, owing to the cylinder firing pressure transmitting power from the pistons to the crankshaft.

Crankshaft: changes the movement of the pistons and the connecting rod. There are eccentric offset rod bearings that convert reciprocating motion into rotating motion. The crankshaft must be very strong and machined from forged alloy, carbon steel, or cast iron alloy.

Flywheel: connected on one end of the crankshaft and through its inertia reduces vibration, allows the engine to be bolted to an external load and sometimes has teeth that engage starting motors for startup. Increasing the number of cylinders increases the frequency of the firing strokes, making it possible to use smaller flywheels.

Valves: control the flow of the air–fuel mixture or the air inside and outside the cylinder.

A.3.3 Auxiliary Systems

Air intake system: responsible for providing cool filtered air at the right fuel mixture to be fed to the cylinders. It consists of a fuel tank, channels, an injector pump, filters, and injecting nozzles.

Cooling system: water cooling to transfer waste heat out of a diesel engine block. It is very rare to have air-refrigerated diesel engines.

Lubrication system: contains oil that serves two purposes: to lubricate bearing surfaces and to absorb friction-generated heat. A pressure relief valve maintains oil pressure in the galleries, returning the oil through a filter to the oil pan.

A.4 TERMINOLOGY OF DIESEL ENGINES

Lower point: the lowest point the piston reaches in its descending course.

Upper point: the highest point the piston reaches in its ascending course.

Cylinder capacity: the volume capacity of a cylinder, corresponding to the maximum acceptable volume of air in the cylinder. It is calculated as

$$V = \pi S r^2$$

where S is the piston course and r is the internal radius of the cylinder.

Compression rate: the relationship between the total volume of a cylinder (v_a) and the volume of the compression chamber (v_e), given by

$$\rho_c = \frac{v_a + v_e}{v_e}$$

To increase the output power it is possible to increase the compression rate, reducing v_e or increasing v_a.

A.4.1 Diesel Cycle

In a four-stroke engine there are two possible cycles where the camshaft is geared to rotate at half of the speed of the crankshaft with one event per stroke; in a two-stroke cycle, more events have to be timed per stroke. Four-stroke diesel engines are used in diesel–electric facilities for small loads, such as for powering installations with personal computer, mini power plants, and more recently, small power plants. On the other hand, two-stroke diesel engines are used in large diesel–electric facilities. Since this book is directed more specifically to small power plants, the four-stroke diesel engine is discussed next.

Caloric energy is transformed to mechanical work by the combustion of a fuel (diesel) inside the cylinder. Four processes, illustrated in Figure A.2, describe the operation. The thermal efficiency of the diesel engine is given by

$$\eta_{\mathrm{TD}} = 1 - \frac{1}{r_c^{k-1}} \frac{r_c^k - 1}{k(r_c - 1)}$$

where r_c is the fuel cut rate and k is Boltzmann's constant.

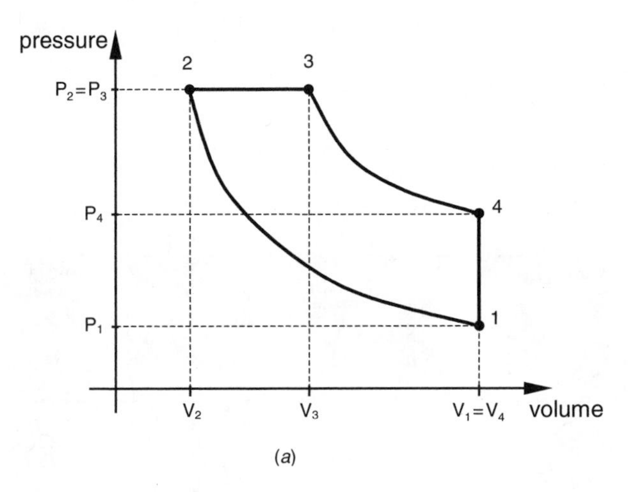

(a)

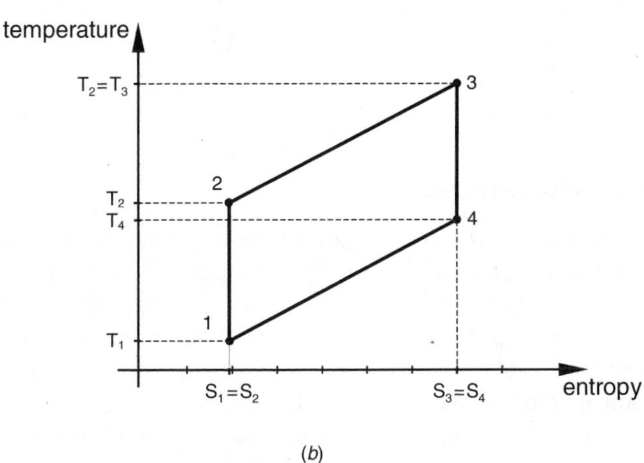

(b)

1 - Isentropic pressure (process 1-2)
2 - Admission (isobaric heat transfer) (process 2-3)
3 - Isentropic expansion (Process 3-4)
4 - Discharge (process 4-1)

Figure A.2 Diesel cycle: (*a*) *P–V* diagram; (*b*) *T–s* diagram.

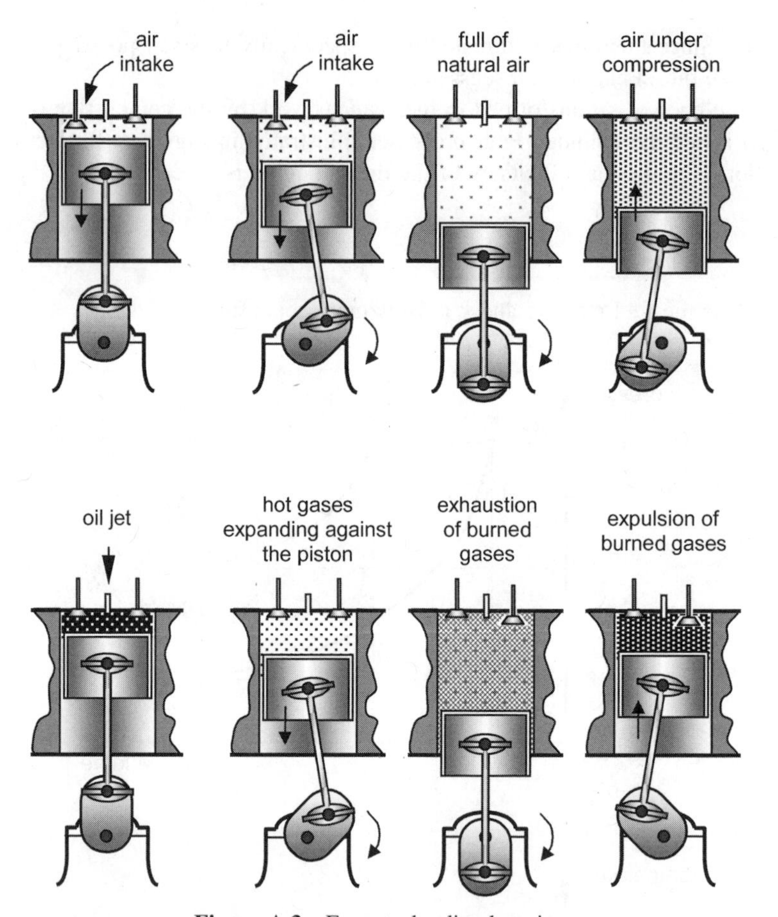

Figure A.3 Four-stroke diesel engine.

A.4.2 Combustion Process

Figure A.3 illustrates the operation of a four-stroke diesel engine. The first cycle begins with admission of air to fill the cylinder completely (which is at a 180° turn of the crank). In the diesel engine, fuel has been injected into the chamber in which air is compressed up with air to 1/16 of its original volume and then heated to about 500° to 600°C. During injection, the air temperature is high enough to provoke self-ignition in the dispersed droplets of fuel in the chamber. Some vaporization of at least part of the fuel is necessary to establish zones of suitable air–fuel composition.

Four-Stroke Diesel Engine Four-stroke models are built for small power sources, up to 4000 hp. Since they are very efficient and have satisfactory durability, they are recommended for heavy-duty applications with a high daily rate of use

[3,4]. High-speed models are predominant because they allow appreciable reduction in weight and cost for a given power level, being a very appropriate issue for electric power generation. For low-speed loads, the advantage of direct coupling (without reducers) justifies the construction of motors using lower rotating speed, especially for high-power applications. Spark ignition engines can easily be adapted to run on ethanol or methanol, and vegetable oils can substitute diesel fuels and are easily integrated with biomass generating systems.

The necessary time to atomize, vaporize, and mix the fuel is called *physical delay*. Soon after the physical delay, a reaction chain begins for the combustion process, called *chemical delay*. The total delay should be as small as possible, because fuel accumulates in the chamber as it is being injected.

At the onset of combustion, the fuel burns and causes pressure elevation. In general, the injection continues beyond that point, and the remaining fuel burns as it is being injected. The combustion ends only after having traveled part of the expansion course. As a result, combustion timing in a diesel engine is defined by four phases:

1. Delay (physical and chemical)
2. Fast increase of pressure at the initial combustion
3. Controlled combustion, to limit the maximum pressure
4. Burning during the expansion, at constant pressure

A.5 DIESEL ENGINE CYCLE

Diesel engines may operate at a very fast combustion rate, approaching constant volume for most of the fuel. Such an operation is obtained when the delay period is long enough that the fuel injected is quite well mixed and most is evaporated before the combustion. However, such an operation is undesirable, due to the high resulting maximum pressures and the high rates of pressure elevation.

In practical operation, the fuel injection system and operating conditions are carefully selected to limit the rates of pressure elevation and maximum pressures to those below the maximum attainable. The equivalent fuel–air cycle at constant volume represents the maximum power and efficiency supplied in the diesel engine and can be used for comparison purposes. However, with intentional limitation of maximum pressure, the equivalent fuel–air cycle is chosen as the basis for the evaluation of diesel cycles.

The equivalent fuel/air cycle at limited pressure has the following characteristics in common with the current cycle: (1) compression rate, (2) fuel–air rate, (3) maximum pressure, (4) point of mixture density, and (5) composition of the mixture at all points.

A.5.1 Relative Diesel Engine Cycle Losses

The losses relative to the cycle of a diesel engine are very variable because they are dependent on the operating conditions of the fuel, the design, the injection system, and the combustion chamber. Whereas in spark ignition motors the burn speed of the mixture at the beginning is relatively low and then accelerates, eventually reaching a maximum, in the diesel engine the opposite occurs. Because the available supply of oxygen decreases as burning progress, the mixture process and burning process tend to reduce only in the final instants.

Most of the combustion processes have a relatively constant temperature. Changes in the amount of fuel per cycle (fuel–air rate) are partially responsible for the constant temperature of the process. In addition, they seem to have a minor effect on the maximum temperature and are not related to the crank angular position.

A.5.2 Classification of Diesel Engines

Diesel engines can be used as motive prime movers for electrical generators. They are of various types and characteristics, and a systematic classification is very difficult. In this appendix they are classified according to their construction, cycle, speed, and overall operation, since those features are more relevant for generating systems. A diesel engine can be vertical or horizontal in construction according to disposition of the cylinders, whereas according to its rated speed, the diesel engine can be fast, medium, or slow.

Fast motors have a relatively low weight/power ratio, high specific power, and reduced dimensions, with an upper maximum rotation above 1200 rpm. They are more economical because of high speed, but technical difficulties limit diesel engine construction for the medium- and small-power range. They are used as prime movers in the to drive auxiliary services of power plants and are mostly suitable for driving electrical generators.

Medium-speed diesel engines operate at a rotating range between 600 and 1000 rpm. They are used for quite high-power applications and a relatively low weight/power ratio. To have good durability characteristics, these motors should operate at full power and at constant rotating speed for long periods of time.

The slow diesel engine is frequently used for auxiliary reserve power plant applications where the weight/power ratio is not of fundamental importance. These motors are bulky, operating between 400 and 450 rpm, and are built for maximum power up to 10,000 hp. They do not use much the cylinder capacity to weight, but their operation in a low rpm translates in better mechanical reliability, a very important requirement for reserve prime movers in utility power plants. Those engines are usually adapted to burn heavy liquid fuels of several characteristics.

According to the operational piston mode, an engine can be classified as:

- *A diesel engine of single effect*: the combustion pressure acts on only one face of the piston.
- *A diesel engine of double effect*: the combustion pressure acts alternately on each face of the piston.

As for the operational cycle, the diesel engine can be:

- *A four-stroke diesel engine:* the operating cycle follows four stages: aspiration, compression, combustion, and escape.
- *A two-stroke diesel engine:* the operating cycle is reduced to aspiration–compression and combustion–escape.

Automatic control techniques are used to control speed in diesel engines for grid-connected applications. The regulator continually acts directly, or through a servomotor, on the command of the valve that varies the fuel flow.

A.6 TYPES OF FUEL INJECTION PUMPS

Injector pumps are built for dosing the amount of fuel, adjusting it to the load, in agreement with the control set point from the centrifugal regulator. According to the way the regulator acts on the pump, they are classified into three types.

Pump with Regulation by Free Retreat In those pumps, the path traveled by the piston is variable. The injection of the fuel uses only part of the path traveled by the piston; fuel injected after that is returned to the aspiration stage of the chamber.

Pump with Regulation by Course Control In this pump type, regulation of the amount of fuel injected takes place by varying the course of the piston through a sloped output. In such a system, the piston is impelled during its entire course. The pump with regulation by course control has the advantage of simplicity of its piston and sleeve but presents the inconvenience of fast wear-out of the contact surface, due to the high specific pressure on the contact points. The exit shaft moves easily in a sense, but requires strong effort to move backward.

Pump with Regulation by Strangled Retreat The pump operates in accordance with the principle of constant piston course, as in regulation by free retreat. Regulation of the amount of fuel impelled connects the cylinder of the pump with the aspiration through an appropriate conduit. A small valve commanded by the regulator controls the conduit. If the communication orifice is opened completely by the valve, all fuel is injected up to the pump cylinder capacity. If it is necessary to reduce the amount of fuel injected, the valve controls that fuel, allowing the excess to return to the aspiration conduit. Therefore, varying the valve aperture also varies the relationship between aspirated and injected fuel, where the maximum aperture corresponds to a null impulse, because all the impelled fuel returns to the aspiration conduit.

A.7 ELECTRICAL CONDITIONS OF GENERATORS DRIVEN BY DIESEL ENGINES

All reciprocating engines have flywheels to compensate for the oscillations and vibrations. For electric power generation it is important to smooth out the

TABLE A.1 Minimum Moment of Inertia to Avoid Resonance (50 and 60 Hz) in Four-Stroke Diesel Engines

	Moment of Inertia ($kg \cdot m^2/kVA$)	
Speed (rpm)	50 Hz	60 Hz
150	350.00	243.06
167	228.00	158.33
188	144.00	100.00
214	84.50	58.68
250	45.50	31.60
300	21.90	15.21
375	9.00	6.25
428	5.30	3.68
500	2.90	2.01
600	1.40	0.97
750	0.57	0.40
1000	0.18	0.13

mechanical shaft movement; otherwise, voltage and frequency oscillations may affect the electric performance. This situation is worse in power plants of similar size coupled in parallel, because the voltage and power fluctuations may impair and disable the parallel connection [3–5].

Parallel operation is satisfactory only if the oscillation period of the generator is far apart from the fundamental oscillation of the force pulses of the prime mover. The minimum moments of inertia free of any resonance necessary to damp the movement of 50/60-Hz parallel generators driven by four-stroke diesel engines can be taken from Tables A.1 and A.2. The values in these tables depend on the

TABLE A.2 Minimum Moment of Inertia to Avoid Resonance (50 and 60 Hz) in Two-Stroke Diesel Engines

	Moment of inertia ($kg \cdot m^2/kVA$)	
Speed (rpm)	50 Hz	60 Hz
150	87.50	60.76
167	57.00	39.58
188	36.00	28.00
214	21.20	14.72
250	11.40	7.92
300	5.50	3.82
375	2.25	1.56
428	1.32	0.92
500	0.72	0.50
600	0.35	0.24
750	0.14	0.01
1000	0.06	

short-circuit characteristics of the generators and vary according to their constructing features. To obtain values for the necessary inertia moments, one should multiply by the value of nominal power in kVA. In Tables A.1 and A.2 the values are valid only for parallel operation with machines of the same driving class and speed. Operation of different operating classes, such as coupling of a generator driven by a diesel engine with another driven by a steam turbine, should be designed carefully, because the different speeds and reverse influences might cause mechanical resonance.

The flywheel is always coupled to the generator rotor. For slow generators, the rotor itself can act as the flywheel if its weight and diameter are well designed. In small groups of medium speed (500 to 1500 rpm), the flywheel should also be coupled to the prime mover, or additional flywheels must be installed between coupling brakes of the prime mover and the generator. The effect of this flywheel on the motor is to reduce vibration by smoothing out the power stroke as each cylinder fires, as a mounting surface to bolt the engine up to its load sometimes is used with gear teeth to allow starting the motor to engage and crank the diesel.

Another aspect to be taken into account is the possibility of overloading diesel engines. It is recommend not to exceed 5 to 10% of its full load. It is possible to count on this overload capacity in frequent intervals or in operation periods with certain duration. Therefore, instead of using very heavy flywheels to regularize load variations, the diesel engine should use special devices preferentially for load regulation. The required mechanical power of a diesel engine to drive an electrical generator can be analyzed as follows.

As electrical generators easily support an overload of up to 5%, a 5% lower power can be adopted for the generator, resulting in

$$N_g = 0.95 N_{\mathrm{di}} \eta_g (0.746)$$

where N_g is the power of the generator (kW), N_{di} the power of the diesel engine in hp, and η_g the efficiency of the generator.

When a diesel engine–powered electric generator works in parallel with other generators driven by prime movers of various types, the operation is such that the diesel groups must operate at constant load even if the turbogenerators compensate for the overloads. As an example, the overload values used by some manufacturers of diesel engine are 6% of overload for 30 minutes, 12.5% of overload for 5 minutes, and up to 20% for temporary load oscillations.

REFERENCES

[1] C. F. Taylor, *Analysis of Internal Combustion Motors*, Edgard Blücher, São Paulo, Brazil.

[2] A. G. Domschke and O. Garcia, *Internal Combustion Motors*, Department of Mechanical Engineering, Polytechnic School of the Universidade de São Paulo, São Paulo, Brazil, 1968.

[3] J. R. Vázquez, *Maquinas motrices y generadores de energia elétrica*, Ceac, Barcelona, Spain.

[4] E. J. Kates, *Diesel and High-Compression Gas Engines Fundamentals*, American Technical Society, Chicago, 1965.

[5] http://www.engineersedge.com/power transmission/power transmission menu.shtml.

APPENDIX B

GEOTHERMAL ENERGY

B.1 INTRODUCTION

The Earth's crust is formed from enormous slabs—the tectonic plates—from the original ball of liquid and gas it was billions of years ago. The tectonic plates are actually moving very slowly over a massive layer of very hot rock, separating from, crushing into, or sliding under one another. The tectonic plate movement is also the way the geothermal resources were formed in our planet. The manifestations of this movement are the volcanoes that result from the high levels of heat energy found in the Earth's core. The Earth core is a very hot mixture of incandescent matter under its thin, crustlike mantle, and together with the tectonic plate movement is responsible for hot springs, steam vents, and geysers. *Geo* means Earth, and *thermal* means heat; so *geothermal* represents Earth heat. Once available, the geothermal energy can be used directly or indirectly, as discussed in this chapter.

Geothermal energy is advantageous because it is renewable, reliable, and efficient, and as a group, geothermal power plants can generate power more than 95% of the time. These plants are seldom off-line for maintenance or repairs and are the highest-capacity factors of all power plants. The disadvantages of geothermal energy include its limited use due to site availability (requiring 1 to 8 acres of

Integration of Alternative Sources of Energy, by Felix A. Farret and M. Godoy Simões

land per megawatt) and its environmental impact due to drilling. Its current prices also range from 5 to 8 cents per kilowatthour [1].

For every 100 m (about 328 ft) below ground, the temperature of the rock increases about 3°C (about 5.4°F). In other words, at about 3 km below ground, the temperature of the rock would be hot enough to boil water. Deep beneath the surface, water sometimes makes its way close to the hot rock and turns into boiling water or steam. The hot water can reach temperatures of more than 150°C. This is actually hotter than boiling water (100°C). It does not turn into steam because it is not in contact with the air. When this hot water comes up through a crack in the Earth, it is called a *hot spring* (e.g., Emerald Pool at Yellowstone National Park) or sometimes explodes into the air as a *geyser* (e.g., Old Faithful Geyser).

Although the upper layer of the Earth, close to the surface, is not very hot, it gets much hotter with depth below ground, working like an isolation layer. Almost everywhere across the planet, the upper 3 m below ground level stays at the same temperature, between 10 and 16°C. This is the case in a basement of a building or in a cavern below ground, where the temperature of the area is almost always cool.

About 10,000 years ago, Paleo-Indians used hot springs in North America for cooking. Areas around hot springs were neutral zones where warriors of fighting tribes would bathe together in peace. Every major hot spring in the United States can be associated with Native American tribes. California hot springs, such as the Geysers in the Napa area, were vital and sacred areas to tribes from that area. In other places around the world, people used hot springs for rest and relaxation. Hot springs in Japan are plentiful, attracting crowds of tourists throughout the year, and the Japanese have enjoyed natural hot springs for centuries. The ancient Romans built elaborate buildings to enjoy hot baths, such as those in the town of Bath, England. At present, areas with hot springs are being looked at as possible sources of geothermal energy. South Fork, Colorado, a town in the San Luis Valley that lags behind the rest of the state in economic development, recently conducted a preliminary feasibility study that indicated the presence of a source of geothermal energy.

B.2 GEOTHERMAL AS A SOURCE OF ENERGY

Heat flows outward from the Earth's interior, causing a convective motion in the mantle, which, in turn, drives plate tectonics. *Plate tectonics* is the movement of plates within the Earth, plates that collide periodically usually causing other events to take place. When two plates collide, one plate is subducted beneath the other and the subducted plate slides downward, reaching pressures and temperatures that cause the plate to melt, forming magma. The magma ascends on buoyant forces, pushing plates into the crust of the Earth and causing large amounts of heat to be produced. At the surface, magma forms volcanoes; below the surface, magma creates subterranean regions of hot rock. Faults and cracks at the Earth's surface allow water seepage into the ground. The water is then heated by the hot rock and can naturally circulate back to the surface, forming hot springs, geysers, and mud pots. Assuming that the water is not allowed to recirculate but is trapped by impermeable rock, the water fills the rock pores and cracks, forming a geothermal reservoir.

The direct use of geothermal energy is achieved primarily by using shallower reservoirs of lower temperatures. Hot springs and spas are very popular around the world and are utilized when the sulfur content is relatively low. Other applications involve use of the reservoirs for agricultural production, aquacultures, industrial applications, and heating.

Geothermal resources can be used either for power generation or direct consumption. There are four methods of power generation: direct-steam power generation, flash-steam power generation, binary cycle power generation, and combined/hybrid power generation. Direct-steam power generation is the least common type of generation and is characterized by vapor-dominated and dry steam reservoirs. Most common uses of shallower reservoirs of lower temperatures are for bathing and spas, agriculture, aquaculture, industrial use, and district heating. The industrial use is related to the drying of fish, fruits, vegetables, and timber; wool washing; cloth dying; paper manufacture; milk pasteurization; and piping under sidewalks and roads to keep them from icing over in freezing weather.

Flash-steam generation, the most common type of generation, is used in high-temperature applications; an example is shown in Figure B.1. It is characterized by liquid-dominated reservoirs, flashing the water into steam to turn turbines.

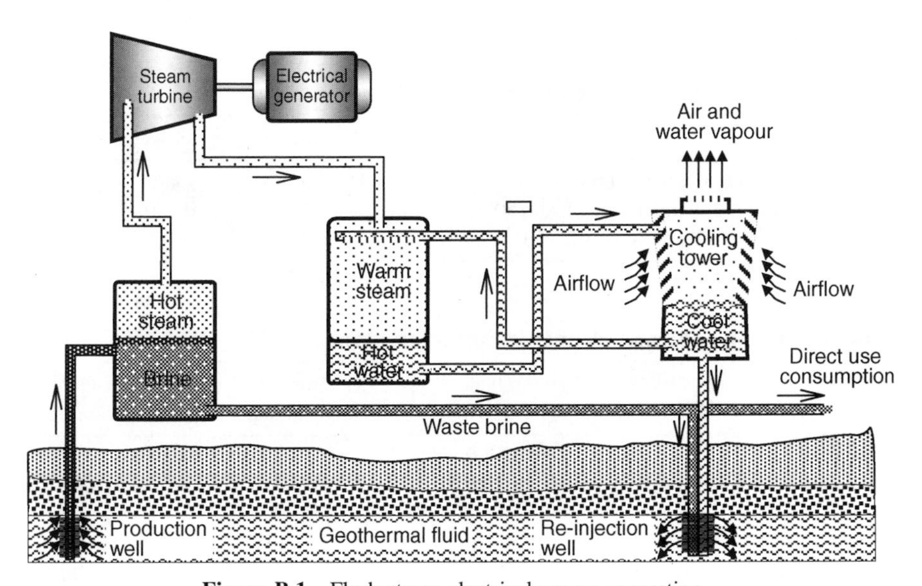

Figure B.1 Flash-steam electrical power generation.

Binary-cycle power generation is also very common, but is the most expensive, as it utilizes a working fluid. The binary cycle can be used with lower-temperature reservoirs, as the water heats the working fluid, and the working fluid steam actually turns the turbine. Finally, combined or hybrid generation is not used very widely and is currently being researched to combine the foregoing techniques to obtain the most efficient geothermal system.

B.2.1 Geothermal Economics

Today, it is common to use geothermally heated water in swimming pools and health spas. The hot water from below ground can also warm buildings for growing plants. In San Bernardino in southern California, hot water from below ground is used to heat buildings during the winter; the hot water runs through miles of insulated pipes to dozens of public buildings. City Hall, animal shelters, retirement homes, state agencies, a hotel, and a convention center are some of the San Bernardino buildings heated this way. In Iceland, many of the buildings and swimming pools in the capital of Reykjavik and elsewhere are heated with geothermal hot water. Iceland has at least 25 active volcanoes and many hot springs and geysers.

There are many advantages of geothermal energy, but there are also some disadvantages. To determine whether or not the advantages outweigh the disadvantages, it is important to conduct a feasibility study for the area in question.

The implementation cost of geothermal power depends on the depth and temperature of the resource, well productivity, environmental compliance, project infrastructure, and economic factors such as the scale of development and project financing costs. Tables B.1, B.2, and B.3 illustrate the costs associated with geothermal applications.

TABLE B.1 Geothermal Resources Cost ($/tonne)

	Steam	Hot water
High temperature ($>150\,°C$)	3.5–6.0	n/a
Medium temperature (100–150 °C)	3.0–4.5	0.2–0.4
Low temperature ($<100\,°C$)	n/a	0.1–0.2

TABLE B.2 Geothermal Plant Unit Cost (cents/kWh)

	High-Quality Resource	Medium-Quality Resource	Low-Quality Resource
Small plants, <5 MW	5.0–7.0	5.5–8.5	6.0–10.5
Medium plants, 5–30 MW	4.0–6.0	4.5–7	Normally not suitable
Large plants, >30 MW	2.5–5.0	4.0–6.0	Normally not suitable

TABLE B.3 Direct Capital Costs of Large Plant Exploration ($/kW Installed Capacity)

Large Plants (>30 MW)	100–200	100–400
Steam field	$300–450	$400–700
Power plant	$750–1100	$850–1100
Total	$1150–1750	$1350–2200

As an example, the overall cost of a geothermal plant in South Fork, Colorado, with a capacity of about 300 MW, is approximately $39 million [2]. This cost does not include the exploration costs, as the initial feasibility study/exploration has already been completed. This cost would remain virtually unchanged for 25 to 30 years, as there is little maintenance and repair associated with geothermal systems.

B.2.2 Geothermal Electricity

As noted earlier, hot water or steam from below ground can also be used to make electricity in a geothermal power plant. Steam carries noncondensable gases of variable concentration and composition. In California, there are 14 areas where geothermal energy is used to make electricity. Some are not yet used because the resource is too small, too isolated, or the water temperatures are not hot enough to make electricity [2]. California's main sources are:

- The Geysers area north of San Francisco
- The northwest corner of the state near Lassen Volcanic National Park
- The Mammoth Lakes area, the site of a huge ancient volcano
- The Coso Hot Springs area in Inyo County
- The Imperial Valley in the south

Some areas have enough steam and hot water to generate electricity. Holes are drilled into the ground and pipes lowered into the hot water, like a drinking straw in a soda. The hot steam or water comes up through these pipes from below ground, like the Geysers Unit 18 located in the Geysers geothermal area of California. California's geothermal power plants produce about one-half of the world's geothermally generated electricity. These geothermal power plants produce enough electricity for about 2 million homes.

A geothermal power plant is like a regular electrical power plant except that no fuel is burned to heat water into steam. The Earth heats the steam or hot water in a geothermal power plant. It goes into a special turbine whose blades spin, and the shaft from the turbine is connected to a generator to make electricity. The steam is then cooled in a cooling tower. The white "smoke" rising from the plants is steam

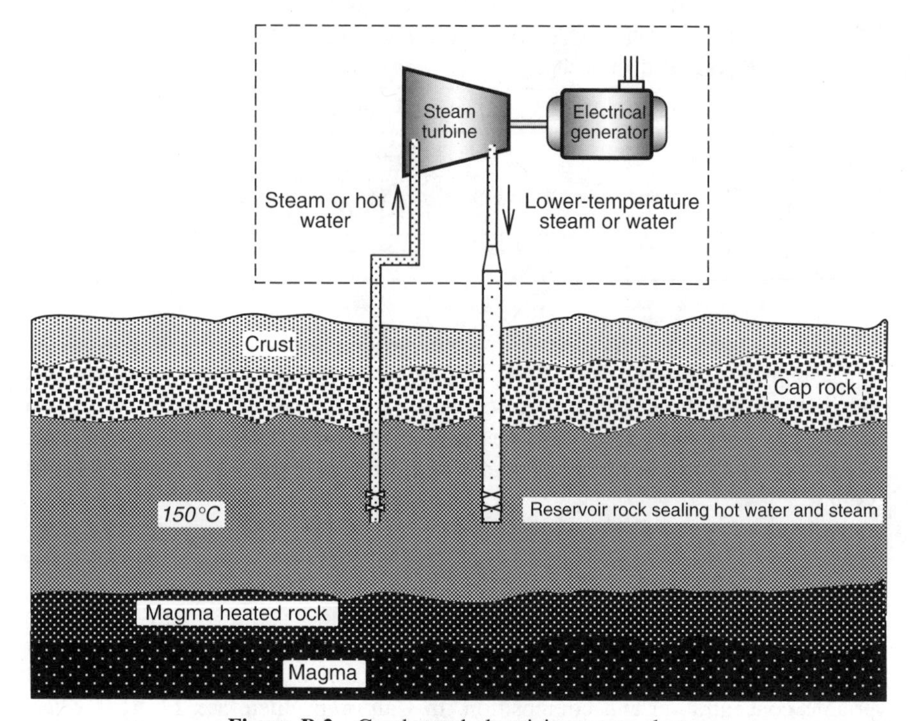

Figure B.2 Geothermal electricity power plant

given off in the cooling process. The cooled water can then be pumped back below ground to be reheated by the Earth.

Figure B.2 gives a glimpse into the inside of a power plant using geothermal energy. Hot water flows into and out of the turbine, and this hot water circulation turns the generator. The electricity goes out to the transformer and then to huge transmission lines that link the power plants to homes, schools, and businesses.

B.2.3 Geothermal/Ground Source Heat Pumps

A geothermal or ground source heat pump system can use the constant temperature under the ground to heat or cool a building or drying crops. Pipes are buried in the ground near the building as depicted in Figure B.3. Inside these pipes, a fluid is circulated to harness the heat and move it to the interior of the ambient to be warmed. In winter, heat from the warmer ground goes through the heat exchanger of a heat pump, which sends warm air into the home or business. During hot weather, the process is reversed. Hot air from inside the building goes through the heat exchanger and the heat is passed into the relatively cooler ground. Heat removed during the summer can also be used to preheat water [3].

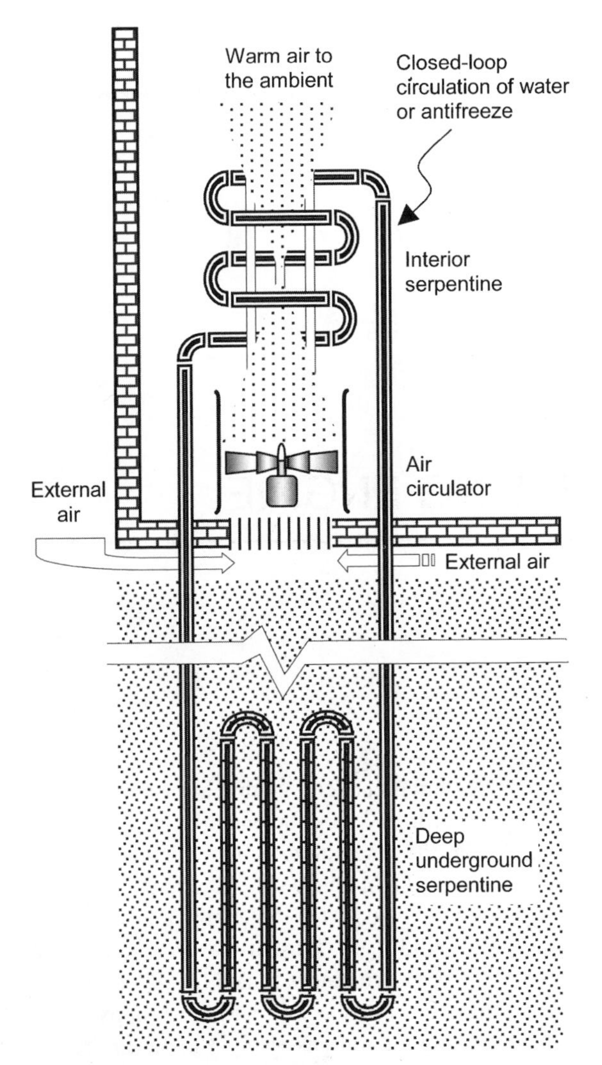

Figure B.3 Direct extraction of geothermal heat.

REFERENCES

[1] http://www.energyquest.ca.gov/story/chapter11.html.

[2] *Renewable Energy Potential in South Fork*, Internal Report for Colorado School of Mines, Engineering Division, Golden, CO, 2003.

[3] http://www.ghpc.org/about/movie.htm.

APPENDIX C

THE STIRLING ENGINE

C.1 INTRODUCTION

The Stirling engine is an external heat engine that is vastly different from the internal combustion engines of ordinary cars. Invented by Reverend Robert Stirling in 1816, the Stirling engine has the potential to be much more efficient than a gasoline or diesel engine. It was invented to create a safer alternative to the steam engines of the time, whose boilers often exploded due to the high pressure of the steam and the primitive materials used to build them. Presently, to generate electricity for homes and businesses, research Stirling generators fueled by either solar energy or natural gas have been tested. They run on solar power during sunny weather and convert automatically to clean burning natural gas at night or when the weather is cloudy. However, today's Stirling engines are used only in some very specialized applications, such as in submarines, intelligent buildings, or auxiliary power generators for yachts, where quiet operation is important. As an example of industrial interest, between 1958 and 1970 General Motors began projects involving Stirling engines in automotive applications [1,2].

In broad terms, Stirling engines can be useful only in places where silence is required. When there is time for slow warm-up, there are some plentiful heating and cooling sources. When there is a demand for a low-speed motor, constant output power, and no power surges, there are even some abundant sources of fuel that

Integration of Alternative Sources of Energy, by Felix A. Farret and M. Godoy Simões

are already available or easily feasible. Such restrictions interest energy researchers because of their potential capabilities for clean, quiet, and diverse sources, and efficient exchange of energy, although there has not been a successful mass-market application for Stirling engines [1,2].

Stirling engines are known to run on low temperature differences, so they tend to be rather large for the amount of power they put out. However, this may not be a significant drawback since these engines can largely be manufactured from lightweight and inexpensive materials such as plastics. These engines can be used for applications such as irrigation and remote water pumping. For wider temperature differences, they can also be used in revertible applications if mechanical work is input into the engine by connecting an electric motor to the power output shaft. As a result, one end will get hot and the other end will get cold. In a correctly designed Stirling cooler, the cold end will get extremely cold. Stirling coolers (built for research use) can cool below 10 K [3].

C.2 STIRLING CYCLE

A Stirling engine uses the Stirling cycle, which is unlike the cycles used in internal-combustion engines. Three major points should be observed:

1. The gases used inside a Stirling engine never leave the engine. There are no exhaust valves that vent high-pressure gases, as in gasoline or diesel engines, and no explosions are taking place. Thus, Stirling engines are very quiet.
2. A flywheel or propeller with gentle spin is needed to start the engine.
3. The Stirling cycle uses an external heat source, which can be anything from sunshine, geothermal heat, gasoline, and solar energy to the heat produced by decaying plants. No combustion takes place inside the cylinders of the engine.

There are several ways to construct a Stirling engine, but here we describe two different configurations of this engine's operation. The key principle of a Stirling engine is that a fixed amount of a gas is sealed inside the engine. Stirling engines are external combustion engines, since the heat is supplied to the air inside the engine from a source outside the cylinder instead of being supplied by fuel burning inside the cylinder. But experiments with hydrogen and helium as working fluids led to vast improvements and modifications of Stirling's invention, due to their lower specific heats. Using gases with different properties led to a reduction in the size of the engine as well as to new applications for the machine [4–6].

The Stirling cycle involves a series of events that change the pressure of the gas inside the engine, causing it to do work. It utilizes the four basic thermodynamic processes of rotating energy conversion: compression, expansion, heat addition, and heat rejection. The Stirling cycle is an idealized cycle very similar to the Carnot cycle, with some key process specifications. The first process is an isothermal compression, which occurs in the cold space, where heat is transferred from the working fluid to the surroundings to maintain the temperature during compression

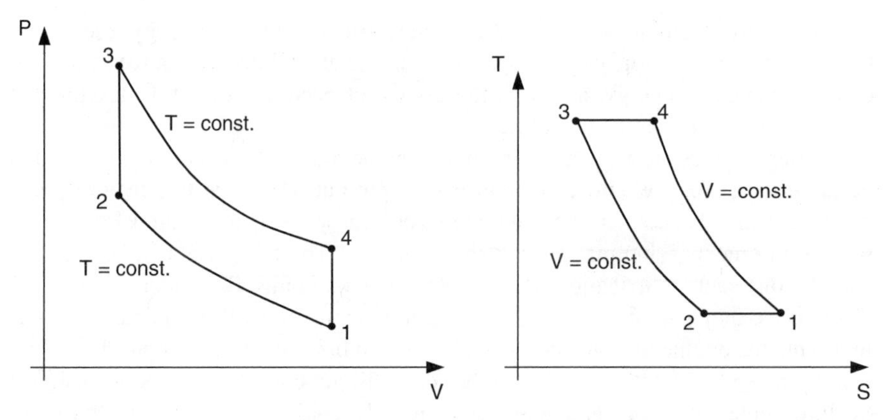

Figure C.1 *P–v* and *T–s* diagrams for the ideal stirling cycle.

(see Figure C.1). The second process is a constant-volume heat addition. This is followed by the isothermal expansion of the working fluid, where heat is added to the system during the expansion to maintain temperature. The last process is heat rejected from the working fluid at constant volume. All four processes can be seen in the *P–v* and *T–s* diagrams in Figure C.1.

There are several properties of gasses that are critical to the operation of Stirling engines:

1. If there is a fixed amount of gas in a fixed volume of space, and the temperature of that gas is raised, the pressure will increase according to the *Gay–Lussac gas law*, given by

$$\frac{p_1 V_1}{T_1} = \frac{p_2 V_2}{T_2} = nR \tag{C.1}$$

   where p, V, and T refer, respectively, to the pressure, volume, and temperature inside a cylinder and the subscripts 1 and 2 to two different states of these variables, n is the number of molecules $\cdot$ gram, and $R = 0.8207$ L $\cdot$ atm/mol $\cdot$ K is the universal gas constant.
2. If a fixed amount of gas is compressed (decreasing the volume of its space), the temperature of that gas will increase, but equation (C.1) holds for every intermediary state.

To understand how the Stirling cycle works, it is best to examine a simplified model of this engine using two crossed cylinders, one with a full piston and the other with a foam half piston, as depicted in Figure C.2. The upper half part of the lower cylinder is heated by an external source (such as fire), and the other is cooled by an external cooling source (such as ice). The gas chambers of the two cylinders are connected, and the upper piston and foam half piston are connected to each other mechanically by a linkage that determines how they will move in rela-

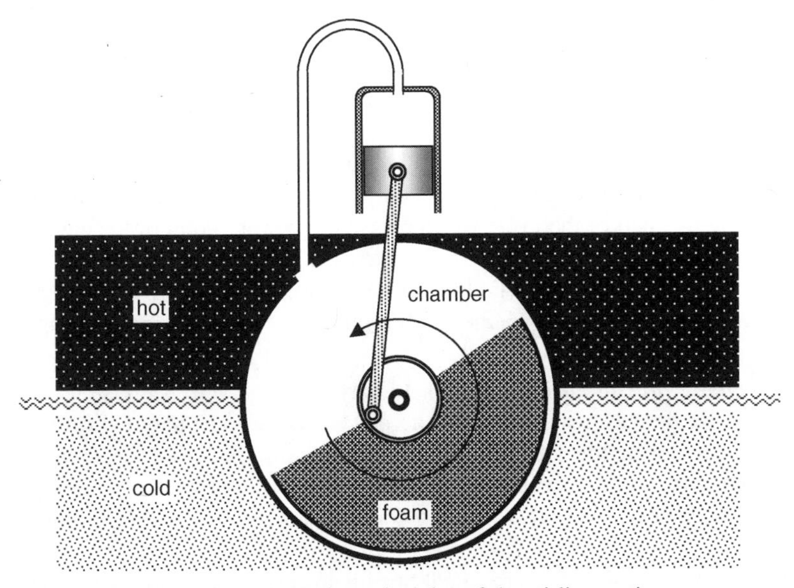

Figure C.2 Operating principles of the stirling engine.

tion to one another [6–8]. There are four parts to the Stirling cycle with the piston and half piston accomplishing all of the parts of the cycle. Heat is added to the gas in the top half of the cylinder causing pressure to build up. This forces the foam half piston down, producing work in steps as follows [9–11]:

1. Heat is added to the gas inside the heated cylinder (top half), causing pressure to build. This forces the full piston to move down. This part of the cycle can be also done via a crankshaft, with two rods at 90° phase shift. This is the part of the Stirling cycle that actually does the work. This part of the cycle continues until most of the air is still in the hot upper portion, expanding, and pushing the piston fully down.
2. The spinning foam half piston moves up while the top piston, connected to the same crankshaft, moves down. Most of the air is now in the cold lower portion, contracting and sucking the piston up. This pushes the hot gas into the cooled cylinder, which quickly cools the gas to the temperature of the cooling source, lowering its pressure. This makes it easier to compress the gas in the next part of the cycle.
3. The foam half piston in the cooled cylinder (bottom) starts to compress the gas. Heat generated by this compression is removed by the cooling source.
4. The half piston moves up while the top piston moves down at some 90° phase lag. This forces the gas into the heated portion of the cylinder, where it heats up quickly and builds pressure. At this point the cycle repeats.

The Stirling engine creates power only during the first part of the cycle. The amount of work produced is represented by integration of the area between the higher and lower temperatures in the Stirling cycle diagram. Its movement is smoothed out by a flywheel connected to the crankshaft [10,11]. There are two principal ways to increase the power output of a Stirling cycle:

1. *Increase power output in stage 1.* In part 1 of the cycle, the pressure of the heated gas pushing against the piston performs work. Increasing the pressure during this part of the cycle will increase the power output of the engine. One way to increase the pressure is by increasing the temperature of the gas as predicted by equation (C.1). In Section C.4 we will see how a device called a regenerator can improve the power output of the engine by temporarily storing heat.
2. *Decrease pressure use in stage 3.* In part 3 of the cycle, the pistons perform work on the gas, using some of the power produced in part 1. Lowering the pressure during this part of the cycle can decrease the power used during this stage of the cycle (effectively increasing the power output of the engine). One way to decrease the pressure is to cool the gas to a lower temperature.

In this section we described the ideal Stirling cycle. Actual working engines vary the cycle slightly because of the physical limitations of their design, as dealt with in the following sections.

C.3 DISPLACER STIRLING ENGINE

Instead of having two pistons, a displacer-type engine has one piston and a displacer. The displacer serves to control when the gas chamber is heated and when it is cooled [8–10], as in Figure C.3. This type of Stirling engine, sometimes used in classroom demonstrations, can run using as little energy as human body heat. In order to run, the engine requires a temperature difference between the top and the bottom of the large cylinder. One part of the engine is kept hot while another part is kept cold, separated by a displacer. This displacer is mounted carefully so that it does not touch the walls of the cylinder. A mechanism then moves the air back and forth through and around the displacer, between the hot side and the cold side, as in Figure C.3. When the air is moved to the hot side, it expands and pushes up on the piston, and when the air is moved back to the cold side, it contracts and pulls. After the hot air expands and pushes the piston as far as the connecting rod allows, the air still has quite a bit of heat energy left in it. In this case, the difference between the temperature of your hand and the air around it is enough to run the engine.

In Figure C.3, two pistons can be seen:

1. *The power piston.* This is the smaller piston at the top of the engine. It is a tightly sealed piston that moves up as the gas inside the engine expands.
2. *The displacer.* This is the large piston in the drawing. This piston is very loose in its cylinder, so air can move easily between the heated and cooled sections of the engine as the piston moves up and down.

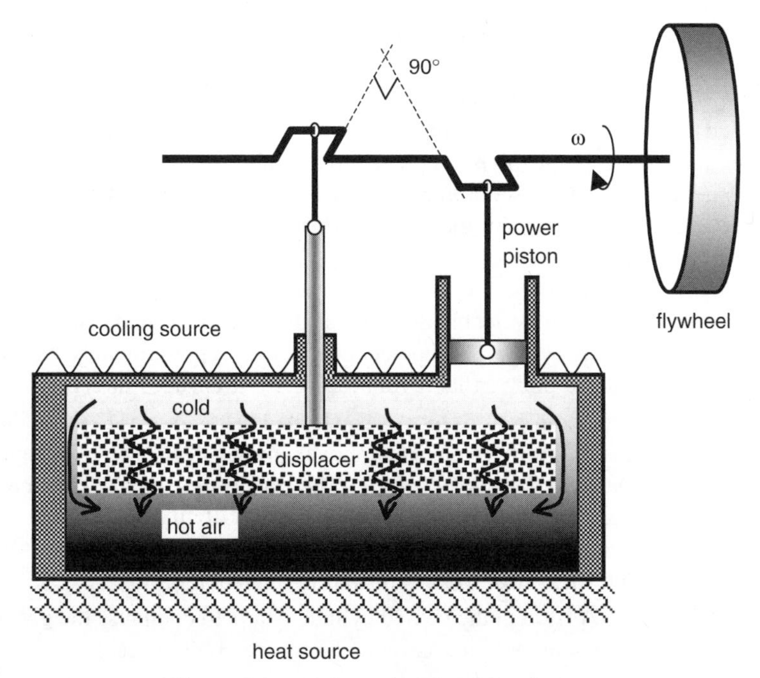

Figure C.3 Stirling principles with displacer.

The displacer moves up and down to control whether the gas in the engine is being heated or cooled. There are two positions:

1. When the displacer is near the top of the large cylinder, most of the gas inside the engine is heated by the heat source and it expands. Pressure builds inside the engine, forcing the power piston up.
2. When the displacer is near the bottom of the large cylinder, most of the gas inside the engine cools and contracts. This causes the pressure to drop, making it easier for the power piston to move down and compress the gas.

The engine repeatedly heats and cools the gas, extracting energy from the gas's expansion and contraction with some waste heat. Some Stirling engines are made to store some of the waste heat by making the air flow through economizer tubes that absorb some of the heat from the air. This precooled air is then moved to the cold part of the engine, where it cools very quickly, and as it cools it contracts, pulling down on the piston. Next, the air is moved mechanically back through the preheating economizer tubes to the hot side of the engine, where it is heated even further, expanding and pushing up on the piston. This type of heat storage, used in many industrial processes, is currently called *regeneration*. Stirling engines do not require regenerators to work, but, well-designed engines will run faster and put out more power if they have a regenerator [3,12].

C.4 TWO-PISTON STIRLING ENGINE

In a two-piston Stirling engine, the heated cylinder is heated by an external flame. The cooled cylinder is air-cooled and has fins on it to aid in the cooling process. A rod stemming from each piston is connected to a small disk, which is, in turn, connected to a larger flywheel (see Figure C.4). This keeps the pistons moving when no power is being generated by the engine. The flame heats the bottom cylinder continuously [1,2,10].

1. In the first part of the cycle, pressure builds, forcing the heated piston to move to the left and do work. The cooled piston stays approximately stationary because it is at the point in its revolution where it changes direction.
2. In the next stage, both pistons move. The heated piston moves to the right and the cooled piston moves up. This moves most of the gas through the regenerator and into the cooled piston. The regenerator is a device that can store heat temporarily. It might be wire mesh that the heated gases pass through. The large surface area of the wire mesh quickly absorbs most of the heat. This leaves less heat to be removed by the cooling fins.
3. Next, the cooled piston in the cooled cylinder starts to compress the gas. Heat generated by this compression is removed by the cooling fins.
4. In the last phase of the cycle, both pistons move; the cooled piston moves down and the heated piston moves to the left. This forces the gas across the regenerator (where it picks up the heat that was stored there during the previous cycle) and into the heated cylinder. At this point, the cycle begins again.

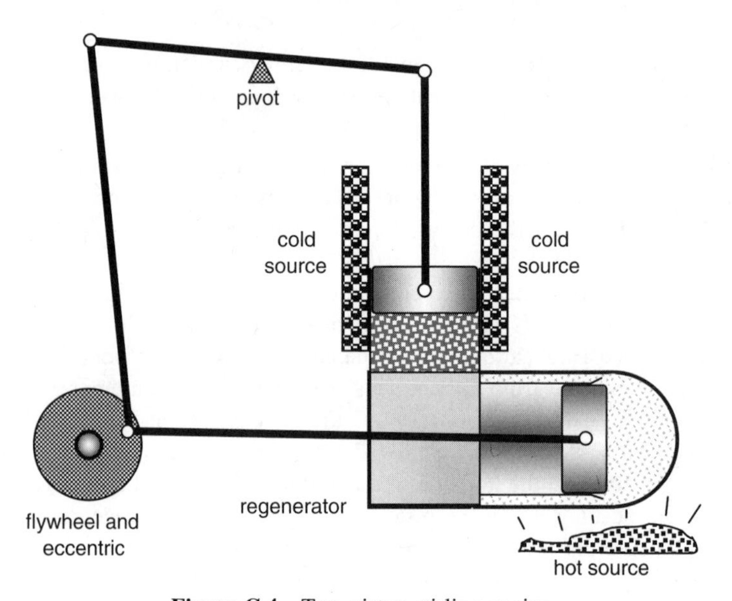

Figure C.4 Two-piston stirling engine.

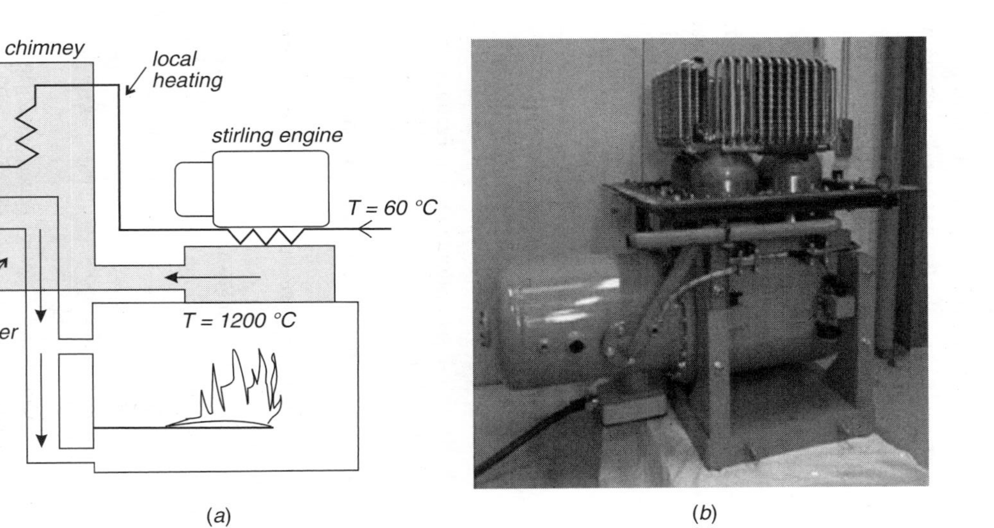

Figure C.5 CHP plant using a Stirling engine: (*a*) CHP plant described by H. Carlsten; (*b*) 35-kWe prototype for biomass combustion plants. (Courtesy of BIOS Energy & H. Carlsten.)

The possibilities for mass-market applications of Stirling engines are still limited. There are a couple of key characteristics that make Stirling engines impractical for use in many applications, including most cars and trucks. Because the heat source is external, it takes a little time delay for the engine to respond to changes in the amount of heat being applied to the cylinder; it takes time for the heat to be conducted through the cylinder walls and into the gas inside the engine. This means:

- The engine requires a long time to warm up before it can produce useful power.
- The engine cannot change its power output quickly.

These shortcomings prevent this engine from replacing automotive internal-combustion engines. However, a Stirling-engine-powered hybrid car might be feasible, as in the process diagram of a CHP plant demonstrated by H. Carlsten and the newly developed 35-kWe prototype for biomass combustion plants shown in Figure C.5.

REFERENCES

[1] N.W. Lane and W.T. Beale, Free-Piston Stirling Design Features. Presented at Eighth International Stirling Engine Conference, May 27–30, 1997, University of Ancona, Italy.

[2] C. M. Hargreaves, *Phillips Stirling Engine*, Elsevier Science, Amsterdam, The Netherlands, 1991.

[3] M. J. Collie, (Ed.), *Stirling Engine: Design and Feasibility for Automotive Use*, Noyes Data Corporation, Park Ridge, NJ, 1979.

[4] G. T. Reader and C. Hooper, *Stirling Engines*, E. & F.N. Spon, New York, 1983.

[5] Volunteers in Technical Assistance, http://idh.vita.org/pubs/docs/stirling.html.

[6] J. Lewis, New simplified heat engine, http://www.emachineshop.com/engine/animation.htm, accessed Mar., 2005.

[7] Stirling Energy Systems, leader in alternative or green energy, http://www.stirlingenergy.com.

[8] *Stirling Engine Reference, Guide and Catalog*, American Stirling Company, http://www.stirlingengine.com/ and http://www.stirlingtech.com/, 2002.

[9] C. D. West, *Principles and Applications of Stirling Engines*, Van Nostrand Reinhold, New York, 1986.

[10] J. R. Senft, *An Introduction to Low Temperature Differential Stirling Engines*, Moriya Press, River Falls, WI, 1996.

[11] G. Walker, *Stirling Engines*, Clarendon Press, Oxford, 1980.

[12] http://www.pureenergysystems.com/os/StirlingEngine/photologie/sitetranslation/blue/.

INDEX

Integration of Alternative Sources of Energy, by Felix A. Farret and M. Godoy Simões